Geopedology

Joseph Alfred Zinck
Graciela Metternicht • Gerardo Bocco
Héctor Francisco Del Valle

Editors

Geopedology

An Integration of Geomorphology and
Pedology for Soil and Landscape Studies

 Springer

Editors
Joseph Alfred Zinck
Faculty of Geo-Information Science
  and Earth Observation (ITC)
University of Twente
Enschede, The Netherlands

Institute of Environmental Studies
University of New South Wales
Sydney, New South Wales, Australia

Gerardo Bocco
Centro de Investigaciones en Geografía
  Ambiental (CIGA)
Universidad Nacional Autónoma
  de México (UNAM)
Morelia, Michoacán, Mexico

Graciela Metternicht
Institute of Environmental Studies
University of New South Wales
Sydney, New South Wales, Australia

Héctor Francisco Del Valle
Consejo Nacional de Investigaciones
  Científicas y Técnicas (CONICET)
Centro Nacional Patagónico (CENPAT),
  Instituto Patagónico para el Estudio de
  los Ecosistemas Continentales (IPEEC)
Puerto Madryn, Chubut, Argentina

ISBN 978-3-319-19158-4   ISBN 978-3-319-19159-1 (eBook)
DOI 10.1007/978-3-319-19159-1

Library of Congress Control Number: 2015958768

Springer Cham Heidelberg New York Dordrecht London
© Springer International Publishing Switzerland 2016

Printed on acid-free paper

Springer International Publishing AG Switzerland is part of Springer Science+Business Media (www.springer.com)

Preface

My Dilemmas with Soil-Landscape Relationships

In 1952 I graduated and was employed full time as a soil scientist with the USDA Soil Conservation Service. That winter I went back to Iowa State College to attend a course studying Hans Jenny's 1941 book – all that good information and discussion with other students and soil scientists. The big dilemma was that it was not possible to solve the soil forming equation; it merely was a stimulus to guide our thinking about soils and their distributions in time and space. Now it is 60+ years later, and we still can't solve the axiom of pedology; but, oh my, we have learned a tremendous lot about "reading soil landscapes."

A dilemma may be considered an undesirable choice suggesting reluctance to make a decision. Often it occurs because we believe we do not have enough information to make the right one or at least a better decision. Most of life is this way; we make judgments all the time, for example, much of each day involves evaluating choices, putting them into classes that separate them from each other or grouping them into populations of similarity.

The articles in this book are about prototypes. They are perceptions of what has been selected as starting points in classifying. They are the basis for creating groups of objects, entities, and even ideas that enable us to separate the complexity of the world about us into manageable formats. You likely were a young adult when someone thought you ought to know something about landforms, maybe even soils, and by then you already had developed some prototypes of what those objects were based on where you lived and how you grew up in a family and a community. It is highly possible that soils and their relationships with landscapes were mainly introduced when you went to a university. There you were introduced to many new prototypes, and the teachers were anxious that you accepted them and that they became part of your archives of working knowledge.

Your whole life is built around prototypes and classifications and what you do with them. Those which you accept eventually become the basis for "aha moments" when you comprehend what they seem to mean to you. In this book the authors hope you will have some good "aha moments" as soon as possible and that you continue using and learning more about geoforms and soils as seen in the field. They are sharing with you what they mean (information) and want you to accept what they say (intent). Each of us is unique in what we have been exposed to all of our life; thus, our experiences are not the same, our prototypes are not the same, and our "aha moments" are quite different. What you do with the information provided is up to you.

We tend to like causal relationships as they enable us to interpret the stimuli we receive through our senses as well as our thought processes. And this becomes a big dilemma as we try to comprehend soil-landscape relationships. Regression analyses and correlations do not prove causal relationships. We want them to, so we often take them as positive evidence of causal relationships. Let us not be pessimistic. There is too much excitement and joy in pedology to be negative; however, I want to tell you some dilemmas in my ability to read landscapes.

Can we say what we mean and mean what we say? I would like to share with you some dilemmas of mine about understanding soil-landscape relationships. I will discuss nine dilemmas that have faced me along pathways I did not always antici-pate. They are classifying and classifications, scales of seeing and presenting infor-mation, properties and their interpretations, sampling, building mental models, applying models, evaluating relationships, presenting our understanding, and the future. I suggest you read "Advancing the frontiers of soil science towards a geosci-ence" by Larry Wilding and Henry Lin, published by *Geoderma* in 2006. It is a nice summary of the past and looks to the future.

Classifying and Classifications

The first is the process of making decisions, and the other is a means of organizing information; thus, one is doing and the other is having. As humans we do not seem to have the choice of segregating and grouping entities; it obviously was a matter of survival and dealing with conditions and situations every day. To help us do this, we develop prototypes that eventually represent large populations. Everything we see, smell, taste, hear, and touch is classified. In pedology we have developed many standards to assist us in selecting prototypes and many of their properties. This enables us to communicate better with one another. I think that if we can't classify, we may be brain dead.

A major dilemma in classifying is agreeing on what are the objects (entities) that we want to recognize as the individuals of a larger population. For some it is a pedon or profile, both of which are small volumes. If you accept a pedon, there are trillions in the soil populations of our pedosphere. We commonly select small volumes as samples of a soil, but for me they are not soils themselves. That decision is influ-enced by our beliefs of what a soil is.

Classifications, and more particularly taxonomies, are multi-categorical systems to organize classes recognized at each categorical level according to a set of requirements. The purpose of taxonomy is to better understand the relationships among members of smaller groupings based on definitions at each categorical level. The classes of higher levels are divided into small groups at each lower level, and the classes at the lowest category are similar to individual entities. The diagnostic features at higher categories accumulate through the system and determine boundaries for classes at lower levels.

Naming systems are applied to taxonomies. Every country or culture has developed their own soil classification; however, most have related them to the names of the US Soil Taxonomy or the World Reference Base of IUSS. These two systems rely on concepts of soil formation and evolution for defining categories and classes within. Similar efforts continue today as we search for improved global communications.

The dilemma in classifying and in having taxonomies is directly tied to our perceptions associated with scales. It is difficult to change old habits.

Scales of Observation and Presenting Information

As we consider soil-landscape relationships, we immediately are faced with the scales at which we observe soils and the landscapes in which they occur. Do we think about pedons, or is something larger and more inclusive relevant to what we visualize and want to convey to others with maps? It seems to come down to what you believe soils are and how they are distributed in a landscape. This is what soil survey is all about, and it is particularly important for maps at detailed scales, e.g., 1:10,000–1:30,000. Why? At these scales the smallest delineations cover areas much larger than individual kinds of soils; thus, they have inclusions of other soils and landscape features. For example, a wet area in a larger field of similar soils can be an inclusion or it can be depicted with a defined spot symbol. Thus, we have a spatial dilemma but also have options, and their use depends on the purpose of that survey. What is acceptable in one region may not be satisfactory in another.

There are names and descriptions of many landforms. They are often perceived as what we observe where we are, standing in a field or looking at satellite images. This may easily become a dilemma. I previously thought that in the USA it would be desirable to have a standardized set for a scale of 1:24,000 which was common for many topographic maps. Landforms can be based on specified geometric forms of components or as concepts of landscape formation and evolution. Both are relevant when they satisfy the purpose; otherwise, they may create a dilemma for a user.

When pedogenesis is of interest, then time scales need to also be considered. The same is true for landform evolution which generally has a longer time frame than soil property development because of our axiom of soils. Very few soil landscapes are older than Pleistocene, and their surface layers are usually much younger. Many are now modified by human interactions making recognition, description, and classification more problematic.

Presentations explaining or hypothesizing soil-landscape relationships are never quite satisfying because the applications of space and time scales are complex. But we teach and learn by communicating with others and try to understand their opinions and conclusions which are crucial for advancement of our field of science. It is a never-ending struggle, and each of us is a product of the progress associated with these struggles.

Properties and Their Interpretations

Properties are those features of soils and landforms which we measure, commonly in the field with rather simple tools. For soil profiles we have standards for colors and their patterns, texture, structure, consistence, coarse fragments of stones and wood, thickness and boundaries of horizons, nature of materials in layers, and uncommon inclusions often related to animals and insects.

In a landscape there are external features such as positions of slopes, their steepness, shape, size, extent, and surfaces may have rock outcrops, scree, or even evidence of prior anthropic uses. Tools of measurement may be simple and with guidelines for recording the observations provided. As technology provides higher precision instruments and products, our measurements have increased, are made more easily, and provide data not previously available. Different kinds of imagery often provide clues related to features of interest. For example, infrared photography colors are associated with vegetation health and vigor. Normal colors may highlight small differences of plant growth, moisture status, and irregularity of surface features.

Combinations of properties are used as diagnostics for taxonomies; however, they may differ in national taxonomies. The US Soil Taxonomy and the World Reference Base have many similar diagnostics (often with different names) and some striking differences. For example, US ST accepts soil moisture and temperature regimes as diagnostic properties at high categorical levels, whereas WRB does not. Alternatives to provide such information are dilemmas that must be considered.

For those interested mainly with pedogenesis, the soil properties to describe, measure, and interpret usually exist at larger scales. For example, concretions, salts, cutans, pores, and their spatial distributions are relevant to ascertaining the processes affecting such properties. Meso and micro features of landforms, both surficial and internal, such as microtopography or internal stratification patterns are used to support concepts of formation and evolution. Scales of space and time need to be considered when interpreting how and when landforms have developed and been modified. Most of these decisions relate to identifying and classifying soils.

When land is used to produce agricultural crops, pastures or forestry, the users want to know about qualities of soils and how well they will perform. These are complex interpretations of current and future behavior and functions which benefit from the expertise of other disciplines and on different time scales.

An early interpretive classification organized soil information into a land capability system, mainly for agricultural uses. As soil survey organizations worked more closely with agronomists, engineers, geologists, and extension personnel, functional interpretations became ever more popular. In the USA soil potential ratings made in cooperation with land users were efforts to directly work with owners and operators. In a recent published survey in California, there are 500+ pages devoted to 30 kinds of interpretations for 155 soil map units.

Soil surveys open doors for people to better understand the complexity of soils in landscapes. Conflicts, different points of view, and other dilemmas are common, normal, and part of our learning processes.

Sampling Soil-Landscape Relationships

When soils are sampled as part of pedogenesis research, the individuals selected are usually small volumes such as pedons or profiles. Depth samples are taken to enable vertical differences to be detected. These depth distributions of properties provide data that are recognized as layering, often as lithological discontinuities, suggesting changes in a landscape that have influenced soil properties and their distributions in space or time or both.

Should samples be by horizons or by equal depth increments? This depends on the questions you are asking! Over time we have learned it is useful to have bulk density measurements to determine weight per unit volume rather than only weight per unit weight (usually as percent). If you only have weight measurements, then depth functions would relate to different thicknesses in a profile. We use wt/vol data because for most soils there are general linear trends of volume in each material present in a profile. For materials that shrink and swell with moisture changes, or have obvious accumulations of soluble salts, there may also be changes of volume in addition to weight changes. Dilemmas, of course, are constantly testing the way you make choices! The deeper you go below a soil profile, as in critical zone sites, the more geomorphic and geologic properties you will encounter.

Perhaps more common is recognition of the components in map unit delineations. Hopefully most of them will be other soils of similar nature and not dominated by non-soil entities. Rocky surfaces, small areas of coarser textures (like drifting sands), small wet depressions, small bedrock outcrops, and abandoned building sites or excavations all are possibilities depending on where you are.

There have been many schemes proposed and used to estimate map unit composition. Some transects are perpendicular to hill slopes to identify changes of soils from summits across upper, mid, and lower backslopes and then into or across footslopes and into toeslopes. Sedimentation patterns differ among these segments of a slope providing clues to the erosional-depositional evolution of that landscape. Various statistical schemes and procedures support such estimates.

In the 1950s–1970s studies were made by Bob Ruhe and colleagues in Iowa, New Mexico, Oregon, and North Carolina in the USA. It was usual to prepare geomorphic and soil maps independently but at the same scale and then evaluate them together. This was grist for many models of soil-landscape relationships. Dilemmas were common but slowly they were resolved. Similar research throughout the world has collected data relevant for pedogenesis and for geomorphic evolution of landscapes. A lot of the data has been, and can be, utilized in interpreting functions of soil landscapes and their behavior.

Building Mental Models

We sample to get more data with which we can piece together both spatial and temporal features in soil landscapes. Even soil taxonomies and geomorphic taxonomies are mental abstractions of what we think we know at a given point in time. Extrapolating from samples to devise mental models is always challenging. We have precise measurements of properties, but how accurate are they? Are you sure? Mathematical procedures are used to support our perceptions. Some deal with extrapolations, and we may or may not overlay them on landscape maps.

It is easy to gloss over gaps and areas of uncertainty in models. Obviously models search for central concepts of properties and relationships that we are interested in. A useful tool is to propose several hypotheses to explain the data sets used to develop our explanations and opinions. Our chances to know the truth seem very remote, but multiple working hypotheses give us opportunities to eliminate details that do not seem to contribute or support the model.

Visual diagrams, graphs, distributions, statistics, 3D block diagrams of soil landscapes, and stages of evolution of both landscapes and soils in those landscapes promote learning of new models and the processes thought to be associated with them. I like 3D diagrams. Be imaginative, work at being more creative, stretch your mind, collaborate with others, and be a dreamer of dreams!

Applying Models

There likely are many taxonomies of landforms in different environments such as tropical, humid, tundra, and arctic or as descriptive hierarchies of landforms by processes of formation and eventual evolution. Some taxonomies such as the one by Alfred Zinck may take us into uncharted generalizations that lead us into new ways of thinking. The world is big; its stories are many and involve different scales to imagine such schemes. Because we start by standing in a landscape and its ecosystem, we may need to visualize going up and down scales to comprehend what is being generalized. In some places three meters takes you into another microworld; in others like the Russian steppe, they force us to ignore certain components to gain the complex reality and visualize the grandeur of such associations.

In the USA catenas have generally been known as wetness classes in similar parent materials and climatic regions. It is one thing to describe the models of moisture classes and quite another to see them as landscape components. They fit nicely in some regions and not well at all in others. Does this sound like a dilemma? Good, you are getting a sense about some undesirable choices that occur when trying to read landscapes and build working models.

How far can we extrapolate a working model of a population of mental models in space? This implies that specific soil-landscape relationships have spatial boundaries. Detailed scale soil maps generally do not cross such boundaries. The concept of components is allowed to stretch a bit more than actually searching for those spatial boundaries. It is another dilemma, of course, but where populations are detected to join or merge at smaller scales, eventually someone tries to resolve the conflict. Some have been undesirable choices, and the larger-scale maps may not yet have been corrected or modified.

Evaluating Relationships

Evaluations open the door to a different side of the coin. Are you looking for the central concepts and confidence limits of the composition of a group of the same named map units delineated on maps? You automatically want more data, recent updated data based on more samples than you likely will ever have. In these situations we feel trapped. We would prefer not to be the decision maker.

It is very important that the reasons for an evaluation are stated and agreed on. As a provider you have a sense of the utility of the applied models, and the user or potential user may have a different sense about the utility. Sometimes estimating the upper and lower confidence limits of the precision (and hopefully the accuracy) helps distinguish the provider and user degrees of acceptance for the application of the models in the area of interest.

It is readily apparent that map scales influence our perceptions of specific soil-landscape relationships. The dilemma occurs as a choice between scientific integrity and the usefulness of practical interpretations provided for the map users. The age and evolution of a soil landscape goes beyond the needs or wants of the decision maker who wants to maximize effectiveness and profitability of managing these same soil-landscape relationships.

Who is responsible for using mental models for soil-landscape relationships? Is there respect both among providers and among recipients? Respect by each group is critical to understanding the set of dilemmas that exist for both groups. Will statistics, pedometrics, or other evaluation procedures help us achieve our goals? We have a lot of work to do, and time is short!

Presenting Our Understanding

Soil maps illustrate our models as we perceive them in landscapes. The base maps have often been airphotos; consequently, we interpret the patterns as vegetation, buildings, roads, communities, and our choices of segments of the landscape. Delineation of the map units is named commonly as phases of taxonomic classes, such as surface textures, slope ranges, and surface rockiness or stoniness. The soils represented are not taxonomic classes; rather, they are associations of individuals (such as polypedons of a specific soil) with inclusions of very similar soils called taxadjuncts and also different kinds of soils. The description of soil map units depends on the choices we have made to show our comprehension of the variability that exists in different landscapes. Some landscape units contain less variability than others; thus, attention must be given to the illustrations and descriptions of the map units associated with the subdivisions of landforms. Sampling by transects or by sizes and shapes of delineations is seldom adequate from a statistical perspective; however, results may be given and explained for map users. Sometimes inset maps at larger scales illustrate the complexity that is being generalized as the final map. Glaciated landscapes and those of broad river plains are often very complicated. Because such landscapes may be young geomorphologically, there are obvious differences that have not been subdued or modified. Plants, animals, and insects respond to these differences in the manner which permits them to survive. Each of these results in micro-changes; some are recognized and described and others missed or ignored. Does this sound like more dilemmas? Change the purpose of having knowledge, and the map units and their distributions will also change.

Although standards and guidelines for their use are prepared and promoted, the actual description and delineation of soil landscapes are tempered by the surveyors as interpreters making the maps. Each of us has had different experiences, and what we learn and accept influence our perceptions and concepts that we use in recognizing and applying our mental models.

An interesting phenomenon about soil landscape maps is that we cannot smoothly go from one scale to another. The spatial links are imperfectly understood. As a result we make maps at many scales sometimes to satisfy our curiosity of what and how generalizations will be made. Even when we make maps by slightly increasing or decreasing scales, we have not accepted what is distracting our perceptions and ability to predict from one scale to another.

Power laws are scale invariant and also contain indicators of fractal or near-fractal patterns. At detailed soil map scales, often sizes of delineations of the same map units show differences of fractal dimensions. What does this mean? We still are not sure. Many other scientific disciplines cannot readily move from scale to scale because the composition of classes at each level is not well known.

Why do we like map generalizations? It may be that as natural classifiers, we try to minimize having so many groups of individuals. This enables us to have fewer prototypes to recognize and use in our daily endeavors. They are attempts to simplify the world around us and find those features that support our perceptions of reality.

Maps and supporting texts are the records of how our discipline of pedology has progressed over the past century and may provide some starting places for the future.

The Future

Perhaps the biggest dilemma we face in understanding and using soil-landscape relationships is that created by the need to find practical solutions to safely use and conserve soils for future civilizations. Our history has dealt with many discouraging choices in providing adequate information that affects national policies for taking care of natural resources. The Dust Bowl in the USA in the 1930s was an example of not having adequate information and policies to avoid the disasters that resulted. Why is this a dilemma? It is a mismatch of understanding and accepting the long-term rates of processes that form soils and promote their evolution and the short-term needs for additions such as fertilizers, water, protection from erosion, the use of insecticides, and numerous other manipulations. We are not satisfied with soils as we find them, only as we modify them to meet our desires.

Groups within the International Union of Soil Sciences are working on many aspects of understanding soil resources. In addition there is more collaboration among and across disciplines to develop concepts and data sets to help all of us become better stewards of our natural resources rather than controllers of nature.

What have we learned? We need to build on the strengths from the past, stand on the shoulders of giants, and share data, concepts, models, and applications. There are many innovation and creativity reserves to be utilized. The main issues of sustainability are not technological; they are issues of human rights and moral values. And these are dilemmas of critical undesirable choices. Will we make them – in time?

Concluding Comment

Throughout my ramblings above I have not mentioned how you begin to see and read landscapes. I think that is because we accept a method and then put it out of our mind. Most landscapes have involved the movement of water at various times in their evolution, so we start by looking at the rivers and major streams that exist today in landscapes. As we follow them upstream, they branch into lower-order stream segments, and if you stay with it eventually you get to the little order 1 drainage ways where water collects and starts its journey to the seas and oceans. That sounds simple enough, doesn't it?

Okay now that you are at the top of some mountain, or on a broad nearly level plain, or even on some intermediate landscape that is of current interest, you slowly follow the stream channel and its little alluvial accumulations along the sides.

The careful observer notices that not all of those little terraces are parallel to the current stream profile and wonders why. These are "aha" moments when you realize that the base levels have changed and are markers of alterations in the adjacent uplands. These cove positions are accumulators, whereas the more active changes are on nose slopes and the side slopes where erosion is stripping off the surface layers time after time after time. When I look at a 3D block diagram of a landscape, the coves with order 1 stream segments are not usually emphasized; it is assumed that they are understood.

As you move farther downstream, the terraces of alluvial fills are more and more complex and become landscapes themselves with hidden stories of the periodic episodes of landscape formation and evolution. Remember that sand dunes and desert environments are built on foundations of older landscapes which usually were molded and modified by running water. Tropical landscapes are often very complex combinations of stepped landscapes, yet the stream systems hold the beginning clues for us to unravel. Arctic environments obviously have a different system of how water collects, moves, and dissipates, and so those clues must be learned. Being able to imagine sequences of events of landform evolution enables us to develop working models (prototypes) we test and evaluate as we learn how to "read landscapes". This process is simple, imaginative, practical, and generally not mentioned!

Once I was trying to explain some characteristics of pedologists. I mentioned honesty, optimism, meaningful, and enthusiasm, that is, being honest, being positive, being relevant, and being dedicated. If you put them together they spell HOME, and for pedologists that means "helping our mother Earth." We can describe principles, methods, and techniques "ad nauseam," but can we make a difference in the use of soil resources? HOME "is" the mission of pedologists. The foundation of HOME is learning to "read the landscapes," visually and mentally.

Former Director, Soil Survey Division,   Richard W. Arnold
USDA-NRCS,
Washington, DC, USA

Contents

The original version of the Table of Contents was revised. An erratum can be found at
DOI 10.1007/978-3-319-19159-1_34

Chapter 1
Presentation

J.A. Zinck, G. Metternicht, H.F. Del Valle, and G. Bocco

Abstract Geopedology aims at integrating geomorphology and pedology to analyze soil-landscape relationships and map soils as they occur on the landscape. This book on geopedology fills a knowledge gap, presenting a proven approach for reliable mapping of soil-landscape relationships to derive value-added information for policy making, planning, and management at scales ranging from local to national to continental. This chapter introduces the structure and contents of the book.

Keywords Geomorphology • Pedology • Soil mapping • Soilscape

Soil is a vital resource for society at large, and an important determinant of the economic status of nations (Daily et al. 1997). Soils are used for many purposes, ranging from agricultural to engineering to sanitary, and provide a broad range of ecosystem services. However, soils have commanded lesser consideration and attention than other components of the natural capital, such as water and forests. They are increasingly exposed to degradation through erosion, salinization,

J.A. Zinck (✉)
Faculty of Geo-Information Science and Earth Observation (ITC), University of Twente,
Enschede, The Netherlands

Institute of Environmental Studies, University of New South Wales, Sydney, NSW, Australia
e-mail: alfredzinck@gmail.com

G. Metternicht
Institute of Environmental Studies, University of New South Wales, Sydney,
NSW, Australia
e-mail: g.metternicht@unsw.edu.au

H.F. Del Valle
Consejo Nacional de Investigaciones Científicas y Técnicas (CONICET), Centro Nacional
Patagónico (CENPAT), Instituto Patagónico para el Estudio de los Ecosistemas Continentales
(IPEEC), Puerto Madryn, Chubut, Argentina
e-mail: delvalle@cenpat-conicet.gob.ar

G. Bocco
Centro de Investigaciones en Geografía Ambiental (CIGA), Universidad Nacional Autónoma
de México (UNAM), Morelia, Michoacán, Mexico
e-mail: gbocco@ciga.unam.mx

© Springer International Publishing Switzerland 2016   1
J.A. Zinck et al. (eds.), *Geopedology*, DOI 10.1007/978-3-319-19159-1_1

compaction, and/or pollution. It takes nature centuries, even millennia to form a few centimeters of soil, while billions of tons of arable land are eroded every year.

The soil patrimony remains largely unknown at scales appropriate for practical uses. Traditionally, soil data were collected by systematic soil surveys organized by national government agencies, but the latter have decreased considerably over the last decades because of global economic recession and a tendency of planners to disregard soil information. The multiplication of (pseudo)-natural disasters, including landslides, gullying, flooding, and competing uses of land for food and bio-fuels have contributed to create public awareness about the relevant role the pedosphere plays in the natural and anthropogenic environments. Recent papers and global initiatives such as the Global Soil Partnership of the FAO and the GlobalSoilMap.net show a renewed interest in soil research and its applications to improved planning and management of this fragile and finite resource (Hartemink 2008; Sanchez et al. 2009; McBratney et al. 2014). Furthermore, the United Nations have acknowledged the global role of soils in declaring the International Year of Soils 2015.

Traditional soil surveys remain an expensive piece of information, and least developed countries lack human and financial resources to undertake detailed soil mapping. To make soil survey cost-effective and more attractive to users, technological and methodological innovations for data gathering and conversion into information have been developed through increased use of information technology in the areas of remote sensing, geographic information systems, spatial modelling, and spatial statistics. These technological advances have facilitated the development of digital soil mapping. However, digital soil cartography is still mainly limited to terrain/land surface properties, in contrast to the 3D soil body that farmers, engineers, planners, and extension officers manage for decision making.

The use of geomorphology, integrated with pedology, has proven to speed up and improve soil inventory. The external geomorphic terrain features (i.e. morphometric and morphographic attributes) help delineate natural soil distribution units, while the internal features of the geomorphic material (i.e. morphogenic and morphochronologic attributes) contribute to understand soil formation. Geomorphology and pedology are conceptually and practically related. In terms of soil mapping, geomorphic units (i.e. geoforms) provide cartographic frames for soil delineations, while pedologic descriptions supply the soil information of the delineations (e.g. soil properties, use, classification). Geomorphology alone provides information on three of the five soil forming factors (i.e. relief/topography, parent material, and age). Thus the combination of geomorphology and pedology is a mutually beneficial endeavor, and fits nicely with the underlying principles of international standard taxonomic soil classifications such as the World Reference Base for Soil Resources, largely based on soil morphology as an expression of soil formation conditions (Dominati et al. 2010).

Geopedology aims at integrating geomorphology and pedology to analyze soil-landscape relationships and map soils as they occur on the landscape. The geopedologic view is similar to the frequently used expression of "soil geomorphology". A few decades ago, several reference books and seminal papers focused on soil geomorphology; however, the most recent one dates back to 2005 (Schaetzl and

Anderson). This new book on geopedology fills a knowledge gap, presenting a proven approach for reliable mapping of soil-landscape relationships to derive value-added information for policy making, planning, and management at scales ranging from local to national to continental. The book presents the theoretical and conceptual framework of the geopedologic approach and a bulk of applied research showing its use and benefits for knowledge generation relevant to geohazard assessment and prediction, land use planning and conflict mitigation, and landscape management.

Part I introduces the theoretical framework. Basic geopedologic concepts are described, with emphasis on the construction of a hierarchic system organizing geoforms into six categories to serve soil mapping at different levels of detail. The geopedology approach to soil survey combines pedologic and geomorphic criteria to establish soil map units. Geomorphology provides the contours of the map units ("the container"), while pedology provides the soil components of the map units ("the content"). Therefore, the units of the geopedologic map are more than soil units in the conventional sense of the term, since they also contain information about the geomorphic context in which soils have formed and are distributed. In this sense, the geopedologic unit is an approximate equivalent of the soilscape unit, but with the explicit indication that geomorphology is used to define the landscape. This is usually reflected in the map legend, which shows the geoforms as entry point to the legend and their respective pedotaxa as descriptors.

Part II includes a set of papers showing how geopedology relates to a variety of research fields in which the relationships between soil and landscape are approached in different ways. The papers have in common the analysis of soil distribution patterns on the landscape (i.e soilscape) using geopedology from various points of view and for different objectives, such as indigenous soilscape perception, soilscape ecology, soilscape history, soilscape diversity, and soilscape complexity and heterogeneity.

Part III introduces case studies that use digital techniques (GIS-based spatial analysis and modeling, remote sensing) to derive morphometric parameters, specifically from digital elevation models (DEM), describing topographic and drainage features of the geomorphic landscape. This section shows complementarity between the geopedologic and the digital soil mapping approaches. A digital soil map is essentially a spatial database of soil properties, based on a statistical sample of landscapes (Sanchez et al. 2009). The geopedologic approach can act as the conceptual framework that 'guides' digital soil mapping. For instance, the segmentation of the landscape into geomorphic units provides spatial frames in which digital terrain models combined with remote-sensed data and geostatistical analyses can be applied to assess detailed spatial variability of soils and geoforms, better framing digital mapping over large territories. Geopedology provides information on the structure of the landscape in hierarchically organized geomorphic units, while digital techniques supply relevant information that helps characterize the geomorphic units, mainly the morphographic and morphometric terrain surface features.

In Parts IV and V, case studies at local, regional, and national scales show the use of geopedologic information in multi-purpose applications. Of all components of the natural landscape, the geomorphic component is the most integrating one. The features and origin of the geoforms reflect the influence of the geologic substratum and internal geodynamics, while being modelled under the influence of climate and exogenous geodynamics. By integrating geoform and soil information, geopedology supplies an appropriate framework for geohazard studies, land degradation assessment, land use conflict analysis, land use planning, and land suitability evaluation.

References

Daily G, Matson P, Vitousek P (1997) Ecosystem services supplied by soils. In: Daily G (ed) Nature's services: societal dependence on natural ecosystems. Island Press, Washington, DC
Dominati E, Patterson M, Mackay A (2010) A framework for classifying and quantifying the natural capital and ecosystem services of soils. Ecol Econ 69:1858–1868
Hartemink AE (2008) Soils are back on the global agenda. Soil Use Manage 24:327–330
McBratney A, Field DJ, Koch A (2014) The dimensions of soil security. Geoderma 213:203–213
Sanchez P, Ahamed S, Carré F et al (2009) Digital soil map of the world. Science 325:680–681
Schaetzl R, Anderson S (2005) Soils: genesis and geomorphology. Cambridge University Press, New York

Part I
Foundations of Geopedology

Chapter 2
Introduction

J.A. Zinck

Abstract This chapter introduces to the foundations of geopedology. Geopedology, as it is considered in this book, refers to the relations between geomorphology and pedology, with emphasis on the contribution of the former to the latter. More specifically, geopedology is in the first instance a methodological approach to soil inventory, while providing at the same time a framework for geographic analysis of soil distribution patterns. The prefix *geo* in geopedology refers to the earth surface – the geoderma – and as such covers, in addition to geomorphology, concepts of geology and geography.

Keywords Geopedology • Geoforms • Geography • Soil • Pedology

Geopedology, as it is considered here, refers to the relations between geomorphology and pedology, with emphasis on the contribution of the former to the latter. More specifically, geopedology is in the first instance a methodological approach to soil inventory, while providing at the same time a framework for geographic analysis of soil distribution patterns. The prefix *geo* in geopedology refers to the earth surface – the *geoderma* – and as such covers, in addition to geomorphology, concepts of geology and geography. Geology intervenes through the influence of tectonics in the geoforms of structural origin, and through the influence of lithology in the production of parent material for soils as a result of rock weathering. Geography relates to the analysis of the spatial distribution of soils according to the soil forming factors. However, in the concept of geopedology, emphasis is on geomorphology as a major structuring factor of the pedologic landscape and, in this sense the term geopedology is a convenient contraction of geomorphopedology. Geomorphology covers a wide part of the physical soil forming framework through the relief, the surface morphodynamics, the morphoclimatic context, the unconsolidated or weathered materials that serve as parent materials for soils, and the factor

J.A. Zinck (✉)
Faculty of Geo-Information Science and Earth Observation (ITC), University of Twente, Enschede, The Netherlands

Institute of Environmental Studies, University of New South Wales, Sydney, NSW, Australia
e-mail: alfredzinck@gmail.com

© Springer International Publishing Switzerland 2016
J.A. Zinck et al. (eds.), *Geopedology*, DOI 10.1007/978-3-319-19159-1_2

time. Geopedology underpins the argument of Wilding and Lin (2006) that the frontiers of soil science would benefit from moving towards a geoscience.

The relationship between geomorphology and pedology can be considered within the context of landscape ecology. With its integrative approach, landscape ecology tries to bridge the gap between related disciplines, both physical and human, that provide complementary perceptions and visions of the structure and dynamics of natural and/or anthropized landscapes. Landscape ecology, as a discipline of integration, has holistic vocation, but it is often practiced de facto as parts of a whole. For instance, one stream emphasizes the ecosystem concept as the basis of the biotic/ecological landscape (Forman and Godron 1986); while another stream stresses the concept of land as the basis of the cultural landscape (Zonneveld 1979; Naveh and Lieberman 1984); and still another one puts emphasis on the concept of geosystem as the basis of the geographic landscape (Bertrand 1968; Haase and Richter 1983; Rougerie and Beroutchachvili 1991). Geomorphology and pedology participate in this concert, and their respective objects of study, i.e. geoform and soil, constitute an essential, inseparable pair of the landscape.

Geoforms or terrain forms sensu lato are the study object of geomorphology. Soils are the study object of pedology, a branch of soil science. The relations between both objects and between both disciplines are intimate and reciprocal. Geoforms and soils are essential components of the earth's epidermis (Tricart 1972), sharing the interface between lithosphere, hydrosphere, biosphere, and atmosphere, within the framework of the noosphere as soils are resources subject to use decisions by human individuals or communities. It is not a mere static juxtaposition; there are dynamic relationships between the two objects, one influencing the behavior of the other, with feedback loops. Moreover, in nature, it is sometimes difficult to categorically separate the domain of one object from the domain of the other, because the boundaries between the two are fuzzy; geoforms and soils interpenetrate symbiotically. This integration of the geoform and soil objects, that coexist and coevolve on the same land surface, has fostered the study of the relations between the two. As it often happens, the interface between disciplines is a frontier area where new ideas, concepts, and approaches sprout and develop.

The analysis of the relationships and interactions between geoforms and soils and the practical application of these relationships in soil mapping and geohazard studies have received several names such as soil geomorphology, pedogeomorphology, morphopedology, and geopedology, among others, denoting the transdisciplinarity of the approaches. By the position of the terms in the contraction word, some authors want to point out that they put more emphasis on one object than on the other. For instance, Pouquet (1966) who has been among the first ones to use the word geopedology, emphasizes the pedologic component and implements geopedology as an approach to soil survey and to erosion and soil conservation studies. In contrast, Tricart (1962, 1965, 1994) who has possibly been one of the first authors to use the word pedogeomorphology, puts the accent on the geomorphic component.

Chapter 3 of this book illustrates the variety of modalities implemented to address the relationships between geomorphology and pedology. The applied context in which geopedology was developed is different from other ways of visualizing the relationships between both disciplines; this specificity of geopedology is described in Chap. 4. The geopedologic approach focuses on the inventory of the soil resource. This means logically addressing themes such as soil characterization, formation, classification, mapping, and evaluation. Chapter 5 summarizes relevant aspects of these themes with emphasis on the hierarchic structure of the soil material, which allows highlighting that geomorphology is involved at various levels. The application of geomorphology in soil survey programs at various scales, from detailed to generalized, requires to establish a hierarchic taxonomy of the geoforms, so that the latter can serve as cartographic frames for soil mapping and, additionally, as genetic frames to help interpret soil formation. These aspects are addressed in Chap. 6 (criteria for classifying geoforms), Chap. 7 (geoform classification), and Chap. 8 (geoform attributes).

This text is partially drawn from the lecture notes used in a course on geopedology under the heading of *Physiography and Soils* (Zinck 1988), taught by the author at the International Institute for Geo-Information Science and Earth Observation (ITC, Enschede, The Netherlands) as part of an annual postgraduate course in soil survey in the period 1986–2003. Shorter versions of the course were also taught by the author on several occasions between 1970 and 2003 in various countries of Latin America, especially in Venezuela and Colombia.

References

Bertrand G (1968) Paysage et géographie physique globale. Esquisse méthodologique. Rev Géogr Pyrénées et SO 39(3):249–272
Forman RTT, Godron M (1986) Landscape ecology. Wiley, New York
Haase G, Richter H (1983) Current trends in landscape research. Geo J (Wiesbaden) 7(2):107–120
Naveh Z, Lieberman AS (1984) Landscape ecology. Theory and application. Springer, Munich
Pouquet J (1966) Initiation géopédologique. Les sols et la géographie. SEDES, Paris
Rougerie G, Beroutchachvili N (1991) Géosystèmes et paysages. Bilan et méthodes. Armand Colin, Paris
Tricart J (1962) L'épiderme de la terre. Esquisse d'une géomorphologie appliquée. Masson, Paris
Tricart J (1965) Principes et méthodes de la géomorphologie. Masson, Paris
Tricart J (1972) La terre, planète vivante. Presses Universitaires de France, Paris
Tricart J (1994) Ecogéographie des espaces ruraux. Nathan, Paris
Wilding LP, Lin H (2006) Advancing the frontiers of soil science towards a geoscience. Geoderma 131:257–274
Zinck JA (1988) Physiography and soils. Lecture notes. International Institute for Aerospace Survey and Earth Sciences (ITC), Enschede
Zonneveld JIS (1979) Land evaluation and land(scape) science. International Institute for Aerospace Survey and Earth Sciences (ITC), Enschede

Chapter 3
Relationships Between Geomorphology and Pedology: Brief Review

J.A. Zinck

Abstract The relationships between geomorphology and pedology, including the conceptual aspects that underlie these relationships and their practical implementation in studies and research, have been referred to under different names, the most common expression being *soil geomorphology*. Definitions and approaches are reviewed distinguishing between academic stream and applied stream. There is consensus on the basic relationships between geomorphology and pedology: geomorphic processes and resulting landforms contribute to soil formation and distribution while, in return, soil development has an influence on the evolution of the geomorphic landscape. However, a unified body of doctrine is yet to be developed, in spite of a clear trend toward greater integration between the two disciplines.

Keywords Approaches • Evolution • Integration • Classification • Conceptual

3.1 Introduction

The relationships between geomorphology and pedology, including the conceptual aspects that underlie these relationships and their practical implementation in studies and research, have been referred to under different names. Some of the most common expressions are *soil geomorphology* (Daniels et al. 1971; Conacher and Dalrymple 1977; McFadden and Knuepfer 1990; Daniels and Hammer 1992; Gerrard 1992, 1993; Schaetzl and Anderson 2005; among others), *soils and geomorphology* (Birkeland 1974, 1990, 1999; Richards et al. 1985; Jungerius 1985a, b), *pedology and geomorphology* (Tricart 1962, 1965a, b, 1972; Hall 1983), *morphopedology* (Kilian 1974; Tricart and Kilian 1979; Tricart 1994; Legros 1996),

J.A. Zinck (✉)
Faculty of Geo-Information Science and Earth Observation (ITC), University of Twente, Enschede, The Netherlands

Institute of Environmental Studies, University of New South Wales, Sydney, NSW, Australia
e-mail: alfredzinck@gmail.com

© Springer International Publishing Switzerland 2016
J.A. Zinck et al. (eds.), *Geopedology*, DOI 10.1007/978-3-319-19159-1_3

geopedology (Principi 1953; Pouquet 1966), and *pedogeomorphology* (Conacher and Dalrymple 1977; Elizalde and Jaimes 1989), without mentioning the numerous publications that treat the subject but do not explicitly use one of these terms in their title. Due to this diversity of expressions, it is convenient to first define what the relations between geomorphology and pedology cover, and subsequently analyze the nature of the relationships.

3.2 Definitions and Approaches

Soil geomorphology, sometimes called pedologic geomorphology or pedogeomorphology, is the term most frequently found in English-published literature, with the word geomorphology being a noun and the word soil being an adjective that qualifies the former. According to this definition, the center of interest is geomorphology, with the contribution of pedology. However, under the same title of soil geomorphology, there are research works in which the roles are reversed. Therefore, in practice, the relationship between geomorphology and pedology goes both ways. The emphasis given to one of the two disciplines depends on a number of factors including, among others, the context of the study, the purpose of the research, and the primary discipline of the researcher.

The relations between geomorphology and pedology as scientific disciplines, and between geoform and soil as study objects of these disciplines can be viewed in two ways depending on the focus and weight given to the leading discipline. In one case, emphasis is on the study of the geoforms, while soil information is used to help resolve issues of geomorphic nature, as for example, characterizing the geoforms or estimating the evolution of the landscape. Literally, this approach corresponds to the expression of soil geomorphology or pedogeomorphology. In the other case, focus is on the formation, evolution, distribution, and cartography of the soils, with the contribution of geomorphology. Literally, this approach corresponds to the expression of geomorphopedology, or its contraction as geopedology. In practice, the various expressions have been used interchangeably, showing that the distinction between the two approaches is fuzzy. Based on this apparent dichotomy, two streams, initially separated, have contributed to the development of the relations between geomorphology and pedology: (1) an academic stream, oriented towards the investigation of the processes that take place at the geomorphology-pedology interface, and (2) a more practical stream, applied to soil survey and cartography. The first one flourished more in hillslope landscapes, which offer propitious conditions to conduct toposequence (catena) and chronosequence studies, whereas the second one developed more in depositional, relatively flat landscapes, with conditions suitable for the use of soils for agricultural or engineering purposes.

3.2.1 Academic Stream

The academic stream consists of research conducted mainly at universities for scientific purposes. It is based on detailed site and transect studies to identify features of interdependence between geoforms and soils without preset paradigm. In general, this stream seeks to use geomorphology and pedology for analyzing, in a concomitant way, the processes of formation and evolution of soils and landscapes. This current covers in fact a variety of approaches, as illustrated by the definitions given by various authors with regard to their conceptions of the relationships between geomorphology and pedology and the study domains covering these relationships. Hereafter, some definitions of soil geomorphology are presented in chronological order.

- The analysis of the balance between geomorphogenesis and pedogenesis and the terms of control of the former on the latter in soil formation (Tricart 1965a, b, 1994).
- The use of pedologic research techniques in studies of physical and human geography (Pouquet 1966).
- The study of the landscape and the influence of the processes acting in the landscape on the formation of the soils (Olson 1989).
- The study of the genetic relationships between soils and landscapes (McFadden and Knuepfer 1990).
- The assessment of the genetic relationships between soils and landforms (Gerrard 1992).
- The application of geologic field techniques and ideas to soil investigations (Daniels and Hammer 1992).
- The study of soils and their use in evaluating landform evolution and age, landform stability, surface processes, and past climates (Birkeland 1999).
- The scientific study of the origin, distribution, and evolution of soils, landscapes, and surficial deposits, and of the processes that create and modify them (Wysocki et al. 2000).
- The scientific study of the processes of evolution of the landscape and the influence of these processes on the formation and distribution of the soils on the landscape (Goudie 2004).
- A field-based science that studies the genetic relationships between soils and landforms (Schaetzl and Anderson 2005).
- A subdiscipline of soil science that synthesizes the knowledge and techniques of the two allied disciplines, pedology and geomorphology, and that puts in parallel the genetic relationships between soil materials and landforms and the commensurate relationships between soil processes and land-forming processes (Thwaites 2007).
- The study that informs on the depositional history in a given locality, and also takes into account the postdepositional development processes in the interpretation of the present and past hydrological, chemical, and ecological processes in the same locality (Winter 2007).

This short review, which is far from being exhaustive, shows the diversity of concepts and conceptions encompassed in the expression *soil geomorphology*. From the above definitions, several main approaches may be derived:

- Geologic approach, with geomorphology as a subdiscipline of geology; this reflects the times when soil surveyors' basic training was in geology.
- Geomorphic approach, considering pedology as a discipline that gives support to geomorphology; etymologically, this approach could be called pedogeomorphology.
- Pedologic approach, considering geomorphology as a discipline that gives support to pedology; etymologically, this approach could be called geomorphopedology.
- Integrated approach, based on the reciprocal relations between both disciplines.
- Elevation of soil geomorphology at the level of a science, exhibiting therefore a status higher than that of a simple approach or type of study.

3.2.2 Applied Stream

The applied stream is related to soil survey and consists in using geomorphology for soil cartography. Historically, the analysis of the spatial relationships between geomorphology and pedology and the implementation of the soil-geoform duo were born out of practice. Soil survey has been the field laboratory where the modalities of applying geomorphology to soil cartography were formulated and tested. The structure of the geomorphic landscape served as background to soil mapping, while the dynamics of the geomorphic environment helped explain soil formation, with feedback of the pedologic information to the geomorphic knowledge.

Originally, different modalities of combining geomorphology and pedology were used for cartographic purposes, including the preparation of separate maps, the use of geomorphology to provide thematic support to soil mapping, and various forms of integration. Some authors and schools of thought advocated the procedure of antecedence: first the geomorphic survey (i.e. the framework), then the pedologic survey (i.e. the content), carried out by two different teams (Tricart 1965a; Ruhe 1975). In other cases, there was more integration, with mixed teams making systematic use of the interpretation of aerial photographs (Goosen 1968). Already in the 1930s, the soil survey service of the USA (National Cooperative Soil Survey) had an area of study in soil geomorphology (parallel mode), which was later on formalized with the mission of establishing pedogeomorphic relation models at the regional level to support soil survey (Effland and Effland 1992). The contribution of Ruhe (1956) meant a breakthrough in the use of geomorphology for soil survey in the USA. Ruhe was in favour of completely separating the description of the soils from the study of geomorphology and geology in a work area. Only after completing the disciplinary studies, could the interpretation of the relationships between soil characteristics and landforms be undertaken (Effland and Effland 1992). In the second half of the twentieth century, progress in systematic soil cartography, especially in developing countries, and progress in soil cartography to support agricultural development projects in a variety of countries have led to various forms of integration, with mixed teams of geomorphologists and pedologists. Work performed by French agencies such as ORSTOM (now IRD) and IRAT provide examples of this kind of soil cartography.

The need to boost agricultural production to support fast population growth has led many developing countries in the middle of the last century, especially in the tropics, to initiate comprehensive soil inventory programs. These were carried out mostly by public entities (ministries, soil institutes) and partly by consultancy agencies. In Venezuela, for instance, soil inventory began in the 1950–1960s as local and regional projects to support the planning of irrigation systems in the Llanos plains and, subsequently, as a nationwide systematic soil inventory. These surveys implemented an integrated approach based on the paradigm of the geopedologic landscape, which is closely related to the concepts of pedon, polypedon, and soilscape as entities for describing, sampling, classifying, and mapping soils. The integration between geomorphology and pedology took place all along the survey process, from the initial photo-interpretation up to the elaboration of the final map. The integration was reflected in the structure of the legend with two columns, a column for the geomorphic units that provide the cartographic frames, and a column for the soil units that indicate the soil types. This kind of approach is more appropriate for technical application than for scientific investigation. However, applied research underlies always the survey process, as new soil-geofom situations and relationships might occur and require analysis that goes beyond the strict survey procedure. This is a relatively formalized and systematic approach that can be applied with certain homogeneity by several soil survey teams working at various scales. One of the major requirements to make the implementation of geomorphology more effective in this kind of integrated survey is to apply a system of geoform taxonomy.

A novel way of integration can be found in the morphopedologic maps, based on the concept of the morphogenesis/pedogenesis balance (Tricart 1965b, 1994). Integration takes place not only at conceptual level, but also at the cartographic level (i.e. mapping procedure). The map distinguishes between stable and dynamic elements. The relatively stable geologic substratum, including lithology and structural settings, forms the map background, on which the geomorphic units are superimposed. Each map unit is characterized in the legend by the dominant pedogenic and dominant geomorphogenic processes. This information is used to derive a balance between pedogenesis and geomorphogenesis, which serves as a basis for identifying limitations to soil use.

The implementation of geomorphology in soil survey has strengthened the link between geomorphology and pedology. This practical cooperation has contributed more than academic studies in small areas or at site locations to enhance understanding of their reciprocal relations. These developments were closely related to the golden period of soil inventories during the second half of the twentieth century, particularly in emerging countries that needed soil information at various scales for ambitious agricultural development and irrigation projects. By mid-century, the systematic use of photo-interpretation revolutionized the soil survey procedure, and made the contribution and mediation of geomorphology indispensable for identifying and delineating the surficial expression of soil units on the landscape. The rise of the liberal economy and the globalization of the economic relations over the last decade of the past century resulted in letting the market laws decide on the occupation and use of the territory. This meant the suspension of many land-use planning projects and, by the same token, the cancellation of the supporting soil inventory and land evaluation programs (Zinck 1990; Ibáñez et al. 1995). More recently, a growing societal awareness with regard to soil degradation and erosion is calling the attention on the threats affecting the soil resource, while creating new initiatives and opportunities for soil mapping (Hartemink and McBratney 2008; Sánchez et al. 2009).

Contemporaneously, the multiplication of GIS-related databases to store and manage the variety of data and information provided by the inventories of natural resources revealed the need for a unifying criterion able to structure the entries to the databases: geomorphology showed it could provide this structuring frame (Zinck and Valenzuela 1990). Hence the importance of having a classification system of the geoforms, preferably with hierarchic structure, to serve as comprehensive entry to the various information systems on natural resources, their evaluation, distribution, and degradation hazards.

In recent years, emphasis has been drawn to digital soil mapping based on remote-sensed data, together with the use of a variety of spatial statistics and geographical information systems (McBratney et al. 2003; Grunwald 2006; Lagacherie et al. 2007; Boettinger et al. 2010; Finke 2012; among others). The combination of remote sensing techniques and digital elevation models (DEM) allows improving predictive models (Dobos et al. 2000; Hengl 2003), but tends to see the soil as a surface rather than a three-dimensional body. Remote sensors provide data on individual parameters of the terrain surface and the surficial soil layer. There are also techniques and instruments able to detect soil property variations with depth via proximal sensing (e.g. frequency-domain electromagnetic methods FDEM, ground-penetrating radar GPR, among others), but their use is still partly experimental. Digital elevation models allow relating these parameters with relief variations, but the contribution of geomorphology is generally limited to geomorphometric attributes (Pike et al. 2009).

Some authors put emphasis on improving the precision of the boundaries between cartographic units as compared with a conventional soil map (Hengl 2003), or predicting spatial variations of soil properties and features such as for example the thickness of the solum (Dobos and Hengl 2009), or comparing the cartographic accuracy of a conventional soil map with that of a map obtained by expert system (Skidmore et al. 1996). In all these cases, morphometric parameters are mobilized along with pre-existing soil information (soil maps and profiles). The essence of the soil-geomorphology paradigm, in particular the genetic relationships between soils and geoforms and their effect on landscape evolution, is not sufficiently reflected in the current digital approach. It is difficult to find any theoretical or conceptual statement on soil-geoform relationships, except the reference that is usually made to classic models such as the hillslope model of Ruhe (1975) and the soil equation of

Jenny (1941, 1980). Technological advances in remote sensing and digital elevation modelling are mainly used to explore and infer soil properties and their distribution in the topographic space. From an operational point of view, digital soil mapping is still mostly limited to the academic environment and essentially consists in mapping attributes of the soil surface layer, not full soil bodies that are actually the units managed by users (e.g. farmers, engineers). In official entities in charge of soil surveys, digital cartography is frequently limited to digitizing existing conventional soil maps (Rossiter 2004). There are few examples of national or regional agencies that have adopted automated methods for the production of operational maps (Hengl and MacMillan 2009).

3.3   Nature of the Relationships and Fields of Convergence

There is a collection of books on soil geomorphology that deal with the topic from different points of view according to the area of expertise of each author (Birkeland 1974, 1999; Ruhe 1975; Mahaney 1978; Gerrard 1981, 1992; Jungerius 1985a; Catt 1986; Retallack 1990; Daniels and Hammer 1992; Schaetzl and Anderson 2005; among others). These works are frequently quite analytical, recording benchmark case studies and describing exemplary situations that illustrate some kind of relationship between geomorphology and pedology. An epistemological analysis of the existing literature is needed to highlight the variety of points of view and enhance broader trends. Synthesis essays can be found in some scientific journal articles. What follows here is based on a selection of journal papers and book chapters, which provide a synthesis of the matter at a given time and constitute milestones that allow evaluating the evolution of ideas and approaches over time.

3.3.1   Evolution of the Relationships

The purely geologic conception of Davis (1899) on the origin of landforms as a function of structure, process and time, excluded soil and biota in general as factors of formation (Jungerius 1985b). For half a century, the denudation cycle of Davis has influenced the approach of geomorphologists, more inclined to develop theories than observe the cover materials on the landscape and to give preference to the analysis of erosion features rather than depositional systems. By contrast, the paradigm of soil formation, born from the pioneer works of Dokuchaiev and Sibirzew, and subsequently formalized by Jenny (1941, 1980), was based on a number of environmental factors including climate, biota, parent material, relief, and time. These original conceptual differences have led geomorphologists and pedologists to ignore each others for a long time (Tricart 1965a), although Wooldridge (1949) had already written an early essay on the relationships between geomorphology and pedology. McFadden and Knuepfer (1990) note that soils have historically been neglected by many geomorphologists, who gave preference to the analysis of sedimentological and stratigraphic relations or morphometric studies. The situation changed by mid-twentieth century when it was recognized that the two models could be combined based on interrelated common factors (geologic structure, parent material, relief, time, and stage of evolution) and complementary factors (processes, climate, biota). This has allowed researchers to use the concepts and methods of both disciplines in varying combinations and for various purposes.

Tricart (1965a) has been one of the first researchers to draw the attention on the mutual relations uniting geomorphology and pedology. According to this author, geomorphology provides a framework for soil formation as well as elements of balance for pedogenesis, while pedology provides information about the soil properties involved in morphogenesis. Jungerius (1985b) shows that, although geomorphology and pedology have different approaches, the study objects of these two disciplines, i.e. landforms and soils, share the same factors of formation; the same author also highlights the fact that the relationships are two-way, generating mutual contributions. Since the early works of synthesis, which focused on what one discipline could bring to the other, the field of soil geomorphology has evolved toward greater integration, variable according to the topics, with simultaneous use of geomorphology and pedology and less consideration for the conventional boundaries that separate both disciplinary domains. In some universities there are now departments that house the two disciplines under the same roof (e.g. Department of Geomorphology and Soil Science, Technical University of Munich, Freising, Germany).

3.3.2   Mutual Contributions

Since the relationships between geomorphology and pedology are multiple, the spectrum of the areas and topics of interdisciplinary research is wide and varied, and the preferences depend on the orientation of each researcher. In the absence of a formal body of themes, here is how some authors have synthesized the content of soil geomorphology.

Already half a century ago, Tricart in his treatise on *Principes et Méthodes de la Géomorphologie* (Tricart 1965a) showed the reciprocal relationship of the two disciplines.

• Geomorphology contributes to pedology providing morphogenic balances that reflect the translocation of materials at the earth's surface. The concept of morphogenic balance is well illustrated in the case of the soil toposequences or catenas, where the removal of materials at the slope summit causes soil truncation, while the accumulation of the displaced materials at the footslope causes soil burying. Another example of balance between antagonistic processes that control soil development on slopes is the difference of intensity between the weathering of the substratum and the ablation of debris on the terrain surface. In active alluvial areas, the morphology of the soil results from the balance between the deposition rate of the sediments and their incorporation in the soil by the pedogenic processes.

- Geomorphology also provides a natural setting in which soil formation and evolution take place. The geomorphic environment, by way of integrating the factors of parent material, relief, time, and surface processes, constitutes an essential part of the spatial and temporal framework in which soils originate, develop, and evolve. Tricart argued that geomorphic mapping should precede soil mapping, and was not in favor of integrating both activities.

- In return, pedology provides information on soil properties such as texture, structure, aggregate stability, iron content, among others, which play an important role in the resistance of the surface materials to the morphogenic processes. Privileging his own discipline, Tricart suggests that pedology ought to be a branch of geomorphology, for the reason that pedology studies specific features of the phenomena taking place at the contact between lithosphere and atmosphere, in particular in the stratum where living beings modify a surficial part of the lithosphere, while geomorphology covers the greater part of the earth's epidermis. This view is shared by other authors such as, for example, Gerrard (1992) or Daniels and Hammer (1992). Tricart, however, recognizes that the most important thing is actually to intensify the ties of cooperation between both disciplines.

The volume on *Pedogenesis and Soil Taxonomy* published in 1983 (Wilding et al.) has been a reference book in its time, the main purpose of which was to provide a balance between soil morphology and genesis to help understand and use the comprehensive classification system of Soil Taxonomy (Soil Survey Staff 1975). The chapter written by Hall (1983) on geomorphology and pedology is an interesting inclusion in a work specifically oriented towards soil taxonomy. The above author shows that the soil is more than an object of classification and tries to reconcile soil and landscape, an aspect largely ignored in Soil Taxonomy. Hall emphasizes that it is necessary to map soils and geomorphic surfaces independently and establish correlations later, a point of view that coincides with positions previously defended by Tricart (1965a) and Ruhe (Effland and Effland 1992). He says that it is not allowed, in a new study area, to predict soils from their location on the landscape or infer the geomorphic history of the area only on the basis of soil properties. Despite this somewhat old-fashioned position, Hall acknowledges the lack of clear boundaries between geomorphic and pedologic processes, and that interdisciplinary studies are needed to explain the features that both sciences address.

In a supplement of the CATENA journal dedicated to *Soils and Geomorphology* Jungerius (1985a, b) presents the results of a broad literature review from the first works of the mid-twentieth century until the publication date of the supplement, with emphasis on papers published in CATENA. The author adopts a dichotomous approach, similar to Tricart's approach, to show the mutual contributions between both disciplines, but with emphasis on the contribution of pedology to geomorphology.

- To illustrate the significance of the landform studies for pedology, it is pointed out that pedologic processes such as additions, losses, translocations, and transformations (Simonson 1959) are under geomorphic control. Subsequently, reference is made to recurring themes in the literature that emphasize the role of relief as a factor of soil formation and geography. Highlighted topics address, for instance, the effect of the terrain physiography on the spatial distribution and cartography of soils, the effect of the topography on the genesis and catenary distribution of soil profiles, and the effect of landscape evolution on soil differentiation.
- The significance of the soil studies for geomorphology is analyzed in more detail. After showing how such studies contribute to prepare geomorphic and soil erosion maps, there is emphasis on two types of study that benefit substantially from the contribution of pedology: the studies of geomorphogenic processes and the paleogeomorphic studies. To investigate the nature of the processes that operate on a slope requires knowing the present soil system, with its spatial and temporal variations. Many of the authors cited by Jungerius (1985b) insist on the importance of the control that the horizon types exert on the geomorphic processes. A key differentiation is made between A horizons and surface crusts and their impact on the patterns of runoff and infiltration, on the one hand, and B horizons and subsurface pans and their impact on the formation of pipes and tunnels, gullies, and mass movements, on the other hand. With respect to the paleogeomorphic studies, these emphasize the importance of the paleosoils as indicators of a landscape stability phase, with the possibility of inferring factors and conditions that prevailed in the same period. The interpretation of the paleosoils helps the geomorphologist reconstruct past climate and vegetation conditions, infer the evolution time of a landscape, detect changes in a landscape configuration, and investigate past geomorphic processes.

3.3.3   Trend Towards Greater Integration

A pioneer work focusing on soils as landscape units is that of Fridland (1974, 1976). Fridland shows that soils are distributed on the landscape according to patterns that shape the structure of the soil mantle. Although the term geomorphology does not appear in his texts, he sets relationships between genetic and geometric soil entities and landforms. Ten years later, Hole and Campbell (1985) adopted Fridland's approach in their analysis of the soil landscape. Contemporaneously to the work of Fridland, Daniels et al. (1971) used the superposition of soil mantles to determine relative ages and sequences of events in the landscape, laying the foundations of pedostratigraphy.

More recent synthesis articles focus on showing how the concepts and methods of the two disciplines have been integrated to investigate interface features, rather than identifying the specific contribution of each discipline individually. Modern studies of soil geomorphology transgress the boundaries between the two sciences of origin and integrate parts of the doctrinal body of both. This new research domain constitutes an interface discipline, or "border country" as it is called by Jungerius (1985b), which gains in autonomy and maturity, with its own methodological approach and topics of interest. This has led Schaetzl and Anderson (2005) to qualify soil geomorphology as a full-fledged science. Hereafter, reference is made to some key articles that attempt to formalize the domain of soil geomorphology.

Olson (1989) considers that a study in soil geomorphology should have three main components, including (1) the recognition of the surface stratigraphy and the parent materials present in an area; (2) the determination of the geomorphic surfaces in space and time; and (3) the correlation between soil properties and landscape features. This approach is in accordance with the definition that Olson (1989) gives of soil geomorphology as the study of the landscape and the influence of landscape processes on soil formation. There is integration of the two disciplines, but geomorphology plays the leading role. In a subsequent publication (Olson 1997), the same author notes that the patterns or models of soil-geomorphology can be applied in a consistent and predictable manner in soil survey and considers that the pedologist should acquire the ability to use the pedogeomorphic patterns to interpolate within a study area or extrapolate to similar geographic areas.

The journal *Geomorphology 3* (1990) published the proceedings of a symposium dedicated to soil geomorphology (Proceedings of the 21st Annual Binghamton Symposium in Geomorphology, edited by Knuepfer and McFadden 1990). In addition to numerous articles analyzing case studies in a variety of sites and conditions, the journal contains two introductory papers that present an overview of the trends in this area in the late 1980s. McFadden and Knuepfer (1990) analyze the link between pedology and surface processes. In a short historical account, they show how pioneer work of some geologists, geomorphologists, and pedologists, concentrating on the study of the genetic relationships between soils and landscapes, resulted in the creation of the soil-geomorphology stream. The authors refer to three topics they consider central to the development of soil geomorphology. First, they point out the significance of the fundamental equation of Jenny (1941) to show the relevance of geomorphology in pedologic research through the factors of climate change, time, and relief. In particular, the study of chronosequences has contributed enormously to understanding geomorphic processes and landscape evolution, especially in river valleys with systems of nested terraces. The theme of fluvial terraces is an outstanding area of convergence, because understanding the genesis of the terraces is important to interpret the soil data. Secondly, the authors take up the issue of modelling and simulation. They contrast the conceptual models, such as those of Jenny (1941) and Johnson et al. (1990), with numerical models designed to simulate the behavior of complex systems, and consider that modelling is still limited by the poor definition of basic concepts such as polygenetic soils, soil-forming intervals, and rates of soil development, among others. Finally, the authors mention some of the problems that the investigation in soil geomorphology faces when dealing with complex landscapes. Hillslopes are a typical example of complex landscape, where the current morphogenic processes sometimes have no or little relationship with the formation of the slope itself, and often there is no clear relationship between slope gradient and degree of soil development. In synthesis, McFadden and Knuepfer (1990) consider that the soil-landform relationship is one of interaction and mutual feedback. The better we understand soils, including the speed at which the formation processes operate and the variations caused by the position of the soils on the landscape, the deeper will be our understanding of the processes that originate the landforms. Reciprocally, whenever we better understand the evolution of the landscape at variable spatial and temporal scales, we will be able to elucidate complex pedologic problems.

In the same special issue of *Geomorphology 3*, Birkeland (1990) points out that it is difficult to work in one specific field of soil geomorphology without using information from the others. He illustrates this need to integrate information by analyzing various types of chronosequence and chronofunction in arid, temperate, and humid regions. Generalizing, Birkeland considers that, in the majority of cases, the studies of soil geomorphology pursue one of the four following purposes: (1) establishing a soil chronosequence that can be used to estimate the age of the surface formations; (2) using the soils, on the basis of relevant properties of diagnostic horizons, as indicators of landscape stability in the short or long term; (3) determining relationships between soil properties that allow inferring climate changes; and (4) analyzing the interactions between soil development, infiltration and runoff, and erosion on slopes.

Following the same order of ideas, Gerrard (1993) considers that the challenge of soil geomorphology is to integrate elements from the four research areas recognized by Birkeland (1990), to develop a conceptual framework of landscape evolution. The author describes several convergent conceptual models, such as those addressing the relationship between thresholds and changes of the soil landscape, the formation of soils on aggradation surfaces, soil chronosequences, and the relationship between soil development and watershed evolution.

The book of Schaetzl and Anderson (2005) on *Soils, Genesis and Geomorphology*, contains an extensive section devoted to soil geomorphology (pp. 463–655). The authors raise soil geomorphology to the level of a discipline that deals specifically with the two-way relations between geomorphology and pedology. The relationships emerge from the fact that soils are strongly related to the landforms on which they have developed. The authors emphasize that soil geomorphology is a science based primarily in field studies. They take up again, with new examples of more or less integrated studies, the three themes that soil geomorphology has been favoring: soil catena studies, soil chronosequences, and reconstruction of landscape evolution through the study of paleosoils. As a relevant attempt to get closer to a definition of the basic principles of the discipline, Schaetzl and Anderson recognize six main topics that comprise the domain of soil geomorphology: (1) soils as indicators of environmental and climatic changes; (2) soils as indicators of geomorphic stability and landscape stability; (3) studies of soil genesis and development (chronosequences); (4) soil-rainfall-runoff relationships, especially with regard to slope processes; (5) soils as indicators of current and past sedimentological and depositional processes; and (6) soils as indicators of the stratigraphy and parental materials of the Quaternary.

This outline is similar, in more detail, to the list of objectives previously proposed by Birkeland (1990). This shows that certain conceptual and methodological coherence has been achieved.

3.4   Conclusion

Several authors have produced books and synthesis articles on soil geomorphology, with extensive lists of references that readers are suggested to consult for more information. This has contributed to make soil geomorphology a discipline in its own right. There is consensus on the basic relationship between geomorphology and pedology: geomorphic processes and resulting landforms contribute to soil formation and distribution while, in return, soil development and properties have an influence on the evolution of the geomorphic landscape. The research themes that have received more attention (in the literature) are chronosequence and toposequence (catena) studies. These two kinds of study provide the majority of the examples used to illustrate the relationships between geomorphology and pedology. Some authors favor the chronosequences as integrated study subjects including pedostratigraphy and paleopedology. Many others emphasize the study of soil distribution and evolution within the framework of the catena concept popularized by the hillslope models of Wood (1942), Ruhe (1960, 1975), and Conacher and Dalrymple (1977). Some articles point out general principles, but there is still no unified body of doctrine. There are few references in international journals that provide some formal synthesis on how to carry out integrated pedogeomorphic mapping.

References

Birkeland PW (1974) Pedology, weathering and geomorphological research. Oxford University Press, New York
Birkeland PW (1990) Soil-geomorphic research – a selective overview. Geomorphology 3:207–224
Birkeland PW (1999) Soils and geomorphology, 3rd edn. Oxford University Press, New York
Boettinger JL, Howell DW, Moore AC, Hartemink AE, Kienast-Brown S (eds) (2010) Digital soil mapping: bridging research, environmental application, and operation, vol 2, Progress in soil science. Springer, New York
Catt JA (1986) Soils and quaternary geology. Clarendon, Oxford
Conacher AJ, Dalrymple JB (1977) The nine-unit landscape model: an approach to pedogeomorphic research. Geoderma 18:1–154
Daniels RB, Hammer RD (1992) Soil geomorphology. Wiley, New York
Daniels RB, Gamble EE, Cady JG (1971) The relation between geomorphology and soil morphology and genesis. Adv Agron 23:51–88
Davis WM (1899) The geographical cycle. The genetic classification of land-forms. Geogr J (Wiley-Blackwell)14:481–504

Dobos E, Hengl T (2009) Soil mapping applications. In: Hengl T, Reuter HI (eds) Geomorphometry: concepts, software, applications, vol 33, Developments in soil science. Elsevier, Amsterdam, pp 461–479

Dobos E, Micheli E, Baumgardner MF, Biehl L, Helt T (2000) Use of combined digital elevation model and satellite radiometric data for regional soil mapping. Geoderma 97(3–4):367–391

Effland ABW, Effland WR (1992) Soil geomorphology studies in the U.S. soil survey program. Agric Hist 66(2):189–212

Elizalde G, Jaimes E (1989) Propuesta de un modelo pedogeomorfológico. Revista Geográfica Venezolana XXX:5–36

Finke PA (2012) On digital soil assessment with models and the pedometrics agenda. Geoderma 171–172(Entering the Digital Era: Special Issue of Pedometrics 2009, Beijing):3–15

Fridland VM (1974) Structure of the soil mantle. Geoderma 12:35–41

Fridland VM (1976) Pattern of the soil cover. Israel Program for Scientific Translations, Jerusalem

Gerrard AJ (1981) Soils and landforms, an integration of geomorphology and pedology. Allen & Unwin, London

Gerrard AJ (1992) Soil geomorphology: an integration of pedology and geomorphology. Chapman & Hall, New York

Gerrard AJ (1993) Soil geomorphology. Present dilemmas and future challenges. Geomorphology 7(1–3):61–84

Goosen D (1968) Interpretación de fotos aéreas y su importancia en levantamiento de suelos, vol 6, Boletín de Suelos. FAO, Roma

Goudie AS (2004) Encyclopedia of geomorphology, vol 2. Routledge, London

Grunwald S (ed) (2006) Environmental soil-landscape modeling: geographic information technologies and pedometrics. CRC/Taylor & Francis, Boca Raton

Hall GF (1983) Pedology and geomorphology. In: Wilding LP, Smeck NE, Hall GF (eds) Pedogenesis and soil taxonomy. I concepts and interactions. Elsevier, Amsterdam, pp 117–140

Hartemink AE, McBratney A (2008) A soil science renaissance. Geoderma 148:123–129

Hengl T (2003) Pedometric mapping. Bridging the gaps between conventional and pedometric approaches. ITC dissertation 101, Enschede, The Netherlands

Hengl T, MacMillan RA (2009) Geomorphometry: a key to landscape mapping and modelling. In: Hengl T, Reuter HI (eds) Geomorphometry: concepts, software, applications, vol 33, Developments in soil science. Elsevier, Amsterdam, pp 433–460

Hole FD, Campbell JB (1985) Soil landscape analysis. Rowman & Allanheld, Totowa

Ibáñez JJ, Zinck JA, Jiménez-Ballesta R (1995) Soil survey: old and new challenges. In: Zinck JA (ed) Soil survey: perspectives and strategies for the 21st century, vol 80, FAO world soil resources report. FAO-ITC, Rome, pp 7–14

Jenny H (1941) Factors of soil formation. McGraw-Hill, New York

Jenny H (1980) The soil resource. Origin and behaviour. Ecological studies, vol 37. Springer, New York

Johnson DL, Keller EA, Rockwell TK (1990) Dynamic pedogenesis: new views on some key soil concepts, and a model for interpreting Quaternary soils. Quat Res 33:306–319

Jungerius PD (ed) (1985a) Soils and geomorphology. Catena supplement, vol 6. Catena Verlag, Cremlingen

Jungerius PD (1985b) Soils and geomorphology. In: Jungerius PD (ed) Soils and geomorphology, vol 6, Catena supplement. Catena Verlag, Cremlingen, pp 1–18

Kilian J (1974) Etude du milieu physique en vue de son aménagement. Conceptions de travail. Méthodes cartographiques. L' Agronomie Tropicale XXIX(2–3):141–153

Knuepfer PLK, McFadden LD (1990) Soils and landscape evolution. Proceedings of the 21st Binghamton symposium on geomorphology. Geomorphology 3(3–4):197–578

Lagacherie P, McBratney AB, Voltz M (eds) (2007) Digital soil mapping: an introductory perspective, vol 31, Developments in soil science. Elsevier, Amsterdam

Legros JP (1996) Cartographies des sols. De l'analyse spatiale à la gestion des territoires. Presses Polytechniques et Universitaires Romandes, Lausanne

Mahaney WC (ed) (1978) Quaternary soils. Geo Abstracts, Norwich

McBratney AB, Mendonça Santos ML, Minasny B (2003) On digital soil mapping. Geoderma 117(1–2):3–52

McFadden LD, Knuepfer PLK (1990) Soil geomorphology: the linkage of pedology and surficial processes. Geomorphology 3:197–205

Olson CG (1989) Soil geomorphic research and the importance of paleosol stratigraphy to Quaternary investigations, midwestern USA, vol 16, Catena supplement. Catena Verlag, Cremlingen, pp 129–142

Olson CG (1997) Systematic soil-geomorphic investigations: contributions of RV Ruhe to pedologic interpretation. Adv Geoecol 29:415–438

Pike RJ, Evans IS, Hengl T (2009) Geomorphometry: a brief guide. In: Hengl T, Reuter HI (eds) Geomorphometry: concepts, software, applications, vol 33, Developments in soil science. Elsevier, Amsterdam, pp 3–30

Pouquet J (1966) Initiation géopédologique. Les sols et la géographie. SEDES, Paris

Principi P (1953) Geopedologia (Geologia Pedologica). Studio dei terreni naturali ed agrari. Ramo Editoriale degli Agricoltori, Roma

Retallack GJ (1990) Soils of the past. Unwin Hyman, Boston

Richards KS, Arnett RR, Ellis S (eds) (1985) Geomorphology and soils. Allen & Unwin, London

Rossiter DG (2004) Digital soil resource inventories: status and prospects. Soil Use Manage 20:296–301

Ruhe RV (1956) Geomorphic surfaces and the nature of soils. Soil Sci 82:441–455

Ruhe RV (1960) Elements of the soil landscape. Trans 7th Intl Congr Soil Sci (Madison) 4:165–170

Ruhe RV (1975) Geomorphology. Geomorphic processes and surficial geology. Houghton Mifflin, Boston

Sanchez PA, Ahamed S, Carré F, Hartemink AE, Hempel J, Huising J, Lagacherie P, McBratney AB, McKenzie NJ, Mendonça-Santos ML, Minasny B, Montanarella L, Okoth P, Palm CA, Sachs JD, Shepherd KD, Vågen TG, Vanlauwe B, Walsh MG, Winowiecki LA, Zhang G (2009) Digital soil map of the world. Science 325:680–681

Schaetzl R, Anderson S (2005) Soils: genesis and geomorphology. Cambridge University Press, New York

Simonson RW (1959) Outline of a generalized theory of soil genesis. Soil Sci Soc Am Proc 23:152–156

Skidmore AK, Watford F, Luckananurug P, Ryan PJ (1996) An operational GIS expert system for mapping forest soils. Photogram Eng Remote Sens 62(5):501–511

Soil Survey Staff (1975) Soil taxonomy. A basic system of soil classification for making and interpreting soil surveys. USDA Agric Handbook 436. US Gov Print Of, Washington

Thwaites RN (2007) Development of soil geomorphology as a sub-discipline of soil science. http://id.loc.gov/authorities/subjects/sh2007010089

Tricart J (1962) L'épiderme de la terre. Esquisse d'une géomorphologie appliquée. Masson, Paris

Tricart J (1965a) Principes et méthodes de la géomorphologie. Masson, Paris

Tricart J (1965b) Morphogenèse et pédogenèse. I Approche méthodologique: géomorphologie et pédologie. Science du Sol 1:69–85

Tricart J (1972) La terre, planète vivante. Presses Universitaires de France, Paris

Tricart J (1994) Ecogéographie des espaces ruraux. Nathan, Paris

Tricart J, Kilian J (1979) L'éco-géographie et l'aménagement du milieu naturel. Editions Maspéro, Paris

Wilding LP, Smeck NE, Hall GF (eds) (1983) Pedogenesis and soil taxonomy. I concepts and interactions. Elsevier, Amsterdam

Winter SM (2007) Soil geomorphology of the Copper River Basin, Alaska, USA. http://id.loc.gov/authorities/subjects/sh2007010089

Wood A (1942) The development of hillside slopes. Geol Ass Proc 53:128–138

Wooldridge SW (1949) Geomorphology and soil science. J Soil Sci 1:31–34

Wysocki DA, Schoeneberger PJ, LaGarry HE (2000) Geomorphology of soil landscapes. In: Sumner ME (ed) Handbook of soil science. CRC Press, Boca Raton, pp E5–E39

Zinck JA (1990) Soil survey: epistemology of a vital discipline. ITC J 1990(4):335–351

Zinck JA, Valenzuela CR (1990) Soil geographic database: structure and application examples. ITC J 1990(3):270–294

Chapter 4
The Geopedologic Approach

J.A. Zinck

Abstract The relationships between geomorphology and pedology can be analyzed from different perspectives: conceptual, methodological, and operational. Geopedology (1) is based on the conceptual relationships between geoform and soil which center on the earth's epidermal interface, (2) is implemented using a variety of methodological modalities based on the three-dimensional concept of the geopedologic landscape, and (3) becomes operational primarily within the framework of soil inventory, which can be represented by a hierarchic scheme of activities. The approach focuses on the reading of the landscape in the field and from remote-sensed imagery to identify and classify geoforms, as a prelude to their mapping along with the soils they enclose and the interpretation of the genetic relationships between soils and geoforms. There is explicit emphasis on the geomorphic context as an essential factor of soil formation and distribution.

Keywords Concept • Geopedologic landscape • Method • Geopedologic integration • Implementation • Contribution to soil survey

4.1 Introduction: Definition, Origin, Development

The first one to use the term *geopedology* was most probably Principi (1953) in his treatise on *Geopedologia (Geologia Pedologica); Studi dei Terreni Naturali ed Agrari*. In spite of the prefix *geo*, the relationships between pedology and geology and/or geomorphology are not specifically addressed, except for the inclusion of three introductory chapters on unconsolidated surface materials, hard rocks, and rock minerals, respectively, as sources of parent material for soil formation. Principi's *Geopedologia* is in fact a comprehensive textbook on pedology. Following

J.A. Zinck (✉)
Faculty of Geo-Information Science and Earth Observation (ITC), University of Twente, Enschede, The Netherlands

Institute of Environmental Studies, University of New South Wales, Sydney, NSW, Australia
e-mail: alfredzinck@gmail.com

© Springer International Publishing Switzerland 2016   27
J.A. Zinck et al. (eds.), *Geopedology*, DOI 10.1007/978-3-319-19159-1_4

the pioneer work of Principi, the term geopedology continues being used in Italy to designate the university programs dealing with soil science in general.

The geopedologic approach, as formulated hereafter, is based on the fundamental paradigm of soil geomorphology, i.e. the assessment of the genetic relationships between soils and landforms and their parallel development, but with a clearly applied orientation and practical aim. The approach puts emphasis on the *reading of the landscape* in the field and from remote-sensed documents to identify and classify geoforms, as a prelude to their mapping along with the soils they enclose and the interpretation of the genetic relationships between soils and geoforms (geoform as defined below). As such, geopedology is closely related to the concept of pattern and structure of the soil cover developed by Fridland (1974, 1976) and taken up later by Hole and Campbell (1985), but with explicit emphasis on the geomorphic context as an essential factor of soil formation and distribution.

It is common acceptance that there are relationships between soils and landscapes, but often without specifying the nature or type of the landscape in consideration (e.g. topographic, ecological, biogeographic, geomorphic). The use of landscape models has shown that the elements of the landscape are predictable and that the geomorphic component especially controls a large part of the non-random spatial variability of the soil cover (Arnold and Schargel 1978; Wilding and Drees 1983; Hall and Olson 1991). Wilding and Drees (1983), in particular, stress the importance of the geomorphic features (forms and elements) to recognize and explain the systematic variations in soil patterns. Geometrically, the geomorphic landscape and its components, which often have characteristic discrete boundaries, are discernible in the field and from remote-sensed documents. Genetically, geoforms make up three of the soil forming factors recognized in Jenny's equation (1941), namely the topography (relief), the nature of the parent material, and the relative age of the soil-landscape (morphostratigraphy). Therefore, the geomorphic context is an adequate frame for mapping soils and understanding their formation.

Geopedology aims at supporting soil survey, combining pedologic and geomorphic criteria to establish soil map units and analyze soil distribution on the landscape. Geomorphology provides the contours of the map units (i.e. the container), while pedology provides their taxonomic components (i.e. the content). Therefore, the geopedologic map units are more comprehensive than the conventional soil map units, since they also contain information about the geomorphic context in which soils are found and have developed. In this sense, the geopedologic unit is an approximate equivalent of the soilscape concept (Buol et al. 1997), with the particularity that the landscape is basically of geomorphic nature. This is reflected in the legend of the geopedologic map, which combines geoforms as entries to the legend and pedotaxa as components.

The geopedologic approach, as described below, was developed in Venezuela with the systematic application of geomorphology in the soil inventory programs that this country carried out in the second half of the twentieth century at various scales and different orders of intensity. In a given project, the practical implementation of geomorphology began with the establishment of a preliminary photo-interpretation map prior to fieldwork. This document oriented the distribution of the observation points, the selection of sites for the description of representative pedons, and the final mapping. As a remarkable feature, geoforms provided the headings of the soil map legend. The survey teams included geomorphologists and pedologists, who were trained in soil survey methodology including basic notions of geomorphology. This kind of training program had started in the Ministry of Public Works (MOP), responsible for conducting the basic soil studies for the location and management of irrigation and drainage systems in the alluvial areas of the country. It was subsequently extended and developed in the Commission for the Planning of the Hydraulic Resources (COPLANARH) and the Ministry of the Environment and Renewable Natural Resources (MARNR). From this experience was generated a first synthesis addressing the implementation of geomorphology in alluvial environment, basically the Llanos plains of the Orinoco river where large soil survey projects for the planning of irrigation schemes were being carried out (Zinck 1970). Later, with the extension of soil inventory to other types of environment, the approach was generalized to include landscapes of intermountain valleys, mountains, piedmonts, and plateaux (Zinck 1974).

Subsequently, the geopedologic approach was formalized as a reference text under the title of *Physiography and Soils* within the framework of a postgraduate course for training specialists in soil survey at the International Institute for Aerospace Survey and Earth Sciences (ITC), now Faculty of Geo-Information Science and Earth Observation, University of Twente, Enschede, The Netherlands (Zinck 1988). For over 20 years, were formed geopedologists originating from a variety of countries of Latin America, Africa, Middle East, and Southeast Asia, who contributed to disseminate and apply the geopedologic method in their respective countries. In these times, the ITC also participated in soil inventory projects within the framework of international cooperation programs for rural development. This in turn has contributed to spreading the geopedologic model in many parts of the intertropical world. In certain countries, this model has received support from official agencies for its implementation in programs of natural resources inventory and ecological zoning of the territory (Bocco et al. 1996).

The geopedologic approach was developed in specific conditions, where the implementation of geomorphology was requested institutionally to support soil survey programs at national, regional, and local levels. Originally, the first demand emanated from the Division of Edaphology, Direction of Hydraulic Works of the Ministry of Public Works in Venezuela. This institutional framework has contributed to determining the application modalities of geomorphology to semi-detailed and detailed soil inventories in new areas for land use planning in irrigation systems and for rainfed agriculture at regional and local levels. The same thing happened later with the small-scale land inventory carried out by COPLANARH as input for the water resources planning at national level. In order to simplify logistics and lower the operation costs, geomorphology was directly integrated into the soil inventory. Hence, *geopedology* turned out to be the term that best expressed the relationship between the two disciplines, with geomorphology at the service of pedology, specifically to support soil mapping. Geomorphology was considered as a tool to improve and accelerate soil survey, especially through geomorphic photo-interpretation.

Geopedology is one of several ways described in Chap. 3 that study the relationships between geomorphology and pedology or use these relationships to analyze and explain features of pedologic and geomorphic landscapes. Compared to other approaches, geopedology has a more practical goal and could be defined as the soil survey discipline, including characterization, classification, distribution, and mapping of soils, with emphasis on the contribution of geomorphology to pedology. Geomorphology especially intervenes to understand soil formation and distribution by means of relational models (for instance, chronosequences and toposequences) and to support mapping. The central concept of geopedology is that of the soil in the geomorphic landscape. The geopedologic landscape is the paradigm.

The application of geomorphology to soil inventory requires hierarchic geoform taxonomy, suitable to be used at various categorial levels according to the degree of detail of the soil inventory and cartography. In Table 4.1, the general structure and main components of such a geoform classification sytem are presented. In this context, the word *geoform* refers to all geomorphic units regardless of the taxonomic levels they belong to in the classification system, while *landform/terrain form* is the generic concept that designates the lower level of the system. The *geoform* concept includes at the same time relief features and cover formations. The vocable *landform* may lead to confusion, because it is used with different meanings in geomorphology,

Table 4.1 Synopsis of the geoform classification system

Level	Category	Generic concept	Short definition
6	Order	Geostructure	Large continental portion characterized by a given type of geologic macro-structure (e.g. cordillera, geosyncline, shield)
5	Suborder	Morphogenic environment	Broad type of biophysical environment originated and controlled by a style of internal and/or external geodynamics (e.g. structural, depositional, erosional, etc.)
4	Group	Geomorphic landscape	Large portion of land/terrain characterized by given physiographic features: it corresponds to a repetition of similar relief/molding types or an association of dissimilar relief/molding types (e.g. valley, plateau, mountain, etc.)
3	Subgroup	Relief/molding	Relief type originated by a given combination of topography and geologic structure (e.g. cuesta, horst, etc.)
			Molding type determined by specific morphoclimatic conditions and/or morphogenic processes (e.g. glacis, terrace, delta, etc.)
2	Family	Lithology/facies	Petrographic nature of bedrocks (e.g. gneiss, limestone, etc.) or origin/nature of unconsolidated cover formations (e.g. periglacial, lacustrine, alluvial, etc.)
1	Subfamily	Landform/terrain form	Basic geoform type characterized by a unique combination of geometry, dynamics and history

Zinck (1988)

pedology, landscape ecology, and land evaluation, among others. The expression *terrain form* would be preferable.

The relationships between geomorphology and pedology can be analyzed from various points of view: conceptual, methodological, and operational. Geopedology (1) is based on the conceptual relationships between geoform and soil which center on the earth's epidermal interface, (2) is implemented using a variety of methodological modalities based on the three-dimensional concept of the geopedologic landscape, and (3) becomes operational primarily within the framework of soil inventory, which can be represented by a hierarchic scheme of integrated activities.

4.2   Conceptual Relationships

Geoform and soil are natural objects that occur along the interface between the atmosphere and the surface layer of the terrestrial globe. They are the only objects that occupy integrally this privileged position. Rocks (lithosphere) lie mostly underneath the interface. Living beings (biosphere) can be present inside or below, but essentially occur above. Air (atmosphere) can penetrate into the interface, but is mostly over it. Figure 4.1 highlights the central position of the geoform-soil duo in the structure of the physico-geographical environment. The geoform integrates the concepts of relief/molding and cover formation.

4.2.1   Common Forming Factors

Because geoform and soil develop along a common interface in the earth's epidermis, a thin and fragile envelope called earth's critical zone where soils, rocks, air, and water interact, they share forming factors that emanate from two sources of matter and energy, one internal and another external.

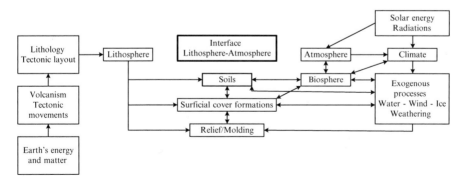

Fig. 4.1 The position of the geoform-soil duo at the interface between atmosphere and lithosphere (Adapted from Tricart 1972)

- The endogenous source corresponds to the energy and matter of the terrestrial globe. The materials are the rocks that are characterized by three attributes: (1) the lithology or facies that includes texture, structure, and mineralogy; (2) the tectonic arrangement; and (3) the age or stratigraphy. The energy is supplied by the internal geodynamics, which manifests itself in the form of volcanism and tectonic deformations (i.e. folds, faults, fractures).
- The exogenous source is the solar energy that acts through the atmosphere and influences the climate, biosphere, and external geodynamics (i.e. erosion, transportation, and sedimentation of materials).

Geoform and soil are conditioned by forming factors derived from these two sources of matter and energy that act through the lithosphere, atmosphere, hydrosphere, and biosphere. The boundaries between geoform and soil are fuzzy. The geoform has two components: a terrain surface that corresponds to its external configuration (i.e. the epigeal component) and a volume that corresponds to its constituent material (i.e. the hypogeal component). The soil body is found inserted between these two components. It develops from the upper layer of the geomorphic material (i.e. weathering products – regolith, alterite, saprolite – or depositional materials) and is conditioned by the geodynamics that takes place along the surface of the geoform (e.g. aggradation, degradation, removal). Many soils do not form directly from hard rock, but from transported detrital materials or from weathering products of the substratum. These more or less loose materials correspond to the surface formations that develop at the interface lithosphere-atmosphere, with or without genetic relationship with the substratum, but closely associated with the evolution of the relief of which they are the lithological expression (Campy and Macaire 1989). The surficial cover formations constitute, in many cases, the parent materials of the soils. The nature and extent of these surface deposits often determine the conditions and limits of the interaction between processes of soil formation (Arnold and Schargel 1978).

The fact that geoform and soil share the same forming factors generates complex cause-effect relationships and feedbacks. One of the factors, namely the relief that corresponds to the epigeal component of the geoforms, belongs inherently to the geomorphology domain. Another factor, the parent material, is partially geomorphic and partially geologic. Time is a two-way factor: the age of the parent material (e.g. the absolute or relative age of a sediment) or the age of the geoform as a whole (e.g. relative age of a terrace) informs on the likely age of the soil; conversely, the dating of a humiferous horizon or an organic layer informs on the stratigraphic position of the geoform. Therefore, the relationships between these three forming factors are both intricate and reciprocal, the geoform being a factor of soil formation and the soil being a factor of morphogenesis (e.g. erosion-accumulation on a slope). Biota and climate influence both the geoform and the soil, but in a different way. In the case of the biota, the relationship is complex, since part of the biota (the hypogeal component) lives within the soil and is considered part of it.

The geoform alone integrates three of the five soil forming factors of the classic model of Jenny (1941), while reflecting the influence of the other two factors.

This gives geomorphology a role of guiding factor in the geoform-soil pair. Its importance as a structuring element of the landscape is reflected in the geomorphic entries to the geopedologic map legend. Figures 4.2 and 4.3 provide an example of this kind of integrated approach, showing each soil unit in its corresponding geomorphic landscape unit.

The geomorphic map of Fig. 4.2 represents the graben of Punata-Cliza in the eastern Andes of Bolivia, close to the city of Cochabamba. For some time this tectonic depression was occupied by a lake that dried up into a lagunary environment. Subsequently, detrital sediments coming from the mountain borders formed fans and glacis in the margins of the depression, leaving uncovered relict lagunary flats in the center of the depression. Photo-interpretation and fieldwork allowed segmenting the alluvial fans in proximal, central, and distal sectors. The geomorphic structure of the depression bottom resulting from this evolution during the Quaternary provides the basic framework for soil formation and spatial distribution. This is reflected in the coupled geomorphic-pedologic legend of Fig. 4.3. The sequential partitioning of the geomorphic environment into landscape, relief, facies, and landform units allowed identifying and mapping geomorphic units with their respective soil taxa, forming thus geopedologic units.

4.2.2   The Geopedologic Landscape

Geoform and soil fuse to form the geopedologic landscape, a concept similar to that of soilscape (Buol et al. 1997), to designate the soil on the landscape. Geoform and soil have reciprocal influences, being one or the other alternately dominant according to the circumstances, conditions, and types of landscape. In flat areas, the

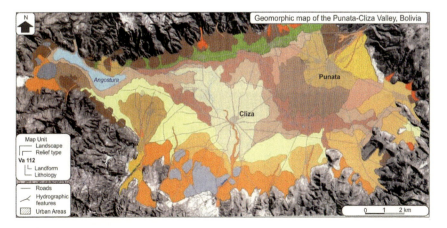

Fig. 4.2 Geomorphic map of the Punata-Cliza tectonic depression, eastern Andes of Bolivia (Metternicht and Zinck 1997)

GEOPEDOLOGIC LEGEND					
LANDSCAPE	RELIEF TYPE	FACIES	LANDFORM	CODE	SOILS
PIEDMONT	Dissected-depositional glacis	Alluvial	Proximal	Pi 111	*Association:* Typic Calciorthids / Typic Camborthids
			Central	Pi 112	*Consociation:* Typic Camborthids (ca)* / Ustochreptic Camborthids
			Distal	Pi 113	*Association:* Ustalfic Haplargids / Ustochreptic Camborthids
	Depositional glacis	Colluvio-alluvial	Distal	Pi 213	*Consociation:* Ustochreptic Camborthids / Typic Camborthids
	Active fans	Alluvial	Active channels	Pi 411	*Miscellaneous land type:* Mixed Alluvial
			Inactive channels	Pi 412	*Consociation:* Typic Torrifluvents / Typic Torriorthents
	Recent fans	Colluvio-alluvial		Pi 51	*Association:* Ustic Torriorthents / Typic Torrifluvents
	Old dissected fans	Glacio-alluvial	Proximal	Pi 661	*Association:* Typic Camborthids / Typic Haplargids
			Central	Pi 612	*Consociation:* Ustochreptic Camborthids (ca)*
			Distal	Pi 613	*Consociation:* Ustochreptic Camborthids
	Hills	Quartzitic sandstones		Pi 71	*Consociation:* Lithic Torriorthents
		Marls sandstones limestones		Pi 72	*Consociation:* Typic Calciorthids / Lithic Calciorthids
VALLEY	Lagunary depressions	Alluvio-lagunary	Higher lagunary flats	Va 111	*Association:* Fluventic Camborthids / Ustochreptic Camborthids
			Middle lagunary flats	Va 112	*Association:* Ustalfic Haplargids / Ustochreptic Camborthids
			Lower lagunary flats	Va 113	*Association:* Ustalfic Haplargids (saso)* / Ustochreptic Camborthids (sa)*
		Lagunary	Playas	Va 124	*Association:* Typic Salorthids / Natric Camborthids
* Phases: (ca) calcareous (saso) saline-alkaline (sa) saline					

Fig. 4.3 Geopedologic legend of the map shown in Fig. 4.2, referring to the Punata-Cliza tectonic depression, eastern Andes of Bolivia (Metternicht and Zinck 1997)

geopedologic landscapes are mainly constructional, while they are mainly erosional in sloping areas.

4.2.2.1 Flat Areas

In flat constructional areas, the sedimentation processes and the structure of the resulting depositional systems control often intimately the distribution of the soils, their properties, the type of pedogenesis, the degree of soil development and, even,

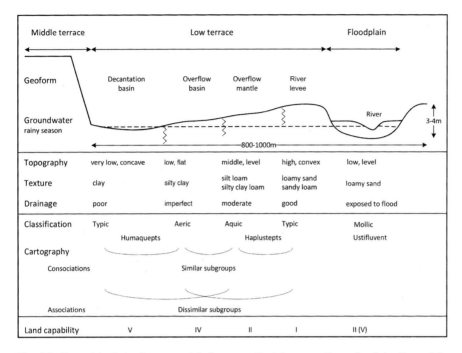

Fig. 4.4 Geopedologic landscape model of a young fluvial terrace. Example of the Guarapiche river valley, northeast of Venezuela; pedotaxa refer to the dominant soil type in each geoform

the use potential of the soils. The valley landscape offers good examples to illustrate these relationships. Figure 4.4 represents a transect crossing the lower terrace built by the Guarapiche river in the north-east of Venezuela during the late Pleistocene (Q1). In the wider sectors of the valley, the river activity produced a system that consists of a sequence of depositional units including river levee, overflow mantle, overflow basin, and decantation basin, in this order across the valley from proximal positions close to the paleo-channel of the river, to the distal positions on the fringe of the valley.

The relevant characteristics of the four members of the depositional system are as follows (pedotaxa refer to dominant soils):

- River levee (or river bank): highest position of the system, convex topography, narrow elongated configuration; textures with dominant sandy component (loamy sand, sandy loam, sometimes sandy clay loam); well drained; Typic Haplustepts (or Fluventic); land capability class I.
- Overflow mantle: medium-high position, flat topography, wide configuration; textures with dominant silty component (silt loam, silty clay loam); moderately well drained; Aquic Haplustepts (or Fluvaquentic); land capability class II.
- Overflow basin: low position, flat to slightly concave topography, wide oval configuration; mainly silty clay texture; imperfectly drained; Aeric Humaquepts; land capability class IV.

- Decantation basin: lowest position of the system, concave topography, closed
  oval configuration; usually very fine clay texture; poorly drained; Typic
  Humaquepts, sometimes associated with Aquerts; land capability class V.

The transitions between geomorphic positions are very subtle to imperceptible
on the terrain surface. External markers such as slight undulations of field border
fences and changes in color or compaction of dirt road trails help presume changes
of positions. Unit boundaries and kinds were tentatively recognized by photo-
interpretation on the basis of tone nuances, but definitively identified by field observations along transects. Parent material must be qualified to identify geoforms. The
total relief amplitude between levee and decantation basin is approximately 2 m
over a distance of about 600 m (0.3 % transversal slope).

The soil classes referred to in this example correspond to the dominant soils in
each geomorphic unit. Major soils are generally accompanied by subordinate soils
that may have common taxonomic limits with the dominant soils in the classification system (i.e. similar soils) and some inclusions that are usually not contrasting.
The geoform, with its morphographic, morphometric, morphogenic and morphochronologic features, controls a number of properties of the corresponding soil unit
(e.g. topography, texture, drainage) and relates to its taxonomic classification and
land use capability. The geoform also guides the composition of the cartographic
unit, with the possibility of mapping soil consociations on the basis of similar subgroups (e.g. Aquic Haplustepts and Aeric Humaquepts) or soil associations on the
basis of dissimilar subgroups (e.g. Typic Haplustepts and Aeric Humaquepts),
according to the soil distribution pattern and the mapping scale. The geomorphic
framework, which controls the determination and delineation of the soil map units,
makes that these units are relatively homogeneous, allowing for a reasonably reliable soil interpretation for land use purposes.

The Guarapiche valley example is an ideal textbook model, rather unfrequent in
its full expression. The complete sequence in the right depositional order occurs
mainly in the largest sections of the valley that have been sedimentologically stable
over some time (see Fig. 4.9 in Sect. 4.3.3.1). In narrow sections, some of the geomorphic positions are usually missing, with for instance the levee running parallel
to the basin. In other places, the river axis has been shifting over the depositional
area, moving for instance during a heavy flood event from unstable channel between
high levees to the low-lying marginal basin position. This results in less organized
spatial geomorphic structures and more complex geopedologic units with contrasting sediment stratifications and superpositions.

The soil sequence in a given geopedologic landscape can also vary, for instance,
according to the prevailing bioclimatic conditions (e.g. Mollisols sequence in a
moister climate) or according to the age of the terrace (e.g. Alfisols sequence on a
Q2 terrace and Ultisols sequence on a Q3 terrace). Post-depositional perturbations
in flat areas, through fluvial dissection of older terraces or differential eolian
sedimentation-deflation, for example, may cause divergent pedogenesis and increase
variations in the soil cover that are often not readily detectable. The resulting geopedologic landscapes are often much more complex than the initial constructed ones

(Ibáñez 1994; Amiotti et al. 2001; Phillips 2001; among others). McKenzie et al. (2000) mention the case of strongly weathered sesquioxidic soils in Australia that were formed under humid and warm climates during the Late Cretaceous and Tertiary and are now persisting under semiarid conditions, showing the imprints from successive environmental changes.

4.2.2.2   Sloping Areas

In sloping areas and other ablational environments, the relationships between geoform and soil are more complex than in constructed landscapes. The classic soil toposequence is an example of geopedologic landscape in sloping areas. The lateral translocation of soluble substances, colloidal particles, and coarse debris on the terrain surface and within the soil mantle results in the formation of a soil catena, whose differentiation along the slope is mainly due to topography and drainage. Typically, the summit and shoulder of a hillslope lose material, which transits along the backslope and accumulates on the footslope. This relatively simple evolution usually results in the formation of a convex-concave slope profile with shallow soils at the top and deep soils at the base. When the translocation process accelerates, for instance after removal of the vegetation cover, soil truncation occurs on the upper slope facets, while soil fossilization takes place in the lower section because pedogenesis is no longer able to digest all the incoming material via continuous soil aggradation/cumulization. Such an evolution reflects relatively clear relationships between the geomorphic context and the soil cover, which can be approximated using the slope facet models. The segmentation of the landscape into units that are topographically related, such as the facet chain along a hillside, provides a sound basis for conducting research on spatial transfers of soil components (Pennock and Corre 2001). However, this idealized soil toposequence model might not be that frequent in nature.

On many hillsides, soil development, properties, and distribution are less predictable than in the case of the classic toposequence. Sheet erosion controlled by the physical, chemical and biological properties of the topsoil horizons, along with other factors, causes soil truncation of variable depths and at variable locations. Likewise, the nature of the soil material and the sequence of horizons condition the morphogenic processes that operate at the terrain surface and underneath. For instance, the difference in porosity and mechanical resistance between surficial horizons, subsurficial layers and substratum controls the formation of rills, gullies and mass movements on sloping surfaces, as well as the hypodermic development of pipes and tunnels. The geopedologic landscapes resulting from this active geodynamics can be very complex. Their spatial segmentation requires using geoform phases based on terrain parameters (e.g. slope gradient, curvature, drainage, micro-relief, local erosion features, salinity spots, etc).

Paleogeographic conditions may have played an important role in hillslope evolution and can explain a large part of the present slope cover formations. Slopes are complex registers of the Quaternary climate changes and their effect on vegetation, geomorphic processes, and soil formation. The resulting geopedologic landscapes are polygenic and have often an intricate, sometimes chaotic structure. The superimposition or overlapping of consecutive events causing additions, translocations, and obliterations, with large spatial and temporal variations, makes it often difficult to decipher the paleogeographic terrain history and its effect on the geopedologic relationships.

The following example shows that an apparently normal convex-concave slope can conceal unpredictable variations in the covering soil mantle. The case study is a soil toposequence along a mountain slope between 1100 masl and 1500 masl in the northern Coastal Cordillera of Venezuela (Zinck 1986). Soils have developed from schist under dense tropical cloud forest, with 1850 mm average annual rainfall and 19 °C average annual temperature. Slope gradient is 2–5° at slope summit, 40–45° at the shoulder, 30–40° along the backslope, and 10–25° at the footslope. By the time of the study, no significant erosion was observed. However, several features indicate that the current soil mantle is the result of a complex geopedologic evolution, with alternating morphogenic and pedogenic phases, during the Holocene period.

- Except at the slope summit, soils have formed from detrital materials displaced along the slope, and not directly from the weathering in situ of the geologic substratum.
- There is no explicit correlation between slope gradient and soil properties. For instance, shoulder soils are deeper than backslope soils, although at higher slope inclination.
- Many soil properties such as pedon thickness and contents of organic carbon, magnesium and clay show discontinuous longitudinal distribution along the slope (Figs. 4.5 and 4.6). The most relevant interruption occurs in the central stretch of the slope, around 1300 m elevation.
- Soils in the upper part of the slope have two Bt horizons (a sort of bisequum) that reflect the occurrence of two moist periods favoring clay illuviation, separated by a dry phase.

Pollen analysis of sediments from a nearby lowland lake reveals that, by the end of the Pleistocene, the regional climate was semi-arid, vegetation semi-desertic, and soils probably shallow and discontinuous (Salgado-Labouriau 1980). From the beginning of the Holocene when the cloud forest started covering the upper ranges of the Cordillera, deep Ultisols developed. During the Holocene, dry episodes have occurred causing the boundary of the cloud forest to shift upwards and leaving the lower part of the slope, below approximately 1350–1300 masl, exposed to erosion. The presence, in the nearby piedmont, of thick torrential deposits dated 3500 BP and 1500 BP indicates that mass movements have episodically occurred upslope during the upper Holocene. This would explain why soil features and properties show a clear discontinuity at mid-slope, around 1300 masl.

The alternance of morphogenic and pedogenic activity along mountain and hill slopes causes geopedologic relationships to be complex in sloping areas, in general

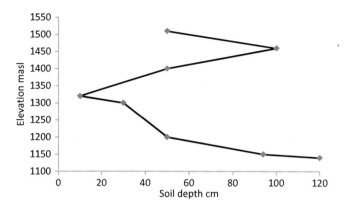

Fig. 4.5 Variation of soil depth with elevation along a mountain slope in the northern Coastal Cordillera of Venezuela (Zinck 1986)

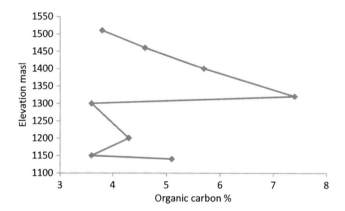

Fig. 4.6 Variation of organic carbon content (0–10 cm) with elevation along a mountain slope in the northern Coastal Cordillera of Venezuela (Zinck 1986)

more complex than in flat areas. The older the landscape, the more intricate are the relationships between soil and geoform because of the imprints left by successive environmental conditions.

4.3   Methodological Relationships

The methodological relationships refer to the modalities used to analyze the spatial distribution and formation of the geofom-soil complex. Geomorphology contributes to improving the knowledge of soil geography, genesis, and stratigraphy. In return, soil information feeds back to the domain of geomorphology by improving the knowledge on morphogenic processes (e.g. slope dynamics). The above needs the integration of geomorphic and pedologic data in a shared structural model to iden- tify and map geopedologic units.

4.3.1 Geopedologic Integration: A Structural Model

Figure 4.7 shows the data structure of the geoform-soil complex in the view of the geopedologic approach (Zinck and Valenzuela 1990). Soil survey data are typically derived from three sources: (1) visual interpretation and digital processing of remote-sensed documents, including aerial photographs, radar and multi-spectral images, and terrain elevation models; (2) field observations and instrumental mea- surements, including biophysical, social, and economic features; and (3) analytical determinations of mechanical, physical, chemical, and mineralogical properties in the laboratory. The relative importance of these three data sources varies according to the scale and purpose of the soil survey. In general terms, the larger is the scale of the final soil map, the more field observations and laboratory determinations are required to ensure an appropriate level of information.

As soils and geoforms are three-dimensional bodies, external and internal (rela- tive to the terrain surface) features are to be described and measured to establish and delimit soil map units. The combination of data and information provided by sources (1) and (2) serves to describe the environmental conditions and areal dynamics (e.g. erosion, flooding, aggradation of sediments, changes in land uses, etc) and to delin- eate the map units. At this level, the implementation of geomorphic criteria through interpretation of remote-sensed documents and field prospection plays a relevant role for the identification and characterization of the soil distribution patterns and

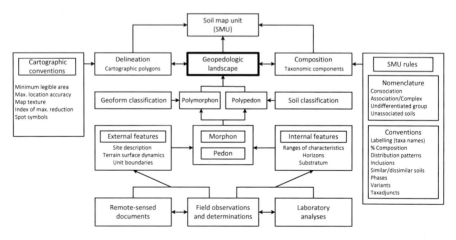

Fig. 4.7 Conceptual-structural model of the geopedologic approach (Zinck and Valenzuela 1990)

the understanding of their spatial variability. The interpretation of remote-sensed documents (photo, image, DEM) can benefit from applying a stepwise procedure of features identification using the geoform hierarchy to highlight the nested structure of the landscape (see Table 4.1). The sequence of steps includes photo/image reading, identification of master lines, sketching the structure of the landscape to select representative cross sections, pattern recognition along the cross sections, delimitation of the geomorphic units via interpolation and extrapolation, and establishing a preliminary geomorphic interpretation legend for field verification.

The combination of data and information provided by sources (2) and (3) allows characterizing and quantifying the properties of the pedologic materials, geomorphic cover formations, and geologic substrata. The horizon (or layer) is the basic unit of data collection. Horizon and substratum information is aggregated in observation profiles, modal pedons, and modal morphons. Pedon and polypedon are described and established according to the criteria of Soil Taxonomy (Soil Survey Staff 1999). The morphon is the geomorphic equivalent of the pedon. It is described at the same site as the pedon but without fixed size standards. Conventionally, the areal size of a pedon varies from 1 m² for horizontally layered soils to 10 m² for soils having cyclic horizons. The extent of a morphon is obviously larger to capture the variations of the terrain surface. The description of the morphon includes internal and external features. The internal features correspond to the characteristics and properties of the geomaterial in the substratum, thus the parent material of the soil. The external features cover the conditions and dynamics of the terrain area at the site of description and its surroundings. The pedologic material (i.e. the solum) occupies the volume between the substratum and the terrain surface. As in the case of the pedon, the morphon is the description and sampling site. Therefore, pedon and morphon are two fundamentally related entities. This is nothing new, since the description of the pedon has always included that of the parent material and surface features. However, the contribution of the geomorphic analysis methods improves the characterization of the geomaterials in the substratum and that of the surface geodynamics. The methodological integration can be achieved by experts skilled in both geomorphology and pedology or by interdisciplinary teams.

The concepts of polypedon and polymorphon are significantly different from each other. The polymorphon corresponds to a whole geoform and is therefore a more comprehensive unit than the polypedon. A polymorphon can include more than one polypedon, and this is actually often the case, especially at the upper levels of the geoform classification system. The foregoing is reflected in the taxonomic composition of the map units: a relatively homogeneous geoform may correspond to a consociation of similar soils, while a less homogeneous geoform may correspond to an association of dissimilar soils. The identification and description of the polymorphon follow the criteria set out in Chaps. 6, 7 and 8, which deal with the taxonomy and attributes of the geoforms. Variations among identification profiles by comparison with a modal profile (pedon or morphon) are expressed in terms of ranges of characteristics for each taxon present in a map unit.

At this stage, the available data consist of: (1) geopedologic point observations, with additional information on the spatial variations of the characteristics, and (2) a framework of spatial units based essentially on external geomorphic criteria (i.e. characteristics of the terrain surface). The combination of the two results in a map of geopedologic units.

For mapping purposes, both objects – soil and geoform – are given identification names (i.e. taxonomic names) that are supplied by their respective classification systems. Assemblies of contiguous similar soils, forming polypedons, are classified by comparison with taxonomic entities established in soil classification systems, such as Soil Taxonomy (Soil Survey Staff 1999), the WRB classification (IUSS 2007), or any national classification. A similar procedure is used for the classification of the geomorphic units, moving from the description and sampling unit (morphon) to the classification entity (polymorphon). A basic geomorphic unit (polymorphon) can contain one or more polypedons. For instance, Entisols (e.g. Mollic Ustifluvents) and Mollisols (e.g. Fluventic Haplustolls) can occur intermixed with contrasting inclusions in a recent river levee position. The combination in the landscape of a polymorphon with the associated polypedons constitutes a geopedologic landscape unit.

Due to the inherent spatial anisotropy of the pedologic material, which is generally more pronounced than the anisotropy of the geomorphic material, soil delineations are usually heterogeneous. This requires that the taxonomic components of a map unit be named and their respective proportions quantified using conventional rules of soil cartography (Soil Survey Staff 1993). The delimitation of polygons follows a number of cartographic conventions that assure a good readability of the soil map. In this way, the geopedologic landscape units, cartographically and taxonomically controlled, as unique combinations of geomorphic polygons and their pedologic contents, result being the soil map units.

This theoretical-methodological model of the geoform-soil complex can be implemented to design the structure of an integrated geopedologic database, such as shown in Zinck and Valenzuela (1990).

4.3.2 Geopedologic Integration: Soil Geography, Genesis, and Stratigraphy

Within the framework of the previously described geopedologic model, themes such as soil geography, genesis, and stratigraphy can benefit substantially from the integration of pedologic and geomorphic methods.

4.3.2.1 Soil Geography

Soil survey generates information on the spatial distribution of soils. The implementation of geomorphic criteria in soil survey improves the identification and delimitation of the soils. At the same time, the rationality of the geopedologic approach contributes to compensate or partially replace what Hudson (1992) called the acquisition of tacit knowledge for the application of the soil-landscape paradigm. The integrated geopedologic analysis facilitates the reading of the landscape, because the geomorphic context controls, in a large proportion, the soil types that are found associated in a given kind of landscape such as, for instance, the sequence of levee-mantle-basin in an alluvial plain or the sequence of summit-shoulder-backslope-footslope along a hillside. These models of geopedologic associations that are genetically related and produce characteristic spatial patterns, are the components (i.e soil combinations) of what Fridland (1974) calls *the structure of the soil cover* and Schlichting (1970) formulates as *Bodensoziologie* (i.e. pedosociology). Geopedologic spatial patterns depict the landscape and its elements the same way they can be seen in nature, in contrast to the artificial delineations shown on some geostatistically-based soil maps. This is why geopedologic maps are easy to read, even for non-specialists. For instance, on Fig. 4.2 it is easy to recognize the triangular shape of the alluvial fans.

- *Soil identification* is based on the description of the soils in the field, which leads to their characterization and classification. Geomorphology contributes to this activity through the selection of the description sites. The use of geomorphic criteria facilitates the choice of representative sites, regardless of the implemented sampling scheme. In oriented sampling, the observation sites are pre-selected based on geomorphic criteria within units delimited by interpretation of aerial photos or satellite images. Random sampling only makes sense if it is applied within the framework of units previously established with geomorphic criteria. A random sampling scheme is more objective and appropriate for statistical data analysis, but frequently generates a number of little representative profiles and, for this reason, is more expensive.

  Grid-based systematic sampling is difficult to apply as an operational technique to an entire soil survey project because it would be too costly. It is useful when applied locally to estimate the spatial variability of the soils within and between selected map units and to establish their degree of purity. Bregt et al. (1987) compare two thematic soil maps, one derived from a conventional soil map and another one obtained by kriging of grid point data. The average purity of the map units, determined on the basis of three criteria including thickness of the A horizon, depth to gravel, and depth to boulder clay, is 77 % in both cases, with less dispersion in the first case (72–82 %) than in the second (69–85 %). The interpretation of geostatistical data is probably more meaningful when geomorphic criteria are used.

- *Soil delimitation* is based on the interpretation of aerial photos and satellite images, the use of digital elevation models, and fieldwork. The features detected by remote sensing are essentially ground surface features, which are often of geomorphic nature. Therefore, what is observed or interpreted in remote-sensed documents are characteristics of the epigeal part of the geoforms and soils. The hypogeal part is still largely inaccessible and some of its features can be detected at distance only with special techniques (e.g. GPR). This is efficient when a three-dimensional representation of the geomorphic landscape is available, which can be obtained by stereoscopic interpretation of aerial photos or satellite images or based on a combination of images and elevation or terrain models.

In this context, geomorphology contributes to the following tasks related to soil delimitation: (1) the selection of sample areas, transects, and traverses; (2) the drawing of the soil map unit boundaries based on the conceptual relations between geoforms and soils (common forming factors; geopedologic landscape); and (3) the identification, temporal monitoring, and explanation of the spatial variability of the soils.

- *Soil variability* is partly controlled by the geomorphic context, especially systematic variability (Wilding and Drees 1983). Landform and soil patterns match often on a one-to-one correspondence (Wilding and Lin 2006). Geomorphology provides criteria for segmenting the soilscape continuum into discrete units that are relatively homogeneous. Such units are suitable frameworks for estimating the spatial variability of soil properties using geostatistical analysis (Saldaña et al. 1998; Kerry and Oliver 2011). They have been used also as reference units to apply spatial analysis metrics, including indices of heterogeneity, diversity, proximity, size and configuration, for quantitatively describing soil distribution patterns at various categorial levels of geoform (i.e. landscape, relief, terrain form) (Saldaña et al. 2011; Toomanian 2013).

The mapping scale and observation density influence the relationship between geoform and soil, as the spatial variability of the geomorphic and pedologic properties are not the same magnitude. In general, at large scales the latter vary more than the former, especially at short distances. Therefore, the geopedologic approach may perform better at smaller than at larger scales. Rossiter (2000) considers that the approach is adequate for semi-detailed studies (scales 1:35,000 to 1:100,000). Esfandiarpoor Borujeni et al. (2009) analyzed the effect of three observation point intervals (125, 250, and 500 m) on the results of applying the geopedologic approach to soil mapping and concluded that this approach works satisfactorily in reconnaissance or exploratory surveys. To increase the accuracy of the geopedologic results at large scales, they suggest adding a category of landform phase. The geoform classification system already includes the concept of phase for any practical subdivision of a landform or of any geoform class at other categorial levels (Zinck 1988). Using statistical and geostatistical methods, Esfandiarpoor Borujeni et al. (2010) show that the means of the soil variables in similar landforms within their study area were comparable but not their variances. They conclude that the geopedologic soil mapping approach is not completely satisfactory for detailed mapping scales (1:10,000 to 1: 25,000) and suggest, as above, the use of landform phases to increase the accuracy of the geopedologic results.

Similarly, the geoform-soil integration facilitates the extrapolation of information obtained in sample areas to unvisited areas or areas of difficult access, using artificial neural networks and decision trees, among other techniques (Moonjun et al. 2010; Farshad et al. 2013). Using a set of terrain parameters extracted from a digital elevation model, Hengl and Rossiter (2003) show that supervised landform classification allowed extrapolating geopedologic information obtained from photo-interpretation of selected sample areas over a large hill and plain region with about 90 % reproducibility.

The geomorphic context is far from embracing the full span of soil variability. However, its contribution to soil cartography decreases in general the amplitude of variation of the soil properties within map units enough to make practical interpretations and decisions for land use planning. Systematic soil surveys using the geopedologic approach in large areas have performed satisfactorily when used for general land evaluation. Specific applications such as precision farming or site engineering need to be supported by very detailed soil information.

4.3.2.2 Soil Genesis and Stratigraphy

Geomorphic processes and environments are used, respectively, as factors and spatial frameworks to explain soil formation and evolution. The geomorphic context, through parent material (weathering products or depositional materials), relief (slope, relative elevation, aspect), drainage conditions, and morphogenesis, controls a large part of the soil forming factors and processes. In return, the soil properties influence the geomorphic processes. There is co-evolution between the pedologic and geomorphic domains. At the same time, the geomorphic history controls soil stratigraphy, while soil dating (i.e. chronosequences) helps reconstruct the evolution of the geomorphic landscape. The use of geomorphic research methods and techniques contributes to elucidate issues in soil genesis and stratigraphy.

Figure 4.8 shows a model of geopedologic relationships in a chronosequence of nested alluvial terraces, in the Guarapiche river valley, Venezuela (Zinck 1970). The geoform, here at the categorial level of terrain form (see Table 4.1), controls soil

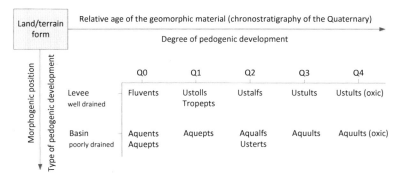

Fig. 4.8 Model of geopedologic relationships in alluvial soils, Guarapiche river valley, Venezuela (Zinck 1970)

formation in two directions. On the one hand, the relative age of the geomorphic material, i.e. the parent material of the soils, from Holocene (Q0) to lower Pleistocene (Q4), directly influences the *degree* of pedogenic development from the level of Entisol to that of Ultisol. On the other hand, the nature of the geomorphic position closely influences the *type* of pedogenic development, distinguishing between well drained soils with ustic regime in levee position and poorly drained soils with aquic regime in basin position.

4.3.3   Geopedologic Integration: A Test of Numerical Validation

4.3.3.1   Materials and Method

The contribution of geomorphology to soil knowledge and, in particular, to the spatial distribution of soils can be considered efficient if, among other things, it facilitates and improves the grouping of the soils into relatively homogeneous cartographic units. To substantiate the geopedologic integration and validate quantitatively the relationships between geoform and soil, the technique of numerical classification was implemented, as the latter allows comparing the performance of an object classification system in relation to a reference system (Sokal and Sneath 1963).

A numerical classification test of the geopedologic units supplied by a semi-detailed soil survey (1:25,000) of the Guarapiche river valley, northeast of Venezuela (Zinck and Urriola 1971), was run to estimate the efficiency of both the soil classification and the geoform classification in building consistent groups by comparison with the phenetic groups of the numerical classification (Zinck 1972). The geopedologic units belong to a chronosequence of nested terraces, spanning the Quaternary from the lower Pleistocene (Q4) to the Holocene (Q0). Soils have formed mostly from longitudinal alluvial deposits, coming from the upper catchment area of the river, and secondarily from local colluvial deposits (Fig. 4.9).

Twenty-six pairs of modal pedons-morphons, representative of the soil series mapped in the survey area, were chosen, and 24 mechanical, physical and chemical properties were selected to characterize the pedologic material (solum) and the geomorphic material (parent material). Soil units classified at subgroup level (Soil Survey Staff 1960, 1967) and geomorphic units classified by depositional facies and relative age were compared. Data handling implemented techniques and methods available in the 1960s when the essay was performed: (1) the method of Hole and Hironaka (1960) for estimating the index of similarity between pairs of units and elaborating the similarity matrix, and (2) the method using unweighted pair-groups with arithmetic mean as described in Sokal and Sneath (1963) to cluster the units, construct the dendrogram represented in Fig. 4.10, and calculate the average similarities.

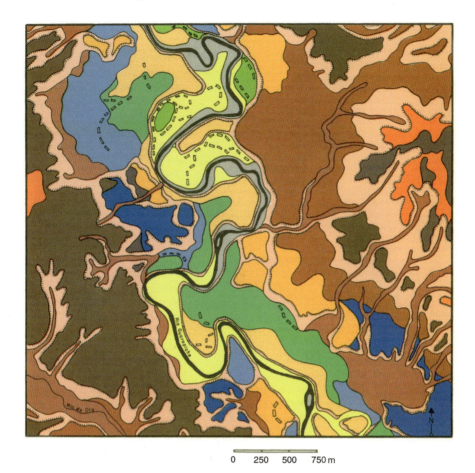

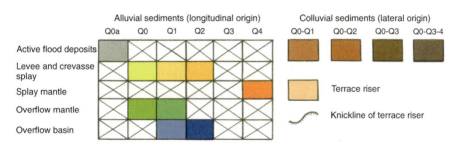

Fig. 4.9 Portion of the Guarapiche river valley, northeast of Venezuela, showing a chronosequence of nested terraces covering the Quaternary period (from Q0 to Q4). The boundaries of the cartographic units are essentially of geomorphic nature, while their contents are of pedologic nature (consociations and associations of soil series, not shown here). Extract of the original soil map at 1:25,000 scale (Zinck and Urriola 1971)

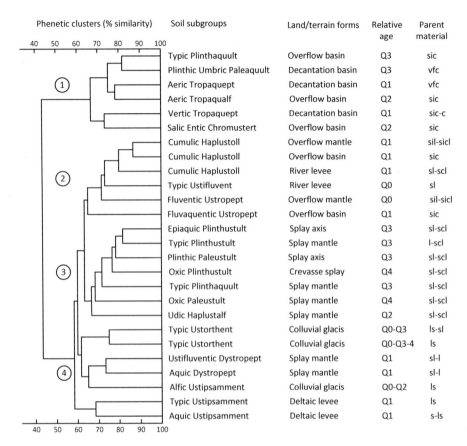

Phenetic clusters (% similarity)	Soil subgroups	Land/terrain forms	Relative age	Parent material
	Typic Plinthaquult	Overflow basin	Q3	sic
	Plinthic Umbric Paleaquult	Decantation basin	Q3	vfc
①	Aeric Tropaquept	Decantation basin	Q1	vfc
	Aeric Tropaqualf	Overflow basin	Q2	sic
	Vertic Tropaquept	Decantation basin	Q1	sic-c
	Salic Entic Chromustert	Overflow basin	Q2	sic
	Cumulic Haplustoll	Overflow mantle	Q1	sil-sicl
	Cumulic Haplustoll	Overflow basin	Q1	sic
	Cumulic Haplustoll	River levee	Q1	sl-scl
②	Typic Ustifluvent	River levee	Q0	sl
	Fluventic Ustropept	Overflow mantle	Q0	sil-sicl
	Fluvaquentic Ustropept	Overflow basin	Q1	sic
	Epiaquic Plinthustult	Splay axis	Q3	sl-scl
	Typic Plinthustult	Splay mantle	Q3	l-scl
	Plinthic Paleustult	Splay axis	Q3	sl-scl
③	Oxic Plinthustult	Crevasse splay	Q4	sl-scl
	Typic Plinthaquult	Splay mantle	Q3	sl-scl
	Oxic Paleustult	Splay mantle	Q4	sl-scl
	Udic Haplustalf	Splay mantle	Q2	sl-scl
	Typic Ustorthent	Colluvial glacis	Q0-Q3	ls-sl
	Typic Ustorthent	Colluvial glacis	Q0-Q3-4	ls
	Ustifluventic Dystropept	Splay mantle	Q1	sl-l
④	Aquic Dystropept	Splay mantle	Q1	sl-l
	Alfic Ustipsamment	Colluvial glacis	Q0-Q2	ls
	Typic Ustipsamment	Deltaic levee	Q1	ls
	Aquic Ustipsamment	Deltaic levee	Q1	s-ls

Fig. 4.10 Dendrogram showing four groups of geopedologic units; Guarapiche river valley, Venezuela (Zinck 1972). Soil classification according to Soil Survey Staff (1960, 1967). Relative age of the geomorphic material (i.e. soil parent material) by increasing order from Q0 (Holocene) to Q4 (lower Pleistocene). Texture of the parent material: *s* sand, *l* loam, *si* silt, *c* clay, *vf* very fine

4.3.3.2 Results

The numerical classification generated four phenetic groups with a variable number of geopedologic units (i.e. soil-geoform combinations). The soils are reported as subgroup classes. Geoforms are identified by their sedimentary position at the terrain form level, their relative age, and the texture of the depositional material (i.e. the parent material of the soils).

• Group 1: six geopedologic units that share the following characteristics: low topographic positions of overflow basin (three) or decantation basin (three), poorly drained (five units with aquic regime), and fine-textured (silty clay or clay), regardless of the chronostratigraphy of the parental materials (relative age varying from Q1 to Q3) and the degree of soil development (one Vertisol, two Inceptisols, one Alfisol, two Ultisols).

- Group 2: six geopedologic units that share the following characteristics: medium to high topographic positions of levee (two), overflow mantle (two), and overflow basin (two), well drained, textures mostly loamy and silty, soils of incipient to moderate development (one Entisol, two Inceptisols, three Mollisols), all formed from recent to relatively recent materials (Q0 and Q1).
- Group 3: seven geopedologic units that share the following characteristics: medium to high topographic positions of splay axis, splay mantle and crevasse splay, moderately well to well drained, textures sandy loam and sandy clay loam, soils of advanced development (one Alfisol, six Ultisols), all formed from old materials (Q3 and Q4).
- Group 4: seven geopedologic units with predominantly sandy textures (loamy sand and sandy loam) that restrict soil development to an incipient stage (five Entisols including three Psamments, two Inceptisols); the soils occur in a variety of depositional sites (deltaic levee, splay mantle, colluvial glacis) and chronostratigraphic units (from Q0 to Q4; the colluvial deposits being of continuous, diachronic formation).

In all cases, the factor that most closely controls the grouping of the geopedologic units is of geomorphic nature, with specific leading factors clustering the soils in each group:

- Group 1: basin depositional facies and low position in the landscape.
- Group 2: relatively recent age of the parental materials (late Pleistocene to Holocene).
- Group 3: advanced age of the parental materials (lower to early middle Pleistocene).
- Group 4: coarse textures of the parent materials.

4.3.3.3 Conclusion

Mean similarities of great soil groups (73 %) and terrain forms (75 %) are comparable to the average similarity of the numerical groups (75 %), indicating that the three classification modes are relatively efficient in generating consistent groupings. Groups 2 and 3 are more homogeneous than groups 1 and 4. The factors that most contribute to differentiate the four groups and generate differences within the heterogeneous groups are attributes of the geoforms, in particular their depositional origin (with their particle size distribution), their position in the landscape, and their relative age. These factors basically correspond to three of the five soil forming factors: i.e. parent material, topography-drainage, and time, which together highlight the contribution of geomorphology to pedology and constitute the foundation of geopedology.

4.4 Operational Relationships

4.4.1 Introduction

The conceptual and methodological relationships between geoform and soil can be implemented basically in two ways: (1) through studies at representative sites, usually of limited extent, to analyse in detail the genetic relationships between geoforms and soils (scientific studies, mostly in the academic domain), and (2) through the inventory of the soils as a resource to establish the soil cartography of a territory (project area, region, entire country) and assess their use potential and limitations (practical studies, in the technical domain).

The operational relationships are examined here in the framework of the soil inventory, from the generation of the geopedologic information through field survey to its interpretation through land evaluation for multi-purpose uses. In this process, geomorphology can play a relevant role. The operational importance of geomorphology refers to the value added to the soil survey information when geomorphology is incorporated into the successive stages of the survey operation.

Soil survey is an information system, which can be represented by a model that describes its structure and functioning using systems analysis, and which allows to estimate the efficiency of the contribution of geomorphology to the soil survey. The opportunity to conduct a trial of this nature was given by a semi-detailed soil survey project to be carried out in the basin of Lake Valencia, Venezuela (Zinck 1977). This is a region of approximately 1000 km² of flat land bordered by mountains, traditionally used for intensive irrigated agriculture, but increasingly exposed to land-use conflicts as a result of fast, uncontrolled urban-industrial sprawling. The size of the study area, the level of detail of the survey, the diversity of objectives to meet, and the number of personnel involved, were decisive factors in the design of the study. A reference framework was needed to plan the survey activities, establish the timetable for implementation, and select the variety of soil interpretations required to supply the necessary information for land-use planning and contribute to mitigate the land-use conflicts.

4.4.2 The Structure of the Soil Survey

Proceeding by iteration, a model structure with five categorial levels was obtained, as represented in Fig. 4.11. The three lower levels comprise the domain proper of the soil survey – its internal area – where the information is produced. The two upper levels represent the sphere of influence of the soil survey – its external area – where the information generated is implemented. Each level responds to a generic concept and, at each level, a series of tasks is performed (Tables 4.2, 4.3, 4.4, 4.5, and 4.6).

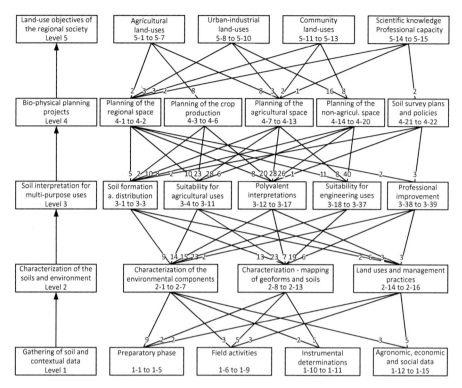

Fig. 4.11 Graph representing the soil survey as an information system, with production, interpretation, and dissemination of data and information, Lake Valencia project (Zinck 1977). The numbers in the boxes refer to the themes labelled in Tables 4.2, 4.3, 4.4, 4.5, and 4.6. The numbers inserted in the arrows indicate the amount of critical pathways through which information circulates from a given level to the following one

The numbers in the boxes refer to the themes labelled in Tables 4.2, 4.3, 4.4, 4.5, and 4.6. The numbers inserted in the arrows indicate the amount of critical pathways through which information circulates from a given level to the following one.

- Level 1: elementary tasks, which consist in the generation of the basic data, including the interpretation of aerial photos, satellite images and DEM, soil description and sampling, laboratory determinations, and gathering of agronomic, social, and economic data.
- Level 2: intermediate tasks, which consist in the synthesis of the information, including the characterization of the environmental components, characterization and mapping of the geoforms and soils, and description of the land-use types and management practices.
- Level 3: final tasks, which consist in the interpretation of the information for multiple purposes, including the genetic interpretation of the soils and their formation environments, land evaluation for agricultural, engineering, sanitary, recreational and aesthetic purposes, and professional improvement of the geopedologists.

Table 4.2 Level 1 themes: elementary soil study tasks; information collection

1-1 Collection and analysis of existing no-pedologic information
1-2 Photo-field exploration, analysis of existing soil information, identification soil legend
1-3 Generalized 1: 50,000 photo-interpretation, identification of the physical-natural macro-units
1-4 Selection of the sample areas
1-5 Detailed 1: 25,000 photo-interpretation, identification of the geoforms, location of the sample areas
1-6 Survey of the sample areas
1-7 Control observations, photo-interpretation adjustments
1-8 Composition of the cartographic units, descriptive soil legend
1-9 Description of representative pedons
1-10 Physical field determinations and measurements
1-11 Laboratory determinations
1-12 Survey of crop yields, production costs, and development costs
1-13 Survey of irrigation practices
1-14 Survey of cultivation and conservation practices
1-15 Evaluation of deforestation, levelling, drainage, stone-removal costs

Lake Valencia project (Zinck 1977)

Table 4.3 Level 2 themes: intermediate soil study tasks; synthesis of the information on soil and environment characterization

2-1 Characterization of the climate
2-2 Characterization of the surface hydrology and hydrography
2-3 Characterization of existing hydraulic works
2-4 Characterization of the water quality
2-5 Characterization of the topography
2-6 Characterization of the geology and hydrogeology
2-7 Characterization of the geomorphology and hidrogeomorphology
2-8 Geopedologic mapping and soil map preparation
2-9 Morphologic characterization of the soils
2-10 Chemical characterization of the soils
2-11 Mineralogical characterization of the soils
2-12 Physical characterization of the soils
2-13 Mechanical characterization of the soils
2-14 Survey of current land-uses
2-15 Survey of management practices and levels
2-16 Evaluation of required improvements and their feasibility

Lake Valencia project (Zinck 1977)

Table 4.4 Level 3 themes: final soil study tasks; multi-purpose interpretations	3-1 Overall characterization of the natural environment (integrated study)
	3-2 Spatial distribution of the soils (soil chorology)
	3-3 Genesis and taxonomic classification of the soils
	3-4 Land suitability for rainfed agriculture
	3-5 Land suitability for irrigated agriculture
	3-6 Land suitability for ornamental plants and garden vegetables
	3-7 Agricultural productivity (productivity of the land)
	3-8 Development costs for agricultural land-use
	3-9 Current soil fertility
	3-10 Soil salinity
	3-11 Limitations of the land for the use of mechanized farm implements
	3-12 Characterization of the natural drainage
	3-13 Drainability of the land
	3-14 Current morphodynamics (erosion, sedimentation)
	3-15 Erodibility of the land
	3-16 Land irrigation requirements
	3-17 Water availability
	3-18 Sources of material for topsoil
	3-19 Sources of sand and gravel
	3-20 Sources of material for road filling
	3-21 Constraints for road network design
	3-22 Limitations for road cuts
	3-23 Limitations for placement of cables and pipes
	3-24 Limitations for foundations of low buildings and houses
	3-25 Limitations for embankment foundations
	3-26 Limitations for residential areas
	3-27 Limitations for streets and parking lots
	3-28 Limitations for excavation of channels
	3-29 Limitations for construction of farm ponds
	3-30 Limitations for construction of dikes
	3-31 limitations for septic filtration areas
	3-32 Limitations for oxidation ponds
	3-33 Limitations for waste disposal areas
	3-34 Limitations for recreation areas (picnic, play grounds)
	3-35 Limitations for lawns, golf courses, landscaping
	3-36 Limitations for camping sites
	3-37 Limitations for sports fields
	3-38 Training of the technical personnel
	3-39 Publications, conferences, education
	Lake Valencia project (Zinck 1977)

Table 4.5 Level 4 themes:
regional planning and
development projects,
designed and executed by
official and private entities

4-1 Soil correlation
4-2 Land-use zoning in the regional space (arbitration between competitive uses)
4-3 Ecological zoning of crops
4-4 Selection of crop and rotation systems
4-5 Substitution of crops in time and space
4-6 Increase of land productivity (yields)
4-7 Determination of agricultural plot sizes
4-8 Irrigation planning and management
4-9 Improvement of poorly drained soils
4-10 Improvement of saline soils
4-11 Management of heavy soils (clay soils)
4-12 Soil conservation techniques
4-13 Agricultural extension
4-14 Urban and peri-urban planning (master zoning plan)
4-15 Supply of water and gas
4-16 Control of soil and water pollution
4-17 Disposal or recycling of industrial, urban, and agricultural wastes
4-18 Channelling and excavation of effluents
4-19 Planning of communication routes
4-20 Tourism development
4-21 Professional training and improvement
4-22 Expanding basic knowledge in geomorphology and pedology
Lake Valencia project (Zinck 1977)

- Level 4: primary external objectives, which correspond to biophysical planning in the local and regional contexts, including territorial zoning, planning of the agricultural and non-agricultural areas, planning of the agricultural production, and formulation of soil survey policies and plans.
- Level 5: final external objectives, which correspond to the concerns, perceptions, and priorities of the regional (or national) society in terms of agricultural land-use, urban-industrial land-use, use of community spaces, and creation of scientific knowledge and improvement of professional skills.

4.4.3   The Functioning of the Soil Survey

The operation of the system refers to the information flows that circulate through the soil survey. To identify the direction of the information flows and evaluate their intensity, several matrices relating the themes of the consecutive layers of the model

Table 4.6 Level 5 themes:
relevant technical issues
faced by the regional (or
national) community

5-1 Marginal agriculture
5-2 Land reform
5-3 Intensification processes of agriculture
5-4 Incorporation of new areas to agricultural activities
5-5 Supply of agricultural products for human consumption
5-6 Supply of special agricultural products (flowers, out-of-season crops)
5-7 Supply of raw agricultural materials for the industry
5-8 Creation of industrial zones
5-9 Urbanization processes (cities, towns, secondary residences)
5-10 Transport of people, products, energy, and information
5-11 Areas for recreation and tourism (water bodies, areas for outdoor activities and sports)
5-12 Protected areas (parks, reserves, green areas)
5-13 Environmental conservation, protection, and improvement
5-14 Enlargement of the technical capacity of the regional community
5-15 Increase in basic scientific knowledge

Lake Valencia project (Zinck 1977)

were built. The matrices were subjected to the judgement of a team of ten experts in soil survey, who identified the relationships between the themes of pairs of levels and assessed the intensity of these relationships through a rating procedure using two score ranges: 0–9 for the internal area and 0–2 for the external area. The individual estimates were averaged to get the direction and intensity of the information flows. This resulted in a complex graph of flows that is shown simplified in Fig. 4.11. The graph indicates the orientation and the amount of flows (critical pathways) that connect each theme with others. The combination of the two criteria of orientation and number of flows allowed establishing a ranking of the soil survey tasks according to their importance in generating or transmitting information.

4.4.4   The Contribution of Geomorphology to Soil Survey

The direct contribution of geomorphology takes place at levels 1 and 2.

- Level 1: geomorphology contributes to the tasks of photo-interpretation, selection of sample areas, identification of representative sites, and delineation of the geopedologic units.

- Level 2: geomorphic synthesis is one of the most prolific themes of the system by the number of flows issued and the number of themes reached at level 3 (30 themes). Based on this performance, the geomorphic synthesis ranked as the most efficient theme of level 2, along with the topography theme.

Thus, the incorporation of geomorphology helps streamline, speed up and improve the soil survey. Unfortunately, nowadays soil inventory is not given priority on political agendas, despite the severe risks of degradation of the soil resource.

4.5 Conclusions

In addition to promoting integration between geomorphology and pedology, geopedology focuses on the contribution of the former to the latter for soil mapping and understanding of soil formation. This contribution is based on the following.

- The geoforms and other geomorphic features, including processes of formation, aggradation and degradation, can be recognized by direct observation in the field and by interpretation of remote-sensed documents (aerial photographs and satellite images) and products derived therefrom (e.g. DEM). Documents that allow stereoscopic vision have the advantage of providing the third dimension of the geoforms in terms of volume and topographic variations. In this regard, aerial photographs are still the more faithful and explicit documents for the interpretation of the relief at large and medium scales.
- Many geoforms have relatively discrete boundaries, facilitating their delimitation. This is particularly the case of constructed geoforms in depositional systems (e.g. geoforms of alluvial, glacial, and eolian origin) and, to a lesser extent, those built in morphogenic systems controlled by endogenous processes (e.g. geoforms of volcanic and structural origin). By contrast, hillsides frequently show continuous variations, which can be approximated using the slope facet models.
- Geoforms are generally distributed in landscape systems controlled by a dominant forming agent (e.g. water, ice, wind). The foregoing results in families of geoforms associated in characteristic patterns that repeat in the landscape. This allows interpolating/extrapolating information in mapping areas and predicting the occurrence of geopedologic units at unvisited sites.
- Geoforms are relatively homogeneous at a given categorial level and with respect to the properties that are diagnostic at this level. The hypogeal component, corresponding to the morphogenic and morphostratigraphic features of the material, is usually more homogeneous than the epigeal component, corresponding to the morphographic and morphometric features of the terrain surface. The nonrandom, systematic variations of the soil mantle are frequently of geomorphic nature.
- The geomorphic context is an important framework of soil genesis and evolution, covering three of the five classic soil forming factors, namely the features of the

relief-drainage compound, the nature of the parent material, and the age of the geoform. Many soils have not formed directly from the hard bedrock, but rather from the geomorphic cover material (e.g. unconsolidated sediments, slope materials in translation, regolith, weathering layers).

- To sum up the foregoing, geomorphic analysis enables segmenting the continuum of the physiographic landscape into spatial units that are frameworks for (1) interpreting soil formation along with the influence of biota, climate and human activity, (2) composing the soil cartographic units, and (3) analyzing the spatial variations of the soil properties.

The geopedologic approach is essentially descriptive and qualitative. Geoforms and soils are considered as natural bodies, which can be described by direct observation in the field and by interpretation of aerial photos, satellite images, topographic maps, and digital elevation models. The approach relies on a combination of basic knowledge in geomorphology and pedology, incremented by working experience, in particular the experience gained from the practice of field observation and landscape reading. Expert knowledge, the acquisition and development of which constitute an inherent process in human societies in evolution, represents a source of cognitive richness that is nowadays attempted to be formalized before it disappears. Expert knowledge has been considered as a factor of subjectivity (Hudson 1992) and personal bias (McBratney et al. 1992) in the conventional practice of soil survey, in contrast to the pedometric (digital) soil mapping which would be more objective (Hengl 2003). Geopedology is a conventional approach with the particularity and advantage that bias and subjectivity can be minimized or compensated by the systematic and integrated use of geomorphic criteria. Geoforms provide a comprehensive cartographic framework for soil mapping, which goes beyond the mere morphometric terrain characterization. However, both modalities, the qualitative and the quantitative, can be usefully combined. Geopedologic units are reference units for more detailed geostatistical studies and for the spatial control of the digital data that are used to measure soil and geoform attributes. "The full potential of (digital) terrain analysis in soil survey will be realized only when it is integrated with field programs with a strong emphasis on geomorphic and pedologic processes" (McKenzie et al. 2000).

References

Amiotti N, Blanco MC, Sanchez LF (2001) Complex pedogenesis related to differential aeolian sedimentation in microenvironments of the southern part of the semiarid region of Argentina. Catena 43:137–156

Arnold R, Schargel R (1978) Importance of geographic soil variability at scales of about 1: 25,000. Venezuelan examples. In: Drosdoff M, Daniels RB, Nicholaides III JJ (eds) Diversity of soils in the tropics. ASA special publication, 34. ASA, Madison, pp 45–66

Bocco G, Velázquez A, Mendoza ME, Torres MA, Torres A (1996) Informe final, subproyecto regionalización ecológica, proyecto de actualización del ordenamiento ecológico general del territorio del país. INE-SEMARNAP, México

Bregt AK, Bouma J, Jellineck M (1987) Comparison of thematic maps derived from a soil map and from kriging of point data. Geoderma 39:281–291

Buol SW, Hole FD, McCracken RJ, Southard RJ (1997) Soil genesis and classification, 4th edn. Iowa State University Press, Ames

Campy M, Macaire JJ (1989) Géologie des formations superficielles. Géodynamique, faciès, utilisation. Masson, Paris

Esfandiarpoor Borujeni I, Salehi MH, Toomanian N, Mohammadi J, Poch RM (2009) The effect of survey density on the results of geopedological approach in soil mapping: a case study in the Borujen region, Central Iran. Catena 79:18–26

Esfandiarpoor Borujeni I, Mohammadi J, Salehi MH, Toomanian N, Poch RM (2010) Assessing geopedological soil mapping approach by statistical and geostatistical methods: a case study in the Borujen region, Central Iran. Catena 82:1–14

Farshad A, Shrestha DP, Moonjun R (2013) Do the emerging methods of digital soil mapping have anything to learn from the geopedologic approach to soil mapping or vice versa? In: Shahid SA, Taha FK, Abdelfattah MA (eds) Developments in soil classification, land use planning and policy implications: innovative thinking of soil inventory for land use planning and management of land resources. Springer, Dordrecht, pp 109–131

Fridland VM (1974) Structure of the soil mantle. Geoderma 12:35–41

Fridland VM (1976) Pattern of the soil cover. Israel Program for Scientific Translations, Jerusalem

Hall GF, Olson CG (1991) Predicting variability of soils from landscape models. In: Mausbach MJ, Wilding LP (eds) Spatial variabilities of soils and landforms. SSSA Special Publication, 28. SSSA, Madison, pp 9–24

Hengl T (2003) Pedometric mapping. Bridging the gaps between conventional and pedometric approaches. ITC dissertation 101, Enschede, The Netherlands

Hengl T, Rossiter DG (2003) Supervised landform classification to enhance and replace photo-interpretation in semi-detailed soil survey. Soil Sci Soc Am J 67:1810–1822

Hole FD, Campbell JB (1985) Soil landscape analysis. Rowman & Allanheld, Totowa

Hole FD, Hironaka M (1960) An experiment in ordination of some soil profiles. Soil Sci Soc Am Proc 24(4):309–312

Hudson BD (1992) The soil survey as paradigm-based science. Soil Sci Soc Am J 56:836–841

Ibáñez JJ (1994) Evolution of fluvial dissection landscapes in Mediterranean environments: quantitative estimates and geomorphic, pedologic and phytocenotic repercussions. Z Geomorphol 38:105–119

IUSS (2007) World reference base for soil resources. World soil resources report 103. IUSS Working Group WRB/FAO, Rome

Jenny H (1941) Factors of soil formation. McGraw-Hill, New York

Kerry R, Oliver MA (2011) Soil geomorphology: identifying relations between the scale of spatial variation and soil processes using the variogram. Geomorphology 130:40–54

McBratney AB, de Gruijter JJ, Brus DJ (1992) Spatial prediction and mapping of continuous soil classes. Geoderma 54:39–64

McKenzie NJ, Gessler PE, Ryan PJ, O'Connell DA (2000) The role of terrain analysis in soil mapping. In: Wilson JP, Gallant JC (eds) Terrain analysis. Principles and applications. Wiley, New York, pp 245–265

Metternicht G, Zinck JA (1997) Spatial discrimination of salt- and sodium-affected soil surfaces. Intl J Remote Sens 18(12):2571–2586

Moonjun R, Farshad A, Shrestha DP, Vaiphasa C (2010) Artificial neural network and decision tree in predictive soil mapping of Hoi Num Rin sub-watershed, Thailand. In: Boettinger JL, Howell DW, Moore AC, Hartemink AE, Kienast-Brown S (eds) Digital soil mapping: bridging research, environmental application, and operation. Springer, New York, pp 151–163

Pennock DJ, Corre MD (2001) Development and application of landform segmentation procedures. Soil Tillage Res 58:151–162

Phillips JD (2001) Divergent evolution and the spatial structure of soil landscape variability. Catena 43:101–113

Principi P (1953) Geopedologia (Geologia Pedologica). Studio dei terreni naturali ed agrari. Ramo Editoriale degli Agricoltori, Roma
Rossiter DG (2000) Methodology for soil resource inventories. Lecture notes, 2nd revised version. International Institute for Aerospace Survey and Earth Sciences (ITC), Enschede, The Netherlands
Saldaña A, Stein A, Zinck JA (1998) Spatial variability of soil properties at different scales within three terraces of the Henares valley (Spain). Catena 33:139–153
Saldaña A, Ibáñez JJ, Zinck JA (2011) Soilscape analysis at different scales using pattern indices in the Jarama-Henares interfluve and Henares River valley, Central Spain. Geomorphology 135:284–294
Salgado-Labouriau ML (1980) A pollen diagram of the Pleistocene-Holocene boundary of Lake Valencia, Venezuela. Rev Palaeobot Palynol 30:297–312
Schlichting E (1970) Bodensystematik und bodensoziologie. Z Pflanzenernähr Bodenk 127(1):1–9
Soil Survey Staff (1960) Soil classification: a comprehensive system. 7th approximation. US Government Printing Office, Washington
Soil Survey Staff (1967) Supplement to soil classification system (7th approximation). US Soil Conservation Service, Washington
Soil Survey Staff (1993) Soil survey manual. US Department of Agriculture handbook 18. US Government Printing Office, Washington, DC
Soil Survey Staff (1999) Soil taxonomy. US Department of Agriculture handbook 436. US Government Printing Office, Washington, DC
Sokal RR, Sneath PHA (1963) Principles of numerical taxonomy. Freeman, San Francisco
Toomanian N (2013) Pedodiversity and landforms. In: Ibáñez JJ, Bockheim J (eds) Pedodiversity. CRC Press/Taylor & Francis Group, Boca Raton, pp 133–152
Tricart J (1972) La terre, planète vivante. Presses Universitaires de France, Paris
Wilding LP, Drees LR (1983) Spatial variability and pedology. In: Wilding LP, Smeck NE, Hall GF (eds) Pedogenesis and soil taxonomy. I concepts and interactions. Elsevier, Amsterdam, pp 83–116
Wilding LP, Lin H (2006) Advancing the frontiers of soil science towards a geoscience. Geoderma 131:257–274
Zinck JA (1970) Aplicación de la geomorfología al levantamiento de suelos en zonas aluviales. Ministerio de Obras Públicas (MOP), Barcelona
Zinck JA (1972) Ensayo de clasificación numérica de algunos suelos del Valle Guarapiche, Estado Monagas, Venezuela. IV Congreso Latinoamericano de la Ciencia del Suelo (resumen). Maracay
Zinck JA (1974) Definición del ambiente geomorfológico con fines de descripción de suelos. Ministerio de Obras Públicas (MOP), Cagua
Zinck JA (1977) Ensayo sistémico de organización del levantamiento de suelos. Ministerio del Ambiente y de los Recursos Naturales Renovables (MARNR), Maracay
Zinck JA (1986) Una toposecuencia de suelos en el área de Rancho Grande. Dinámica actual e implicaciones paleogeográficas. In: Huber O (ed) La selva nublada de Rancho Grande, Parque Nacional "Henri Pittier". El ambiente físico, ecología vegetal y anatomía vegetal. Fondo Editorial Acta Científica Venezolana y Seguros Anauco CA, Caracas, pp 67–90
Zinck JA (1988) Physiography and soils. Lecture notes. International Institute for Aerospace Survey and Earth Sciences (ITC), Enschede
Zinck JA, Urriola PL (1971) Estudio edafológico Valle Guarapiche, Estado Monagas. Ministerio de Obras Públicas (MOP), Barcelona
Zinck JA, Valenzuela CR (1990) Soil geographic database: structure and application examples. ITC J 1990(3):270–294

Chapter 5
The Pedologic Landscape: Organization of the Soil Material

J.A. Zinck

Abstract The soil material is organized from structural, geographic, and genetic points of view. Structurally, the soil material is multiscalar with features and properties specific to each scale level. The successive structural levels are embedded in a hierarchic system of nested soil entities or holons known as the holarchy of the soil system. At each hierarchic level of perception and analysis of the soil material, distinct features are observed that are particular to the level considered. The whole of the features describes the soil body in its entirety. Each level is characterized by an element of the soil holarchy, a unit (or range of units) measuring the soil element perceived at this level, and a means of observation or measurement for identifying the features that are diagnostic at the level concerned. The levels are labelled based on a connotation with the proper dimension of the soil element into consideration at every level: nano, micro, meso, macro, and mega. The holarchy of the soil system allows highlighting relevant relationships between soil properties and geomorphic response at different hierarchic levels. These relationships form the conceptual essence of geopedology.

Keywords Hierarchic levels • Soil reactions • Micromorphologic components • Soil horizons • Pedon • Polypedon

5.1 Introduction

The soil material is organized from structural, geographic, and genetic points of view. Structurally, the soil material is multiscalar with features and properties specific to each scale level. The successive structural levels are embedded in a

J.A. Zinck (✉)
Faculty of Geo-Information Science and Earth Observation (ITC), University of Twente, Enschede, The Netherlands

Institute of Environmental Studies, University of New South Wales, Sydney, NSW, Australia
e-mail: alfredzinck@gmail.com

© Springer International Publishing Switzerland 2016
J.A. Zinck et al. (eds.), *Geopedology*, DOI 10.1007/978-3-319-19159-1_5

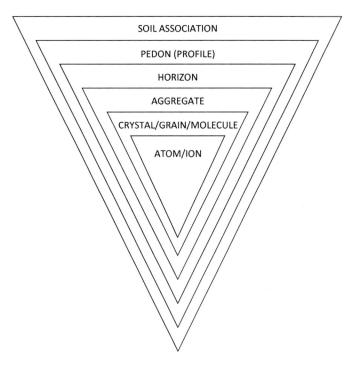

SOIL ASSOCIATION

PEDON (PROFILE)

HORIZON

AGGREGATE

CRYSTAL/GRAIN/MOLECULE

ATOM/ION

Fig. 5.1 The holarchy of the soil system (Adapted from Haigh 1987)

hierarchic system of nested soil entities, or holons, that Haig (1987) has called the holarchy of the soil system (Fig. 5.1). Geographically, the soil material is not randomly distributed on the landscape; instead, it is organized according to spatial distribution patterns under the control of the soil forming factors (Fridland 1974, 1976; Hole and Campbell 1985). Genetically, the soil material is formed and develops as an open system of exchanges and transformations of matter and energy (Jenny 1941; Simonson 1959).

Hereafter, a model similar to Haigh's holarchy is used to introduce some basic soil notions and analyze their relationships with the geopedologic approach at various scalar levels (Table 5.1). This scheme of nested holons is a condensate of pedology ranging from molecular reactions to the (geo)pedologic landscape. At each hierarchic level of perception and analysis of the soil material, distinct features are observed that are particular to the level considered. The whole of the features describes the soil body in its entirety. At each level correspond an element of the soil holarchy, a unit (or range of units) measuring the soil element perceived at this level, and a means of observation or measurement for identifying the features that are diagnostic at the level concerned. The levels are labelled based on a connotation with the proper dimension of the soil element into consideration at every level: nano, micro, meso, macro, and mega (Table 5.1).

Table 5.1 Hierarchic levels of the soil system

Level	Unit	Concept	Soil feature
Nano	nm-μm	Particle	Basic soil reactions
Micro	μm-mm	Aggregate	Micromorphologic structure
Meso	mm-cm-dm	Horizon	Differentiation of the soil material
Macro	m	Pedon	Soil volume for description and sampling
Mega	m-km	Polypedon	Soil classification and mapping – (geo)pedologic landscape

(Zinck 1988)

5.2 Nano-level

At the nano-level, the soil material is considered in its elementary form of molecules and combinations of molecules into particles, which can be either identified through chemical reactions, or observed using an electron microscope, or determined by X-ray diffraction. At this level take place the basic reactions of the soil material: chemical, mechanical, and physico-chemical. These reactions control processes and features such as rock weathering and soil formation, but also mass movements and other erosion phenomena that have the particularity of manifesting and taking visual expression at coarser levels of perception.

5.2.1 *Chemical Reactions*

The chemical reactions, which take place in the soil material as well as in the parent material (hard rock or unconsolidated sediment) to transform the latter into soil material, operate in two modalities: (1) by solubility changes of the chemical compounds in the salts, carbonates, and silicates, and (2) by structural changes in the oxide minerals.

- Solution (salts): $NaCl + H_2O \Leftrightarrow Na^+ + Cl^- + H_2O$
- Carbonation (carbonates): $CO_2 + H_2O \Rightarrow HCO_3^- + H^+$
  $CaCO_3 + (HCO_3^- + H^+) \Rightarrow Ca(HCO_3)^2$
- Hydrolysis (silicates): $KAlSi_3O_8 + HOH \Rightarrow HAlSi_3O_8 + KOH$
- Hydration (oxides): $2Fe_2O_3 + 3H_2O \Rightarrow 2Fe_2O_3 * 3H_2O$
- Oxido-reduction (oxides): $4FeO + O_2 \Leftrightarrow 2Fe_2O_3$

The performance of these reactions depends on the bioclimatic conditions, the nature of the substratum, and the type of relief and associated drainage conditions, among other factors. These are basic processes of rock weathering, alteration of unconsolidated materials, and formation of pedogenic material. Some processes operate only in specific geopedologic environments. For instance, the dissolution, concentration and, eventually, (re)crystallization of salts and the resulting geoforms are typical of halomorphic conditions in coastal and dry inland areas. Likewise, the dissolution of carbonates into bicarbonates and the mobilization of the latter are typical of calcimorphic conditions and responsible, in particular, for the formation of karstic relief. The hydrolysis of potassium feldspar, favored by high humidity and high temperature in tropical environment, results in the formation of acid clay together with potassium hydroxide that is lost by lixiviation. Hydration makes iron oxide more fragile. Oxydo-reduction is a reversible process typical of the intertidal zone.

5.2.2   Mechanical Reactions

The mechanical reactions depend on the way particles are arranged and associated. Coarse particles have the tendency to pile up into different kinds of packing, while the behavior of the fine particles depends on the intensity of their agglomeration into various kinds of fabric. In general terms, these mechanical reactions of nano-level determine the susceptibility of the materials to mass movements, the geomorphic expression of which is visible on the landscape at coarser levels of perception (from meso to mega).

5.2.2.1   Types of Packing

Coarse particles including sand and coarse silt grains (2–0.02 mm) cluster in piles, the structure of which varies according to the degree of roundness of the grains. Rounded grains (e.g. sand grains of marine or eolian origin) usually present a cubic arrangement with limited contact surface and high porosity. This allows water to penetrate readily in the pore space, resulting in water pressure in the pores that tends to separate the grains. For this reason, the cubic packing is in general an unstable arrangement, which facilitates the process of moving sands (quicksands). Less rounded grains (e.g. sand grains of alluvial or colluvial origin) generally show a tetrahedral type of packing, with greater contact surface and lower porosity, which is a more stable arrangement. Irregular grains and rock fragments tend to be tightly interlocked, with large friction surface that ensures greater stability of the material.

5.2.2.2   Types of Fabric

The fabric arrangement of the fine particles, including clay and fine silt (<0.02 mm), depends on the mode and intensity of the contacts between particles in the soil solution. Various modes of particle association in clay suspensions are recognized, with four basic types of micro-mechanical fabric, ranging from the total absence of agglomeration (i.e. deflocculated state) to a strongly agglomerated condition (i.e. flocculated state), and a series of combinations of these basic types (Mitchell 1976) (Fig. 5.2).

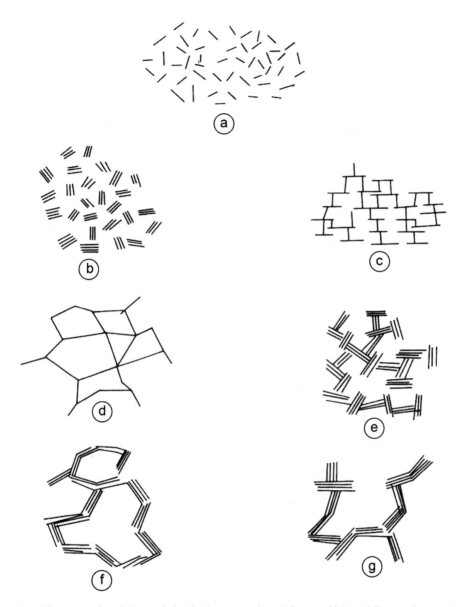

Fig. 5.2 Modes of particle association in clay suspensions (After van Olphen 1963): (**a**) dispersed and deflocculated; (**b**) aggregated but deflocculated; (**c**) edge-to-face flocculated but dispersed; (**d**) edge-to-edge flocculated but dispersed; (**e**) edge-to-face flocculated and aggregated; (**f**) edge-to-edge flocculated and aggregated; (**g**) edge-to-face and edge-to-edge flocculated and aggregated (Adapted from Mitchell 1976)

The fabric types are related to the moisture content in the soil, which determines the mechanical state of the material, from liquid to solid, and the consistence limits (i.e. Atterberg limits) between mechanical states. Obviously, the fabric depends also on other factors such as the type of clay, organic matter content, and the presence of salts, among others.

In geopedologic terms, the fabric of the soil material plays an important role in the generation of mass movements (Table 5.2).

- Deflocculated state: all particles are individually in suspension in the soil solution, without interaction between particles. This fabric condition favors the occurrence of mudflows.
- Dispersed state: there are elementary associations between individual particles, essentially contacts between particle edges and faces. This fabric condition creates a risk of solifluction.
- Aggregated state: there are associations between particle clusters, creating a situation that favors the potential occurrence of landslides.
- Flocculated state: all kinds of contact between faces and between edges and faces take place, generating the most stable arrangement of particles in the soil solution and resulting in high soil strength and stability.

5.2.3 Physico-chemical Reactions

The physico-chemical reactions are based on the colloidal properties of clay and humus. Both compounds have electronegative charges at the edges of the layers and in the space between layers. The electronegative charges attract cations with decreasing intensity according to the lyotropic sequence of preferential adsorption, which reflects the number of charges and the hydrated size of the cations: $Al^{+++} > Ca^{++} > Mg^{++} > K^+ = NH_4^+ > Na^+$. Divalent cations play an important role in establishing bridges between clay particles, which is a basic process for the formation of aggregates. The physico-chemical reactions that take place at the nanolevel control soil fertility, aggregation, structural stability and its influence on soil susceptibility to erosion.

Table 5.2 Influence of the fabric type and the consistence of the soil material in the generation of mass movements (the most likely to occur)

Fabric type	State of the material	Mass movement
Deflocculated	Liquid	Mudflow
Dispersed	Plastic	Solifluction
Aggregated	Semi-solid	Landslide
Flocculated	Solid	Metastability
Organization of the soil material	Soil property (consistence, Atterberg limits)	Morphogenic process (geomorphic response)

5.2.4 Relationship with Geopedology

The reactions taking place at the nano-level determine the fundamental processes of soil formation, evolution, differentiation, as well as degradation. The production of regolith through rock weathering, the alteration of the unconsolidated cover formations, and the transformation of these loose materials into soil material largely depend on the chemical and physico-chemical reactions that operate in the substratum – inherently the domain of geomorphology. The different mechanical reactions that take place in the soil material and regolith, according to variations in moisture content, control the morphogenesis by mass movements, the impact of which is directly visible in the landscape.

5.3 Micro-level

At the micro-level, the object of interest is the soil aggregate, which can be observed with the use of a petrographic microscope. This is the investigation domain of micromorphology. The observation of an aggregate in thin section under the petrographic microscope allows characterizing the micromorphologic structure of the soil matrix, both in its solid component and porous component, and identifying features derived from the addition of material and transformation of the matrix. Some of these micromorphologic characteristics are shown schematically in Fig. 5.3 and summarized in Table 5.3.

5.3.1 The Micromorphologic Components

At the micro-level, the soil material is divided into two main components: the soil matrix, which corresponds to the soil material in situ, and pedologic features. Each of these two components is subdivided into elements that play important roles in the functioning of the soil, including plasma, pore space, skeleton grains, and pedologic features (Table 5.3).

5.3.1.1 Skeleton Grains

The skeleton grains consist of:

- Mineral grains, essentially sand and silt grains, which constitute the inert soil material, without colloidal properties, that dominates in coarse-grained soils.
- Organic fragments, which are pieces of undecomposed organic material, essentially fragments of leaves, twigs, and branches (folic material), that dominates in the litter.

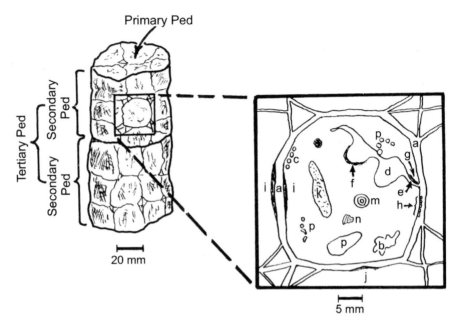

Fig. 5.3 Micropedologic features. Voids: (*a*) packing voids, (*b*) vugh, (*c*) vesicles, (*d*) chamber, (*e*) channel. Cutans: (*f*) chamber cutan, (*g*) channel cutan, (*h*) skeletans, (*i*) argillan or sesquan, (*j*) stress cutan. Other features: (*k*) pedotubule, (*l*) nodule, (*m*) concretions, (*n*) papule. Note that the S-matrix is the mass of plasma, skeleton grains (*p*), and voids (Adapted from Buol et al. 1997)

Table 5.3 Micromorphologic organization of the soil material

Soil material	Soil matrix (S-matrix) (soil material in situ)	Solids	Skeleton grains (coarse material)
			Plasma (fine material)
		Pore space (voids, pores)	Vesicles
			Chambers, vughs
			Channels
			Planes
	Pedologic features (addition to or transformation of soil material)	Cutans	
		Glaebules	
		Tubules	
		Plasma separations	

5.3.1.2 Plasma

The plasma is the active phase of the solid material, where the chemical and physico-chemical reactions take place and which controls the mechanical mobility of the fine particles. The plasma is endowed with relevant properties, among others:

- Colloidal property that provides the clay minerals and the humus with electronegative charges.
- Solubility property that allows salts and carbonates to be converted into ions.
- Chelation property, thanks to which insoluble compounds (e.g. Fe and Al sesquioxides) can migrate in association with organic molecules.

5.3.1.3 Pores

Pores vary in configuration and location within and between aggregates, and for this reason fulfill different functions. Packing voids, vesicles, and chambers are examples of pore differentiation in the soil.

- Packing pores are located around the aggregates and control the permeability, with its influence on drainage, and the adhesion between aggregates.
- Vesicles are closed empty spaces, without active function.
- Chambers are pores open on one extremity, which retain moisture even when the soil appears to be dry; these are places where the microfauna (e.g. bacteria) responsible for the decomposition of the organic matter tends to concentrate, and where the oxido-reduction mechanisms responsible for hydromorphism take place.

5.3.1.4 Pedologic Features

Micromorphologic soil features derive essentially from the addition of new material to the soil and/or the transformation of the soil material in situ.

- The additions can be traced by the coatings (cutans) that form when fine particles move within the soil solution from eluvial horizons and deposit in the pores or on the surface of the aggregates in the underlying illuvial horizons. According to the nature of the constituents, different types of cutan are recognized, including clay cutans (argillans), iron cutans (ferrans), manganese cutans (manganans), etc.
- The transformations can be (1) physical: e.g. pressure faces (stress cutans) on the surface of the aggregates caused by contraction-expansion; (2) chemical: e.g. local concentration of chemical compounds (Fe_2O_3, $CaCO_3$, SiO_2) in the form of nodules and concretions; and (3) biological: e.g. fecal nodules, pedotubules.

5.3.2 Relationship with Geopedology

The micromorphologic characteristics represent an important source of information for the genetic interpretation of the soils and for inferring soil properties and qualities that control geomorphic processes.

- The pedologic features, which refer to the additions and transformations that take place in the soil material, are indicators of soil formation and evolution. The translocation of substances (e.g. clay illuviation) is a particularly good example that reveals a type of pedogenic dynamics. The micromorphologic analysis also allows identifying paleo-environmental influences in polygenic soils (Jungerius 1985) and correlatively in the evolution of the geomorphic landscape.
- The soil matrix has influence on geomorphogenesis. The nature of the plasma conditions the aggregate stability, which plays a relevant role in the processes of soil erosion by water and wind. Porosity controls the movement of water and air in the soil. The microporosity determines the capacity of water retention in the soil, while the macroporosity determines the surface runoff, the infiltration, and the percolation of water through the soil. An imbalance between these different terms of the water dynamics on the surface of and within the soil causes susceptibility to sheet erosion and mass movement.

5.4 Meso-level

At the meso-level, the organization entity of the soil material is the horizon, which usually consists of a mass of aggregates, except when the material is single-grain (sandy soil) or compact (clay soil). Horizons result from the differentiation of the parent material by pedogenic processes. The mode of analysis is direct observation and description in the field.

5.4.1 Horizon Definition and Designation

A horizon is a layer of soil material with a unique combination of properties, different from the properties of the soil in the horizons above and below that horizon (e.g. color, texture, structure). The concept of horizon refers to the pedogenic material and is therefore different from the concept of stratum that refers to the geogenic material (in the C layer). Soil horizons are identified at three successive levels using a designation nomenclature of letters and numbers.

5.4.1.1 Primary Divisions: The Master Horizons

The primary divisions reflect the effect of the basic soil forming processes, resulting in the differentiation of the soil material in master horizons. These are identified by capital letters (O, A, E, B, C, R). At this level, the horizons are distinguished according to the nature of the material and according to their position in the soil profile.

The distinction of the material according to its nature allows separating the organic material from the mineral material. A material is considered to be organic (O horizon) when it complies with the following contents of organic carbon (OC):

- In well drained soils: OC >20 %.
- In poorly drained soils: OC $\geq$18 %, if clay $\geq$60 %; OC $\geq$12 %, if clay=0 %; proportional percentages of OC for intermediate clay contents.

The distinction of the material according to the position in the profile leads to separate four kinds of horizon/layer: surficial horizon (topsoil), subsurface horizon, subsoil, and substratum.

- Topsoil horizons: A and E horizons

    - *A horizon*: layer where the incorporation of organic matter occurs and where the biologic activity shows its maximum expression; there may also be some downwashing of constituents.
    - *E horizon*: layer that loses soil material through eluviation according to the degree of solubility of the constituents. A generalized sequence by order of decreasing susceptibility to leaching includes: salts, carbonates, bases, clay, OM, Fe and Al sesquioxides. In an extremely leached situation, only SiO_2 remains in situ, giving the horizon a whitish color (albic horizon).

- Subsurface horizons: B horizons
  The nature of the B horizon varies according to the process of formation, which can operate by weathering of the parent material (consolidated or loose), illuviation of chemical compounds (salts, carbonates, clay, OM, sesquioxides, etc.), and neoformation of clay minerals.
- Subsoil: C layer=parent material.
- Substratum: R layer=bedrock.

5.4.1.2   Secondary Divisions: Specific Genetic Features

The secondary divisions inform on specific genetic features of the horizons, using lowercase letters:

- Degree of decomposition of the organic material:

  i=slightly decomposed organic material (**Fi**brist).
  e=moderately decomposed organic material (**He**mist).
  a=strongly decomposed organic material (**Sa**prist).

- Degree of weathering of the mineral material: w (Bw), r (Cr).
- Accumulation: z, y, k, n, t, h, s, q, in order of decreasing mobility of the chemical compounds, referring respectively to salts more soluble than calcium sulphate, gypsum, carbonates, sodic clay, clay, humus, sesquioxides, and silica.
- Concentration: c, o, v, referring respectively to concretions, no-concretionary nodules, and plinthite.

- Transformation: f, g, m, p, x, b, d, referring respectively to frozen soil, gleization, compaction, plowpan, fragipan, buried horizon, and densified horizon.

5.4.1.3   Tertiary Divisions

The tertiary divisions are concerned with a variety of unrelated features, using arabic numerals:

- Subdivision of genetic horizons based on differences in color and/or texture, among other criteria (e.g. Bt1-Bt2) (numerical suffixes).
- Lithologic discontinuity based on textural contrasts indicating several successive depositional phases that result in the superposition of layers or profiles (e.g. Bt-2Bt-2C) (numerical prefixes).
- Bisequum that reflects the superimposition or imprint of a recent soil within a soil formed previously under different bioclimatic conditions, vegetation cover, or land-use. For instance, a Spodosol developing under pine plantation that invades the upper part of an Alfisol previously formed under deciduous forest (e.g. O-A-E-Bs-E′-Bt′-C).

5.4.2   Relationship with Geopedology

The designation symbols are information vectors that summarize the relevant characteristics of a horizon, including properties, mode of formation, and position in the profile. The nomenclature is used to identify genetic horizons based on the qualitative inference of the process(es) responsible for their formation. For instance, a Bw horizon reflects weathering of primary minerals, whereas a Bt horizon reflects clay illuviation. To be diagnostic for taxonomic classification of the soils, genetic horizons must comply with quantitative requirements (e.g. color, depth, thickness, % content, etc.) specified by the taxonomic system that is implemented. For this reason, it can be stated that all argillic horizons are Bt horizons, but not all Bt horizons are argillic horizons.

The soil information describing the nature of the horizons and, especially, their sequence in the profiles is very useful in geomorphic research on the susceptibility of the soils and cover formations to erosion processes. As highlighted by Jungerius (1985), A and B horizons exert different control on the geomorphic processes. The difference in strength between surficial horizons (A) and subsurface horizons (Bt) often determines the depth of soil truncation by sheet erosion. Similarly, differences in physico-mechanical properties between consecutive horizons may cause shear planes that can activate surface mass movements. Suffusion, piping, and tunnelling processes also depend on the sequence of and contrast between horizons.

5.5   Macro-level

5.5.1   Definition

At the macro-level, the basic concept is the pedon, which is defined as the minimum soil volume for describing and sampling a soil body (Soil Survey Staff 1975, 1999). Conventionally, the pedon is represented with a hexagonal configuration (Fig. 5.4). It covers a large part of the lateral and vertical variations of a soil body. The normal

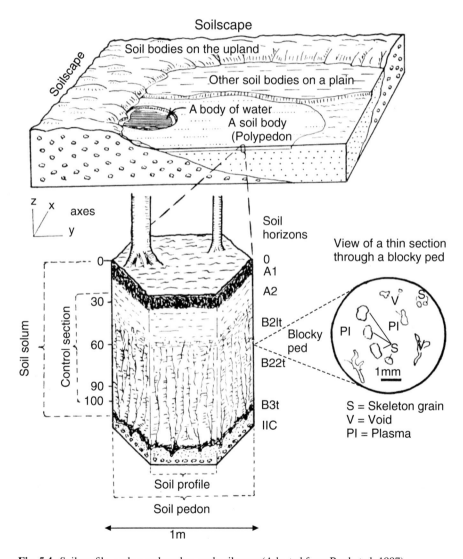

Fig. 5.4 Soil profile, pedon, polypedon, and soilscape (Adapted from Buol et al. 1997)

size of the area is 1 m^2 in the case of a soil with approximately parallel horizons and isotropic spatial variations. The maximum size of the area is 10 m^2 when horizons show cyclic variations. The theoretical depth is down to the parent material of the soil, but for practical reasons it is usually limited to the upper 2 m.

5.5.2 Related Concepts

Several other concepts that characterize the soil body are related with the pedon concept, such as soil profile, solum, and control section.

- Soil profile is a face of the pedon including the entire sequence of horizons, commonly used to describe and sample. Statistical trials have shown that, when collecting material of a horizon laterally in all faces of the pedon to obtain a composite sample, probable mean errors can be divided approximately by two for most of the physical and chemical parameters (Wilding and Drees 1983).
- Solum includes soil horizons O + A + E + B, the C and R layers being excluded.
- Control section is the specific depth of the pedon within which selected soil characteristics need to occur to be considered diagnostic for taxonomic classification. For instance, for most of the soils, the family of particle-size distribution is determined within the depth of 25–100 cm. Likewise, to be diagnostic, plinthite should be present at <125 cm depth at great group level (e.g. Plinthustult) and at <150 cm depth at subgroup level (e.g. Plinthic Paleustult).

5.5.3 Relationship with Geopedology

Geomorphic literature does not provide any criteria or norms that specify the size of the minimum area for description and sampling. In practice, there is no space limitation for the description of the epigeal component of the geoform, since processes and features of the terrain surface are directly observable. However, defining a minimum observation area can be useful for comparison between sites and for generalization of field information. With respect to the hypogeal component of the geoform, thus the proper geomorphic material (i.e. regolith, depositional material) that constitutes the C layer of the soils, it is not directly accessible to observation, description and sampling, except when there are natural or artificial exposures. Therefore, geomorphic research faces an issue of minimum volume for description and sampling similar to the one that has been solved in pedology with the concept of pedon. As the geopedologic survey integrates the description of the geoform and that of the soil in one place, the size criteria of the pedon may also apply to the morphon. The morphon covers the features of both the terrain surface and the subsoil/substratum, while the pedon covers the volume of the intermediate material that corresponds to the solum. In the geopedologic practice, the two are inseparable and their distinction may be regarded as superfluous.

The above comments apply primarily to the lower level of the hierarchic classification of the geoforms i.e. that of landform/terrain form (see Chap. 7). They are less pertinent at the higher categories of the system, since the external features of the geoform often allow inferring the nature of the substratum.

5.6 Mega-level

5.6.1 Definition

The polypedon is the basic concept at the mega-level. It is an extended soil body formed by adjacent similar pedons that fit within the range of variation of a single taxonomic unit (e.g. soil series) (Soil Survey Staff 1975, 1999). It is a real physical soil body, limited by "no-soils" (e.g. rock outcrops, water bodies, built areas, etc.) or by pedons that exhibit dissimilar characteristics. The minimum area is 2 m^2 (i.e. two pedons), but there is no specification of maximum area. The concepts of soil body and soil individual are synonymous with polypedon. In similar terms, Boulaine (1975) proposed the concept of *genon* to designate the soil volume of all pedons that have the same structure and characteristics and that result from the same pedogenesis.

5.6.2 Relationship with Geopedology

- The polypedon constitutes the fundamental link between the actual soil volume (i.e. pedon) and the taxonomic unit in the classification system. It is the concept used to taxonomically classify the soil bodies. A polypedon comprises all contiguous pedons of equal classification.
- The polypedon provides the pedologic content of the cartographic unit. A polypedon is a concrete soil individual (i.e. soil body) on the landscape. Polypedon and landscape together form the soilscape. Polypedons can constitute (1) relatively pure map units with one dominant polypedon per unit (consociacion), or (2) composite map units comprising more than one dominant polypedon (association, complex).
- The polypedon correlates with the geomorphic unit (polymorphon), especially at the lower taxonomic level (landform/terrain form). In its simplest expression, a polypedon together with the corresponding geomorphic frame forms a geopedologic landscape unit. However, the geopedologic landscape is usually more complex, because a single geoform often comprises more than one polypedon.

5.7   Conclusion

The holarchy of the soil system allows highlighting relevant relationships between soil properties and geomorphic response at different hierarchic levels. These relationships form the conceptual essence of geopedology. A notable phenomenon refers to the cause-effect relationships between reactions that occur in the soil material at micro-scale, thus not directly perceptible, and their geomorphic expression in the landscape at macro-scale. This is especially the case of landscape shaping by mass movements, which are controlled by micro-mechanical reactions in the soil fabric. With respect to soil cartography, the most conspicuous relationship takes place at the mega-level, where polypedon and polymorphon integrate to form a geopedologic landscape unit.

References

Boulaine J (1975) Géographie des sols. Presses Universitaires de France, Paris
Buol SW, Hole FD, McCracken RJ, Southard RJ (1997) Soil genesis and classification, 4th edn. Iowa State University Press, Ames
Fridland VM (1974) Structure of the soil mantle. Geoderma 12:35–41
Fridland VM (1976) Pattern of the soil cover. Israel Program for Scientific Translations, Jerusalem
Haigh MJ (1987) The holon: hierarchy theory and landscape research. Catena Suppl 10:181–192. CATENA Verlag, Cremlingen
Hole FD, Campbell JB (1985) Soil landscape analysis. Rowman & Allanheld, Totowa
Jenny H (1941) Factors of soil formation. McGraw-Hill, New York
Jungerius PD (1985) Soils and geomorphology. In: Jungerius PD (ed) Soils and geomorphology. Catena Suppl 6:1–18. CATENA Verlag, Cremlingen
Mitchell JK (1976) Fundamentals of soil behavior. Wiley, New York
Simonson RW (1959) Outline of a generalized theory of soil genesis. Soil Sci Soc Am Proc 23:152–156
Soil Survey Staff (1975) Soil taxonomy. A basic system of soil classification for making and interpreting soil surveys. USDA agriculture handbook 436. US Government Printing Office, Washington, DC
Soil Survey Staff (1999) Soil taxonomy. USDA agriculture handbook 436. US Government Printing Office, Washington, DC
van Olphen H (1963) An introduction to clay colloid chemistry. Wiley, New York
Wilding LP, Drees LR (1983) Spatial variability and pedology. In: Wilding LP, Smeck NE, Hall GF (eds) Pedogenesis and soil taxonomy. I concepts and interactions. Elsevier, Amsterdam, pp 83–116
Zinck JA (1988) Physiography and soils. ITC soil survey lecture notes. International Institute for Aerospace Survey and Earth Sciences, Enschede

Chapter 6
The Geomorphic Landscape: Criteria for Classifying Geoforms

J.A. Zinck

Abstract Combining the basic criteria to build a taxonomic system with the hierarchic arrangement of the geomorphic environment determines a structure of nested categorial levels. Five of these levels are essentially deduced from the epigeal physiographic expression of the geoforms. To substantiate the relationship between geoform and soil, it is necessary to introduce in the system information on the internal hypogeal component of the geoforms, namely the constituent material, which is in turn the parent material of the soils. As a result of the foregoing, an additional level is needed to document the lithology in the case of bedrock substratum or the facies in the case of unconsolidated cover materials. This leads finally to a system with six categorial levels, identified by their respective generic concepts, including from upper to lower level: geostructure, morphogenic environment, geomorphic landscape, relief/molding, lithology/facies, and the basic landform or terrain form. Such a system with six categories complies with *Miller's Law*, which postulates that the capacity of the human mind to process information covers a range of seven plus or minus two elements.

Keywords Geomorphic classifications • Classification system structure • Levels of landscape perception • Geoform taxonomy • Geomorphometry

6.1 Introduction

Unlike other scientific disciplines, geomorphology still lacks a formally structured taxonomic system to classify the forms of the terrestrial relief, hereafter designated as *geoforms*. There is some consensus for grouping the geoforms according to the

J.A. Zinck (✉)
Faculty of Geo-Information Science and Earth Observation (ITC), University of Twente, Enschede, The Netherlands

Institute of Environmental Studies, University of New South Wales, Sydney, NSW, Australia
e-mail: alfredzinck@gmail.com

© Springer International Publishing Switzerland 2016   77
J.A. Zinck et al. (eds.), *Geopedology*, DOI 10.1007/978-3-319-19159-1_6

families of processes that operate on given geologic substrata or in given biocli-matic zones. Examples of the former are the karstic forms generated by the dissolu-tion of calcareous rocks, desert forms shaped by wind, glacial forms resulting from the activity of ice, or alluvial forms controlled by the activity of the rivers. However, these geoforms are not integrated in a structured hierarchic scheme. It is necessary to create a system that allows accommodating and organizing the geoforms accord-ing to their characteristics and origin, and considering also their hierarchic relation-ships. This requires a multicategorial framework.

Geoform is the generic concept that designates all types of relief form regardless of their origin, dimension, and level of abstraction, similarly to how the concept of soil is used in pedology or the concept of plant in botany (Zinck 1988; Zinck and Valenzuela 1990). The term of geoform, with generic meaning, has been introduced recently in the Spanish version of the FAO Guidelines for soil description (FAO 2009). Geoforms have an internal (hypogeal) component and an external (epigeal) component in relation to the terrain surface. The internal component is the material of the geoform (the content), the characteristics of which convey genetic and strati-graphic (i.e. chronological) information. The external component of the geoform is its shape, its "form" (the container), which expresses a combination of morpho-graphic and morphometric characteristics. The external component is directly accessible to visual perception, proximal or distal, either human or instrumental. Ideally, the classification of the geoforms should reflect features of both compo-nents, i.e. the constituent material and the physiographic expression. The external appearance of the geoforms is very relevant for their direct recognition and cartog-raphy. For this reason, a system of geoform classification must necessarily combine perception criteria of the geomorphic reality and taxonomic criteria based on diag-nostic attributes.

Seemingly, geoform taxonomy has not fomented the same interest as plant tax-onomy and soil taxonomy did. This might be due to the fact that more importance has been given to the analysis of the morphogenic processes than to geomorphic mapping which requires some kind of classification of the geomorphic units. There are few countries that have had, at some time, a systematic program of geomorphic mapping similar to those carried out in several Eastern European countries after the Second World War or in France in the second part of the last century (Tricart 1965; CNRS 1972).

Soil map legends often ignore the geomorphic context that, however, largely controls soil formation and distribution. Usually, the legend of the soil maps shows only the pedotaxa, without mentioning the landscapes where the soils are found, although the concept of "soilscape" is considered to provide the spatial framework for mapping polypedons (Buol et al. 1997). A mixed legend, showing the soil in its geomorphic landscape, facilitates the reading, interpretation, and use of the soil map by nonspecialists working in academic and practitioner environments (see the example in Fig. 4.2, Chap. 4). With the use of GIS, the geomorphic context is emerging as the structuring element of a variety of legends, including legends of taxonomic maps, interpretive maps, and land-use planning maps, among others.

6.2 Examples of Geomorphic Classification

Geomorphologists have always shown some interest in classifying geoforms, but the criteria used for this purpose have changed over the course of time and are still very diverse. After mentioning some geomorphic classification approaches, the structure of a taxonomic system for geoform classification is described. This has been developed from geopedologic surveys in Venezuela and later used in the ITC (Enschede, The Netherlands) to train staff from a variety of countries in Latin America, Africa, Middle East, and Southeast Asia (Zinck 1988; Farshad 2010).

6.2.1 Classification by Order of Magnitude

The dimensional criterion has been used by several authors to classify the geomorphic units (Tricart 1965; Goosen 1968; Verstappen and Van Zuidam 1975; among others). These classifications are hierarchic, with emphasis on structural geomorphology in the upper levels of the systems. The classification proposed by Cailleux-Tricart (Tricart 1965) in eight temporo-spatial orders of magnitude is a representative example of this approach (Table 6.1). The spatial dimension and the temporal dimension of the geomorphic units vary concomitantly from global to local and from early to recent. Tricart (1965) considers that the dimension of the geomorphic objects (facts and phenomena) intervenes not only in their classification, but also in the selection of the study methods and in the nature of the relationships between geomorphology and neighboring disciplines.

   With a similar but less elaborate approach, Lueder (1959) distributes the geoforms in three orders of magnitude. The first order includes continents and ocean basins. Mountain ridges are an example of second order. The third order includes a variety of forms such as valley, depression, crest, and cliff.

Table 6.1 Taxonomic classification of the geomorphic units by Cailleux-Tricart

Order	Unit types	Unit examples	Extent (km^2)	Time (years)
I	Configuration of the earth's surface	Continent, ocean basin	10^7	10^9
II	Large structural assemblages	Shield, geosyncline	10^6	10^8
III	Large structural units	Mountain chain, sedimentary basin	10^4	10^7
IV	Elementary tectonic units	Serranía, horst	10^2	10^7
V	Tectonic accidents	Anticline, syncline	10	10^6–10^7
VI	Relief forms	Terrace, glacial cirque	10^{-2}	10^4
VII	Microforms	Lapies, solifluction	10^{-6}	10^2
VIII	Microscopic features	Corrosion, disaggregation	10^{-8}	–

Summarized from Tricart (1965)

6.2.2 Genetic and Genetic-Chorologic Classifications

There are variants of genetic classification of the geoforms based on the conventional division of geomorphology as a scientific discipline in specialist areas concerned with different types of geoforms (Table 6.2).

The genetic-chorologic classification of geoforms is based on the concept of morphogenic zone. The latitudinal and altitudinal distribution of the morphogenic zones parallels the division of the earth's surface in large bioclimatic zones, generating a series of morphoclimatic domains, each with a specific association of geoforms: glacial, periglacial, temperate (wet, dry), mediterranean, subtropical, and tropical (wet, dry). The classification combines origin and geographic distribution of the geoforms. It is often used to present and describe the geoforms by chapters in textbooks on geomorphology. This type of classification is based on some kind of hierarchic structure and leads to a typology of the geoforms, but does not provide a clear definition of the criteria used in the ranking and typology. There is tendency to emphasize one type of attributes of the geoforms to the detriment of others: for instance, the dimension, or the genesis, or the geographic distribution.

The project of the Geomorphic Map of France (CNRS 1972) establishes a hierarchy of geomorphic information in five levels, called *terms*, as reference frames to gather the data, represent them cartographically, and enter them in the map legend. The five terms are in descending order: the location, the structural context (type of structural region, lithology, tectonics), the morphogenic context (age, morphogenic system), surface formations (origin of the material, particle-size distribution, consolidation, thickness, morphometry), and finally the forms. The last term contains the entire collection of recognized forms, with grouping into classes and subclasses according to the origin of the forms. Each form is given a definition and a symbol for its cartographic representation. Two main groups of forms are distinguished: (1) the endogenous forms (volcanic, tectonic, structural), and (2) the forms originated by external agents (eolian, fluvial, coastal, marine, lacustrine, karstic, glacial, periglacial and nival forms, and slope and interfluve forms).

For the purpose of soil mapping, Wielemaker et al. (2001) proposed a hierarchic terrain objects classification, qualified as morphogenic by the authors, which includes five nested levels, namely region, major landform, landform element, facet, and site. This system was derived from the analysis of a concrete case study located in Southern Spain, using a methodological framework to formalize expert knowledge on soil-landscape relationships and an interactive GIS procedure for sequential disaggregation of the landscape (de Bruin et al. 1999).

Table 6.2 Families of geoforms as per origin

Study fields of geomorphology	Types of geoforms
Structural geomorphology: types of relief	Cuesta, fold, shield reliefs, etc.
Climatic geomorphology: types of molding	Glacial, periglacial, eolian moldings, etc.
Azonal geomorphology: types of form	Alluvial, lacustrine, coastal forms, etc.

A variant of genetic-chorologic classification is the ordering of landscapes and geoforms in the context of a given country (Zinck 1974; Elizalde 2009). This type of classification combines physico-geographic units at the higher levels of the system with taxonomic units at the lower levels. The physico-geographic units belong to a specific regional context and, therefore, cannot be generalized or extrapolated to other regional situations. The division of a country into physiographic provinces and natural regions is an example of this type of nomenclature. Instead, the taxa of the lower categories (e.g. landscape types or relief types) convey sufficient abstraction to be recognizable on the basis of differentiating features in a variety of regional contexts.

6.2.3 Morphometric Classification

First attempts of morphometric relief characterization go back to mid-nineteenth century in the Germanic countries. However, it was only after the Second World War that systematic use of morphometric techniques was made to describe features of the topography, parameters of the hydrographic network, drainage density, and other measurable attributes of the relief (Tricart 1965). In recent decades, the technology of the digital elevation models (DEM) has given a new impulse to morphometry and automated extraction of morphometric information (Pike and Dikau 1995; Hengl and Reuter 2009). Geomorphometry focuses on the quantitative analysis of the terrain surface with two orientations: a specific morphometry that analyzes the discrete features of the terrain surface (e.g. landforms/terrain forms), and a general morphometry that deals with the continuous features. In its present state, geomorphometry pursues essentially the characterization and digital analysis of continuous topographic surfaces (Pike et al. 2009).

The use of DEM has allowed measuring and extracting attributes that describe topographic features of the landscape (Gallant and Wilson 2000; Hutchinson and Gallant 2000; Olaya 2009). The most frequently measured parameters include altitude, slope, exposure, curvature, and roughness of the relief, among others. The spatial distribution of these parameters allows inferring the variability of hydrologic, geomorphic, and biological processes in the landscape. The combination of data derived from DEM and satellite images contributes to improve predictive models (Dobos et al. 2000).

There are attempts to classify landforms and model landscapes using morphometric parameters (Evans et al. 2009; Hengl and MacMillan 2009; Nelson and Reuter 2012). Idealized geometric primitives (Sharif and Zinck 1996) and ideal elementary forms (Minár and Evans 2008) have been used to segment the landscape and approximate the representation of a variety of terrain forms. The implementation of automated algorithms to classify landforms has facilitated the mapping of landform elements and relief classes (Pennock et al. 1987; MacMillan and Pettapiece 1997; Ventura and Irvin 2000; Meybeck et al. 2001; Iwahashi and Pike 2007; MacMillan and Shary 2009). Ventura and Irvin (2000) analyzed different methods of automated landform classification for soil landscape studies, but the experiments were basically restricted to slope situations according to the classic models of Ruhe (1975) and Conacher and Dalrymple (1977).

The use of quantitative parameters allows describing continuous variations of topographic features with the support of fuzzy sets techniques (Irwin et al. 1997; Burrough et al. 2000; MacMillan et al. 2000). However, this approach may be less efficient in identifying differentiating characteristics of geoforms that have discrete boundaries, as is frequent in erosional (e.g. gullies, solifluction features) and depositional areas (e.g. alluvial or eolian systems). The DEM-based analysis leads to a classification of topographic features of the relief and contributes to the morphometric characterization of the terrain forms, but does not generate a terrain form classification in the geomorphic sense of the concept. The classification of slope facets by shape and gradient is essentially a descriptive classification which does not convey information on the origin of the relief. However, this kind of classification results in an organization of the relief features that allows formulating hypotheses about their origin (Small 1970). Compared with the multiplication of tests carried out in rugged areas, the possibilities of digital mapping in flat areas, especially areas of depositional origin, have been so far less explored.

In the FAO Guidelines for soil description (2006), landforms are described by their morphology and not by their origin or forming processes. The proposed landform classification in a two-level hierarchy is based mainly on morphometric criteria. At the first level, three classes called, respectively, level land, sloping land, and steep land, are considered. These classes are subdivided according to three morphometric attributes including slope gradient, relief intensity, and potential drainage density. Applying this procedure to the level-land class, for instance, four subclasses are recognized, namely plain, plateau, depression, and valley floor. Sloping-land and steep-land include plain, valley, hill, escarpment zone, and mountain subclasses, differentiated by the above morphometric features.

6.2.4 Ethnogeomorphic Classification

Indigenous people in traditional communities use topographic criteria, before taking the soils into consideration, to identify ecological niches suitable for selected crops and management practices. Their approach to segment a hillside into relief units is similar to the slope facet models of Ruhe (1975) and Conacher and Dalrymple (1977). Likewise in depositional environments, where the topographic variations are often subtle and less perceptible, farmers clearly recognize a variety of landscape positions, as for instance the characteristic *banco-bajio-estero* trio (bank-depression-backswamp) for pasture management in the Orinoco river plains. Trials of participatory mapping, with the collaboration of local land users and technical staff, show that the mental maps of the farmers visualize the relief using a detailed nomenclature, which allows converting them into real maps that are very similar to the geomorphic maps prepared by specialists (Barrera-Bassols et al. 2006,

2009). The two maps in Fig. 6.1 show cartographic as well as taxonomic similarities: main unit delineations coincide, and taxa recognized by scientists and local famers are comparable (e.g. gently sloping lava flow vs tzacapurhu meaning lava flow of stony land).

Indigenous soil classifications usually include the relief at the top level of the classification system, forming the basis of ethnogeopedology. In their perception of the environment, indigenous farmers use the relief, along with other features of the landscape, as a main factor for identifying, locating, and classifying soils. Because

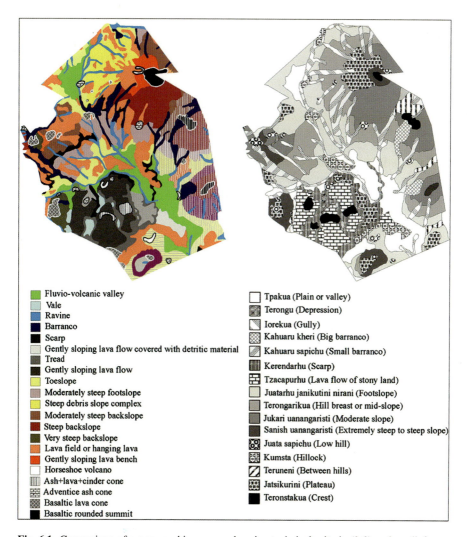

Legend (left):
- Fluvio-volcanic valley
- Vale
- Ravine
- Barranco
- Scarp
- Gently sloping lava flow covered with detritic material
- Tread
- Gently sloping lava flow
- Toeslope
- Moderately steep footslope
- Steep debris slope complex
- Moderately steep backslope
- Steep backslope
- Very steep backslope
- Lava field or hanging lava
- Gently sloping lava bench
- Horseshoe volcano
- Ash+lava+cinder cone
- Adventice ash cone
- Basaltic lava cone
- Basaltic rounded summit

Legend (right):
- Tpakua (Plain or valley)
- Terongu (Depression)
- Iorekua (Gully)
- Kahuaru kheri (Big barranco)
- Kahuaru sapichu (Small barranco)
- Kerendarhu (Scarp)
- Tzacapurhu (Lava flow of stony land)
- Juatarhu janikutini nirani (Footslope)
- Terongarikua (Hill breast or mid-slope)
- Jukari uanangaristi (Moderate slope)
- Sanish uanangaristi (Extremely steep to steep slope)
- Juata sapichu (Low hill)
- Kumsta (Hillock)
- Teruneni (Between hills)
- Jatsikurini (Plateau)
- Teronstakua (Crest)

Fig. 6.1 Comparison of a geomorphic map made using technical criteria (*left*) and a relief map drawn up according to the indigenous Purhépecha nomenclature (*right*) of the territory of San Francisco Pichátaro, Michoacán, in the volcanic belt of Central Mexico (Adapted from Barrera-Bassols et al. 2006)

of the importance that both disciplines give to the relief factor, ethnopedology and geopedology are strongly related.

6.3 Bases for a Taxonomic Classification System of the Geoforms

6.3.1 Premises and Basic Statements

A set of assumptions is formulated hereafter as a basis for structuring a taxonomic system of the geoforms and improving the traditional approaches to geomorphic classification.

- The object to be classified is a unit of the geolandscape or subdivision thereof that can be recognized by its configuration and composition. The most commonly used term to designate this entity in English-written geomorphic literature is *landform*. The same term is indistinctly used by geomorphologists, geologists, pedologists, agronomists, ecologists, architects, planners, contemplative and active users of the landscape, among others, but there is no standard definition accepted by everybody. In the FAO Guidelines for soil description (2006), the concept of *major landform* is considered to refer to the morphology of the whole landscape. Way (1973) provides a satisfactory definition in the following terms: "*Landforms* are terrain features formed by natural processes, which have a defined composition and a range of physical and visual characteristics that occur wherever the form is found and whatever is the geographic region". This statement poses two basic principles: (1) a landform is identified using internal constituents as well as external attributes, and (2) a landform is recognized by its intrinsic characteristics and not according to the context in which it occurs. In Spanish language, landform literally means *forma de tierra(s)*, a term that has an agricultural or agronomic connotation. *Land* in landscape ecology includes not only the physical features of the landscape, but also the biota and the human activities (Zonneveld 1979, 1989). The term *terrain form* is more appropriate to designate the elementary relief form, while the term *geoform* is the generic concept that encompasses the geomorphic units at all categorial levels. *Terrain form* is etymologically equal to terms with similar geomorphic meaning used in other languages, such as *forma de terreno* in Spanish and *forme de terrain* in French.
- The objects that are classified are the geoforms, or geomorphic units, which are identified on the basis of their own characteristics, rather than by reference to the factors of formation. Local or regional combinations of criteria such as climate, vegetation, soil, and lithology, which are associated with the geoforms and contribute to their formation, can be referred to in the legend of the geomorphic map, but are not intrinsically part of the classification of the geoforms. The climate factor is implicitly present in the geoforms originated by exogenous morphogenic agents (snow, ice, water, wind).

- Classes of geoforms are arranged hierarchically to reflect their level of member-ship to the geomorphic landscape. For instance, a river levee is a member of a terrace, which in turn is a member of a valley landscape. Therefore, levee, ter-race, and valley shall be placed in different categories in a hierarchic system, because they correspond to different levels of abstraction. Similarly, the slope facets (i.e. summit, shoulder, backslope, and footslope) are members of a hill, which is a member of a hilland type of landscape.
- The genesis of the geoforms is taken into consideration preferably at the lower levels of the taxonomic system, since the origin of the geomorphic units can be a matter of debate and the genetic attributes may be not clear or controversial, or their determination may require a number of additional data. At higher levels, the use of more objective, rather descriptive attributes is privileged, in parallel with the criteria of pattern recognition implemented in photo and image interpretation.
- The dimensional characteristics (e.g. length, width, elevation, slope, etc.) are subordinate attributes and are not diagnostic for the identification of the geo-forms. A geoform belongs to a particular class regardless of its size, provided it complies with the required attributes of that class. For instance, the extent of a dune or a landslide can vary from a few m^2 to several km^2.
- The names of the geoforms are often derived from the common language and some of them may be exposed to controversial interpretation. Priority is given here to those terms that have greater acceptation by their etymology or usage.
- The concepts of physiographic province and natural region, as well as other kinds of chorologic units related to specific geographic contexts, are not taken into account in this taxonomic system, because they depend on the particular conditions of a given country or continental portion, a fact that limits their level of abstraction and geographic repeatability.
- The geographic distribution of the geoforms is not a taxonomic criterion. The chorology of the geoforms is reflected in their cartography and in the structure of the geomorphic map legend.
- Toponymic designations can be used as phases of the taxonomic units (e.g. Cordillera de Mérida, Pantanal Basin).

6.3.2   Prior Information Sources

The development of the geoform classification system uses prior knowledge in terms of concepts, methods, information, and experience.

- Existing geoform typologies, with definitions and descriptive attributes, have been partially taken from the literature. The proposed classification builds on and organizes prior knowledge in a hierarchic taxonomic system. Some of the key documents that were consulted for this purpose are as follows:

- Various classic textbooks of geomorphology: Tricart and Cailleux (1962, 1965, 1967, 1969), Tricart (1965, 1968, 1977), Derruau (1965, 1966), Thornbury (1966), Viers (1967), CNRS (1972), Garner (1974), Ruhe (1975), and Huggett (2011), among others.
- Dictionaries and encyclopedias: Visser (1980), Lugo-Hubp (1989), Fairbridge (1997), and Goudie (2004), among others.
- Manuals of geomorphic photo-interpretation: Goosen (1968), Way (1973), Verstappen and Van Zuidam (1975), Verstappen (1983), and Van Zuidam (1985), among others.

- For the structure of the system, inspiration was taken from the conceptual framework of the USDA Soil Taxonomy (Soil Survey Staff 1975, 1999) with regard to the concepts of category, class, and attribute.
- Development and validation of the system have taken place essentially in Venezuela and Colombia, within the framework of soil survey projects at different scales from detailed to generalized, with the implementation of geomorphology as a tool for soil mapping (applied geomorphology). The system was modified and improved progressively as ongoing field surveys provided new knowledge. Subsequently, the already established system became teaching and training matter in postgraduate courses in soil survey at the ITC (Zinck 1988) for students from different parts of the world, especially Latin America, Africa, Middle East, and Southeast Asia.

6.3.3   Searching for Structure: An Inductive Example

Let's consider the collection of objects included in Fig. 6.2 (Arnold 1968). Squares, triangles, and circles can be recognized. The objects are large or small, green (G) or red (R). Thus the objects are different by shape, size, and color. Based on these three criteria, the objects may be classified in various ways. One option is to sort the objects first by size, then by color, and finally by shape (Fig. 6.3). They can also be sorted successively by shape, color, and size. Six hierarchization alternatives are possible. This simple experiment shows that artificial or natural objects may be classified in various ways. Any alternative is valid, if it meets the objective pursued.

From example in Fig. 6.2, three basic elements of a hierarchic classification system can be induced by effect of generalization: category, class, and attribute.

- The categories are hierarchic levels that give structure to the classification system. Three categories are present, identified by generic criteria (size, color, shape). Several (6) hierarchic arrangements are possible.
- Classes are groups of objects that have one or more differentiating characteristics in common. There are seven differentiating characteristics: large, small, red, green, square, triangular, and circular. The aggregation of characteristics generates an increase of classes from the top to the bottom of the system.
- Attributes are characteristics or properties of the objects, such as red, green, large, small, square, triangular, and circular.

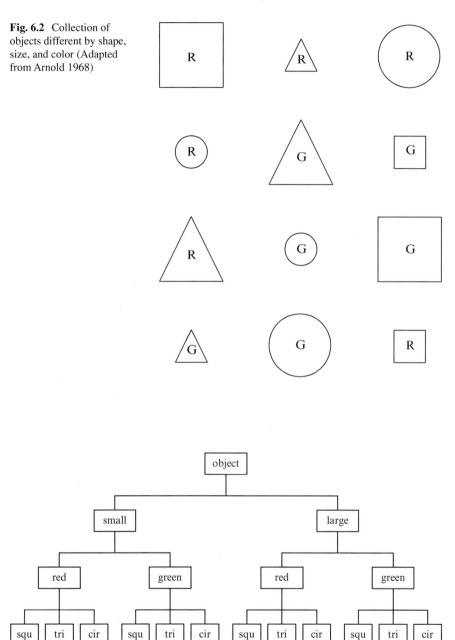

Fig. 6.2 Collection of objects different by shape, size, and color (Adapted from Arnold 1968)

Fig. 6.3 Hierarchic arrangement of the objects displayed in Fig. 6.2 by size (2 classes), color (4 classes), and shape (12 classes) (*squ* square, *tri* triangular, *cir* circular)

6.4   Structure and Elements for Building a Taxonomic System of the Geoforms

A taxonomic system is characterized by its structure (or configuration) and its elements (or components).

6.4.1   Structure

Various configuration models are possible: hierarchic, relational, network, and linear, among others (Burrough 1986). In general, the hierarchic multicategorial model is considered appropriate for taxonomic purposes. Haigh (1987) states that the hierarchic structure is a fundamental property of all natural systems, while Urban et al. (1987) consider that breaking a landscape into elements within a hierarchic framework allows to partially solve the problem of its apparent complexity. Although a hierarchic structure is less efficient than, for instance, a relational system or a network system in terms of automated data handling by computer, it is however particularly suitable for archiving, processing, and retrieving information by the human mind (Miller 1956, 2003).

A system can be compared to a box containing all the individuals belonging to the object that is sought to be classified: for example, all soils, all geoforms. The collection of individuals constitutes the universe that is going to be divided into classes and arranged into categories. The classification results in (1) a segmentation of the universe under consideration (e.g. the soil cover continuum) into populations, groups, and individuals by descending disaggregation, and (2) a clustering of individuals into groups, populations, and universe by ascending aggregation.

6.4.2   Elements

6.4.2.1   Category

A category is a level of abstraction. The higher the level of the category, the higher is the level of abstraction. Each category comprises a set of classes showing a similar level of abstraction. A category is identified by a generic concept that characterizes all classes present in this level (color, size, shape, in Fig. 6.3). For instance, a valley landscape, a fluvial terrace, and a river levee are objects belonging to different levels of abstraction. The levee is a member of the terrace, which in turn is a member of the valley. In a hierarchic system of geoforms, these geomorphic entities shall be placed in three successive categories.

6.4.2.2   Class

A class is a formal subdivision of a population at a given categorial level. A class can be determined using different modalities among which the two following are commonly implemented: (1) the range of variation of a diagnostic attribute or a combination thereof, and (2) a central class concept in relation to which other classes deviate by one or more characteristics.

An example of the first modality is provided by the way the percentage of base saturation is used in soil taxonomy as a threshold parameter to separate Alfisols ($\geq$35 %) and Ultisols (<35 %). Using a similar procedure, the strata dip in sedimentary rocks allows separating several classes of monoclinal relief, including mesa, cuesta, creston, hogback, and bar (Fig. 6.4). A similar approach can be applied to the classification of the geoforms caused by mass movements through segmentation of the continuum between solid and liquid states using the consistence limits (Fig. 6.5). There are very few references in the geomorphic literature where the segmentation of a continuum is used to differentiate related geoforms.

The central typifying concept is used to position a typical class in relation to intergrades and extragrades, which depart from the central class by deviation of some attributes. This is the case, for instance, of the "Typic" as used at subgroup level in the USDA Soil Taxonomy (Soil Survey Staff 1975, 1999). No examples were found in the geomorphic literature implementing formally the central concept to distinguish modal situations from transitional ones.

6.4.2.3   Taxon

A taxon (or taxum) is a concrete taxonomic unit as a member of a class established at a given categorial level. Usually, a particular taxon covers only part of the range of variation allowed in the selected attributes that define the class. For instance, the texture of a river bank, above the basal gravel strata, can vary from gravelly to sandy clay loam. A particular bank can be sandy to sandy loam without covering the entire diagnostic textural range.

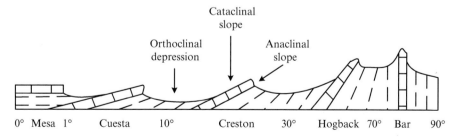

Fig. 6.4 Monoclinal relief classes determined based on strata dip ranges in sedimentary bedrocks (e.g. limestone, sandstone) (Adapted from Viers 1967)

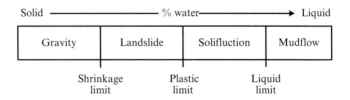

Fig. 6.5 Classes of geoforms originated by different kinds of mass movement

6.4.2.4 Attribute

An attribute is a characteristic (or variable) used to establish the limits of the classes that make up the system and to implement these limits in the description and classification of individuals. There are several kinds of attribute, as for instance:

- Dichotomous: e.g. presence or absence of iron reduction mottles, concentration of carbonates or other salts.
- Multi-state without ranges: e.g. types of soil structure, types of depositional structure.
- Multi-state with ranges: e.g. size of structural aggregates, plasticity and adhesion classes.
- Continuous variation: e.g. base saturation, bedrock dip.

Implementing these basic taxonomic criteria in geomorphology requires (1) the inventory of the known geoforms and their arrangement in a hierarchic system, and (2) the selection, categorization (diagnostic or not), hierarchization, and measurement of the attributes used to identify and describe the geoforms.

6.5 Levels of Perception: Exploring the Structure of a Geomorphic Space

Geomorphology is primarily a science of observation, aiming at the identification and separation of landscapes from topographic maps, digital elevation or terrain models, and remote-sensed documents allowing stereoscopic vision, but mainly by reading the physiographic features in the field. Geoforms can be perceived by human vision or artificial sensors, because they have a physiognomic appearance on the earth's surface (i.e. geolandscape). Physiography describes this external appearance corresponding to the epigeal component of the geoforms. Thanks to their scenic expression, geoforms are the most directly structuring elements of the terrain, more than any other object or natural feature. Even a non-scientific observer can notice that any portion of the earth's crust shows a structure determined by the relief, which allows subdividing it into components. The times that a terrain area can be subdivided into elements depend on the level of perception used for the segmentation. Although the concept of perception level is subjective when the human eye is used, it helps hierarchize the structural components of a terrain surface.

Hereafter, an example is developed that illustrates the effect of the perception scale on the sequential identification of different terrain portions. The example refers to the contact area between the Caribbean Sea and the northern edge of the South American continent in Venezuela (Zinck 1980). The use of successive perception levels, increasingly detailed, materialized by observation platforms of decreasing elevation in relation to the earth's surface, allows dividing the selected portion of continent into classes of geoforms that are distributed over various hierarchic categories (Fig. 6.6 and Table 6.3). An observer mounted on a spaceship at about 800–1,000 km elevation would distinguish two physiographic provinces, namely the east-west oriented coastal mountain chain of the Cordillera de la Costa to the north and the basin of the Llanos Plains to the south. These two macro-units of contrasting relief correspond to two types of geostructure: a folded cordillera-type mountain chain and a geosincline-type sedimentary basin, respectively. From a airplane flying at about 10 km elevation, one can distinguish the two parallel branches of the Cordillera de la Costa, namely the Serranía del Litoral range to the north and the Serranía del Interior range to the south, separated by an alignment of tectonic depressions such as that of Lake Valencia. These units are natural regions that correspond to types of morphogenic environment: the mountain ranges are structural environments undergoing erosion, whereas depressions are depositional environments. When increasing the level of perception as from a helicopter flying at 2 km elevation, a mountain range can be divided into mountain and valley landscapes. A field transect through a valley allows to cross a series of topographic steps with risers and treads that correspond to fluvial terraces. Detailed field observation of the topography and sediments in a given terrace will reveal a sequence of depositional units from the highest, the river levee (bank), to the lowest, the decantation basin (swamp). The results of this exploratory inductive procedure, leading to a sequential segmentation of a portion of the South American continent, are summarized in Table 6.3. This empirical approach generates a hierarchic scheme of geoforms in five nested categorial levels, each identified by a generic concept from general to detailed (Fig. 6.7).

6.6   Structure of a Taxonomic System of the Geoforms

Combining the basic criteria to build a taxonomic system (Sects. 6.3 and 6.4) with the results of the exploration aimed at detecting guidelines of hierarchic arrangement in the geomorphic environment (Sect. 6.5), a structure of nested categorial levels is obtained. Five of these levels are essentially deduced from the epigeal physiographic expression of the geoforms. The units recognized at the two upper levels are identified by local names, because they belong to a particular national or regional context. These are chorologic units which are formalized as taxonomic

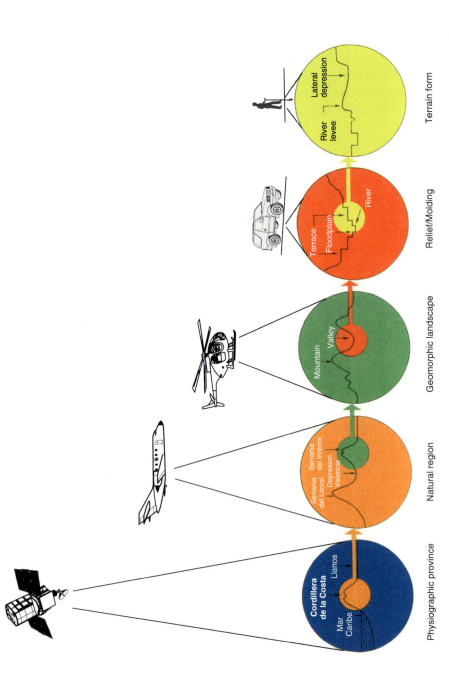

Fig. 6.6 Successive levels of perception of geoforms from different observation elevations. From *left* to *right*: physiographic province (geostructure), natural region (morphogenic environment), geomorphic landscape, relief/molding, terrain form (Zinck 1980). The features referred to are explained in Table 6.3

Table 6.3 Sequential identification of geoforms according to increasing levels of perception

Observation platform	Observation area	Observed features	Criteria used Inferred factors	Resulting geoforms	Derived generic categorial concepts
Satellite	Large continental portion	*Cordillera de la Costa* narrow, longitudinal, high relief mass; abrupt limits	Topography Internal geodynamics (orogenic area)	Cordillera (folded mountain chain)	Geostructure
		Llanos del Orinoco extensive, flat, low relief mass	Topography Internal geodynamics (sinking area)	Geosyncline (sedimentary basin)	
Airplane	Cordillera	*Serranía del Litoral Serranía del Interior* parallel, dissected mountain ranges	Topography Internal/external geodynamics (erosion)	Structural/erosional environment	Morphogenic environment
		Depresión de Valencia low-lying, flat terrain areas; concave margins	Topography Internal/external geodynamics (sedimentation)	Depositional environment	
Helicopter	Structural/erosional environment	Parallel mountain ridges	Topography Tectonics Hydrography	Mountain	Geomorphic landscape
		Narrow longitudinal depressions, parallel or perpendicular to the ridges	Topography Tectonics Hydrography	Valley	

(continued)

Table 6.3 (continued)

Observation platform	Observation area	Observed features	Criteria used Inferred factors	Resulting geoforms	Derived generic categorial concepts
Earth surface	Valley	Topographic step treads separated by risers	Topography	Terrace	Relief/molding
		Valley bottom, river system, riparian forest	Topography Drainage Vegetation	Floodplain	
Terrain surface and subsurface	Terrace	Longitudinal, narrow, convex bank; well drained, coarse-textured	Topography Drainage Morphogenesis	Levee	Terrain form
		Large, concave depression, poorly drained, fine-textured	Topography Drainage Morphogenesis	Basin	

Based on the features observed in Fig. 6.6
Zinck (1988)

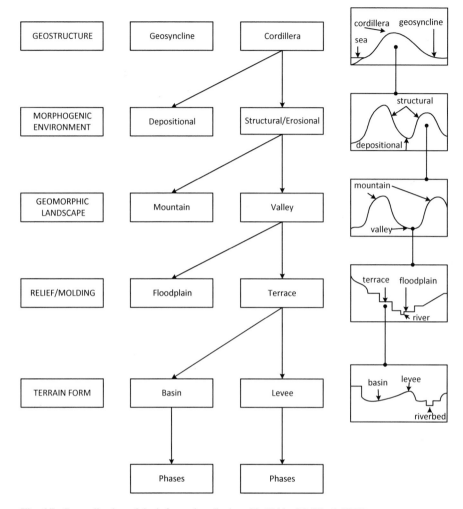

Fig. 6.7 Generalization of the information displayed in Table 6.3 (Zinck 1988)

units under the generic concept of geostructure and morphogenic environment, respectively. To substantiate the relationship between geoform and soil, it is necessary to introduce in the system information on the internal hypogeal component of the geoforms, namely the constituent material, which is in turn the parent material of the soils. As a result of the foregoing, an additional level is needed to document the lithology, in the case of bedrock substratum, or the facies in the case of unconsolidated cover materials. After several iterations, this category was inserted between the level of relief/molding (level 3) and the level of terrain form (level 1). Its inclusion in the lower part of the system is justified by the fact that field data are often needed to supplement or clarify the general information provided by the geologic maps (see Fig. 7.3 and Table 7.2 in Chap. 7). This leads finally to a system

Table 6.4 Synopsis of the geoform classification system

Level	Category	Generic concept	Short definition
6	Order	Geostructure	Large continental portion characterized by a type of geologic macro-structure (e.g. cordillera, geosyncline, shield)
5	Suborder	Morphogenic environment	Broad type of biophysical environment originated and controlled by a style of internal and/or external geodynamics (e.g. structural, depositional, erosional, etc.)
4	Group	Geomorphic landscape	Large portion of land/terrain characterized by given physiographic features: it corresponds to a repetition of similar relief/molding types or an association of dissimilar relief/molding types (e.g. valley, plateau, mountain, etc.)
3	Subgroup	Relief/molding	Relief type originated by a given combination of topography and geologic structure (e.g. cuesta, horst, etc.)
			Molding type determined by specific morphoclimatic conditions and/or morphogenic processes (e.g. glacis, terrace, delta, etc.)
2	Family	Lithology/facies	Petrographic nature of the bedrocks (e.g. gneiss, limestone, etc.) or origin/nature of the unconsolidated cover formations (e.g. periglacial, lacustrine, alluvial, etc.)
1	Subfamily	Landform/terrain form	Basic geoform type characterized by a unique combination of geometry, dynamics, and history

Zinck (1988)

with six categorial levels (Table 6.4), identified by their respective generic concepts that are explained in Chap. 7. It can be noted that obtaining a system with six categories complies with the rule called *Miller's Law*, which postulates that the capacity of the human mind to process information covers a range of seven plus or minus two elements (Miller 1956, 2003).

6.7 Conclusion

Geoforms are the emerging parts of the earth's crust. Their distinct physiognomic features make them directly observable through visual and artificial perception from remote to proximal sensing. Changing the scale of perception changes not only the degree of detail but most significantly the nature of the object observed. For instance, a levee is a member of a terrace which is a member of a valley, thus three geomorphic objects bearing different levels of abstraction. The geolandscape is a hierarchically structured and organized domain. Therefore, a multicategorial system, based on nested levels of perception to capture the information and taxonomic criteria to organize that information, is an appropriate frame to classify geoforms.

References

Arnold R (1968) Apuntes de agrología (documento inédito). Ministerio de Obras Públicas (MOP), Barquisimeto

Barrera-Bassols N, Zinck JA, Van Ranst E (2006) Local soil classification and comparison of indigenous and technical soil maps in a Mesoamerican community using spatial analysis. Geoderma 135:140–162

Barrera-Bassols N, Zinck JA, Van Ranst E (2009) Participatory soil survey: experience in working with a Mesoamerican indigenous community. Soil Use Manage 25:43–56

Buol SW, Hole FD, McCracken RJ, Southard RJ (1997) Soil genesis and classification, 4th edn. Iowa State University Press, Ames

Burrough PA (1986) Principles of geographical information systems for land resources assessment. Clarendon Press, Oxford

Burrough PA, van Gaans PFM, MacMillan RA (2000) High-resolution landform classification using fuzzy k-means. Fuzzy Set Syst 113:37–52

CNRS (1972) Cartographie géomorphologique. Travaux de la RCP77. Mémoires et Documents, vol 12. Editions du Centre National de la Recherche Scientifique, Paris

Conacher AJ, Dalrymple JB (1977) The nine-unit landscape model: an approach to pedogeomorphic research. Geoderma 18:1–154

de Bruin S, Wielemaker WG, Molenaar M (1999) Formalisation of soil-landscape knowledge through interactive hierarchical disaggregation. Geoderma 91:151–172

Derruau M (1965) Précis de géomorphologie. Masson, Paris

Derruau M (1966) Geomorfología. Ediciones Ariel, Barcelona

Dobos E, Micheli E, Baumgardner MF, Biehl L, Helt T (2000) Use of combined digital elevation model and satellite radiometric data for regional soil mapping. Geoderma 97(3–4):367–391

Elizalde G (2009) Ensayo de clasificación sistemática de categorías de paisajes. Primera aproximación, edn revisada. Maracay, Venezuela

Evans IS, Hengl T, Gorsevski P (2009) Applications in geomorphology. In: Hengl T, Reuter HI (eds) Geomorphometry: concepts, software, applications, vol 33, Developments in Soil Science. Elsevier, Amsterdam, pp 497–525

Fairbridge RW (ed) (1997) Encyclopedia of geomorphology. Springer, New York

FAO (2006) Guidelines for soil description, 4th edn. Food and Agricultural Organization of the United Nations, Rome

FAO (2009) Guía para la descripción de suelos, cuarta edn. Organización de las Naciones Unidas para la Agricultura y la Alimentación, Roma

Farshad A (2010) Geopedology. An introduction to soil survey, with emphasis on profile description (CD-ROM). University of Twente, Faculty of Geo-Information Science and Earth Observation (ITC), Enschede

Gallant JC, Wilson JP (2000) Primary topographic attributes. In: Wilson JP, Gallant JC (eds) Terrain analysis: principles and applications. Wiley, New York, pp 51–85

Garner HF (1974) The origin of landscapes. A synthesis of geomorphology. Oxford University Press, New York

Goosen D (1968) Interpretación de fotos aéreas y su importancia en levantamiento de suelos. Boletín de Suelos 6. FAO, Roma

Goudie AS (ed) (2004) Encyclopedia of geomorphology, vol 2. Routledge, London

Haigh MJ (1987) The holon: hierarchy theory and landscape research, Catena Supplement 10. CATENA Verlag, Cremlingen, pp 181–192

Hengl T, MacMillan RA (2009) Geomorphometry: a key to landscape mapping and modelling. In: Hengl T, Reuter HI (eds) Geomorphometry: concepts, software, applications, Developments in Soil Science 33. Elsevier, Amsterdam, pp 433–460

Hengl T, Reuter HI (eds) (2009) Geomorphometry: concepts, software, applications, Developments in soil science 33. Elsevier, Amsterdam

Huggett RJ (2011) Fundamentals of geomorphology. Routledge, London

Hutchinson MF, Gallant JC (2000) Digital elevation models and representation of terrain shape. In: Wilson JP, Gallant JC (eds) Terrain analysis: principles and applications. Wiley, New York, pp 29–50

Irwin BJ, Ventura SJ, Slater BK (1997) Fuzzy and isodata classification of landform elements from digital terrain data in Pleasant Valley, Wisconsin. Geoderma 77:137–154

Iwahashi J, Pike RJ (2007) Automated classifications of topography from DEMs by an unsupervised nested-means algorithm and a three-part geometric signature. Geomorphology 86(3–4):409–440

Lueder DR (1959) Aerial photographic interpretation: principles and applications. McGraw-Hill, New York

Lugo-Hubp J (ed) (1989) Diccionario geomorfológico. Universidad Nacional Autónoma de México, Cd México

MacMillan RA, Pettapiece WW (1997) Soil landscape models: automated landscape characterization and generation of soil-landscape models, Research report 1E. Agriculture and Agri-Food Canada, Lethbridge

MacMillan RA, Shary PA (2009) Landforms and landform elements in geomorphometry. In: Hengl T, Reuter HI (eds) Geomorphometry: concepts, software, applications, Developments in soil science 33. Elsevier, Amsterdam, pp 227–254

MacMillan RA, Pettapiece WW, Nolan SC, Goddard TW (2000) A generic procedure for automatically segmenting landforms into landform elements using DEMs, heuristic rules and fuzzy logic. Fuzzy Set Syst 113:81–109

Meybeck M, Green P, Vorosmarty CJ (2001) A new typology for mountains and other relief classes: an application to global continental water resources and population distribution. Mt Res Dev 21:34–45

Miller GA (1956) The magical number seven, plus or minus two: some limits on our capacity for processing information. Psychol Rev 63(2):81–97

Miller GA (2003) The cognitive revolution: a historical perspective. Trends Cogn Sci 7(3):141–144

Minár J, Evans IS (2008) Elementary forms for land surface segmentation: the theoretical basis of terrain analysis and geomorphological mapping. Geomorphology 95:236–259

Nelson A, Reuter H (2012) Soil projects. Landform classification from EU Joint Research Center, Institute for Environment and Sustainability. http://eusoils.jrc.ec.europa.eu/projects/landform/

Olaya V (2009) Basic land-surface parameters. In: Hengl T, Reuter HI (eds) Geomorphometry: concepts, software, applications, Developments in soil science 33. Elsevier, Amsterdam, pp 141–169

Pennock DJ, Zebarth BJ, De Jong E (1987) Landform classification and soil distribution in hummocky terrain, Saskatchewan, Canada. Geoderma 40:297–315

Pike RJ, Dikau R (eds) (1995) Advances in geomorphometry. Proceedings of the Walter F. Wood memorial symposium. Zeitschrift für Geomorphologie Supplementband 101

Pike RJ, Evans IS, Hengl T (2009) Geomorphometry: a brief guide. In: Hengl T, Reuter HI (eds) Geomorphometry: concepts, software, applications, Developments in soil science 33. Elsevier, Amsterdam, pp 3–30

Ruhe RV (1975) Geomorphology. Geomorphic processes and surficial geology. Houghton Mifflin, Boston

Sharif M, Zinck JA (1996) Terrain morphology modelling. Int Arch Photogramm Remote Sens XXXI(Part B3):792–797

Small RJ (1970) The study of landforms. A textbook of geomorphology. Cambridge University Press, London

Soil Survey Staff (1975) Soil taxonomy. A basic system of soil classification for making and interpreting soil surveys. USDA agriculture handbook 436. US Government Printing Office, Washington, DC

Soil Survey Staff (1999) Soil taxonomy. USDA Agric Handbook 436. US Gov Print Of, Washington

Thornbury WD (1966) Principios de geomorfología. Editorial Kapelusz, Buenos Aires

Tricart J (1965) Principes et méthodes de la géomorphologie. Masson, Paris

Tricart J (1968) Précis de géomorphologie. T1 Géomorphologie structurale. SEDES, Paris

Tricart J (1977) Précis de géomorphologie. T2 Géomorphologie dynamique générale. SEDES-CDU, Paris

Tricart J, Cailleux A (1962) Le modelé glaciaire et nival. SEDES, Paris

Tricart J, Cailleux A (1965) Le modelé des régions chaudes. Forêts et savanes. SEDES, Paris

Tricart J, Cailleux A (1967) Le modelé des régions périglaciaires. SEDES, Paris

Tricart J, Cailleux A (1969) Le modelé des régions sèches. SEDES, Paris

Urban DL, O'Neill RV, Shugart HH Jr (1987) Landscape ecology. A hierarchical perspective can help scientists understand spatial patterns. BioScience 37(2):119–127

Van Zuidam RA (1985) Aerial photo-interpretation in terrain analysis and geomorphological mapping. ITC, Enschede

Ventura SJ, Irvin BJ (2000) Automated landform classification methods for soil-landscape studies. In: Wilson JP, Gallant JC (eds) Terrain analysis: principles and applications. Wiley, New York, pp 267–294

Verstappen HT (1983) Applied geomorphology; geomorphological survey for environmental development. Elsevier, Amsterdam

Verstappen HT, Van Zuidam RA (1975) ITC system of geomorphological survey. ITC, Enschede

Viers G (1967) Eléments de géomorphologie. Nathan, Paris

Visser WA (ed) (1980) Geological nomenclature. Royal geological and mining society of the Netherlands. Bohn, Scheltema & Holkema, Utrecht

Way DS (1973) Terrain analysis. A guide to site selection using aerial photographic interpretation. Dowden, Hutchinson & Ross, Stroudsburg, Pennsylvania

Wielemaker WG, de Bruin S, Epema GF, Veldkamp A (2001) Significance and application of the multi-hierarchical landsystem in soil mapping. Catena 43:15–34

Zinck JA (1974) Definición del ambiente geomorfológico con fines de descripción de suelos. Ministerio de Obras Públicas (MOP), Cagua

Zinck JA (1980) Valles de Venezuela. Lagoven, Petróleos de Venezuela, Caracas

Zinck JA (1988) Physiography and soils, ITC soil survey lecture notes. International Institute for Aerospace Survey and Earth Sciences, Enschede

Zinck JA, Valenzuela CR (1990) Soil geographic database: structure and application examples. ITC J 1990(3):270–294

Zonneveld JIS (1979) Land evaluation and land(scape) science. ITC, Enschede

Zonneveld JIS (1989) The land unit – a fundamental concept in landscape ecology, and its applications. Landsc Ecol 3(2):67–86

Chapter 7
The Geomorphic Landscape: Classification of Geoforms

J.A. Zinck

Abstract This chapter attempts to organize existing geomorphic knowledge and arrange the geoforms in the hierarchically structured system with six nested levels introduced in the foregoing Chap. 6. Geoforms are grouped thematically, distinguishing between geoforms mainly controlled by the geologic structure and geoforms mainly controlled by the morphogenic agents. It is thought that this multicategorial geoform classification scheme reflects the structure of the geomorphic landscape sensu lato. It helps segment and stratify the landscape continuum into geomorphic units belonging to different levels of abstraction. This geoform classification system has shown to be useful in geopedologic mapping, and it could also be useful in digital soil mapping.

Keywords Geotaxa • Geostructure • Morphogenic environment • Geomorphic landscape • Relief/molding • Lithology/facies • Terrain form/landform

7.1 Introduction

The terms used hereafter to name the geoforms have been taken from a selection of textbooks, compendia, and other general books of geomorphology, including among others: Tricart and Cailleux (1962, 1965, 1967, 1969), Tricart (1965, 1968, 1977), Derruau (1965, 1966), Thornbury (1966), Viers (1967), CNRS (1972), Garner (1974), Zinck (1974), Ruhe (1975), Verstappen and Van Zuidam (1975), Visser (1980), Verstappen (1983), Van Zuidam (1985), Lugo-Hubp (1989), Fairbridge (1997), Goudie (2004), and Huggett (2011). Readers may not unanimously agree with the proposed terminology, as some terms can be subject to controversial interpretation or variability of use among geomorphologists, geomorphology schools, and countries.

J.A. Zinck (✉)
Faculty of Geo-Information Science and Earth Observation (ITC), University of Twente, Enschede, The Netherlands

Institute of Environmental Studies, University of New South Wales, Sydney, NSW, Australia
e-mail: alfredzinck@gmail.com

© Springer International Publishing Switzerland 2016   101
J.A. Zinck et al. (eds.), *Geopedology*, DOI 10.1007/978-3-319-19159-1_7

The geomorphic vocabulary, especially vocables referring to landforms, is to a large extent of vernacular origin, derived from terms used locally to describe landscape features and transmitted orally from generation to generation (Barrera-Bassols et al. 2006). Many of these terms, initially extracted from indigenous knowledge by explorers and field geomorphologists, subsequently received more precise definitions and were gradually incorporated into the scientific language of geomorphology. A typical example is the term *karst*, which refers to a mound of limestone fragments in Serbian language, and now applies to the dissolution process of calcareous rocks and the resulting geoforms. Many terms are used with different meanings depending on the country. For instance, the term *estero* (i.e. swamp) used in Spain means salt marsh, or tidal flat, or an elongated saltwater lagoon lying between sandbanks in a coastal landscape. In Venezuela, the same term refers to a closed depression, flooded by rainwater most of the time, in an alluvial plain. This kind of semantic alteration of concepts is common in countries colonized by Europeans, who intended to describe unfamiliar landscapes by similarity with their home experience. This resulted in vocabulary confusions and ambiguities that endure today. There is not yet a standardized terminology to label the geoforms, with additional semantic issues when the terms are translated from one language to another. Hereafter, an amalgam of vocables originating from various sources is used to name and describe the classes of geoforms in the six categories of the classification system.

7.2   The Taxonomy: Categories and Main Classes of Geotaxa

The categories in descending order are as follows (see Table 6.4 in Chap. 6):

- Geostructure
- Morphogenic environment
- Geomorphic landscape
- Relief/molding
- Lithology/facies
- Terrain form/landform

7.2.1   Geostructure

The concept of geostructure refers to an extensive continental portion characterized by its geologic structure, including the nature of the rocks (lithology), their age (stratigraphy), and their deformations (tectonics). These macro-units are related to plate tectonics. They include three taxa: cordillera, shield, and geosyncline.

- *Cordillera*: a system of young mountain chains, including also plains and valleys, which have been strongly folded and faulted by relatively recent orogenesis.

The component ranges may have various orientations, but the mountain chain as a whole usually has one single general direction.

- *Shield*: a continental block that has been relatively stable for a long period of time and has undergone only slight deformations, in contrast to cordillera belts. It has been exposed to long-lasting downwasting and is composed mainly of Precambrian rocks.
- *Geosyncline* (or *sedimentary basin*): wide basin-like depression, usually elongate, that has been sinking deeply over long periods of time and in which thick sequences of stratified clastic sediments, layers of organic material, and sometimes volcanic deposits have accumulated. Through orogeny and folding, geosynclines are transformed into mountain ranges.

7.2.2   Morphogenic Environment

The morphogenic environment refers to a general type of biophysical setting, originated and controlled by a style of internal and/or external geodynamics. It comprises six taxa.

- *Structural environment*: controlled by internal geodynamics through tectonic movements (tilting, folding, faulting, overthrusting of bedrocks) or volcanism.
- *Depositional environment*: controlled by the deposition of detrital, soluble and/or biogenic materials, under the influence of water, wind, ice, mass removal, or gravity.
- *Erosional* e*nvironment* (or denudational): controlled by processes of dissection and removal of materials transported by water, wind, ice, mass movement, or gravity.
- *Dissolutional environment*: controlled by processes of rock dissolution generating chemical erosion (karst in calcareous rocks, pseudokarst in non-calcareous rocks).
- *Residual environment*: characterized by the presence of surviving relief features (e.g. inselberg).
- *Mixed environment*: e.g. a structural environment dissected by erosion.

7.2.3   Geomorphic Landscape

7.2.3.1   Definition

Landscape is a complex concept which covers a variety of meanings:

- In common language: scenery of a portion of land or its pictorial representation.
- In media language: political, financial, intellectual, artistic landscape, etc.

- In scientific language: term used differently in landscape ecology, pedology, bio-geography, geomorphology, architecture, etc.
- In the geomorphic literature: the expression *geomorphic landscape* is used with-out taxonomic connotation or mention of the level of generalization; it can thus correspond to any of the six categories of the system described here.
- Adopted definition: large land surface characterized by its physiographic expres-sion. It is formed by a repetition of similar types of relief/molding or an associa-tion of dissimilar types of relief/molding. For instance, a large active alluvial plain may consist of a systematic spatial repetition of the same molding type, namely a set of adjoining floodplains constructed by a network of rivers. In con-trast, a valley usually shows an association of various molding types, such as floodplain, terrace, fan, and glacis.
- Ambiguity of the concept of landscape: a valley, for instance, can cover three different kinds of spatial frame (Fig. 7.1):

    (1) An area of longitudinal transport and deposition of sediments, including the floodplain and terraces of the valley bottom. This space corresponds to the concept of valley sensu stricto.

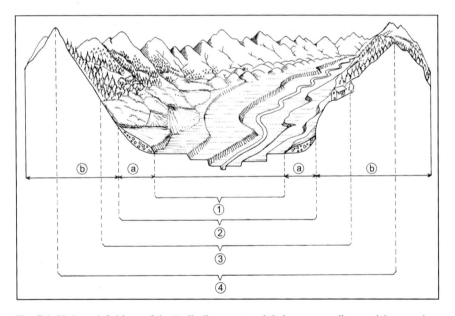

Fig. 7.1 Various definitions of the "valley" concept and their corresponding spatial expressions (Zinck 1980). (*1*) Valley as an area where sediments of longitudinal origin, coming from the catch-ment area of the upper watershed, are deposited in the floodplain and terraces of the valley bottom. (*2*) Valley as an area where longitudinal as well as lateral sediments are deposited, including pied-mont glacis and fans. (*3*) Valley as an area directly influenced by human occupation and activities, including the lower reaches of the surrounding mountain slopes. (*4*) Hydrographic basin delineated by the water divides between adjacent watersheds. (*a*) Piedmont (*b*) Mountain

(2) An area similar to the previous one plus the sectors of lateral deposition forming fans and glacis. This space modeled by side deposits actually corresponds to the concept of piedmont landscape.

(3) An area controlled by human settlements, including the lower parts of the surrounding mountain slopes. This portion of space in fact belongs to the mountain landscape.

There is no consensus on whether restricting the concept of valley to the area covered by longitudinal deposits, or also including one or both of the two other components. An extreme position would be to extend the valley space to the surrounding water divides. In this case, the mountain landscape would vanish.

7.2.3.2 Taxa

The present system of geoform classification recognizes seven taxa at the categorial level of geomorphic landscape: valley, plain, peneplain, plateau, piedmont, hilland, and mountain (Figs. 7.2 and 7.3).

- *Valley*: elongated portion of land, flat, lying between two bordering areas of higher relief (e.g. piedmont, plateau, hilland, or mountain). A valley is usually drained by a single river. Stream confluences are frequent. For recognition, a valley should have a system of terraces which, in its simplest expression, comprises at least a floodplain and a lower terrace. In the absence of terraces, it is merely a fluvial incision, which is expressed on a map by the hydrographic network.
- *Plain*: extensive portion of land, flat, unconfined, low-lying, with low relief energy (1–10 m of relative elevation difference), and gentle slopes, usually less

Fig. 7.2 Types of geomorphic landscape (Zinck 1980). *1* valley; *2* plain; *3* plateau; *4* piedmont; *5* hilland; *6* mountain

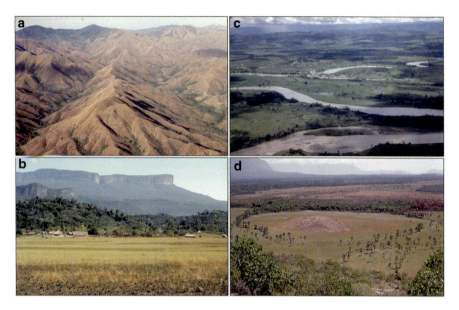

Fig. 7.3 Examples of geomorphic landscapes: (**a**) hilland dissected by two parallel valleys guided by tectonics, mountains in the background, Coastal Cordillera, northern Venezuela; (**b**) sandstone plateau dominating a valley, Guayana Shield, southern Venezuela; (**c**) alluvial plain with meandering river, Llanos basin, central Venezuela; (**d**) peneplain with residual hill in the center, surrounded by an annular glacis and a peripheral circular vale with palm trees, Guayana Shield, southern Venezuela

than 3 %. Several rivers contribute to form a complex fluvial system. Stream diffluences are frequent.

- *Peneplain*: slightly undulating portion of land, characterized by a systematic repetition of low hills, rounded or elongated, with summits of similar elevation, separated by a dense hydrographic network of reticulated pattern. The hills and hillocks have formed either by dissection of a plain or plateau, or by downwasting and flattening of an initially rugged terrrain surface. Often, a peneplain consists of an association of three types of relief/molding: namely hills surrounded by a belt of glacis and, further, by peripheral colluvio-alluvial vales.

- *Plateau*: large portion of land, relatively high, flat, commonly limited at least on one side by an escarpment relating to the surrounding lowlands. It is frequently caused by tectonic uplift of a plain, and the elevated land portion is subsequently subdivided by incision of deep gorges and valleys. The summit topography is table-shaped or slightly undulating, because erosion is mostly linear. The plateau landscape is independent of specific altitude ranges, provided it complies with the diagnostic characteristics of this kind of geoform, such as high position, tabular topography, and escarpments along the edges and the water courses that deeply incise the relief. According to this definition, the table-shaped relief of the Mesa Formation in eastern Venezuela, cut by valleys of variable depth (40–

100 m), makes up a plateau landscape at no more than 200–300 masl, while the Bolivian Altiplano is a plateau landscape lying at 3500–4000 masl.

- *Piedmont*: sloping portion of land lying at the foot of higher landscape units (e.g. plateau, mountain). The internal composition is generally heterogeneous and includes: (1) hills and hillocks formed from pre-Quaternary substratum, exposed by exhumation after the Quaternary alluvial cover has been partially removed by erosion; and (2) fans and glacis, often in terrace position (fan-terrace, glacis-terrace), composed of Quaternary detrital material carried by torrents from surrounding higher terrains. Piedmonts located at the foot of recent mountain systems (cordilleras) usually show neotectonic features, as for example faulted and tilted terraces.
- *Hilland*: rugged portion of land, characterized by a repetition of high hills, generally elongated, with variable summit elevations, separated by a moderately dense hydrographic network and many colluvio-alluvial vales.
- *Mountain*: high portion of land, rugged, deeply dissected, characterized by: (1) important relative elevations in relation to external surrounding lowlands (e.g. plains, piedmonts); (2) strong internal dissection, generating important net relief energy between ridge crests and intramountain valleys.

7.2.4   Relief/Molding

7.2.4.1   Definition

The concepts of relief and molding are based on the definition that is commonly given to both terms in the geomorphic French literature (Viers 1967).

- Relief: geoform that results from a particular combination of topography and geologic structure (e.g. cuesta relief); largely controlled by internal geodynamics.
- Molding: geoform determined by specific morphoclimatic conditions or morphogenic processes (e.g. glacis, fan, terrace, delta); largely controlled by external geodynamics (molding from French word *modelé*).

7.2.4.2   Taxa

Relief and molding include an ample variety of taxa that can be grouped into families according to the dominant forming process: structural, erosional, depositional, dissolutional, and residual (Table 7.1 and Fig. 7.4). In general, the geomorphic literature does not establish a clear differentiation between geoforms of level 3 (relief/molding) and geoforms of level 1 (terrain form/landform). The list of geoforms in Table 7.1 was obtained by iteration, taking into account the possibility to subdivide types of relief and molding into terrain forms/landforms at level 6 of the system. It

Table 7.1 Relief and molding types

Structural	Erosional	Depositional	Dissolutional	Residual
Depression	Depression	Depression	Depression	Planation surface
Mesa (meseta)	Vale	Swale	Dome	Dome
Cuesta	Canyon (gorge)	Floodplain	Tower	Inselberg
Creston	Glacis	Flat (e.g. tidal flat)	Hill (hum)	Monadnock
Hogback	Mesa (meseta)	Terrace	Polje	Tors (boulders field)
Bar	Hill (hillock)	Mesa (meseta)	Blind vale	...
Flatiron	Crest	Fan	Dry vale	
Escarpment	Rafter (chevron)	Glacis	Canyon	
Graben	Ridge	Bay	...	
Horst	Dike	Delta		
Anticline	Trough (glacial)	Estuary		
Syncline	Cirque (glacial)	Marsh		
Excavated anticline	...	Coral reef		
Hanging syncline		Atoll		
Combe		...		
Ridge				
Cone (dome)				
Dike				
...				

Zinck (1988)

is an open-ended collection, which can be improved by the incorporation of additional geoforms.

7.2.5 Lithology/Facies

7.2.5.1 Definition

Level 5 provides information on (1) the petrographic nature of the bedrocks that serve as hard substratum to the geoforms, and (2) the facies of the unconsolidated cover formations that often constitute the internal hypogeal component of the geoforms. In both cases, the information concerns the parental material of the soils.

If the taxonomic system were restricted to depositional geoforms, the present categorial level could result redundant and therefore superfluous, as the lithology would be conveniently covered by the facies of the geomorphic material (i.e. the parent material of the soil) at level 1 of the system (i.e. the terrain form level). However, in areas where the soils are formed directly or indirectly from consolidated

Fig. 7.4 Examples of morphogenic environments and relief/molding types: (**a**) structural: mesetas developed on horizontal sandstone layers, Guayana Shield, southern Venezuela; (**b**) residual: monadnock inselberg rising above a semiarid peneplain landscape, Bahia State, north-eastern Brazil; (**c**) erosional-depositional: glacial landscape with gelifraction crests and cirques in the background, glacial trough in the center, moraines (lateral, ground, and frontal) in the foreground, Venezuelan Andes; (**d**) dissolutional: doline depression in limestone, southern France

geologic material, the system should allow entering information about the lithology of the bedrocks.

In some geomorphic classification systems, the lithology is referred to at high categorial levels. For instance, in the case of the geomorphic map of France, lithology is the second information layer in the structure of the legend, following a first level that deals with the location of the description sites (CNRS 1972).

Analyzing the portion of terrain represented in Fig. 7.5, an observer would recognize successively (hierarchically) the patterns identified in Table 7.2, by reasoning in the field or by photo-interpretation. The example shows that lithology is best positioned below the categorial levels where the concepts of landscape and relief/molding are located, respectively, taking into account criteria such as the hierarchic subdivision mechanism, the level of perception and the degree of resolution through interpretation of aerial photos (API), and the need for field and laboratory data.

7.2.5.2 Taxa

- Bedrocks (according to conventional rock classification):

J.A. Zinck

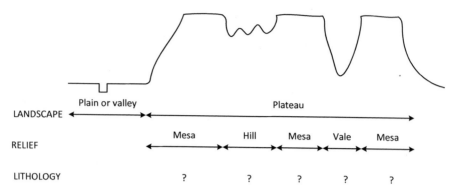

Fig. 7.5 Sequential partition of a plateau landscape into relief patterns to infer the lithology of the substratum (see Table 7.2 for lithology alternatives) (Zinck 1988)

- Igneous rocks, including intrusive rocks (e.g. granite, granodiorite, diorite, gabbro) and extrusive rocks (e.g. rhyolite, dacite, andesite, basalt)
- Metamorphic rocks (e.g. slate, schist, gneiss, quartzite, marble)
- Sedimentary rocks (e.g. conglomerate, sandstone, limolite, shale, limestone)

• Facies of unconsolidated materials:

- Nival (snow)
- Glacial (ice, glacier)
- Periglacial (ice, cryoclastism, thermoclastism)
- Alluvial (concentrated water flow = fluvial = river)
- Colluvial (diffuse, laminar water flow)
- Diluvial (torrential water flow)
- Lacustrine (freshwater lake)
- Lagoonal (brackish water lake)
- Coastal (fringe between continent and ocean; tidal)
- Mass movement (plastic or liquid debris flow; landslide)
- Gravity (rock fall)
- Volcanic (surface flow or aerial shower of extrusive igneous materials)
- Biogenic (coral reef)
- Mixed (fluvio-glacial, colluvio-alluvial, fluvio-volcanic)
- Anthropic (kitchen midden, sambaqui, tumulus, rubble, urban soil, etc.)

Table 7.2 Inference of the substratum lithology related to the plateau landscape depicted in Fig. 7.5

Categorial level	Identification features	Geoform or material inferred	Generic concept	Resolution API Field	
High	Flat summit topography	Plateau	Landscape	+	−
	High position in relation to the surrounding lowlands				
	Abrupt edges (escarpments)				
	Deep river incision				
Intermediate	Summit topography divided into:		Relief/molding	±	±
	(1) Level areas	(1) Mesas			
	(2) Undulating areas	(2) Hills			
Low	(1) If concordance between slope of the terrain surface and dip of the underlying rock layers, then structural surface supported by horizontally-lying rock strata	(1a) Hard sedimentary rocks (e.g. limestone, sandstone) or	Lithology	−	+
		(1b) Hard extrusive igneous rocks (e.g. basalt)			
	(2) If no concordance between terrain surface and rock dip, then erosional surface truncating no-horizontally-lying rock strata	(2a) Tectonized stratified rocks (sedimentary or volcanic) or			
		(2b) Intrusive igneous rocks			

Zinck (1988)
API aerial photo-interpretation

7.2.6   *Terrain Form/Landform*

7.2.6.1   **Definition**

In general, geomorphology textbooks do not establish a formal hierarchic differentiation of geoforms below the level of landscape. The terms *terrain form* and *landform* are often used as a general concept that covers any class of geomorphic unit from landscape level down to the lower levels of the system, without distinction between degrees of abstraction or levels of hierarchy. In this sense, both terms are synonyms of the generic term *geoform*.

Table 7.3 Structural geoforms

Primary relief	Derived relief	Terrain form
Monoclinal		
Cuesta (1–10° dip)	Double cuesta	Relief front (front slope)
Creston (10–30°)	Outlier hill	Scarp (overhang)
Hogback (30–70°)	Flatiron	Debris talus
Bar (70–90°)	Orthoclinal (subsequent) depression	Relief backslope
Flatiron	Cataclinal (consequent) depression	Structural surface
	Anaclinal (obsequent) depression	Substructural surface
		Cataclinal gap
Folded (Jurassian)		
Mont (original anticline)	Excavated anticline	Anticlinal hinge zone
Val (original syncline)	Hanging syncline	Synclinal hinge zone
	Rafter (chevron)	Fold flank
	Creston	Scarp
	Combe	Debris talus
	Cluse	
	Ruz	
Folded (Appalachian)		
	Truncated anticline	Scarp
	Bar	Debris talus
	Hanging syncline	
	Cataclinal gap	
Folded (complex)		
Overthrust nappe	Klippe	Scarp
Overthrust fold	Creston of overturned fold	Debris talus
Box fold	Escarpment of faulted fold	
Diapiric fold	Combe	
Faulted/fractured		
Fault scarp	Faultline scarp	Scarp
Horst	Fault escarpment facet	Debris talus
Graben	Cuesta	
Faults en échelon		
Block-faulted area		

In the present hierarchic system of geoform classification, terrain form/landform is considered as the generic concept of the lower level of the system. It corresponds to the elementary geomorphic unit whose minor internal and/or external variations are signaled by phases. It is characterized by its geometry, dynamics, and history.

The hierarchic arrangement of the collection of geoforms in Tables 7.3, 7.4, 7.5, 7.6, 7.7, 7.8, 7.9, 7.10, and 7.11 is based on expert judgement and field experience (Zinck 1988). Geoforms can be conveniently grouped into: (1) geoforms predominantly controlled by the geologic structure and bedrock substratum (internal

Table 7.4 Volcanic geoforms

Relief	Variety of geoforms
Depression	Crater
	Caldera
	Maar
	Lake
Cone	Ash cone
	Cinder cone
	Lava cone
	Spatter cone
	Stratovolcano
Dome	Cumulo-volcano
	Shield-volcano
	Intrusion dome
	Extrusion dome
	Extrusion cilinder
Flat	Lava flow
	Block lava (aa lava)
	Ropy lava (pahoehoe lava)
	Pillow lava
	Volcanic mudflow (lahar)
	Fluvio-volcanic flow
	Cinder field
	Ash mantle
	Pyroclastic deposit
Mesa Cuesta	Planèze
	Hanging lava flow
	Sill
Bar Dyke Tower Escarpment	Longitudinal dyke
	Annular dyke (ring-dyke)
	Volcano scarp
	Volcanic plug (neck)
	Volcanic chimney (vent)
	Volcanic spine

Table 7.5 Karstic geoforms

Relief	Terrain form
Cockpit karst (dolines)	Karren
Hum karst (hills)	Sima (aven)
Tower karst	Ponor
Cone karst	Doline
Polje (karstic plain)	Uvala
Karrenfeld	
Collapse valley	
Blind valley	
Dry valley	

Table 7.6 Glacial geoforms

Molding	Terrain form
Cirque Trough	Trough threshold
	Cirque threshold
	Trough basin
	Trough shoulder
	Hanging valley (gorge)
	Roches moutonnées
	Ground moraine
	Lateral moraine
	Medial moraine
	Frontal moraine
	Knob-and-kettle till
	Blocks stream
	Dead-ice depression
Flat	Roches moutonnées field
	Drumlin field
	Ground moraine
	Push moraine
	Kame
	Esker
	Fluvio-glacial outwash fan (sandur)

Table 7.7 Periglacial geoforms

Molding	Terrain form
Crest (gelifraction)	Nunatak (horn)
	Debris talus (scree talus)
	Debris fan (scree fan)
Flat	Polygonal ground
	Mud field
	Stone field (pavement)
	Permafrost
	Tundra hummock
	Peatland (moor, bog)
	Dune field
	Loess mantle
Slope	Gravity scree
	Patterned ground
	Striped ground
	Stone stream
	Mud flow (solifluction)

geodynamics), and (2) geoforms predominantly controlled by the morphogenic agents and surface formations (external geodynamics). Section 7.3 provides more details.

Table 7.8 Eolian geoforms

Molding	Terrain form
Flat (dune field, erg)	Barchan
	Nebka
	Parabolic dune
	Longitudinal dune
	Transverse dune
	Pyramidical dune (ghourd)
	Reticulate dune
	Blowout dune (eolian levee)
	Loess cover
	Blowout depression
	Reg (deflation pavement)
	Yardang
Meseta	Hamada (rocky deflation surface)

Table 7.9 Alluvial and colluvial geoforms

Depositional facies/erosion	Terrain form
Overload facies	Scroll bar
	Point bar complex
	River levee
	Distributary levee
	Delta channel levee
	Splay axis
	Splay mantle
	Crevasse splay
	Splay fan
	Splay glacis
	Alluvial fan
Overflow facies	Overflow mantle
	Overflow basin
Decantation facies	Decantation basin
	Backswamp (lateral depression)
	Cut-off meander with oxbow lake
	Infilled channel
Colluvial facies	Colluvial fan
	Colluvial glacis
Water erosion features	Sheet erosion
	Rill
	Gully
	Badland

Table 7.10 Gravity and mass movement geoforms	Process (consistence states)	Terrain form
	Creep (variable consistence)	Creep mantle
		Pied-de-vache
		Terracette
	Flow (plastic/liquid)	Rock flow
		Earth flow
		Debris flow
		Mud flow
		Solifluction sheet
		Solifluction tongue (stripe)
		Solifluction lobe
		Torrential lava
	Slide (semi-solid)	Rotational slide (slump)
		Translational slide (slip)
		Rock slide
		Block slide
		Debris slide
		Landslide
		Landslide scar
	Fall (solid)	Rock fall
		Scree talus

7.2.6.2 Taxa

- Geoforms predominantly controlled by the geologic structure and bedrock substratum

  - Structural (monoclinal, folded, faulted)
  - Volcanic
  - Karstic

- Geoforms predominantly controlled by the morphogenic agents and surface formations

  - Nival, glacial, periglacial
  - Eolian
  - Alluvial and colluvial
  - Lacustrine
  - Gravity and mass movements
  - Coastal

- Banal hillside geoforms

Table 7.11 Coastal geoforms

Formation mode	Terrain form
Mechanical deposition	Beach
	Beachridge (coastal bar)
	Offshore bar (barrier beach)
	Offshore trough
	Baymouth bar (restinga)
	Cuspate bar
	Spit
	Tombolo
	Slikke-schorre (tidal mudflat)
	Lagoon
	Dune
	Sand cay
	Beachrock platform
Biogenic formation	Fringing reef
	Barrier reef
	Reef flat
	Reef front
	Lagoon
Erosion	Cliff
	Wave-cut platform/terrace
	Tidal channel
	Grao

7.3   Classification of the Geoforms at the Lower Levels

7.3.1   Introduction

The geotaxa belonging to the upper and middle levels of the system are defined in the previous section. The present section describes the classification of the geoforms at the lower categorial levels of the system: relief/molding and terrain form. The taxa listings are neither exhaustive nor free of ambiguity. It is mainly an attempt to categorize the existing geotaxa according to their respective level of abstraction and place them either at level 4 or level 6 of the classification system. A variety of synonymous terms can be found in the specialized literature, and the same type of geoform may be referred to with different names. With further progress in geomorphic mapping, probably new types of geoform will be identified and new names will appear. The concepts and terms used here are extracted from general texbooks and treatises in geomorphology. In case of multiple terms for a particular geoform, preference is given to the most commonly used one. Terms borrowed from different languages are kept in their original form and spelling, especially when already internationally accepted.

A criterion often used for grouping the geoforms in families is their origin or formation mode. Hereafter, the concept of origin is used in a broad sense, referring indistinctly to a type of environment (e.g. structural), an agent (e.g. wind), a morphogenic system (e.g. periglacial), or a single process (e.g. decantation).

The concept of origin, as a synonym for formation, is implicitly or explicitly present at all levels of the taxonomic system, but its diagnostic weight increases at the lower levels. The origin controlled by the internal geodynamics is more relevant in the upper categories, while the origin controlled by the external geodynamics is more important in the lower categories. It results from the former that there is a differential hierarchization of the diagnostic attributes according to the origin of the geoforms. For instance, in the case of the structural geoforms, genetic features have maximum weight at the level of the relief type, while in the case of the geoforms caused by exogenous agents (e.g. water, wind, ice), the genetic features have maximum weight at the lower levels of the system (i.e. facies and terrain form).

A morphogenic agent can cause erosional as well as depositional features according to the context in which the process takes place. For this reason, a distinction is made between erosional and depositional terrain forms. Likewise, structural geoforms may have been strongly modified by erosion, a fact which leads to distinguish between original (primary) and derived forms.

A geoform is considered erosional when the erosion process, operating either by areal removal of material or by linear dissection, is responsible for creating the dominant configuration of that geoform. Local modifications caused, for instance, by the incision of rills and gullies or surficial deflation by wind are identified as phases of the affected taxonomic unit. Similarly, point features and phenomena of limited extent are not considered as taxonomic units and are represented by cartographic spot symbols on the maps (e.g. geysers, erratic blocks, pingos, etc).

For the definition of the geoforms whose names are reported in the attached tables, it is recommended to consult the textbooks and dictionaries of geomorphology, namely Derruau (1965), CNRS (1972), Visser (1980), and Lugo-Hubp (1989), among others. The multilingual *Geological Nomenclature* (Visser 1980) is particularly useful, in the current context of unstandardized vocabulary, for short definitions of geoforms and multilingual equivalents. Some geoforms may appear named at both levels of relief/molding and terrain form, because their taxonomic position in the classification system is not yet clearly established.

7.3.2   Geoforms Mainly Controlled by the Geologic Structure

Geostructural control acts through tectonics, volcanism and/or lithology. Therefore, the internal geodynamics is determinant in the formation of this kind of geoforms, in combination with external processes of erosion or deposition in varying degrees. The dissection of primary structural reliefs by mechanical erosion, for instance, results in the formation of derived relief forms. Chemical erosion through limestone dissolution or sandstone disintegration causes the formation of karstic and pseudo-karstic reliefs. Deposition of volcanic ash or scoriae can alter the original configuration of a structural relief.

7.3.2.1 Structural Geoforms Proper

Geoforms directly caused by structural geodynamics (folds and faults) cover a large array of relief types (Table 7.3):

- Monoclinal reliefs: rock layers uniformly dipping up to 90° (see Fig. 6.4 in Chap. 6). Strata of hard rocks (e.g. sandstone, quartzite, limestone) overlie softer rocks (e.g. marl, shale, slate). The duo hard rock/soft rock can be recurrent in the landscape causing the same relief type to repeat several times (e.g. double cuesta).
- Jurassian fold reliefs: symmetrical folds in regular sequences of structural highs (anticlines) and structural lows (synclines) in their original or almost original form; related to important volumes of stratified sedimentary rock layers.
- Appalachian fold reliefs: fold reliefs in advanced stage of flattening and dissection.
- Complex fold reliefs: primary or derived fold reliefs controlled by overthrust tectonics and complex folding.
- Fault reliefs: primary or derived reliefs caused by faults or fractures; the faulting style (i.e. normal, reverse, rotational, overthrust, etc.) controls the type of resulting relief.

7.3.2.2 Volcanic Geoforms

Volcanic materials can constitute the whole substratum or an essential part thereof or be limited to cover formations in a variety of landscapes including mountain, plateau, piedmont, plain, and valley. Volcanic geoforms are of variable complexity, and this makes it difficult to strictly separate relief types and terrain forms. An ash cone, for instance, can be a very simple geoform and constitute therefore an elementary terrain form, while a stratovolcano cone is usually a much more complex geoform with various terrain forms (Table 7.4).

7.3.2.3 Karstic Geoforms

Karst formation operates by chemical erosion of soluble rocks and originates sculpted terrain surfaces and underground gallery systems of complex configuration, characterized by residual geoforms of positive or negative relief. The resulting taxa enter the system essentially at the relief/molding level. The karstic geoforms are both endogenous by the influence of the lithology in their constitution and exogenous by the dissolution process which originates them (Table 7.5).

7.3.3 Geoforms Mainly Controlled by the Morphogenic Agents

Water, wind, and ice are morphogenic agents that cause erosion or deposition according to the prevailing environmental conditions. The resulting geoforms are usually more homogeneous than the geoforms controlled by the internal structure. For this reason, many of the geoforms originated by exogenous agents can be classified at the level of terrain form. Hereafter, six main families of geoforms are distinguished according to their origin.

7.3.3.1 Nival, Glacial, and Periglacial Geoforms

The nival, glacial, and periglacial geoforms have in common the fact that they develop in cold environments (high latitudes and high altitudes) by the accumulation of snow (nival geoforms), alternate freezing-thawing causing gelifraction

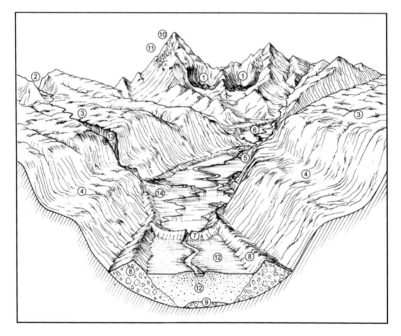

Fig. 7.6 Configuration and components of a glacial valley or glacial trough (Zinck 1980)
Glacial erosion molding: *1* Glacial cirque with lagoon, *2* Glacial diffluence pass, *3* Roches moutonnées (striated surface), *4* Trough shoulder (staircase tread), *5* Threshold with trough narrowing, *6* Basin with trough widening and deepening (lake)
Glacial deposition molding: *7* Frontal moraine barring the water flow (lake), *8* Lateral moraine, *9* Ground moraine
Periglacial molding: *10* Gelifraction horn, *11* Scree talus
Postglacial fluvial molding: *12* Trough filling by fluvial aggradation, *13* Hanging lateral valley with steps, *14* Alluvial fan

(periglacial geoforms), or accumulation of ice mass (glacial geoforms). Some geo-forms result from deposition (e.g. moraines), others from erosion (e.g. glacial cirque) (Fig. 7.6). Some can be recognized and mapped as elementary terrain forms (e.g. a moraine). Others are molding types that consist of more than one kind of ter-rain form. A glacial trough, for instance, can contain different types of moraine (e.g. ground, lateral, frontal), surfaces with "roches moutonnées", hanging valleys, and lagoons, among others (Tables 7.6 and 7.7). Strictly speaking, the nival forms are not terrain forms, since they are covered with snow (e.g. nivation cirque, permanent snowpack, and snow avalanche corridor and fan).

7.3.3.2 Eolian Geoforms

Dry environments, from desert to subdesert, are most favorable to forms arising from the action of the wind. Eolian geoforms occur mainly in coastal or continental plains where the effect of the wind is more pronounced (Table 7.8).

7.3.3.3 Alluvial and Colluvial Geoforms

Alluvial geoforms can occur in almost all types of landscape, but mostly in plains and valleys where they form terraces, floodplains, glacis, and fans. The colluvial geoforms are typical features of the piedmont landscape where they form fans and glacis. The elementary terrain forms are grouped according to the type of deposi-tional process that originates them (Table 7.9).

7.3.3.4 Lacustrine Geoforms

The receding of lake shorelines, which is a common process in drying lakes after the last glaciation, leaves exposed lacustrine material in the form of terraces. In arid and semi-arid environments, stratified fluvio-lacustrine deposits occur in playa-type depressions. In areas emerging from proglacial lakes there are stratified varve deposits.

7.3.3.5 Gravity and Mass Movement Geoforms

The mechanical condition of the material, with continuity from solid state to liquid state, controls the mass movement processes, including creep, flow, slide, and fall, that give rise to the geoforms (Table 7.10).

7.3.3.6   Coastal Geoforms

The most typical coastal geoforms are developed in the coastal lowlands, including molding types such as salt marsh, mangrove marsh, estuary, delta, bay, reef, and atoll. Cliff is the most common form in rocky coasts (Table 7.11).

7.3.4   Banal Hillside Geoforms

Geoforms without remarkable physiographic features have been called *banal* (CNRS 1972). Such geoforms are frequent in soft sedimentary rocks, devoid of structural control (e.g. marls and other argillaceous rocks), and in igneous-meta-morphic rocks without marked schistosity (e.g. granite, gneiss). Their most common physiographic feature is expressed by convex-concave hillslopes that have inspired the slope facet models. This unoriginal but not uncommon topography does not reflect any specific internal or external geodynamics, as is the case of the two other large families of geoforms. Present hills are in general an inheritance of a long geomorphic evolution. They are increasingly exposed to severe slope erosion processes including sheet erosion, rills, gullies, and mass movements.

7.3.4.1   Main Characteristics

- General topography of hills, ridges, and crests, originated by dissection.
- Little or none structural influence, in particular lack of specific control by fold or fault tectonics in the topography.
- Presence of fractures that favor and control the incision and organization of the hydrographic network.
- The drainage pattern has a relevant influence on the configuration of the resulting dissection topography, especially in peneplain and hilland landscapes.
- Homogeneous rock substratum over wide expanses.
- Material of moderate to weak resistance to physical and/or chemical erosion. Banal geoforms are frequent in shale and marl. In warm and moist tropical environments, chemical erosion of granite or gneiss produces also banal geoforms in peneplain landscape.

7.3.4.2   Classes of Banal Hillside Geoforms

Banal geoforms occur at the levels of relief/molding and terrain form in mountain, hilland, peneplain, and piedmont landscapes.

(a) At the level of relief/molding

- The *backbone* configuration consists of an association between a main longitudinal dorsal and a set of perpendicular hills (chevron, rafter, nose) separated by vales (Fig. 7.7). This type of relief is common in fractured sedimentary rocks. Its further evolution generates elongated horseback-shaped hills.
- The *half-orange* configuration consists of a systematic repetition of rounded hills with similar elevation. This type of relief is typical of the peneplain landscape developed in homogeneous but intensively fractured igneous or metamorphic substratum, with reticulate drainage pattern. It is common in the Precambrian shields of the intertropical zone.

(b) At the level of terrain form

Slope segmentation into interrelated facets seems to be the most convenient criterion to subdivide any hilly relief. The slope models such as the nine-unit-land-surface model of Conacher and Dalrymple (1977) or the five-hillslope-element model of Ruhe (1975) can be implemented to this effect. Table 7.12 shows the relationships between slope facet, topographic profile, and dominant morphogenic dynamics according to Ruhe's model (Fig. 7.8). It is worth noting that the toeslope is actually not a slope facet; instead it is a unit that belongs to the adjoining valley or vale, with slope perpendicular to the hillside and with longitudinal deposits.

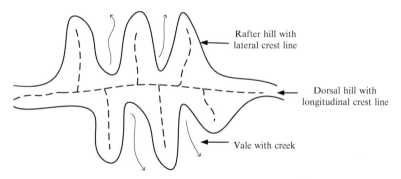

Rafter hill with lateral crest line

Dorsal hill with longitudinal crest line

Vale with creek

Fig. 7.7 Hilland landscape with backbone configuration comprising a longitudinal dorsal and perpendicular rafters

Table 7.12 Slope facet model

Slope facet	Topographic profile	Dominant morphodinamics
Summit	Level/convex	Ablation/erosion
Shoulder	Convex	Erosion
Backslope	Rectilinear-inclined	Material in transit
Footslope	Concave	Lateral accumulation
Toeslope	Concave/level	Longitudinal accumulation

Adapted from Ruhe (1975)

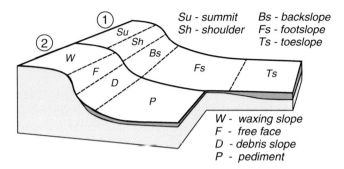

Fig. 7.8 Models of convex-concave "fully developed hillslopes" with lateral deposits (Adapted from Ruhe 1975). (*1*) Ruhe's model (note that the toeslope deposits are of longitudinal origin). (*2*) Model combining elements taken from Wood (1942) and King (1957)

Models are suitable generalizations of real situations. However, the general hillside model with convex-concave profile can be disturbed by irregularities. For instance, the cross section of a hill shows often complications that should be considered in the mapping of the geoforms and soils. These complications can be caused by the heterogeneity of the local geologic substratum or the local morphodynamics. A convex-concave slope can be interrupted by treads and scarps that reflect tectonic influence or lithologic changes. Likewise, the general topographic profile can be locally disturbed or modified by water erosion (e.g. rills and gullies) or mass movements (e.g. terracettes, landslides, solifluction scars and tongues).

7.4   Conclusion

This chapter organizes existing geomorphic knowledge and arranges the geoforms in a hierarchically structured system with six nested levels. It is thought that this multicategorial geoform classification scheme reflects the structure of the geomorphic landscape sensu lato. The approach segments and stratifies the landscape continuum into geomorphic units belonging to different levels of abstraction. This geoform classification system has shown to be useful in geopedologic mapping and could be useful also in digital soil mapping.

References

Barrera-Bassols N, Zinck JA, Van Ranst E (2006) Local soil classification and comparison of indigenous and technical soil maps in a Mesoamerican community using spatial analysis. Geoderma 135:140–162

CNRS (1972) Cartographie géomorphologique. Travaux de la RCP77. Mémoires et Documents, vol 12. Editions du Centre National de la Recherche Scientifique, Paris
Conacher AJ, Dalrymple JB (1977) The nine-unit landscape model: an approach to pedogeomorphic research. Geoderma 18:1–154
Derruau M (1965) Précis de géomorphologie. Masson, Paris
Derruau M (1966) Geomorfología. Ediciones Ariel, Barcelona
Fairbridge RW (ed) (1997) Encyclopedia of geomorphology. Springer, New York
Garner HF (1974) The origin of landscapes. A synthesis of geomorphology. Oxford University Press, New York
Goudie AS (ed) (2004) Encyclopedia of geomorphology, vol 2. Routledge, London
Huggett RJ (2011) Fundamentals of geomorphology. Routledge, London
King LC (1957) The uniformitarian nature of hillslopes. Trans Edinb Geol Soc 17:81–102
Lugo-Hubp J (ed) (1989) Diccionario geomorfológico. Universidad Nacional Autónoma de México, Cd México
Ruhe RV (1975) Geomorphology. Geomorphic processes and surficial geology. Houghton Mifflin, Boston
Thornbury WD (1966) Principios de geomorfología. Editorial Kapelusz, Buenos Aires
Tricart J (1965) Principes et méthodes de la géomorphologie. Masson, Paris
Tricart J (1968) Précis de géomorphologie. T1 Géomorphologie structurale. SEDES, Paris
Tricart J (1977) Précis de géomorphologie. T2 Géomorphologie dynamique générale. SEDES-CDU, Paris
Tricart J, Cailleux A (1962) Le modelé glaciaire et nival. SEDES, Paris
Tricart J, Cailleux A (1965) Le modelé des régions chaudes. Forêts et savanes. SEDES, Paris
Tricart J, Cailleux A (1967) Le modelé des régions périglaciaires. SEDES, Paris
Tricart J, Cailleux A (1969) Le modelé des régions sèches. SEDES, Paris
Van Zuidam RA (1985) Aerial photo-interpretation in terrain analysis and geomorphological mapping. ITC, Enschede
Verstappen HT (1983) Applied geomorphology; geomorphological survey for environmental development. Elsevier, Amsterdam
Verstappen HT, Van Zuidam RA (1975) ITC system of geomorphological survey. ITC, Enschede
Viers G (1967) Eléments de géomorphologie. Nathan, Paris
Visser WA (ed) (1980) Geological nomenclature. Royal geological and mining society of the Netherlands. Bohn, Scheltema & Holkema, Utrecht
Wood A (1942) The development of hillside slopes. Geol Assoc Proc 53:128–138
Zinck JA (1974) Definición del ambiente geomorfológico con fines de descripción de suelos. Ministerio de Obras Públicas (MOP), Cagua
Zinck JA (1980) Valles de Venezuela. Lagoven, Petróleos de Venezuela, Caracas
Zinck JA (1988) Physiography and soils, ITC soil survey lecture notes. International Institute for Aerospace Survey and Earth Sciences, Enschede

Chapter 8
The Geomorphic Landscape: The Attributes of Geoforms

J.A. Zinck

Abstract Attributes are characteristics used for the description, identification, and classification of the geoforms. They are descriptive and functional indicators that make the multicategorial system of the geoforms operational. Four kinds of attribute are used: (1) morphographic attributes to describe the geometry of geoforms; (2) morphometric attributes to measure the dimensions of geoforms; (3) morphogenic attributes to determine the origin and evolution of geoforms; and (4) morphochronologic attributes to frame the time span in which geoforms originated. The morphometric and morphographic attributes apply mainly to the external (epigeal) component of the geoforms, are essentially descriptive, and can be extracted from remote-sensed documents or derived from digital elevation models. The morphogenic and morphochronologic attributes apply mostly to the internal (hypogeal) component of the geoforms, are characterized by field observations and measurements, and need to be substantiated by laboratory determinations.

Keywords Morphography • Morphometry • Morphogenesis • Morphochronology • Attribute classes • Attribute weights

8.1 Introduction

Attributes are characteristics used for the description, identification, and classification of the geoforms. They are descriptive and functional indicators that make the multicategorial system of the geoforms operational. This implies two requirements: (1) select descriptive attributes that help identify the geoforms, and (2) select differentiating attributes that allow classifying geoforms at the various categorial levels of the taxonomic system.

J.A. Zinck (✉)
Faculty of Geo-Information Science and Earth Observation (ITC), University of Twente, Enschede, The Netherlands

Institute of Environmental Studies, University of New South Wales, Sydney, NSW, Australia
e-mail: alfredzinck@gmail.com

© Springer International Publishing Switzerland 2016   127
J.A. Zinck et al. (eds.), *Geopedology*, DOI 10.1007/978-3-319-19159-1_8

To determine a geoform, it is necessary to sequentially perform the following operations:

- Description and measurement, to characterize the properties and constituents
- Identification, to compare the geoforms to be determined with established reference types
- Classification, to place the geoforms to be determined in the taxonomic system

For this purpose, four kinds of attribute are used, following Tricart's proposal with respect to the four types of data that a detailed geomorphic map should comprise (Tricart 1965a, b):

- Geomorphographic attributes, to describe the geometry of geoforms
- Geomorphometric attributes, to measure the dimensions of geoforms
- Geomorphogenic attributes, to determine the origin and evolution of geoforms
- Geomorphochronologic attributes, to frame the time span in which geoforms originated

In order to simplify the expressions, it is customary to omit the prefix *geo* in the denomination of the attributes.

The morphometric and morphographic attributes apply mainly to the external (epigeal) component of the geoforms, are essentially descriptive, and can be extracted from remote-sensed images or derived from digital elevation models. The morphogenic and morphochronologic attributes apply mostly to the internal (hypogeal) component of the geoforms, are characterized from field observations and measurements, and need to be substantiated by laboratory determinations.

8.2 Morphographic Attributes: The Geometry of Geoforms

The morphographic attributes describe the geometry and shape of the geoforms in topographic and planimetric terms. They are commonly used for automated identification of selected geoform features from DEM (Hengl 2003).

8.2.1 Topography

Topography refers to the cross section of a portion of terrain (Fig. 8.1). It can be viewed in two dimensions from a vertical cut through the terrain generating the topographic profile (Table 8.1), and in three dimensions from a terrain elevation model generating the topographic shape (Table 8.2). The characterization of these features is particularly relevant in sloping areas. The shape and the profile of the topography are related to each other, but described at different categorial levels.

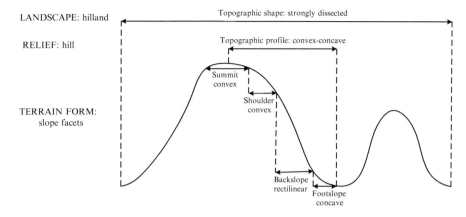

Fig. 8.1 Relationship between topographic attributes and categorial levels of the geoform classification system

Table 8.1 Topographic profile (2D)

Classes	Examples
Level	Mesa, terrace
Concave	Basin, footslope facet
Convex	Levee, summit/shoulder facet
Convex-concave	Slope facet complex
Convex-rectilinear-concave	Slope facet complex
Rectilinear (straight)	Backslope
With intermediate flat step(s)	Slope facet complex
With protruding rock outcrop(s)	Slope facet complex
With rocky scarp(s)	Slope facet complex, cuesta
Asymmetric	Hill, hogback
Irregular	Hillside

Table 8.2 Topographic shape (3D)

Classes	Slope %	Relief amplitude
Flat or almost flat	0–2	Very low
Undulating	2–8	Low
Rolling	8–16	Low
Hilly	16–30	Moderate
Steeply dissected	>30	Moderate
Mountainous	>30	High

The topographic shape attributes are used at landscape level, while the topographic profile attributes are used at the levels of relief and terrain form. The third descriptor, the exposure or aspect which indicates the orientation of the relief in the four cardinal directions and their subdivisions, can be used at any level of the system.

8.2.2   Planimetry

Planimetry refers to the vertical projection of the geoform boundaries on a horizontal plane. It is a two-dimensional representation of characteristic geoform features that closely control the soil distribution patterns. Fridland (1965, 1974, 1976) and Hole and Campbell (1985) were among the first to recognize configuration models that delimit soil bodies and relate these with the pedogenic context. The configuration of the geoform, the design of its contours, the drainage pattern, and the conditions of the surrounding environment are the main attributes described for this purpose.

8.2.2.1   Configuration of the Geoforms

Many geoforms at the levels of relief/molding and terrain form show typical configurations that can be easily extracted from remote-sensed documents, especially air photos. This enables preliminary identification of geoforms based on the covariance between morphographic and morphogenic attributes. For instance, a river levee is generally narrow and elongated, while a basin is wide and massive. The configuration attributes give an idea of the massiveness or narrowness of a geoform (Table 8.3).

8.2.2.2   Contour Design of the Geoforms

The design of the contours describes the peripheral outline of the geoform at the levels of relief/molding and terrain form (Fig. 8.2 and Table 8.4). It can vary from straight (e.g. recent fault scarp) to wavy (e.g. depositional basin) to indented (e.g. scarp dissected by erosion). These variations from very simple linear outlines up to complex convoluted contours that approximate areal configurations, are reflected in variations of the fractal dimension (Saldaña et al. 2011). The attribute of contour design can be used also as an indirect morphogenic indicator. For instance, an

Table 8.3 Configuration of the geoforms

Classes	Examples
Narrow	Levee
Large	Overflow mantle
Elongate	Dike
Massive	Basin
Annular (ring-shaped)	Volcanic ring-dyke
Oval/elliptic	Doline, sinkhole
Rounded	Hill
Triangular	Fan, delta
Irregular	Dissected escarpment

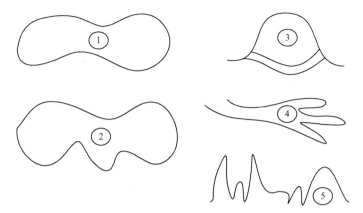

Fig. 8.2 Configuration and contour design of some geoforms (2D). (*1*) Basin with ovate configuration and sinuous contour. (*2*) Basin with ovate configuration and lobulate contour (*lower part*), reflecting the penetration of a digitate crevasse splay fan (see Fig. 8.3). (*3*) Bay closed by an arch-shaped offshore bar. (*4*) Deltaic channel levee with digitate distal extremities. (*5*) Dissected scarp with denticulate contour pattern

Table 8.4 Contour design of geoforms

Classes	Examples
Rectilinear	Escarpment
Arched (lunate)	Coastal bar
Sinuate (wavy)	River levee
Lobulate	Basin
Denticulate	Dissected escarpment
Digitate	Deltaic channel levee (distal sector)
Irregular	Gully, badland

alluvial decantation basin has usually a massive configuration, but the shape of the boundaries can vary according to the dynamics of the neighboring forms. In general, a depositional basin has a sinuous outline, but when a crevasse splay that forms after opening a gap in a river levee in high water conditions penetrates into the basin, the different fingers of the splay create a lobulated distal contour. Thus, a lobulated basin contour can reflect the proximity of a digitate splay fan, with overlap of a light-colored sandy deposit fossilizing the argillaceous gley material of the basin (Fig. 8.3).

8.2.2.3   Drainage Pattern

The drainage pattern refers to the network of waterways, which contributes to enhance the configuration and contour outline of the geoforms. It is mainly controlled by the geologic structure (tectonics, lithology, and volcanism) in erosional

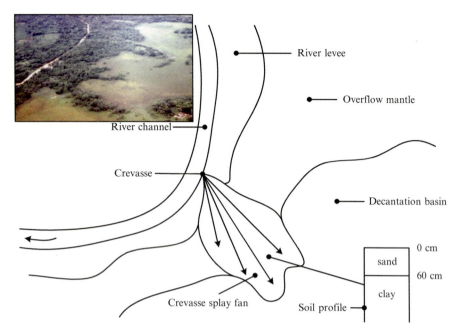

Fig. 8.3 Modification of a basin contour design by the penetration of a crevasse splay fan upon rupture of a levee during high channel water. The intrusion of the fan in the neighboring lateral depression results in the overlaying of sandy cover sediments on top of the clayey basin substratum, creating a lithologic discontinuity at 60 cm depth in this case, with the formation of a buried soil

areas, and by the structure and dynamics of the depositional system in aggradation areas. Representative patterns taken from the Manual of Photographic Interpretation (ASP 1960) are shown in Figs. 8.4 and 8.5: radial pattern of a conic volcano, annular pattern in a set of concentric calderas, dendritic pattern in homogeneous soft sedimentary rocks without structural control, trellis pattern in sedimentary substratum with alternate hard and soft rock layers and with structural control (faults and fractures), parallel pattern in alluvial area, and rectangular pattern in a till plain. The network of waterways creates connectivity between the areas that it crosses and controls the various kinds of flow that traverse the landscape (water, materials, wildlife, vegetation, humans).

8.2.2.4 Neighboring Units and Surrounding Conditions

The geomorphic units lying in the vicinity of a geoform under description shall be mentioned along with the surrounding conditions. This attribute applies at the levels of landscape, relief/molding, and terrain form. According to its position in the landscape, a geoform can topographically dominate another one, be dominated by it, or lie at the same elevation (e.g. a plain dominated by a piedmont). These adjacency

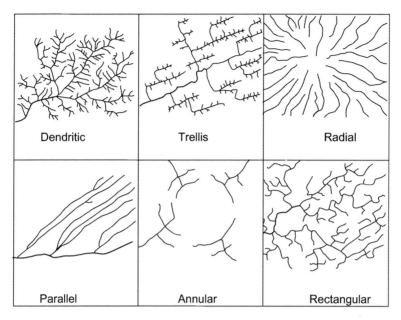

Fig. 8.4 Drainage patterns controlled by features of the geologic and geomorphic structure (see comments in the text) (Adapted from ASP 1960)

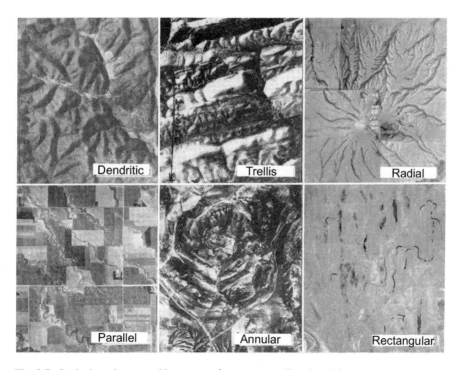

Fig. 8.5 Geologic and geomorphic structure features controlling the drainage patterns (see comments in the text) (Adapted from ASP 1960)

conditions suggest the possibility of dynamic relationships between neighboring geoforms and enable to model them. In a piedmont landscape, for instance, can start water flows that cause flooding in the basins of a neighboring alluvial plain, or material flows that cause avulsion in agricultural fields and siltation in water reservoirs. The segmentation of the landscape into functionally distinct geomorphic units provides a frame for analyzing and monitoring transfers of physical, chemical, mineralogical, and biological components within and between landscapes.

8.2.3   Morphography and Landscape Ecology

The morphographic attributes, in particular the configuration and contour design of the geoforms, have close semantic and cartographic relationships with concepts used in landscape ecology, such as mosaic, matrix, corridor, and patch (Forman and Godron 1986). A deltaic plain is a good example that illustrates the relationship between the planimetry of the geoforms and the metrics used in landscape ecology. A deltaic plain that occupies the distal area in a depositional system is a dynamic entity that receives materials and energy from the medial and proximal sectors of the same system. Delta channels are axes which introduce water and material in the system, conduct them through the system, and distribute them to other positions within the system such as overflow mantles and basins. Channels are elongated, sinuous, narrow corridors that feed the deltaic depositional system. In general, the mantles (overflow or splay) are extensive units that form the matrix of the system. The basins are closed depressions, forming scattered patches in the system (Fig. 8.6).

8.3   Morphometric Attributes: The Dimension of Geoforms

Morphometry covers the dimensional features of the geoforms as derived from a numerical representation of the topography (Pike 1995; Pike and Dikau 1995). Computerized procedures allow the extraction and measurement of a variety of morphometric parameters from a DEM, some being relevant at local scale and others at regional scale, including slope, hypsometry, orientation (aspect), visual exposure, insolation, tangential curvature, profile curvature, catchment characteristics (extent, elevation, slope), and roughness (Gallant and Hutchinson 2008; Olaya 2009). While many of these land-surface parameters are used in topography, hydrography, climatology, architecture, urban planning, and other applied fields, only a few actually contribute to the characterization of terrain forms, in particular the relative elevation, drainage density, and slope gradient. These are subordinate, not diagnostic attributes which can be used at any categorial level with variable weight.

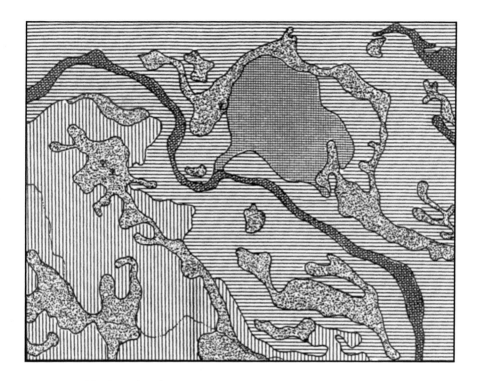

Morphogenic unit	Morphochronologic unit
Delta channel levee	Q1
Crevasse splay fan	Q1
Overflow mantle	Q1
Overflow mantle	Q2
Overflow basin	Q2

0   250   500   750 m

Fig. 8.6 Contact area between two depositional systems differentiated by relative age. Extract from a soil series map of the Santo Domingo river plain, Venezuela; survey scale 1: 25,000 (Adapted from Pérez-Materán 1967) In the *center* and to the *right*, a deltaic alluvial system with relative age Q1 (i.e. upper Pleistocene) fossilizes a previous depositional system of relative age Q2 (i.e. late middle Pleistocene) of which the elongated patches of overflow basin are remnants. The delta channel is the axial unit of the depositional system and functions as a corridor through which water and sediments transit before being distributed within the system. A unit of triangular configuration is grafted on the delta channel, corresponding to a crevasse splay fan that originated upon the opening of a gap in the levee of the channel. The overflow mantles are the matrices of both depositional systems (Q1 and Q2). The basins and the splay fan correspond to patches

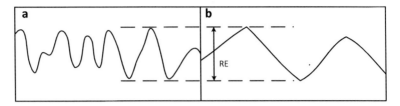

Fig. 8.7 Relationship between drainage density and slope gradient in similar conditions of relative elevation (RE) (Adapted from Meijerink 1988)

Morphometric attributes are interrelated: at a specific range of relative elevation, there is a direct relationship between drainage density and slope gradient; the higher the drainage density, the greater is the slope gradient, and conversely (a and b, respectively, in Fig. 8.7).

8.3.1 Relative Elevation (Relief Amplitude, Internal Relief)

The relative elevation between two geoforms is evaluated as high, medium, or low. Ranges of numerical values (e.g. in meters) can be attributed to these qualitative classes within the context of a given region or project area. Numerical ranges are established on the basis of local or regional conditions and are valid only for these conditions. Relative elevation is a descriptive attribute, and the classes of relative elevation can be differentiating but are not diagnostic. Likewise, the absolute altitude is not a diagnostic criterion, because similar geoforms can be found at various elevations. For instance, the Bolivian Altiplano at 3500–4000 masl, the Gran Sabana area in the Venezuelan Guayana at 800–1100 masl, and the mesetas of eastern Venezuela at 200–400 masl show all three the diagnostic characteristics of the plateau landscape, although at different elevations.

8.3.2 Drainage Density

Drainage density measures the degree of dissection or incision of a terrain surface. Density classes are set empirically for a given region or project area. For instance, Meijerink (1988) determines drainage density classes (called valley density VD) based on the relationship $VD = \Sigma L/A$, where ΣL is the cumulative length of drainage lines in km and A is the area in km^2. Not only the conditions of the region studied but also the study scale affects the numerical values of VD (Fig. 8.8). The FAO Guidelines for soil description (2006) define potential drainage density values based on the number of "receiving" pixels within a window of 10×10 pixels.

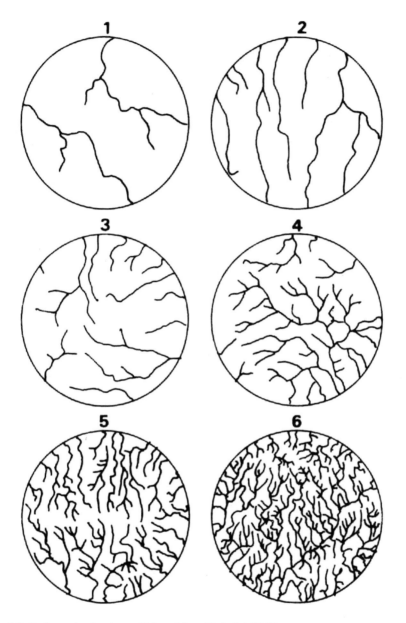

Fig. 8.8 Drainage density classes (Adapted from Meijerink 1988)

8.3.3   Relief Slope

The slope gradient is expressed in percentages or degrees. There are geoforms that have characteristic slopes or specific slope ranges. For instance, a coastal cliff or a young fault escarpment is often vertical and has therefore a slope close to 90°. A debris talus has an equilibrium slope of 30–35°, which corresponds to the angle of repose of the loose debris covering the slope. However, the mere knowledge of these numerical values does not contribute directly to identify the corresponding geoform. The slope gradient is essentially a descriptive attribute, at the most covariant with other attributes of higher diagnostic value. Obviously, a hill has a slope greater than a valley floor.

8.3.4   Terrain and Soil Surface Features

Morphometry is not limited to the extraction of topographic parameters from a DEM. Remote sensing also contributes to the characterization of the epigeal component of the geoforms. A variety of terrain and soil surface features can be identified, measured, and delineated from remote-sensed data. An inventory of parameters that can be characterized from optical and microwave sensors includes mineralogy, texture, moisture, organic carbon, iron oxides, salinity, carbonates, terrain roughness, and erosion features (Wulf et al. 2015). Spectral signatures covariate with laboratory determined or field observed property values. Some of these attributes may perform better al local scale to identify patches of specific surface features such as spatial variations in texture, organic matter content, or soil erosion in a given geomorphic unit. Others can contribute to delineate entire geoforms or associations thereof, for instance, in poorly drained, salt-affected or land degraded areas. Landscapes with no or sparse vegetation cover in dry environments offer the best possibilities for remote-sensed morphometry characterization (Metternicht and Zinck 1997, 2003; del Valle et al. 2010; Metternicht et al. 2010). Del Valle et al. (Chap. 19 in this book) show the capability of the PALSAR L-band to penetrate coarse-textured materials several decimeters below the terrain surface to detect buried geological and geomorphic features. So far less widely used than remote sensors, proximal sensors present promising opportunities to further explore the hypogeal component of the geoforms.

8.3.5   Contribution of Digital Morphometry

With the development of digital cartography, (geo)morphometry is increasingly used to characterize terrain units based on individual numerical parameters that are extracted from a DEM, such as altitude, relative elevation, slope, exposure, and curvature, among others. Attributes such as slope and curvature can present continuous variations in space and are therefore suitable for fuzzy mapping. This is in particular the case of banal hillside reliefs with convex-concave slope profiles according to the model of Ruhe (1975). However, many geoforms have relatively discrete boundaries that reflect their configuration and contour design. This is especially the case of constructed geoforms. In brief, the contribution of digital morphometry resides essentially in the automated estimation of dimensional attributes of the geoforms. However, limiting the description of the geoforms to their morphometric characteristics, just because the latter can be extracted automatically from a DEM, carries the risk of replacing field observation and image reading by numerical parameters which do not reflect satisfactorily the structure and formation of the geomorphic landscape. The scope of the morphometric characteristics to interpret the origin and evolution of the relief is limited, because morphometry covers only part of the external features of the geoforms, their epigeal component.

8.4 Morphogenic Attributes: The Dynamics of Geoforms

Selected geoform attributes reflect forming processes and can therefore be used to reconstruct the morphogenic evolution of an area or infer past environmental conditions. In general, the attribute-process relationship is more efficient for identifying geoforms in depositional environment than in erosional environment. Constructed geoforms are usually more conspicuous than erosional geoforms, except for features such as gullies or karstic erosion forms, for instance. Hereafter, some morphogenic attributes are analyzed by way of examples. Particle size distribution, structure, consistence, mineralogical characteristics, and morphoscopic features are good indicators of the origin and evolution of the geoforms.

8.4.1 Particle Size Distribution

8.4.1.1 Relevance

Particle size distribution, or its qualitative expression of texture, is the most important property of the geomorphic material, as well as of the soil material, because it controls directly or indirectly a number of other properties. The particle size distribution provides basic information for the following purposes:

- Characterization of the material and assessment of its suitabilities for practical uses (e.g. agricultural, engineering, etc.).
- Inference of other properties of the material that closely depend on the particle size distribution (often in combination with the structure of the material), such as bulk density, specific surface area, cohesion, adhesion, permeability, hydraulic conductivity, infiltration rate, consistence, erodibility, CEC, etc.

- Inference and characterization of geodynamic and pedodynamic features such:
  - Transport agents (water, wind, ice, mass movement)
  - Depositional processes and environments
  - Weathering processes (physical and chemical)
  - Soil-forming processes

8.4.1.2   The Information

The particle size distribution of the material is determined in the laboratory using methods such as densitometry or the pipette method to separate the fractions of sand, silt, and clay, and sieves to separate the various sand fractions. The analytical data are used to classify the material according to particle size scales. The most common of these grain size classifications are the USDA classification for agricultural purposes, and the Unified and AASHTO classifications for engineering purposes (USDA 1971). Significant differences between these classification systems concern the following aspects:

- The upper limit of the sand fraction: 2 mm in USDA and AASHTO; 5 mm in Unified.
- The lower limit of the sand fraction: 0.05 mm (50 μm) in USDA; 0.074 mm (74 μm) in Unified and AASHTO (solifluidal threshold).
- The boundary between silt and clay: 0.002 mm (2 μm) in USDA; 0.005 mm (5 μm) in Unified and AASHTO (colloidal threshold).

8.4.1.3   Examples of Inference and Interpretation

Hereafter, some examples are analyzed to show the type of information that can be derived from particle size data to characterize aspects of sedimentology, weathering, and soil formation. The granulometric composition of the material allows inferring and interpreting important features relative to the formation and evolution of the geoforms: for instance, the nature of the agents and processes that mobilize the material, the modalities of deposition of the material and their variations in time and space, the mechanisms of disintegration and alteration of the rocks to form regolith and parent material of the soils, and the differentiation processes of the soil material.

(a) Transport agents

Wind and ice illustrate two extreme cases of relationship between transport agent and granulometry of the transported material.

- Wind is a highly selective transport agent. The competence of the wind covers a narrow range of particle sizes, which usually includes the fractions of fine sand, very fine sand, and coarse silt (250–20 μm). Coarser particles are too heavy, except for saltation over short distances; smaller particles are often

immobilized in aggregates or crusts, a condition that causes mechanical retention in situ. As a result, the material transported by wind is usually
homometric.

- Ice is a poorly selective agent. Glacial deposits (e.g. moraines) include a wide
  range of particles from clay and silt (glacial flour) to large blocks (erratic
  blocks). This results in heterometric material.

(b) Transport processes

Cumulative grain size curves at semi-logarithmic scale, established from the
analytical laboratory data, allow inferring and characterizing processes of transport and deposition, especially in the case of the processes controlled by water
or wind. The granulometric facies of a deposit reflects its origin and mode of
sedimentation (Rivière 1952). According to Tricart (1965a), granulometric
curves are basically of three types, sometimes called canonical curves (Rivière
1952): namely, the sigmoid type, the logarithmic type, and the parabolic type
(Fig. 8.9).

Granulometric curves that correspond to three types of sediments deposited
by a flood event of the Guil river, in southern France, are displayed in Fig. 8.9
(Tricart 1965a).

- The sigmoid or S-shaped curve shows that a large proportion of the sample
  (ca 85 %) lies in a fairly narrow particle size range (150–40 µm), which corresponds mostly to the fractions of coarse silt and very fine sand. This material results from a very selective depositional process, which is common in
  areas of calm, no-turbulent, fluvial overflow sedimentation. In such places,
  the vegetation cover of the soil, especially when it comes to grass, operates an
  effect of sieving and biotic retention mainly of silt and fine sand particles
  (overflow process). Eolian deposits of particles that have been transported
  over long distances, as in the case of loess, generate similar S-shaped curves.
- The logarithmic curve, with a more or less straight slope, reveals that the
  deposit is distributed in approximately equal proportions over all particle size
  classes. This reflects a poorly selective depositional mechanism that is characteristic of the splay process. Glacial moraine sediments can also produce
  logarithmic type curves.
- The parabolic curve shows an abrupt slope inflection in the range of 30–20
  µm. All particles are suddenly laid down upon a blockage effect caused by a
  natural or artificial barrier. For example, a landslide or a lava flow across a
  valley can obstruct the flow of a river and lead to the formation of a lake where
  the solid load is retained.

(c) Depositional terrain forms

A transect across an alluvial valley usually shows a typical sequence of positions built by river overflow. A full sequence may include a sandy to coarse
loamy levee, a silty to fine loamy overflow mantle, and a clayey basin, in this
order from the highest position, closest to the river channel, to the lowest and
farthest position in the depositional system (see Fig. 4.4 in Chap. 4).

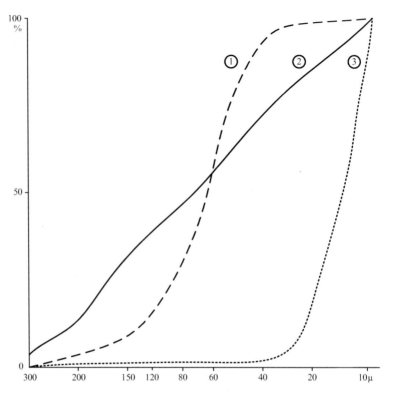

Fig. 8.9 Types of granulometric curves in depositional materials. Sediments of a flood event (June 1957) in the watershed of the Guil river, southern France (Taken from Tricart 1965a). (*1*) Sigmoid curve, characteristic of free sediment accumulation. (*2*) Logarithmic curve, characteristic of a torrential lava flow (in this case) or splay deposits. (*3*) Parabolic curve, characteristic of an accumulation forced by an obstacle obstructing the flow

(d) Lithologic discontinuity

The soil profile included in Fig. 8.3 shows a contrasting change of texture from sand to clay, which constitutes a lithologic discontinuity at 60 cm depth. This particle size change reveals an event of splay deposition following a basin depositional phase.

(e) Weathering processes

- Physical weathering of rocks produces predominantly coarse fragments. This is particularly common in extreme environmental conditions such as the following:

  - Cold environments, where frequent recurrence of freezing and thawing in the cracks and pores causes rock fragmentation. Cryoclastism or gelifraction is common at high latitudes and high altitudes.

- Hot and dry environments, where large thermic amplitudes between day and night favor the repetition of daily cycles of differential expansion-contraction between leucocratic (felsic) minerals and melanocratic (mafic) minerals. Termoclastism is common in desert regions with large daily temperature variations.

- Chemical weathering produces predominantly fine-grained products, especially clay particles that are neoformed upon weathering of the primary minerals of the rocks.

(f) Soil forming processes

A classic example is the comparison of clay content between eluvial and illuvial horizons to infer the process of clay translocation. Soil Taxonomy (Soil Survey Staff 1975, 1999), as well as other soil classification systems, uses ratios of clay content between A and B horizons to recognize argillic Bt horizons. For instance, a B/A clay ratio >1.2 is required for a Bt horizon to be considered argillic, when the clay content in the A horizon is 15–40 %. The B/A clay ratio is also used as an indicator of relative age in chronosequence studies of fluvial terraces.

8.4.2 Structure

8.4.2.1 Geogenic Structure

The geogenic structure refers to the structure of the geologic and geomorphic materials (bedrocks and unconsolidated surface materials, respectively).

(a) Rock structure

The examination of the rock structure allows evaluating the degree of weathering by comparison between the substratum R and the Cr horizon, especially in the case of crystalline rocks (igneous and metamorphic) where the original rock structure can still be recognized in the Cr horizon (saprolite). For instance, gneiss exposed to weathering preserves the banded appearance caused by the alternation of clear stripes (leucocratic felsic minerals) and dark stripes (melanocratic mafic minerals). The weathering of the primary minerals, especially the ferromagnesians, releases constituents, mainly bases, that are lost by washing to the water table. In the Cr horizon, the rock volume remains the same as that of the unweathered rock in the R substratum, but the weight has decreased. For example, the density could decrease from 2.7 Mg m^{-3} in the non-altered rock to 2.2–2.0 Mg m^{-3} in the Cr horizon. This process has received the name of isovolumetric alteration (Millot 1964).

(b) Depositional structures

The sediments show often structural features that reveal the nature of the depositional processes. Rhythmic and lenticular structures are examples of syndepositional structures, while the structures created by cryoturbation and bioturbation are generally postdepositional.

- The rhythmic structure reflects successive depositional phases or cycles. It can be recognized by the occurrence of repeated sequences of strata that are granulometrically related, denoting a process of cyclic aggradation. For example, a common sequence in overflow mantles includes layers with texture varying between fine sand and silt. Consecutive sequences can be separated by lithologic discontinuities.
- The lenticular structure is characterized by the presence of lenses of coarse material within a matrix of finer material. Lenses of coarse sand and/or gravel, several decimeters to meters wide and a few centimeters to decimeters thick, are frequent in overflow as well as splay mantles. They correspond to small channels of concentrated runoff, flowing at a given time on the surface of a depositional area, before being fossilized by a new phase of sediment accumulation.
- Cryoturbation marks result from the disruption of an original depositional structure by ice wedges or lenses.
- Bioturbation marks result from the disruption of an original depositional structure by biological activity (burrows, tunnels, pedotubules).

8.4.2.2 Pedogenic Structure

The soil structure *type* is often a good indicator of how the geomorphic environment influences soil formation. For instance, in a well-drained river levee position, the structure is usually blocky. The structure is massive or prismatic in a basin position free of salts, while it is columnar in a basin position that is saline or saline-alkaline. On the other hand, the *grade* of structural development may reflect the time span of soil formation.

8.4.3 Consistence

Consistence limits, also called Atterberg limits, are good indicators to describe the mechanical behavior, actual or potential, of the geomorphic and pedologic materials according to different moisture contents. In Fig. 8.10, consistence states, limits, and indices, which are relevant criteria in mass movement geomorphology, are related to each other. These relationships are controlled by the particle size distribution and

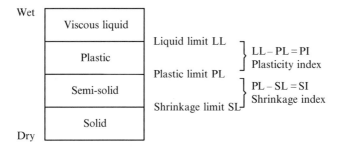

Fig. 8.10 Consistence/consistency parameters

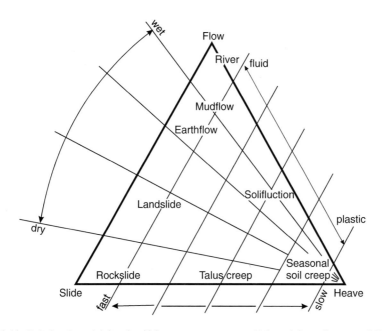

Fig. 8.11 Relational model for classifying mass movements (Adapted from Carson and Kirkby 1972)

mineralogy of the materials. In general, clay materials are mostly susceptible to landsliding, while silt and fine sand materials are more prone to solifluction. A low plasticity index makes the material more susceptible to liquefaction, with the risk of creating mudflows. The graphic model of Carson and Kirkby (1972) shows how continuity solutions in terms of speed, water content, and plasticity that relate the basic mechanisms of swell, slide, and flow, can be segmented for differentiating types of mass movement (Fig. 8.11).

8.4.4   Mineralogy

The mineralogical composition of the sand, silt, and clay fractions in the unconsolidated materials of surface formations is an indicator of the geochemical dynamics
of the environment, as related to or controlled by morphogenic processes, and helps
follow the pathways of tracer minerals. The associations of minerals present in
cover formations allow making inferences about the following features:

– They reflect the dominant lithologies in the sediment production basins.
– They help distinguish between fresh and reworked materials; the latter result
  from the mixing of particles through the surficial translation of materials over
  various terrain units.
– They reflect the morphoclimatic conditions of the formation area: for instance,
  halites in hot and dry environment; kandites in hot and moist environment.
– They reflect the influence of topography on the formation and spatial redistribution of clay minerals along a slope forming a catena of minerals. In humid tropical environment, a catena or toposequence of minerals commonly includes
  kandites (e.g. kaolinite) at hill summit, micas (e.g. illite) on the backslope, and
  smectites (e.g. montmorillonite) at the footslope.

Table 8.5 shows an example of determination of minerals in sand and silt fractions to reconstitute the morphogenic processes acting in the contact area between a
piedmont and an alluvial valley. The sampling sites are located on the lower terrace
of the Santo Domingo river (Barinas, Venezuela) at its exit from the Andean foothills towards the Llanos plain. Sites are distributed along a transect perpendicular to
the valley from the base of the piedmont to the floodplain of the river. Site A is close
to the piedmont, site C is close to the floodplain, and site B is located in an intermediate position.

• Site A: colluvial deposit (reworked material). Rubified colluvium, coming from
  the truncation of a strongly developed red soil lying on a higher terrace (Q3). The
  reworking effect can be inferred from the high contents of clean quartz grains,
  washed during transport by diffuse runoff, and soil aggregates, respectively. The
  absence of rock fragments and micas indicates that colluviation removed fully
  pedogenized material from the piedmont.

Table 8.5 Mineralogy of silt and sand fractions (%); eastern piedmont of the Andes, to the west
of the city of Barinas, Venezuela (Data provided by the Institute of Geography, University of
Strasbourg, France (courtesy J. Tricart))

Site	Clean quartz + feldspars	Ferruginous quartz	Soil aggregates	Rock fragments	Micas	Total
A	40	5	55	0	0	100
B	21	14	22	42	1	100
C	22	0	0	0	78	100

- *Site B*: *mixed deposit, colluvial and alluvial.* Mixture of red colluvium (presence of aggregates) removed from an older soil mantle on a middle terrace (Q2), and recent alluvium (presence of rock fragments) brought by the Santo Domingo river.
- *Site C*: *alluvial deposit.* Holocene alluvial sediments exclusively composed of clean quartz and fresh micas. The high proportion of micas results form the retention of silt particles trapped by dense grass cover.

8.4.5   Morphoscopy

Morphoscopy (or exoscopy) consists of examining coarse grains (sand and coarse silt) under a binocular microscope to determine their degree of roundness and detect the presence of surface features.

- The shape of the grains can vary from very irregular to well rounded:

  - Well rounded grains reflect continuous action by (sea)water or wind.
  - Irregular grains indicate torrential or short-distance transport.

- The brightness of the grains and the presence of surface marks, such as striae, polishing, frosting, chattermarks, gouges, among others, indicate special transport modes or special environmental conditions:

  - Shiny grains: seawater action.
  - Frosted grain surface: wind action.
  - Grains with percussion marks: chemical corrosion or collision of grains transported by wind.

8.5   Morphochronologic Attributes: The History of Geoforms

8.5.1   Reference Scheme for the Geochronology of the Quaternary

The Quaternary period (2.6 Ma) is a fundamental time frame in geopedology, because most of the geoforms and soils have been formed or substantially modified during this period. Pre-Quaternary relict soils exist, but are of fairly limited extent. The Quaternary has been a period of strong morphogenic activity due to climatic changes, tectonic paroxysms, and volcanic eruptions, which have caused destruction, burial, or modification of the pre-Quaternary and syn-Quaternary geoforms and soils, while at the same time new geoforms and soils have developed.

In temperate and boreal zones, as well as in mountain areas, glacial and interglacial periods have alternated several times. In their classic scheme based on observations made in the Alps, Penck and Brückner (1909) considered a relatively limited number of glacial periods (i.e. Würm, Riss, Mindel, Günz). A similar scheme was established for the chronology of the Quaternary period in North America. Recent research shows that the alternations of glacial-interglacial periods were actually more numerous. In Antarctica, up to eight glacial cycles over the past 740,000 years (740 ka) have been recognized. The average duration of climatic cycles is estimated at 100 ka for the last 500 ka and at 41 ka for the early Quaternary (before 1 Ma), with intermediate values for the period from 1 Ma to 500 ka (EPICA 2004). In addition, shorter climate variations have occurred during each glacial period, similar to the Dansgaard-Oeschger events of the last glaciation. Many regions are now provided with very detailed geochronologic reference systems for the Pleistocene and especially for the Holocene. In the intertropical zone, climate change is expressed more in terms of rainfall variations than in terms of temperature variations. Dry periods have alternated with moist periods, in approximate correlation with the alternation between glacial and interglacial periods at mid- and high latitudes.

Quaternary geochronology is conventionally based on the recurrence of climatic periods, which are assumed of promoting alternately high or low morphogenic activity and high or low pedogenic development. Erhart (1956), in his bio-rhexistasis theory, summarizes this dichotomy by distinguishing between (1) rhexistasic periods with unstable environmental conditions, rather cold and dry, conducive to intense morphogenic activity, and (2) biostasic periods with more stable environmental conditions, rather warm and humid, favorable to soil development. The biostasic periods are assumed of having been longer than the rhexistasic periods (Hubschman 1975). Butler's model of K cycles (1959) is based on the same principle of the alternation of stable phases with soil development and unstable phases with predominance of erosion (soil destruction) or sedimentation (soil fossilization). In the context of soil survey, various rather simple geochronologic schemes have been implemented to record the relative age of geoforms and associated soils, using letters such as K (from kyklos), t (from terrace), and Q (from Quaternary), with increasing numerical subscripts according to increasing age of the geopedologic units, assimilated to chronostratigraphic units (Table 8.6). Although these relative chronology schemes have a spatial resolution limited, for instance, to a region or a country, they also allow coarse stratigraphic correlations over larger territories.

Comments on Table 8.6:

- Q identifiers refer to the inferred relative age of the geomorphic material that serves as parent material, thus not directly to the age of the soil derived from this material. In erosional, structural, and residual relief areas, there is often a large gap between the age of the geologic substratum and the age of the overlying soil mantle. In many cases, the bedrock may even not be the parent material of the soil. This occurs in hill and mountain landscapes, where soils often develop from allochthonous slope formations lying atop the rocks in situ. By contrast, in depositional environments, the initiation of soil formation usually coincides fairly well with the end of the period of material accumulation. However, in sedimenta-

Table 8.6 Relative geochronology scheme of the Quaternary

		Rhexistasic periods	Biostasic periods
HOLOCENE			Q0
	Upper	Q1	
			Q1-2
	Late middle	Q2	
PLEISTOCENE			Q2-3
	Early middle	Q3	
			Q3-4
	Lower	Q4	
			Q4-5
PLIO-PLEISTOCENE		Q5	

Zinck (1988)

tion areas of considerable extent, deposition does not stop abruptly or does not stop in all sectors at the same time. For this reason, Q1 deposition in floodplains, for example, can extend locally into Q0 without notable interruption.

- The numerical indices (Q1, Q2, etc.) indicate increasing relative age of the parental materials. Where necessary, the relative scale can be extended (e.g. Q5, etc.) to refer to deposits that overlap the end of the Pliocene (Plio-Quaternary formations).
- Each period can be subdivided using alphabetical subscripts to reflect minor age differences (e.g. Q1a more recent than Q1b).
- Some geoforms, such as for example colluvial glacis, may have evolved over the course of several successive periods. A composite symbol can be used to reflect this kind of diachronic formation (e.g. Q1-Q2; Q1-Q1-2).

8.5.2 Dating Techniques

Idcally, age determination of a geoform or a soil requires finding and sampling a kind of geomorphic or pedologic material that allows using any of the absolute or relative dating techniques available, or a combination thereof, including:

- Carbon-14 (organic soils, charcoal, wood; frequently together with analysis of pollen)
- K/Ar (volcanic materials)
- Thermoluminescence (sediments, e.g. beach sands, loess)

- Dendrochronology (tree growth rings)
- Tephrochronology (volcanic ash layers)
- Varves (proglacial lacustrine layers)
- Analysis of historic and prehistoric events (earthquakes, etc.).

These techniques are relatively expensive and their implementation within the framework of a soil survey project is generally limited for budgetary reasons. On average, a determination of carbon-14 costs 300–350 euros. Some techniques are applicable only to specific kinds of material (e.g. ^{14}C only on material containing organic carbon; K/Ar only on volcanic material). Certain techniques cover restricted ranges of time (e.g. ^{14}C for periods shorter than 50–70 ka; thermoluminescence up to 300 ka). Interpretation errors can result from the contamination of the samples or the residence time of the organic matter (in the case of ^{14}C).

The former suggests that the most common materials in the geomorphic and pedologic context likely to be dated in absolute terms are soil horizons and sedimentary strata containing organic matter. In many situations, this limits practically absolute dating to about 60,000 years BP, a time span that covers the Holocene and a small part of the upper Pleistocene corresponding to half of the last glacial period. This underlines the need for indirect dating means such as those provided by pedostratigraphy.

8.5.3 Relative Geochronology: The Contribution of Pedostratigraphy

8.5.3.1 Definition

Relative geochronology is based on establishing relationships of temporal antecedence between the various geoforms or deposits in a study area and building correlations at several spatial scales. This procedure practically consists in extending the stratigraphic system used in pre-Quaternary geology to the Quaternary period. Geologic maps often provide scarce information about the Quaternary (e.g. Qal for alluvial cover formations, Qr for recent deposits), in comparison with the detailed lithologic information concerning the pre-Quaternary. This information is usually to coarse to determine the temporal frame of soil formation. In contrast, the geopedologic information provided by the proper soil survey can contribute to improving the stratigraphy of the Quaternary.

Pedostratigraphy or soil-derived stratigraphy consists in using selected soil and regolith properties to estimate the relative age of the cover formations and the geoforms on which soils have developed. This makes it possible to determine the chronostratigraphic position of a material or a geoform in a geochronologic reference scheme (Zinck and Urriola 1970; Harden 1982; Busacca 1987; NACSN 2005), with the possibility of recognizing successive soil generations.

Etymologically, pedostratigraphy means the use of soils or soil properties as stratigraphic tracers to contribute establishing the relative chronology of geologic, geomorphic, and pedologic events in a territory. However, according to the definitions provided by the North American Stratigraphic Code (NACSN 2005), the concepts of pedostratigraphy and soil stratigraphy are not strictly synonymous. According to this code, the basic pedostratigraphic unit is the geosol, which differs in various ways from the basic unit of soil stratigraphy, the pedoderm. One of the key differences is that the geosol is a buried weathering profile, while the pedoderm may correspond to a buried soil, a surficial relict soil, or an exhumed soil. Disregarding these definition differences, what is in fact relevant is that soils are recognized as stratigraphic units and, in this sense, the term pedostratigraphy has been used in geomorphology and pedology without complying with the strict definition of geosol. Pedostratigraphy is a privileged area of the geopedologic relationships with mutual contribution of geomorphology and pedology. The chronosequences of fluvial terraces provide illustrative examples of this close interrelation. The relative age of the terraces as determined on the basis of their position in the landscape, the lowest being usually the most recent, generally correlates fairly well with the degree of soil development and conversely. Morphostratigraphy and pedostratigraphy complement each other.

8.5.3.2 Indicators

A variety of pedologic and geomorphic indicators has been used to establish relative chronology schemes of the Quaternary in regions with different environmental characteristics (Mediterranean, tropical, etc.). These criteria include, among others, the following.

- The degree of activity of the geoforms, distinguishing between active geoforms (e.g. dune in formation), inherited geoforms in survival (e.g. hillside locally affected by solifluction), and stabilized geoforms (e.g. coastal bar colonized by vegetation).
- The degree of weathering of the parent material based on the color of the cover formations and the degree of disintegration of stones and gravels (Fig. 8.12). In humid tropical environment, the fragments of igneous and metamorphic rocks found in detrital formations are usually much more altered than most of the sedimentary rock fragments. Quartzite is most resistant in all kinds of climatic condition and often provides the dominant residual fragments in detrital formations of early Quaternary.
- The degree of soil morphological development, inferred from criteria such as color, pedogenic structure, solum thickness, and leaching indices, among others.
  - Color is a good indicator of the relative age of soils, particularly in humid tropical climate, with gradual increase of the red color (rubification/rubefaction) as the weathering of the ferromagnesian minerals in the parent material proceeds.

Fig. 8.12 Quaternary alluvial cover formations differentiated by color resulting from increasing rubefaction through time; materials belonging to (**a**) Holocene to upper Pleistocene (Q0-Q1), western piedmont of Venezuelan Andes; (**b**) late middle Pleistocene (Q2), eastern piedmont of Venezuelan Andes; (**c**) early middle Pleistocene (Q3), eastern piedmont of Venezuelan Andes; (**d**) lower Pleistocene (Q4), mesetas of the eastern Venezuelan Plateau

The possibility of differentiating soil ages by color dims over time in well-developed soils. Red soils can also be recent, when they arise from materials eroded from older rubified soils and redeposited in lower portions of the landscape. Likewise, red soils on limestone can be relatively young.

– The pedogenic structure reflects (1) the conditions of the site and the nature of the parent material which together control the type of structure (e.g. blocky, prismatic, columnar), and (2) the elapsed time that influences the grade of structural development (from weak to strong). The relationship between development grade and time reaches a threshold in well-developed soils, beyond which structure tends to weaken because of the impoverishment in substances that contribute to the cohesion of the soil material (e.g. organic matter, type and amount of clay, divalent cations).

– The thickness of the solum generally increases with the duration of pedogenic development in conditions of geomorphic stability. As in the case of structural development and rubification, solum thickness reaches a threshold over time beyond which increases become gradually insignificant.

– Leaching indices allow evaluating the intensity of the translocation of soluble or colloidal substances from eluvial horizons to the underlying illuvial horizons. The most commonly implemented are the clay and calcium carbonate ratios. The leaching intensity decreases with time as the eluvial horizons become depleted in mobilizable substances, resulting in stabilization of the translocation rates.

• The status of the adsorption complex. In general terms, the adsorption complex of the soil changes quantitatively and qualitatively with increasing time. Soil reaction (pH), cation exchange capacity, and base saturation are among the most sensitive indicators. With the passage of time, soils lose alkaline and alkaline-earth cations, resulting in a decrease or a change of composition (more H^+ and/or Al^{+++}) of the adsorption complex and an increase in acidity of the soil solution.

• Clay mineralogy changes with soil development as a function of time, among other factors. The associations of clay minerals originally present in the Cr or C horizons will be replaced by other associations with increasing time. In general, the 2:1 type clays (e.g. smectites, micas) are going to be replaced by or transformed into 1:1 type clays (e.g. kandites).

8.5.3.3   Combining Indicators

The simultaneous use of several of the above-mentioned soil properties allows determining pedostratigraphic units. To this effect, Harden (1982) established a quantitative index to estimate degrees of soil development and correlate these with dated soil units. The index was originally developed based on a soil chronosequence in the Merced River valley, central California, combining properties described in the field with soil thickness. Eight properties were integrated to form the index, including the presence of clay skins, texture combined with wet consistence, rubification based on change in hue and chroma, structure, dry consistence, moist consistence, color value, and pH. Other properties described in the field can be added if more soils are studied. The occasional absence of some properties did not significantly affect the index. Quantified individual properties and the integrated index were examined and compared as functions of soil depth and age. The analysis showed that the majority of the properties changed systematically within the 3 Ma that span the chronosequence of the Merced River. The index has been applied to other sites with successive adjustments (Busacca 1987; Harden et al. 1991).

Stepped alluvial terrace systems offer usually the possibility to establish illustrative soil chronosequences. Table 8.7 reports data on selected properties of soils that have developed on five consecutive Quaternary terraces in the Guarapiche river valley, northeast of Venezuela. Melanization with mollic horizon and soil structure formation on terrace Q1 corresponds to the first stage of soil development from raw depositional material of Q0. From period Q2 onwards clay illuviation starts upon descarbonation, together with substantial solum deepening. This is followed on level Q3 by important desaturation of the soil complex and soil solution. On the older Q4 terrace kaolinite formation takes place, causing degradation of the adsorption complex. Each terrace is characterized by a different stage of soil development, adding up to further soil evolution. The properties quantifying these consecutive pedogenic stages show value leaps that correspond to pedostratigraphic thresholds. The latter reflect discontinuities in soil formation during the Quaternary. The pedotaxa sequence comprising increasingly developed soils parallels the Quaternary pedogenic evolution, from Entisols to Mollisols to Alfisols to Ultisols to oxic (kanhaplic) Ultisols (Fig. 8.13).

There is no single model describing the relationship between time and soil development. Pedogenic development rates vary according to the considered time segment and the geographic conditions of the studied area. In general, soil development rates decrease when time increases above a given threshold and with increasing aridity (Zinck 1988; Harden 1990).

8.6   Relative Importance of the Geomorphic Attributes

Not all attributes are equally important to identify and classify geoforms. For instance, the particle size distribution of the material is most important, because it has more differentiating power and therefore more taxonomic weight than the relative elevation of a geoform.

8.6.1   Attribute Classes

Following an approach that Kellogg (1959) applied to distinguish between soil characteristics, the attributes of the geoforms can be grouped into three classes according to their weight for taxonomic purposes: differentiating, accessory, and accidental attributes, respectively.

Table 8.7 Pedostratigraphic thresholds; Guarapiche river valley, Venezuela (Zinck and Urriola 1970)

Relative age of parent material	Dominant color	Average solum thickness cm	CaCO$_3$ eq. %	Clay illuviation B/A index	pH 1:1 H$_2$O	Base saturation %	CEC cmol+ kg^{-1} clay	Clay mineral associations	Main taxa
Q0	Grayish- brown	30	>3	–	+/–8	100	80–120	S>K=M	Entisols Inceptisols
Q1	Dark- brown	80	1–3	–	6–7.5	80–100	60–95	S>K>M	Mollisols Inceptisols
Q2	**Reddish-yellow**	**200**	<1	**1.2–1.6**	4.5–6	40–60	40–60	S>K>M	Alfisols Vertisols
Q3	Yellowish- red	250	<0.5	2.1–2.7	**4–5**	**20–40**	40–50	V>K>M	Ultisols
Q4	Red	300	0	2.4–2.5	4–5	<20	**20–30**	**K>>>V>M>HIV**	Ultisols oxic subgr.

Based on the soil subgroups included in the phenetic clusters 1-2-3 of the dendrogram in Fig. 4.10 (Chap. 4)

Properties refer mainly to the solum (mean values or ranges of values)

Properties in bold represent pedostratigraphic thresholds reflecting discontinuities in soil evolution during the Quaternary

Q0 Holocene, Q1 Upper Pleistocene, Q2 Late Middle Pleistocene, Q3 Early Middle Pleistocene, Q4 Lower Pleistocene

S smectite, K kaolinite, M mica, V vermiculite, HIV hydroxy-interlayered vermiculite

Fig. 8.13 Well-drained soil profiles belonging to the chronosequence of the Guarapiche river terrace system, eastern Venezuela (see Table 8.7). All soils have similar parent materials (sandy loam C horizons with mixtures of smectite, kaolinite, and mica minerals) originating throughout the Quaternary from the sedimentary rocks (sandstone, lutite, limestone) of the southern slope of the Coastal Cordillera. The sequence shows the factor time effect on soil formation and differentiation: (**a**) Entisol (Mollic Ustifluvent), Fluvisol, in the floodplain (Q0); (**b**) Mollisol (Cumulic Haplustoll), Phaeozem, on the lower terrace (Q1); (**c**) Alfisol (Typic Haplustalf), Luvisol, on the lower middle terrace (Q2); (**d**) Ultisol ('Kanhaplic' Paleustult), Acrisol, on the upper terrace (Q4) under savanna cover. The Ultisol (Typic Paleustult), Lixisol, of the higher middle terrace (Q3) is not depicted here

8.6.1.1 Differentiating Attributes

A differentiating attribute is one that enables to distinguish one type of geoform from another at a particular categorial level. Therefore, a change in an attribute's state, expressed by a range of values, leads to a change in geoform classification. An attribute that has this property is considered diagnostic. Such an attribute, along with other differentiating attributes, contributes to the identification and classification of the geoforms.

A few examples:

• The dip of the geologic layers is a diagnostic criterion for recognizing monoclinal reliefs and the degree of dipping is a differentiating feature for distinguishing classes of monoclinal reliefs (see Fig. 6.4 in Chap. 6).
• A slope facet should be concave to classify as footslope. In this case, the topographic profile is the differentiating attribute and "concave" is the state of the attribute.
• The material of a decantation basin normally comprises more than 60 % clay fraction. In this case, the particle size distribution is the differentiating attribute, and the attribute state is expressed by the high clay content.

8.6.1.2   Accessory Attributes

An attribute is accessory if it reinforces the differentiating capability of a diagnostic attribute with which it has some kind of correlation (covariant attribute). For instance, the lenticular type of depositional structure can occur in several alluvial facies, but is more common in deposits caused by overload flow accompanied by mechanical friction (river levee, different kinds of splay). By itself, the presence of a lenticular structure is not enough to recognize a type of geoform.

8.6.1.3   Accidental Attributes

An accidental attribute does not contribute to the identification of a particular type of geoform, but provides additional information for its description and characterization. This kind of attribute can be used to create phases of taxonomic units for the purpose of mapping and separation of cartographic units (e.g. slope classes or classes of relative elevation).

8.6.2   Attribute Weight

8.6.2.1   Morphographic Attributes

Morphographic attributes are essentially accessory, sometimes differentiating.

- Accessory weight. For instance, a newly formed river levee has a characteristic morphology (elongated, narrow, sinuous, convex shape), which facilitates its identification in aerial images. An older levee, the contours of which have been obliterated with the passing of time, is more difficult to recognize from its external features. In the case of a levee buried underneath a recent sediment cover, it is possible to reconstruct the configuration and design of the contours by means of perforations. In these last two cases, the identification of the geoform rests primarily on the granulometric composition of material, with accessory support of the morphographic features.
- Differentiating power. In hill and mountain landscapes, the morphographic attributes can be differentiating. For instance, in the case of a convex-concave hillside, the characteristic topographic profile of each single slope facet is in itself differentiating.

8.6.2.2   Morphometric Attributes

Morphometric attributes are predominantly accidental. They contribute to the description of the geoforms, but seldom to their identification. For instance, the difference of elevation (i.e. relative elevation) between the summit surface of a plateau and the surrounding lowlands (e.g. valley or plain landscapes) can be as little as 100–150 m (e.g. the mesetas in eastern Venezuela) or as much as 1000–1500 m (e.g. the Bolivian Altiplano). In both cases, however, the geoform meets the diagnostic plateau attributes at the categorial level of landscape. In general, the dimensional features have low taxonomic weight, but are relevant for the practical use of the geomorphic information, for instance, in evaluation of environmental impacts or land-use planning. To this end, phases of relative elevation, drainage density, and slope gradient can be implemented.

8.6.2.3 Morphogenic Attributes

The morphogenic attributes are essentially differentiating, either individually or in group, especially when they are reinforced by accessory attributes. For instance, the consistence is a diagnostic attribute for assessing the susceptibility of a material to mass movement and for interpreting the origin of the resulting geoforms. The depositional geoforms show always specific ranges of granulometric composition, which is a highly diagnostic attribute in this case.

8.6.2.4 Morphochronologic Attributes

Morphochronologic attributes are mostly differentiating, because the relative age of a geoform is an integral part of its identity. The fact that a river levee has formed during the Holocene (Q0) or during the middle Pleistocene (Q2) probably does not have great effect on its configuration, although the contour design may have been obliterated with the passage of time. However, the chronostratigraphic position of the geoform is differentiating, because it determines a time frame in which the morphogenic processes take place, and which controls the evolution of the soils and their properties.

8.6.3   Attribute Hierarchization

Not all attributes are used at each categorial level of the geoform classification system. Table 8.8 shows an attempt of differential hierarchization of the geomorphic attributes according to their diagnostic weight. This aspect is of growing importance for the automated treatment of the geomorphic information. In Table 8.9 are mentioned the criteria that have guided the hierarchization in terms of attribute amount, nature, function, and implementation at the upper and lower levels of the system, respectively.

Table 8.8 Hierarchization of the geomorphic attributes

Attributes	Landscape	Relief	Lithology	Terrain form
Morphometric				
Relative elevation	+	+	−	o
Drainage density	+	+	−	−
Slope	+	+	−	+
Morphographic				
Topographic shape	+	o	−	−
Topographic profile	−	+	−	+
Exposure	−	+	−	+
Configuration	−	+	−	+
Contour design	−	+	−	+
Drainage pattern	+	+	−	−
Surrounding conditions	+	+	+	+
Morphogenic				
Particle size distribution	−	o	+	+
Structure	−	−	+	+
Consistence	−	−	+	+
Mineralogy	−	−	+	+
Morphoscopy	−	−	+	+
Morphochronologic				
Degree of weathering	−	−	+	+
Degree of soil development	−	−	o	+
Leaching indices	−	−	o	+
Adsorption complex status	−	−	o	+
Clay mineralogy	−	−	+	+

Zinck (1988)
+ Very important attribute
o Moderately important attribute
− Less important attribute

Table 8.9 Relations between geomorphic attributes according to the categories of the system

Attributes	Amount	Nature		Function	Implementation
Upper levels	Few	Descriptive		Generalizing	Interpretation of photos, images, and DEM
		External characterization		Aggregation	
↕	↕	↕		↕	↕
Lower levels	Many	Genetic		Detailing	Field and laboratory
		Internal characterization		Disaggregation	

8.6.3.1 Upper Levels

- Limited number of attributes.
- Preferably descriptive attributes, reflecting external features of the geoforms (i.e. morphographic and morphometric attributes).
- Function of generalizing and aggregating information.
- Information about attributes is mostly obtained by interpretation of aerial photos, satellite images, and digital elevation models.

8.6.3.2 Lower Levels

- Greater number of attributes, resulting from the addition of information.
- Preferably genetic attributes, reflecting internal characteristics of the geoforms (i.e. morphogenic and morphochronologic attributes).
- Function of differentiating and detailing information.
- More field information and laboratory data are required.

8.7 General Conclusion on Geopedology

Geopedology is an approach to soil survey that combines pedologic and geomorphic criteria to establish soil map units. Geomorphology provides the contours of the map units ("the container"), while pedology provides the soil components of the map units ("the content"). Therefore, the units of the geopedologic map are more than soil units in the conventional sense of the term, since they also contain information about the geomorphic context in which soils have formed and are distributed. In this sense, the geopedologic unit is an approximate equivalent of the soilscape unit, but with the explicit indication that geomorphology is used to define the landscape. This is usually reflected in the map legend, which shows the geoforms as entries to the legend and their respective pedotaxa as descriptors.

In the geopedologic approach, geomorphology and pedology benefit from each other in various ways:

- Geomorphology provides a genetic framework that contributes to the understanding of soil formation, covering three of the five factors of Jenny's equation: nature of the parent material (transported material, weathering material, regolith), age and topography (Jenny 1941, 1980). Biota is indirectly influenced by the geomorphic context.
- Geomorphology provides a cartographic framework for soil mapping, which helps understand soil distribution patterns and geography. The geopedologic map shows the soils in the landscape.

- The use of geomorphic criteria contributes to the rationality of the soil survey, decreasing the personal bias of the surveyor. The need of prior experience to ensure the quality of the soil survey is offset by a solid formation in geomorphology.
- Geomorphology contributes to the construction of the soil map legend as a guiding factor. The hierarchic structure of the legend reflects the structure of the geomorphic landscape together with the pedotaxa that it contains.
- The soil cover or soil mantle provides the pedostratigraphic frame based on the degree of soil development, which enables to corroborate the morphostratigraphy (e.g. terrace system).
- The soil cover through its properties (mechanical, physical, chemical, mineralogical, biological) provides data that contribute to assess the vulnerability of the geopedologic landscape to geohazards and estimate the current morphogenic balance (erosion-sedimentation).
- The geopedologic approach to soil survey and digital soil mapping are complementary and can be advantageously combined. The segmentation of the landscape sensu lato into geomorphic units provides spatial frames in which geostatistical and spectral analyses can be applied to assess detailed spatial variability of soils and geoforms, instead of blanket digital mapping over large territories. Geopedology provides information on the structure of the landscape in hierarchically organized geomorphic units, while digital techniques provide information extracted from remote-sensed imagery that help characterize the geomorphic units, mainly the morphographic and morphometric terrain surface features.

This first part of the book addresses the basic concepts and ideas underlying geopedology, with emphasis on the identification, characterization, and classification of geoforms to support soil survey and field soil studies at large. The following parts comprise a variety of studies that implement the geopedologic approach here introduced or other modalities based on soil-landscape relationships, using different methods and techniques, for soil pattern recognition, analysis and mapping, soil degradation assessment, and land use planning.

References

ASP (1960) Manual of photographic interpretation. American Society of Photogrammetry, Washington

Busacca AJ (1987) Pedogenesis of a chronosequence in the Sacramento Valley, California. USA. I Application of a soil development index. Geoderma 41:123–148

Butler BE (1959) Periodic phenomena in landscapes as a basis for soil studies, Soil publication 14. CSIRO, Melbourne

Carson MA, Kirkby MJ (1972) Hillslope form and process. Cambridge University Press, Cambridge

del Valle HF, Blanco PD, Metternicht GI, Zinck JA (2010) Radar remote sensing of wind-driven land degradation processes in northeastern Patagonia. J Environ Qual 39:62–75

EPICA (2004) Eight glacial cycles from an Antarctic ice core. Nature 429(6992):623–628

Erhart H (1956) La genèse des sols en tant que phénomène géologique. Masson, Paris

FAO (2006) Guidelines for soil description, 4th edn. Food and Agricultural Organization of the United Nations, Rome

Forman RTT, Godron M (1986) Landscape ecology. Wiley, New York

Fridland VM (1965) Makeup of the soil cover. Sov Soil Sci 4:343–354

Fridland VM (1974) Structure of the soil mantle. Geoderma 12:35–41

Fridland VM (1976) Pattern of the soil cover. Israel Program for Scientific Translations, Jerusalem

Gallant JC, Hutchinson MF (2008) Digital terrain analysis. In: McKenzie NJ, Grundy MJ, Webster R, Ringrose-Voase AJ (eds) Guidelines for surveying soil and land resources, vol 2, 2nd edn, Australian soil and land survey handbook series. CSIRO, Melbourne, pp 75–91

Harden JW (1982) A quantitative index of soil development from field descriptions: examples from a chronosequence in Central California. Geoderma 28(1):1–28

Harden JW (1990) Soil development on stable landforms and implications for landscape studies. Geomorphology 3:391–398

Harden JW, Taylor EM, Hill C (1991) Rates of soil development from four soil chronosequences in the southern Great Basin. Quat Res 35:383–399

Hengl T (2003) Pedometric mapping. Bridging the gaps between conventional and pedometric approaches. ITC dissertation 101. Enschede, The Netherlands

Hole FD, Campbell JB (1985) Soil landscape analysis. Rowman & Allanheld, Totowa

Hubschman J (1975) Morphogenèse et pédogenèse quaternaires dans le piémont des Pyrénées garonnaises et ariégoises. Thèse de Doctorat, Université de Toulouse-Le-Mirail, Toulouse

Jenny H (1941) Factors of soil formation. McGraw-Hill, New York

Jenny H (1980) The soil resource. Origin and behaviour, vol 37, Ecological studies. Springer, New York

Kellogg CE (1959) Soil classification and correlation in the soil survey. USDA, Soil Conservation Service, Washington

Meijerink A (1988) Data acquisition and data capture through terrain mapping units. ITC J 1988(1):23–44

Metternicht G, Zinck JA (1997) Spatial discrimination of salt- and sodium-affected soil surfaces. Int J Remote Sens 18(12):2571–2586

Metternicht GI, Zinck JA (2003) Remote sensing of soil salinity: potentials and constraints. Remote Sens Environ 85:1–20

Metternicht G, Zinck JA, Blanco PD, del Valle HF (2010) Remote sensing of land degradation: experiences from Latin America and the Caribbean. J Environ Qual 39:42–61

Millot G (1964) Géologie des argiles. Altérations, sédimentologie, géochimie. Masson, Paris

NACSN (2005) North american stratigraphic code. North American Commission on Stratigraphic Nomenclature. AAPG Bull 89(11):1547–1591

Olaya V (2009) Basic land-surface parameters. In: Hengl T, Reuter HI (eds) Geomorphometry: concepts, software, applications, vol 33, Developments in soil science. Elsevier, Amsterdam, pp 141–169

Penck A, Brückner E (1909) Die Alpen im Eiszeitalter. Tauchnitz CH, Leipzig

Pérez-Materán J (1967) Informe de levantamiento de suelos, río Santo Domingo, Venezuela. Ministerio de Obras Públicas (MOP), Caracas

Pike RJ (1995) Geomorphometry: progress, practice, and prospect. Z Geomorphol Suppl 101:221–238

Pike RJ, Dikau R (eds) (1995) Advances in geomorphometry. Proceedings of the Walter F. Wood memorial symposium. Zeitschrift für Geomorphologie Supplementband 101

Rivière A (1952) Expression analytique générale de la granulométrie des sédiments meubles. Indices caractéristiques et interprétation géologique. Notion du faciès granulométrique. Bul Soc Géol de France, 6è Série(II):156–167

Ruhe RV (1975) Geomorphology. Geomorphic processes and surficial geology. Houghton Mifflin, Boston

Saldaña A, Ibáñez JJ, Zinck JA (2011) Soilscape analysis at different scales using pattern indices in the Jarama-Henares interfluve and Henares River valley, Central Spain. Geomorphology 135:284–294

Soil Survey Staff (1975) Soil taxonomy. A basic system of soil classification for making and interpreting soil surveys, USDA agriculture handbook 436. US Government Printing Office, Washington, DC

Soil Survey Staff (1999) Soil taxonomy, USDA agriculture handbook 436. US Government Printing Office, Washington, DC

Tricart J (1965a) Principes et méthodes de la géomorphologie. Masson, Paris

Tricart J (1965b) Morphogenèse et pédogenèse. I Approche méthodologique: géomorphologie et pédologie. Science du Sol 1:69–85

USDA (1971) Guide for interpreting engineering uses of soils. USDA, Soil Conservation Service, Washington, DC

Wulf H, Mulder T, Schaepman ME, Keller A, Jörg PhC (2015) Remote sensing of soils: project report from the Federal Office of the Environment (FOEN/BAFU). University of Zurich

Zinck JA (1988) Physiography and soils, ITC soil survey lecture notes. International Institute for Aerospace Survey and Earth Sciences, Enschede

Zinck JA, Urriola PL (1970) Origen y evolución de la Formación Mesa. Un enfoque edafológico. Ministerio de Obras Públicas (MOP), Barcelona

Part II
Approaches to Soil-Landscape Patterns Analysis

Chapter 9
Linking Ethnopedology and Geopedology: A Synergistic Approach to Soil Mapping. Case Study in an Indigenous Community of Central Mexico

N. Barrera-Bassols

Abstract This chapter conveys findings from an integrated participatory soil-landscape survey in a mountain indigenous community of central Mexico using ethnopedologic and geopedologic approaches. It describes the soil-landscape knowledge that local people use for selecting suitable agro-ecological settings, applying land management practices, and implementing soil conservation measures. Relief and soil maps generated by both procedures, the indigenous and the technical, are compared, and the level of spatial correlation of the map units is assessed. Commonalities, differences, and synergies of both soil knowledge systems are highlighted. Participatory soil survey promotes the collaboration of local farmers and experts and the integration of knowledge systems the leads to better understanding of soil distribution patterns on the landscape and their use potentials.

Keywords Ethno-geopedology • Geopedology • Participatory soil survey • Purhépecha • San Francisco Pichátaro • Mexico

9.1 Introduction

Ethnopedology links technical with cultural aspects related to the land/soil complex. In a broader perspective, it explores soil and land in a cultural and ecological context (Barrera-Bassols 2003). It is founded on a holistic and transdisciplinary approach, linking social and natural sciences with other ways of knowing soil and land, or the "other pedologies", i.e. the soil knowledge of local people (Barrera-Bassols and Zinck 2000). Ethnopedology is rooted into two main scientific domains, ethnoecology and ecological anthropology. Ethnoecology studies traditional knowledge and wisdom systems about nature, exploring the links between the three

N. Barrera-Bassols (✉)
Facultad de Filosofía, Universidad Autónoma de Querétaro, Santiago de Querétaro, México
e-mail: barrera@itc.nl

© Springer International Publishing Switzerland 2016   167
J.A. Zinck et al. (eds.), *Geopedology*, DOI 10.1007/978-3-319-19159-1_9

inseparable spheres of these systems, that is the belief system (or Kosmos), the knowledge system (or Corpus), and the performance system (or Praxis), i.e. the K-C-P complex (Toledo and Barrera-Bassols 2008, 2009). Ecological anthropology studies the symbolic dimensions of nature that are rooted in any cultural system (Viveiros de Castro 2010; Descola 2012).

Ethnopedology focuses on the "other pedologies" theories and practices in an (agro) ecological perspective, and compares the latter with modern soil science at different spatial and temporal scales and operational dimensions of the local context. This synergistic approach allows understanding two models of the same agro-ecological reality, the local model and the researcher's model, or the cultural and the technical models (Barrera-Bassols and Zinck 2000, 2003).

This chapter highlights the links between soil and social sciences by focusing on the cultural dimension of the soil/land complex among non-Western agrarian traditions. A study carried out at local level in a mountain indigenous community of central Mexico shows that synergy between ethnopedology and geopedology helps understand soil-landscape relationships, analyze soil distribution patterns, and improve soil mapping.

9.2 Materials and Method

9.2.1 The Study Area

San Francisco Pichátaro, an indigenous Purhépecha community, is located southwest of the Pátzcuaro lake, in the volcanic highlands of central Mexico (Fig. 9.1). The community territory extends from 2,300 to 3,200 masl along a bioclimatic gradient shifting from temperate subhumid to cold humid as elevation increases (T = 16–12 °C; P = 1000–1500 mm). The configuration of the relief is controlled by a set of Plio-Quaternary basalt cones covered by pyroclasts and separated by small fluvio-volcanic valleys.

All soils are derived from volcanic materials, mainly ash and cinder and, to a lesser extent, basalt lava. Andisols cover about 75 % of the study area; other soils have lost their andic properties because of time and/or climate effect. Soils include (1) Pachic Melanudands on the summits of the highest volcanoes, (2) Typic Haplustands on young volcano slopes, (3) Typic Haplustalfs on older volcano slopes, (4) Humic Haplustands in the higher valleys, and (5) Typic Haplustults in the lower valleys.

Land occupation started as early as 3,500 years ago. The presence of fertile volcanic soils and permanent springs at the foot of the volcanoes and lava flows contributed to making Pichátaro an early center of maize production. Nowadays, Pichátaro is a community of some 4,500 inhabitants that maintain indigenous structures and traditions, including local socio-political institutions, vivid Mesoamerican cultural elements in daily life, syncretic catholic practices, communal landownership, and multiple land use strategy based on maize production. Half of

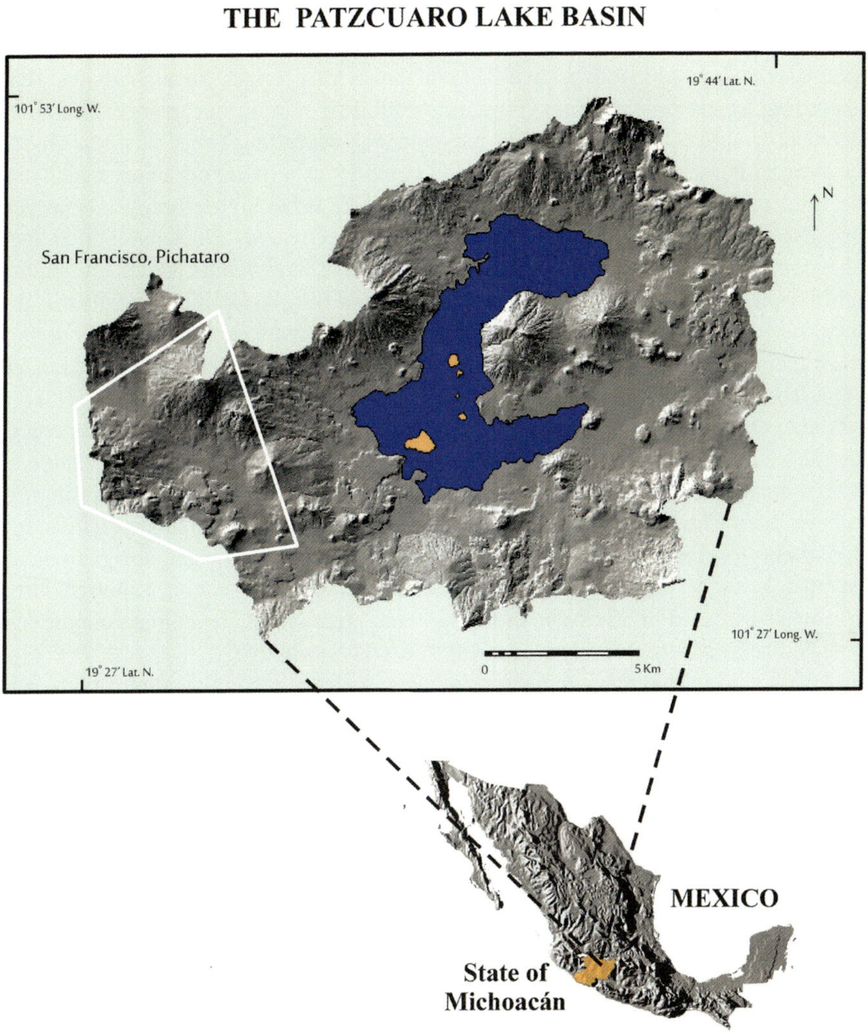

THE PATZCUARO LAKE BASIN

Fig. 9.1 Location of San Francisco Pichátaro within the Pátzcuaro Lake Basin, Michoacán, Mexico

the 10,000 ha territory is used for farming together with some cattle and lamb livestock; the rest is covered by secondary pine and oak forest.

Four main principles govern the local knowledge of land/soil management in the mountain landscapes controlled by the community: (1) land location, (2) land behavior, (3) land resilience, and (4) land quality (Barrera-Bassols and Zinck 2004).

(1) *Land location* is a land management principle based on land characteristics and suitability according to its position in the landscape. Five main landscape types are recognized, including summit and shoulder areas, mid-slope positions, foot-slope positions, valleys, and lava-flow plateaus. (2) *Land movement and behavior* is

the principle that helps farmers understand that the land/soil dominion is dynamic. They recognize, accept, and work with this fact that is reflected in the local expression of "land moves and behaves". Land behavior changes throughout the year according to seasonal rhythms, climatic variability, rainfall occurrence, and management practices. Similarly, land movement is according to its position on the landscape. Farmers consider the land/soil complex as a living organism which can be tired, thirsty, hungry, sick or getting old, like other living beings. However, because soil can grow up again, be rejuvenated, recovered, and rehabilitated, it is also considered fundamentally different from other living beings, which are ineluctably condemned to perish. (3) *Land resilience and restoration* is a principle that the local farmers apply on a regular basis as a practice to improve land quality, but they may also implement exceptional measures to rehabilitate or restore more depleted soils. The way of compromising with nature, by accepting upslope erosion and taking advantage of downslope deposition, is coupled with active sloping land management by means of measures such as sediment trapping, bunds, living fences, deviation of intermittent waterways, terrain leveling, and intensive manuring. (4) *Land quality* reflects a combination of the former three principles, referring to land potential and constraints that result from the position on the landscape, the intensity and periodicity of erosion and deposition of materials, and the management practices applied. Land quality is assessed on the basis of topographic position, micro-climate conditions, selected soil properties, and soil fertility (i.e. soil 'strength').

Using these four management principles, farmers recognize three main land classes, primarily controlled by landscape position and requiring different land care: land on steep slopes, land on valley bottoms, and land in special conditions.

9.2.2 Participatory Soil Survey

A participatory soil survey was carried out with the contribution of the farmers of San Francisco Pichátaro. This included ethnographic (Yin 1994) and ethnopedologic investigations (Shah 1993; Pretty et al. 1995; Sillitoe 1996; Norton et al. 1998; Barrios et al. 2006), complemented by socio-economic and agro-ecological studies (Conway 1985; Farrington 1996; Brussard et al. 2007; Jackson et al. 2007). Participatory soil data gathering was run in parallel with the acquisition of geopedologic information (Fig. 9.2). Both surveys were intentionally not integrated to avoid 'knowledge contamination' through misinterpretation of local environmental knowledge when comparing it with scientific knowledge. Ethnopedologic research was conducted after collecting basic scientific soil data, an approach similar to that of Oudwater and Martin (2003).

9.2.2.1 Ethnographic Survey

Several ethnographic techniques were applied to elicit local environmental knowledge and understanding. Data cross-checking allowed the linkage of conceptual thinking with practical knowledge because local knowledge is not uniform.

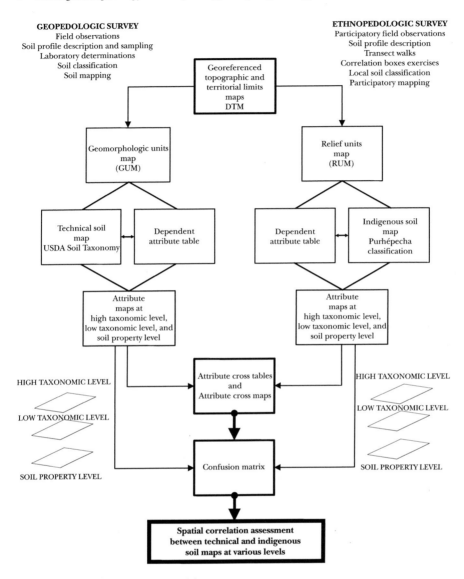

Fig. 9.2 The ethno-geopedologic model

Knowledge differences are related to gender, age, skills and abilities, social prestige, and production specialization. Collective soil knowledge and common land management practices, together with beliefs and symbols, constitute the local social theory of soil and land resources (Toledo 2002; Barrera-Bassols et al. 2006a; Toledo and Barrera-Bassols 2009). This information was complemented by an agronomic survey focusing on crop calendar, farming practices, and production systems. Techniques used included interviews with farmers, soil listing-sorting-ranking, and soil questionnaires (Barrera-Bassols et al. 2009).

Individual and group interviews covering a wide range of issues from soil to socio-economic aspects were carried out. Several techniques were applied to establish soil names and classification as used by local farmers, following the methods implemented in similar case studies (Berlin et al. 1974; Furbee 1989; Bernard 1994; Sandor and Furbee 1996). A list of questions was submitted to 27 farmers, including 6 women, to extract soil information at both village and field levels.

9.2.2.2 Ethnopedologic Survey

Acquisition of ethnopedologic information included knowledge about soil-landscape relationships, soil horizons and profiles, mineral and organic components, soil properties, soil nomenclature, soil taxonomy, spatial distribution of soil classes, soil behaviour, soil quality assessment, soil fertility, soil erosion recognition, and soil and water conservation measures. Also included was knowledge about land use, land-use restrictions and potentials, suitability and limitations for crops, and land management practices. Techniques included soil profile descriptions, soil correlation monoliths, soil-landscape cross sections, and participatory mapping.

The soil profiles described and sampled by the geopedologic survey provided the basis for discussion and exchange of opinions between farmers and technicians on soil nomenclature and properties such as horizons, water holding capacity, internal drainage, stoniness, rooting condition, and biological features. Ethnopedologic information gathered at soil profile sites and along landscape cross sections was contrasted with information acquired in household compounds to test consistency and diversity of soil knowledge. Participatory cross sections were aimed at understanding the landscape structure and local microenvironments. The objective was to analyse farmers' knowledge and experience with respect to identifying soil-relief patterns and boundaries, and assessing soil potentials and limitations for cropping, land-use, and soil water management. The information collected through the various participatory activities allowed translating the farmers' mental representation of soil-landscape patterns into consensual maps. Field data and information were positioned with a GPS and later digitized using the ILWIS software (ITC 2002).

9.3 Soil-Landscape Pattern Recognition, Soil Distribution and Mapping

9.3.1 Comparison of Indigenous and Technical Relief Maps

Comparison of Purhépecha and technical relief maps shows similarities, although indigenous farmers and geomorphologists use different criteria to classify relief forms (Fig. 9.3). Farmers' relief knowledge is based on structural and dynamic characteristics, while specialists rely on relief formation, structure and dynamics.

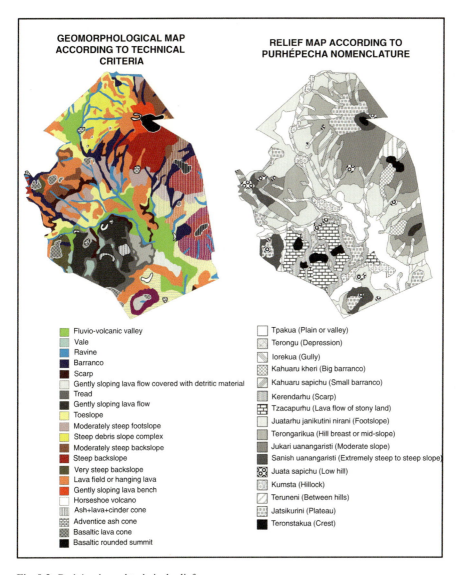

Fig. 9.3 Purhépecha and technical relief maps

Farmers are well aware that the local relief is shaped by volcanism, but volcanism is perceived as a supranatural force. The local relief knowledge is mainly utilitarian, geared towards selecting favorable agro-ecological niches, but it is also inextricably linked to symbolic representations of nature.

Local people divide the relief in three segments: "up or *íotakakhuaru*" (high), "intermediate or *terójkani*" (middle), and "down or *kétsikua*" (low), according to topographic position. This allows them to recognize and manage the spatial distribution of erosion and sedimentation by water during the rainy season and by wind during the dry season, mainly for agricultural purposes. Criteria such as slope, aspect, position, surface lithology, and adjacency or connection to other relief types, are used to describe the configuration of the relief (Fig. 9.4). Additional attributes are implemented to characterize the shape of the topography (e.g. flat, concave, narrow, etc.) and the degree of dissection. Relief is described like a toposequence or catena, in its structural and dynamic content, for practical purposes of slope management. Each relief unit or slope segment is given a local name, which summarizes the environmental conditions and the farming practices required, with emphasis on the seeding of local landraces of maize.

Anthropomorphic terms such as head, shoulder, breast, skirts, and foot are used to recognize the upper, intermediate and lower parts of mountains. Elder people still believe that mountains represent female and male deities, with opposite forces and substances needed for fertility. This belief is rooted in pre-Hispanic Mesoamerican and Purhépecha mythologies. The cult of mountains was originally animistic, but it is still venerated as a syncretic catholic cult, as mountains are believed to be 'life givers' and main water sources.

The ample nomenclature registered shows the criteria used to distinguish relief units. Nomenclature is hierarchic. Mountain range (*juatarhu* or *monte*, or *sierra*), plateau (*jatsikurini* or *mesa*), and valley (*tpakua* or *valle*) are used to distinguish the main local landscapes. Mountains are divided in relief types according to height, position, slope gradient, topographic shape, degree of dissection, and adjacency to other relief types. Mountain slopes (*uanagáristi* or *ladera*) are further subdivided according to relief position, slope gradient, and topographic shape. Gullies (*kauarhu* or *barranco*) are subdivided according to depth, topographic shape, and drainage. Gully dynamics is well understood and permanently assessed for managing runoff and prevent erosion in agricultural lands. People also distinguish several relief features on the lava plateau (*jatsikurini*). Lithology, soil depth, and the amount of rock outcrops are criteria used to recognize relief types and landforms in this local landscape. Finally, valley landforms are distinguished according to their topographic shape, origin of sedimentation, and flooding.

The correlation between the Purhépecha and the technical relief maps, both prepared at the same scale, is high in terms of relief unit identification and spatial distribution. The similarity of the two maps can be grasped visually from Fig. 9.3. The technical map contains more relief types (21) than the local people's map (16). However, there is full correspondence between plateau (*jatsikurini*), valley (*tpakua*), and barranco (*kauarhu*). The indigenous map provides more information on the barranco unit distinguishing between large (*kauarhu kheri*) and small (*kauarhu sapichu*).

Both relief maps, indigenous and technical, were used in the preparation of the geopedologic and ethno-geopedologic maps, respectively. In both cases, relief map units were instrumental not only for soil mapping, but also for understanding soil distribution patterns, soil forming processes, and soil behavior.

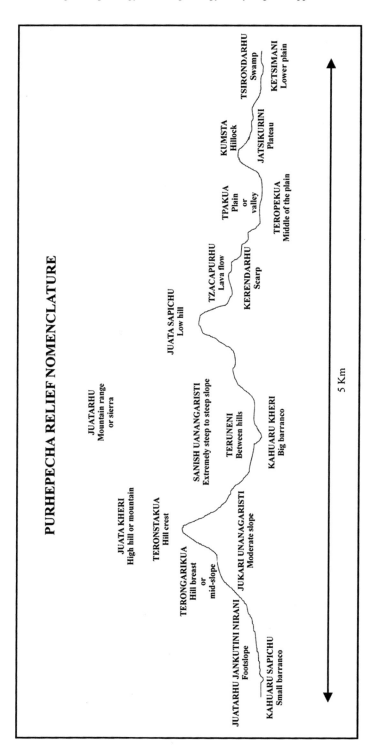

Fig. 9.4 Purhépecha relief nomenclature

9.3.2 Comparison of Indigenous and Technical Soil Maps

To assess the level of consistency and accuracy of both the Purhépecha and the technical soil knowledge systems, a spatial correlation analysis of the soil maps established according to the USDA soil taxonomy (Soil Survey Staff 1999) and the Purhépecha soil classification was carried out (Fig. 9.5). The number of Purhépecha soil classes was higher (21) than the number of classes recognized by the geopedologic survey (19). This shows the fine-tuned local soil classification criteria used by farmers and the contrast with the complex criteria used to determine Andisols according to the USDA Soil Taxonomy (Soil Survey Staff 1999). This is further discussed in Barrera-Bassols (2003). For the purpose of spatial correlation, local soil cartographic units were taken as reference units. Spatial correlation was considered to be high whenever one dominant technical soil class or two similar technical soil classes, one being at least 50 %, occupied 75 % or more of the extent of a local soil cartographic unit.

Using GIS facilities, data were integrated at three levels, including high taxonomic level (order classes), low taxonomic level (subgroup classes), and individual soil property level. At high taxonomic level, 50 % of the Purhépecha and technical soil groups (four groups, respectively) were spatially correlated, while 40 % of them (eight Purhépecha groups and seven technical groups) were spatially correlated at low taxonomic level. These results were obtained with moderate to high taxonomic consistency, showing that both taxonomic systems are themselves spatially robust at the two levels considered, although they use different approaches.

The technical soil taxonomy is based on the recognition of diagnostic properties in all soil horizons, using field descriptions and laboratory determinations, with emphasis on subsurface and subsoil characteristics that are relatively stable over time and reflect main soil forming processes. In contrast, the Purhépecha soil classification is based on the recognition of field-observable topsoil (0–50 cm depth) properties. It uses long-standing farming experience to assess soil functionality and behavior for practical purpose, specifically aimed at the sustainability of rainfed maize cropping in mountain areas. Farmers' monitoring of changes affecting topsoil attributes in space and time is critical for the maintenance of food production, soil management, and soil conservation.

Similarities between Purhépecha and technical soil distribution patterns are related to the nature of the soils in Pichátaro and the way these soils are classified by both systems. The example of Andisols and Dusty soils is illustrative. Both soil groups dominate in more than 75 % of the study area, despite differences in classification criteria. Sixty-five percent of Andisols are spatially correlated with Dusty soils and, conversely, 67 % of Dusty soils are spatially correlated with Andisols. Criteria used to describe Andisols and Dusty soils show commonalities with respect to some physical properties, such as texture, organic matter content, structure development, wet consistence, internal drainage condition, and moisture retention capacity, all of them being field-observable properties. However, many of the andic properties required to classify Andisols according to the technical approach are not observable in the field and must be determined in laboratory. In contrast, the

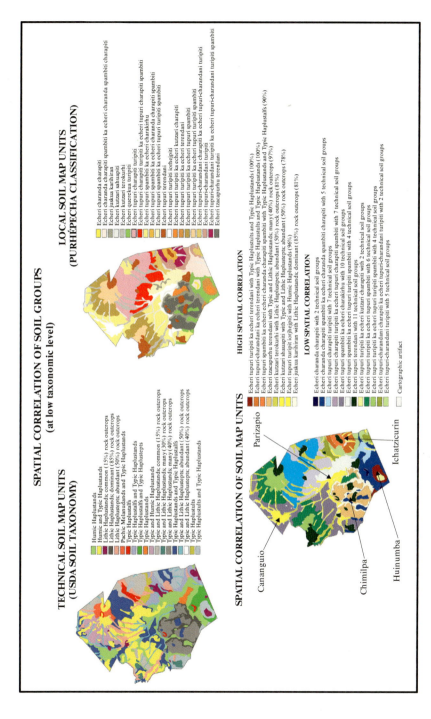

SPATIAL CORRELATION OF SOIL GROUPS
(at low taxonomic level)

TECHNICAL SOIL MAP UNITS
(USDA SOIL TAXONOMY)

Humic Haplustands
Humic and Typic Haplustands
Lithic Haplustands; common (15%) rock outcrops
Lithic Haplustands; dominant (85%) rock outcrops
Lithic Haplustepts; abundant (50%) rock outcrops
Pachic Melanudands and Typic Haplustands
Typic Haplustalfs
Typic Haplustalfs and Typic Haplustands
Typic Haplustalfs and Typic Haplustands
Typic Haplustands
Typic and Humic Haplustands
Typic and Lithic Haplustands; common (15%) rock outcrops
Typic and Lithic Haplustands; many (30%) rock outcrops
Typic and Lithic Haplustands; many (40%) rock outcrops
Typic Haplustands and Typic Haplustalfs
Typic and Lithic Haplustepts; abundant (50%) rock outcrops
Typic and Lithic Haplustepts; abundant (40%) rock outcrops
Typic Haplustults
Typic Haplustults and Typic Haplustands

LOCAL SOIL MAP UNITS
(PURHÉPECHA CLASSIFICATION)

Echeri charanda charapiti
Echeri charanda charapiti spambiti ka echeri charanda spambiti charapiti
Echeri juskua karhíran
Echeri kutzari sahuapiti
Echeri kutzari terokurhi
Echeri querekua turipiti
Echeri tupuri charapiti turipiti
Echeri tupuri charapiti turipiti ka echeri tupuri charapiti spambiti
Echeri tupuri charapiti turipiti ka echeri charakirhu
Echeri tupuri spambiti ka echeri terendani
Echeri tupuri spambiti ka echeri charanda charapiti spambiti
Echeri tupuri spambiti ka echeri tupuri turipiti spambiti
Echeri tupuri terendani
Echeri tupuri turipiti iorhepiti
Echeri tupuri turipiti ka echeri kutzari charapiti
Echeri tupuri turipiti ka echeri terendani
Echeri tupuri turipiti ka echeri tupuri spambiti
Echeri tupuri turipiti ka echeri tupuri turipiti spambiti
Echeri tupuri-charandani charapiti ka echeri tupuri-charandani turipiti
Echeri tupuri-charandani turipiti
Echeri tupuri-charandani turipiti ka echeri tupuri-charandani turipiti spambiti
Echeri tzacapurhu terendani

HIGH SPATIAL CORRELATION

Echeri tupuri turipiti ka echeri terendani with Typic Haplustults and Typic Haplustands (100%)
Echeri tupuri-charandani ka echeri terendani with Typic Haplustults and Typic Haplustands (100%)
Echeri tupuri spambiti ka echeri echeri charanda charapiti spambiti with Typic Haplustalfs and Typic Haplustalfs (90%)
Echeri tzacapurhu terendani with Typic and Lithic Haplustepts; many (40%) rock outcrops (81%)
Echeri kutzari terokurhi with Lithic Haplustepts; abundant (50%) rock outcrops (81%)
Echeri tupuri turipiti iorhepiti with Typic and Lithic Haplustepts; abundant (50%) rock outcrops (78%)
Echeri tupuri shauapiti with Humic Haplustands (96%)
Echeri juskua karhíran with Lithic Haplustands; dominant (85%) rock outcrops (81%)

LOW SPATIAL CORRELATION

Echeri charanda charapiti with 2 technical soil groups
Echeri charanda charapiti spambiti ka echeri charanda spambiti charapiti with 5 technical soil groups
Echeri tupuri charapiti turipiti with 7 technical soil groups
Echeri tupuri charapiti turipiti ka echeri tupuri charapiti spambiti with 7 technical soil groups
Echeri tupuri spambiti ka echeri charakirhu with 10 technical soil groups
Echeri tupuri spambiti ka echeri tupuri turipiti spambiti with 4 technical soil groups
Echeri tupuri terendani with 11 technical soil groups
Echeri tupuri turipiti ka echeri kutzari charapiti with 2 technical soil groups
Echeri tupuri turipiti ka echeri tupuri spambiti with 6 technical soil groups
Echeri tupuri turipiti ka echeri tupuri turipiti spambiti with 4 technical soil groups
Echeri tupuri-charandani charapiti ka echeri tupuri-charandani turipiti with 2 technical soil groups
Echeri tupuri-charandani turipiti with 5 technical soil groups
Cartographic artifact

SPATIAL CORRELATION OF SOIL MAP UNITS

Parizapio
Cananguio
Chimilpa
Huinumba
Ichatzicurin

Fig. 9.5 Spatial correlation of Purhépecha and technical soil map units at low taxonomic level

Purhépecha approach emphasizes observable topsoil properties and agronomic qualities to identify Dusty soils for maize cropping, while also requiring technical determinations to assess the need of chemical fertilizers.

At high and low taxonomic levels, the geopedologic and geo-ethnopedologic approaches, which similarly base soil mapping on relief configuration and variation, constitute synergistic attempts that mobilize the convergence between technical geomorphic and Purhépecha relief knowledge. This is supported by the very high spatial correlation (99 %) between technical and local relief units. Soil-relief relationships proved to be an outstanding factor for farmers and surveyors in soil classification and mapping.

Individual topsoil properties used in both local and technical classifications are spatially better correlated than the full soil map units at high and low taxonomic levels. This confirms that farmers possess an accurate understanding of the functionality and behavior of the arable layer for good crop performance. Soil color and organic matter content show the lowest spatial correlation between both determinations, because of differences in property classes and ranking. Farmers consider that both properties are constantly changing due to land management. Instead of fixed determination and ranking criteria, they use average qualitative estimates. In the technical procedure, topsoil color and organic matter content are often determined over long-time intervals, which do not reflect their short-term variability in space and time as perceived by local people.

When landscapes encompass distinct geomorphic units, both the indigenous and the scientific approaches lead to similar soil distribution patterns and generate comparable soil cartographies. This is the case, for instance, in a clear-cut alluvial terrace landscape or a primary volcanic landscape where discrete relief types and landforms can be easily recognized with well-defined boundaries. Pichátaro is a good example of a well-structured volcanic landscape. In this case, the farmers' relief map and the surveyors' geomorphic map were very similar (Barrera-Bassols et al. 2006b). This explains why the spatial correlation between indigenous soil map units and scientific soil map units was relatively good, in spite of using a restrictive cartographic purity threshold (75 %). Similar results of moderate to high correlation between indigenous and scientific soil map units were obtained by Gobin et al. (1998, 2000, 2001) in Nigeria, Cools et al. (2003) in Syria, Payton et al. (2003) in Bangladesh, and Hillyer et al. (2006) in Namibia. In all these studies, it was recognized that relief and other landscape features were used by local farmers to identify, locate, and classify soils.

9.4   Conclusion: Relevance of Integrating Local and Technical Soil-Landscape Knowledge

The factor showing the greatest contrast between the two classification procedures is soil depth. Technical soil classifications tend to ignore or downplay the diversity of topsoil characteristics, mainly because they can change fairly rapidly under human influence (FAO 1998). In contrast, many non-scientific classifications, such as the Purhépecha system, are based on topsoil characteristics, because the latter control to a large extent soil-related land qualities for food production, soil management, and soil conservation. The simultaneous use of local and technical topsoil classifications helps understand how farmers recognize topsoil variability in space and time at plot, local, and landscape levels for practical purposes. Comparison shows that the local soil quality assessment is multi-dimensional, practical, site-specific, dynamic, and value-laden.

Accordingly, the comprehensive understanding of local soil classifications could contribute to reinforce technical topsoil classification efforts for sustainable land management, as proposed by FAO (1998; see also Sanchez et al. 1982), or the soil quality assessment approach (Romig et al. 1995; Karlen et al. 1997; Lal 1998). There is much need for a fruitful dialogue between soil surveyors, farmers, extensionists and other specialists through exploring, comparing and contrasting rationale and approaches, for instance in soil description, classification, and mapping. This could be done by applying multi-defined soil functions linking crop performance to soil properties, by using classifications that provide useful and practical information, and by avoiding the complex technical language of scientific soil classifications.

Critical to this is the assessment of topsoil characteristics, behavior, and performance throughout the year and between years, as the evaluation of topsoil dynamics is relevant for local land-use decision-making. Some authors still consider "the other pedologies" of limited value, because it is mainly based on the recognition of topsoil characteristics, as compared to scientific systems that use both surface and subsurface horizon features. However, topsoil characteristics are often strongly correlated with subsurface characteristics, especially in areas with the same soil forming factors.

Similarities and differences between making technical and local soil maps reveal synergies that can be further explored to assess soil performance for precision agriculture. Management practices, such as conservation measures, pest and disease control, and crop variety selection, can be tailored to specific farming areas based on soil, relief, and other environmental criteria. Critical understanding and flexible integration of both cognitive systems could benefit from the rationale lying behind indigenous precision agriculture and lead to better crop-specific soil management.

Beyond the soil classification comparison, the spatial correlation analysis at the level of topsoil properties showed that farmers and soil surveyors can talk a common language and together improve the quality of soil research for better soil mapping and for monitoring agronomic and environmental impacts at site, local, and regional levels. A common language requires recognizing that all soil knowledge systems have limitations and that merging technical and local thinking is able to promote sustainable land management schemes. The main benefit of this is correlating and mutually enriching different perceptions about the (soil) world. In other words, ethno-geopedology helps validate scientific soil knowledge to assure that it is not only scientific but also locally relevant and functional. Dialogue should be used as a communication platform to compare and co-validate diverse manners of perceiving, naming, classifying, and mapping soils. Contrasting knowledge systems (e.g. contextual vs general) can be synergist in multi-dimensional ways.

Ethnopedology is geopedologic in essence because farmers use relief units (i.e. geomorphic units), even before relying on soil properties, to select favorable agro-ecological settings on the landscape. Thus the prefix *geo* is implicit in the concept of ethnopedology. Local, indigenous people who have been living for long time in their territory have an acute understanding of the relationships between relief and soil. Local people are innate geopedologists.

References

Barrera-Bassols N (2003) Symbolism, knowledge and management of soil and land resources in indigenous communities: ethnopedology at global, regional and local scales. ITC dissertation series 102, 2 vols. International Institute for Geo-information Science and Earth Observation (ITC), Enschede

Barrera-Bassols N, Zinck JA (2000) Ethnopedology in a worldwide perspective: an annotated bibliography. International Institute for Aerospace Survey and Earth Sciences, Enschede

Barrera-Bassols N, Zinck JA (2003) Ethnopedology: a worldwide view on the soil knowledge of local people. Geoderma 111:171–195

Barrera-Bassols N, Zinck JA (2004) Land moves and behaves: indigenous discourse on sustainable land management in Pichátaro, Pátzcuaro Basin, Mexico. Geogr Ann Ser A 85:266–276

Barrera-Bassols N, Zinck JA, Van Ranst E (2006a) Symbolism, knowledge and management of soil and land resources: ethnopedology at global, regional and local scales. Catena 65:118–137

Barrera-Bassols N, Zinck JA, Van Ranst E (2006b) Local soil classification and comparison of indigenous and technical maps in a Mesoamerican community using spatial analysis. Geoderma 135:140–162

Barrera-Bassols N, Zinck JA, Van Ranst E (2009) Participatory soil survey: experience in working with a Mesoamerican indigenous community. Soil Use Manage 25:43–56

Barrios E, Delve RJ et al (2006) Indicators of soil quality: a south–south development guide for linking local and technical knowledge. Geoderma 135:248–259

Berlin B, Breedlove DE et al (1974) Principles of Tzeltal plant classification. Academic, New York

Bernard HR (1994) Research methodology in anthropology: qualitative and quantitative approaches. Sage, Thousand Oaks

Brussard L, de Ruiter PC et al (2007) Soil biodiversity for agricultural sustainability. Agr Ecosyst Environ 121:233–244

Conway GR (1985) Agroecosystems analysis. Agric Adm 20:31–55

Cools E, De Pauw E et al (2003) Towards an integration of conventional land evaluation methods and farmers' soil suitability assessment: a case study in northwestern Syria. Agr Ecosyst Environ 95:327–342

Descola P (2012) Más allá de naturaleza y cultura. Amorrortu Editores, Buenos Aires

FAO (1998) Topsoil characterization for sustainable land management. Land and Water Development Division. Soil Resources, Management and Conservation Service. FAO, Rome

Farrington J (1996) Socio-economic methods in natural resources research, Natural resources perspectives 9. Overseas Development Institute, London

Furbee L (1989) A folk expert system: soils classification in the Colca Valley, Peru. Anthropol Q 62:83–102

Gobin A, Campling P et al (1998) Integrated toposequence analysis at the confluence zone of the River Ebonyi headwater catchment (southeastern Nigeria). Catena 32:173–192

Gobin A, Campling P et al (2000) Quantifying soil morphology in tropical environments: methods and application in soil morphology. Soil Sci Soc Am 64:1423–1433

Gobin A, Campling P et al (2001) Integrated land resources analysis with an application to Ikem (south-eastern Nigeria). Landsc Urban Plan 53:95–109

Hillyer AEM, McDonagh JF et al (2006) Land-use and legumes in northern Namibia – the value of a local classification system. Agr Ecosyst Environ 117:251–265

ITC (2002) ILWIS user's manual (version 3.11). ITC, Enschede

Jackson LE, Pascual U et al (2007) Utilizing and conserving agrobiodiversity in agricultural landscapes. Agr Ecosyst Environ 121:196–210

Karlen DL, Mausbach MJ et al (1997) Soil quality: a concept, definition, and framework for evaluation. Soil Sci Soc Am J 61:4–10

Lal R (1998) Soil quality and agricultural sustainability. In: Lal R (ed) Soil quality and agricultural sustainability. Ann Arbor Press, Chelsea, pp 3–13

Norton JB, Pawluk RR et al (1998) Observation and experience linking science and indigenous knowledge at Zuni, New Mexico. J Arid Environ 39:331–340

Oudwater N, Martin A (2003) Methods and issues in exploring local knowledge of soils. Geoderma 111:387–402

Payton RW, Barr JJF et al (2003) Contrasting approaches to integrating indigenous knowledge about soils and scientific soil survey in East Africa and Bangladesh. Geoderma 111(3–4):355–386

Pretty J, Guijt T et al (1995) Regenerating agriculture: the agroecology of low-external input and community-based development. In: Kirkby J, O'Keefe P, Timberlake I (eds) The earthscan reader in sustainable development. Earthscan Publications, London, pp 125–145

Romig DE, Garlynd MJ et al (1995) How farmers assess soil health and quality. J Soil Water Conserv 50(3):229–236

Sanchez PA, Couto W et al (1982) The fertility capability classification system: interpretation, applicability and modification. Geoderma 27:283–309

Sandor JA, Furbee L (1996) Indigenous knowledge and classification of soils in the Andes of southern Peru. Soil Sci Soc Am J 60:1502–1512

Shah PB (1993) Local classification of agricultural land in the Jhiku Khola watershed. In: Tamang D, Gill GJ, Thapa GB (eds) Indigenous management of natural resources. Ministry of Agriculture/Winrock International, Kathmandu, pp 159–163

Sillitoe P (1996) A place against time. Land and environment in the Papua New Guinea highlands. Harwood Academic Publishers, Amsterdam

Soil Survey Staff (1999) Keys to soil taxonomy, 8th edn. USDA Natural Resource Conservation Service, Washington, DC

Toledo VM (2002) Ethnoecology: a conceptual framework for the study of indigenous knowledge on nature. In: Stepp JR, Wyndham FS, Zarger R (eds) Ethnobiology and biocultural diversity. International Society of Ethnobiology, University of Georgia Press, Athens, pp 511–522

Toledo VM, Barrera-Bassols N (2008) La memoria biocultural. La importancia ecológica de las sabidurías tradicionales. Icaria, Barcelona

Toledo VM, Barrera-Bassols N (2009) A etnoecologia: uma ciência pós-normal que estuda as sabedorias tradicionais. Desenvolvimento Meio Ambiente 20:31–45

Viveiros de Castro E (2010) Metafísicas caníbales. Líneas de Antropología estructural. Katz Editores, Buenos Aires

Yin RK (1994) Case study research. Design and methods. SAGE Publications, Thousand Oaks

Chapter 10
Diversity of Soil-Landscape Relationships: State of the Art and Future Challenges

J.J. Ibáñez and R. Pérez Gómez

Abstract Pedology and geomorphology are considered independent scientific disciplines, but form in fact a single indivisible system. The diversity analysis of natural resources tries to account for the variety of forms and spatial patterns that display the natural bodies, biotic and abiotic, appearing at the earth's surface. The application of mathematical tools to diversity analysis requires a classification of the universe concerned. Biodiversity studies have a long tradition in comparison to earth sciences. Recently pedologists started paying attention to soil diversity using the same mathematical tools as ecologists use and reaching interesting relations between the spatial patterns of soil and vegetation. So far geodiversity studies are only concerned by the preservation of the geological heritage, bypassing most of the aspects related to its spatial distribution. Vegetation scientists have developed a classification that links climate and plant communities, the so-called syntaxonomic system. The purpose of this chapter is to explore a perspective of joining soils, geoforms, climate, and biocenoses in an integrated and comprehensive approach to describe the structure and diversity of the earth surface systems.

Keywords Geopedology • Pedodiversity • Landscape diversity • Geodiversity • Vegetation patterns

10.1 Introduction

Natural resources vary in the space and time. Throughout centuries naturalists have observed that some landscapes are more heterogeneous than others regardless of the nature of the study object (e.g. biological species, rocks, landforms, soils, etc.) (Ibáñez 2014). It is essential practice in science to identify, categorize, and classify

J.J. Ibáñez (✉)
Centro de Investigaciones sobre Desertificación, CIDE (CSIC-UV), Valencia, Spain
e-mail: choloibanez@hotmail.com

R.P. Gómez
Departamento de Ingeniería, Topográfica y Cartografía, Universidad Politécnica de Madrid (UPM), Madrid, Spain
e-mail: rufino@topografia.upm.es

the heterogeneity of the objects of study, leading to taxonomy. Using variable criteria several classification systems with different numbers of taxa have been proposed. To achieve a common language between experts in a scientific community, the most desirable would be universally accepted taxonomies. It has also been recognized that it is impossible to reach perfect taxonomies, as these are changing over time according to perceptions, scientific progress, and societal information demands (Ibáñez and Montanarella 2013). The taxonomy of a given natural resource in a given time provides an inventory of the global diversity that is accepted by the experts that embrace that mental construct. The tree of the life has been most extensively studied so that biological taxonomies are considered the most elaborate and sophisticated ones (Ibáñez and Montanarella 2013). Ecological studies are ahead several decades compared to other natural resources by providing numerous methodologies and mathematical procedures to estimate biodiversity. In contrast, the analysis of pedodiversity started only in the last decades of the twentieth century (Fridland 1976; Ibáñez et al. 1995). Recently Ibáñez and Bockheim (2013a) and Ibáñez (2014) have synthesized the knowledge on soil diversity. Biodiversity and pedodiversity studies pursue two different but complementary purposes: (a) the analysis of the structure of ecosystems and soil assemblages, respectively, and (b) the preservation of both natural resources. Geodiversity studies focus mainly on the preservation of the geological heritage (Ibáñez and Bockheim 2013b).

According to Gray (2004), geodiversity could be defined as "*the natural range (diversity) of geological (rocks, minerals, fossils), geomorphological (land form, processes) and soil features. It includes their assemblages, relationships, properties, interpretations and systems*". However, so far experts have not proposed any index or mathematical procedure to quantify the diversity of more than one natural resource at a time. Ecologists have not been able either to reach a satisfactory diversity index to include the existing taxonomic distance between biotaxa (Ricotta 2005). It might be therefore advisable to analyze the diversity of each natural system independently. The main problem for the estimation of lithologic and geomorphic diversity derives from the lack of universal classifications. The absence of this kind of taxonomic constructs precludes conducting comparative studies to detect regularities in spatial patterns and laws in different environments.

The purpose of this chapter is to explore a perspective of joining soils, geoforms, climate, and biocenoses in an integrated and comprehensive approach to describe the structure and diversity of the earth surface systems.

10.2   The Concept of Diversity

There is no consensus among experts to define biodiversity and pedodiversity. However, the following proposal is satisfactory from a methodological point of view and does not distinguish between the different objects that can be analyzed. According to Huston (1994, p 65), diversity can be conceptually defined as follows:

> The concept of diversity has two primary components, and two unavoidable value judgments. The primary components are statistical properties that are common to any mixture

of different objects, whether the objects are balls of different colors, segments of DNA that code for different proteins, species or higher taxonomic levels, or soil types or habitat patches on a landscape. Each of these groups of items has two fundamental properties: 1. the number of different types of objects (e.g. species, soil types) in the mixture or sample; and 2. the relative number or amount of each different type of object. The value judgments are: 1. whether the selected classes are different enough to be considered separate types of objects; and 2. whether the objects in a particular class are similar enough to be considered the same type. On these distinctions hangs the quantification of biological diversity.

There are essentially three components of diversity: (a) the variety of taxa (richness); (b) the way in which the individuals are distributed among those taxa (evenness or equitability); and (c) diversity indices that attempt to incorporate both components (a) and (b) in a single value (Magurran 2003; Ibáñez et al. 1990; Ibáñez et al. 2013). In addition, abundance distribution models provide the closest fit to the observed pattern of object abundance (e.g. geometric series, log series, lognormal series, power laws, abundance distribution models). Statistical regression models (e.g. power laws) and other mathematical procedures have been extensively applied also to analyze diversity-area relationships.

Several other methodologies and useful mathematical tools have been proposed to complement the former such as fractals, multifractals, nested subsets theory, neural networks, and some procedures applied by physicists to the study of non-linear systems (see Ibáñez et al. 2013 and references therein).

10.3 Biodiversity, Pedodiversity, Landform Diversity, and Lithological Diversity Patterns

Few studies have been carried out to analyze the relation between pedodiversity and the diversity of other natural resources (Fig. 10.1), being the most frequent those that compare biodiversity and pedodiversity (e.g. Petersen et al. 2010; Williams and Houseman 2013; Ibáñez and Feoli 2013). Ibáñez et al. (2013); Ibáñez (2014) show that biodiversity and pedodiversity are usually positively correlated and have similar spatial patterns. There are few studies that analyze the relations between biodiversity and landforms, lithodiversity and landforms, and pedodiversity and landforms. However, Ibáñez et al. (1994) and Toomanian (2013) have detected positive correlation between pedodiversity and landform diversity, as well as between pedodiversity and lithodiversity.

10.4 Geopedologic and Bioclimatic Approaches

All natural resources are diverse in nature. However, a simpler description of the landscape could be interesting to reduce the division between natural resources by making use of more holistic concepts (e.g. pedodiversity + landform diversity = geopedologic diversity and/or plant community diversity + climate diversity = bioclimatic diversity).

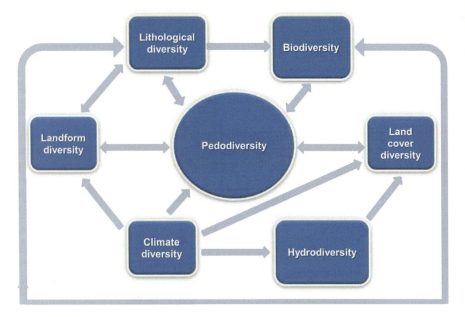

Fig. 10.1 Interrelations between pedodiversity and diversity of soil forming factors

Geopedology proposes a landscape approach that integrates geoforms and soils (Zinck 2013). Because geomorphology takes into account relief and surface morphodynamics in a morphoclimatic context and geoforms are in many aspects conditioned by lithology, this approach is more integrative than pedology and geomorphology applied individually. Zinck (1988) established a hierarchic geopedologic approach to analyze soils in the landscape that includes geostructure, morphogenic environment, geomorphic landscape, relief, lithology, and landform. The geopedologic approach has been applied to soil survey (Zinck 1988) as well as in plant-landforms studies (e.g. Stacey and Monger 2012; Michaud et al. 2013). Geopedology is also a method considered in landscape ecology (e.g. Zonneveld 1989; Saldaña 2013; Zinck 2013). Some geopedologic classifications have been proposed to be applied everywhere (e.g. Zinck 1988, 2013), whereas others have been developed to analyze only specific territories (e.g. Michaud et al. 2013).

In several aspects, the SOTER initiative (e.g. Van Engelen and Dijkshoom 2013) could solve the lack of universal classifications on soils, landforms, slope, surface forms, lithology, etc., if it were accepted at worldwide level and applied at all scales. However, the SOTER methodology does not include the inventory of the mentioned natural resources which must be provided by other initiatives and institutions. This fact, today leads to a dead end. SOTER characterization and classification of landforms is physiographic instead genetic (e.g. geomorphology), not providing helpful information about some relevant aspects (e.g. age and intensity of alteration processes, regolith depth, etc.) for a deeper understanding of landscape genesis and structures. With respect to vegetation, it should be noted that there are many approaches, but few of them as for instance the syntaxonomic system take into account soil features and climate to define phyto-associations and map units. The SOTER approach to vegetation is very descriptive and thus less useful for analyzing the relationships between geoforms and vegetation as we can see in the following paragraphs.

The European school of geobotany termed phytosociology is a discipline focusing on the classification of plant communities (e.g. Westhoff and van der Maarel 1978), with an International Code of Phytosociological Nomenclature (Weber et al. 2000). To make the inventory of, map, and classify plant landscapes according to phytosociology, geobotanists use the so-called syntaxonomic system (e.g. Mirkin 1989). The syntaxonomic approach takes mainly into account plant communities based on the concept of plant natural vegetation (PNV) and climate to develop a classification of plant landscapes in bioclimatic belts (Loidi and Fernández-González 2012). However, the geobotanical school considers also pedologic, geomorphic, and lithological land features for analyzing plant-soil relations at landscape level (Rivas-Martínez 2005). The landscape is divided in units termed tessela and microtessela that correspond to terrestrial areas where the same PNV is present. In the frame of the syntaxonomic classification, the nomenclature of plant assemblages includes terms such as climatophilous (plant communities that only depend on climatic factors), basophilous (plant associations that grow on pedotaxa rich in nutrients; eutric in pedological terms), siliceous (plant associations that grow on pedotaxa poor in nutrients; dystric in pedological terms), calcicolous (plant assemblages associated to the presence of calcium carbonate in soils; calcic and calcaric in soil classifications), edaphophylous (plant associations that depend on specific soil features and properties). There are other terms such as gypsiferous or halophytic (reserved for gypsiferous and halophytic vegetation, associated with gypsum- and salt-rich soils, respectively). The edaphophylous units are divided in edaphoxerophilous (plants adapted to xericity that grow in tessela or microtessela on pedotaxa that store very little water) and edaphohygrophilous (plants species associated to pedotaxa with permanent or seasonal waterlogging). Other terms such as permaseries, geopermaseries, and geoseries are indicators of PNV that occur on sites with abiotic constraints that produce permanent environmental stress for the full development of an ecological succession. Rivas-Martínez (2005), among others, explains concepts and nomenclature. Summarizing, the syntaxonomic system classifies PNV units taking into account all the environmental variables that influence the distribution of plant communities by adding these to the formal nomenclature of each syntaxum. Most of these factors are climatic, but also pedological, geomorphic, and lithological ones are included (Fig. 10.2).

Figure 10.3 shows the richness of PNV in the Almeria province (south-eastern Spain) according to the factors that determine their presence and geographic dispersion in the study area. The geographic distribution (dominant, abundant, common, frequent, rare, endemic) of PNV has been clustered according to the number of bioclimatic belts (BB) (there is a total of seven in the Almeria province) where these plant communities appear. For example abundant PNV means that they appear in most climatic belts, whereas endemic are those that only appear in one of them. It is

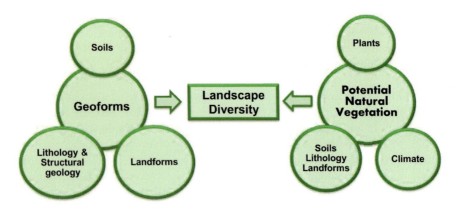

Fig. 10.2 A landscape diversity scheme

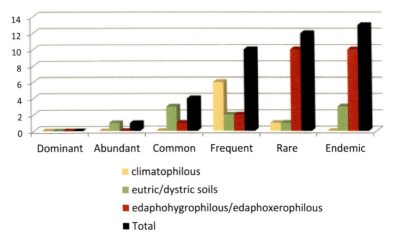

Fig. 10.3 Plant landscape diversity scheme (number of PNV types) and the role of soils in Almeria Province, Spain

noticeable that in the Almeria Province there are much more plant communities determined by soil types and properties than those conditioned by climatic factors only. However, many of these soils are in turn associated to specific landforms.

Thus, in principle it would be possible to analyze the landscape including environmental heterogeneity and diversity using only two classifications: the geopedologic and the syntaxonomic. It is interesting to note that the syntaxonomic system classifies also aquatic vegetation in shallow water bodies, whereas the underlying sediments are also included in the most recent soil taxonomies such as the WRB (IUSS Working Group 2007). These classifications comprise virtually all soil-forming factors, with the exception of humans (Jenny 1941). Nevertheless it is necessary to test whether the combination of both taxonomies improves current landscape analysis. This is a line of research that should be explored.

However, as we stated above, the main concern is that there are no universal classifications widely accepted by experts in lithology and geoforms (Ibáñez et al. 2013), whereas the syntaxonomic approach is only popular in continental Europe.

10.5   Preservation of Geoforms as Part of the Natural Heritage

For many pedologists, soil survey is based on the soil-landscape paradigm (Hudson 1992). The preservation of geodiversity and pedodiversity has generated much interest in recent years. Landform diversity and pedodiversity are part of our natural heritage (Gray 2004; Ibáñez et al. 2013). It is impossible to preserve the soils without preserving the geoforms in which they are formed. Thus irrespective of whether policies or societal demand are intended to preserve pedodiversity or geodiversity, the preservation of geoforms is guaranteed, not needing any additional theoretical scheme.

10.6   Conclusions

All natural resources, biotic and abiotic, are part of the natural heritage. Diversity analysis could be applied to all of them as a mathematical tool to understand their respective spatial patterns in the landscape. The only requirement for implementing these formal procedures is the existence of their respective taxonomies. Obviously universal taxonomies are preferable to national or ad hoc purpose-oriented classifications, as they allow comparing the results by extracting regularities from different regions and environments by different researches. Conventional soil inventories implicitly make use of soil-physiography relationships, the so-called soil-landscape paradigm. The geopedologic approach makes explicit the implicit traditional knowledge in soil survey activities, formalized in a single taxonomy or classification system. Likewise, the syntaxonomic approach plays the same role concerning the plant community-climate relationships. By using both approaches concomitantly it is feasible to achieve a unifying vision of the landscape structure, in the landscape ecology perspective. It is desirable to use the diversity analysis of all natural resources independently but also jointly. The geopedologic approach is a step forward in this direction. The same is true for the preservation of the natural heritage, in view that all natural resources interact with each other, being mutually interdependent.

References

Fridland VM (1976) The soil cover pattern: problems and methods of investigation. Soil combinations and their genesis (Translated from Russian). Keter Publishing House, Jerusalem

Gray M (2004) Geodiversity: valuing and conserving abiotic nature. Wiley, Chichester

Hudson HD (1992) The soil survey as paradigm-based science. Soil Sci Soc Am J 56:836–841. doi:10.2136/sssaj1992.03615995005600030027x

Huston MAH (1994) Biological diversity. Cambridge University Press, Cambridge

Ibáñez JJ (2014) Diversity of soils. In: Warf B (ed) Oxford bibliographies in geography (article online). Oxford University Press, New York

Ibáñez JJ, Bockheim JG (eds) (2013a) Pedodiversity. CRC Press, Boca Raton

Ibáñez JJ, Bockheim JG (2013b) Conclusions. In: Ibáñez JJ, Bockheim JG (eds) Pedodiversity. CRC Press, Boca Raton, pp 229–233

Ibáñez JJ, Feoli E (2013) Global relationships of pedodiversity and biodiversity. Vadose Zone J. doi:10.2136/vzj2012.0186

Ibáñez JJ, Montanarella L (2013) Magic numbers: a metha-analysis for enlarging the scope of a universal soil classification system. JRC Technical Reports, European Commission, Brussels (pdf free available from http://publications.jrc.ec.europa.eu/repository/bitstream/111111111/28069/1/lb-na-25-849-en-n.pdf)

Ibáñez JJ, Jiménez-Ballesta R, García-Álvarez A (1990) Soil landscapes and drainage basins in Mediterranean mountain areas. Catena 17:573–583. doi:10.1016/0341-8162(90)90031-8

Ibáñez JJ, Pérez A, Jiménez-Ballesta R, Saldaña A, Gallardo F (1994) Evolution of fluvial dissection landscapes in Mediterranean environments. Quantitative estimates and geomorphological, pedological and phytocenotic repercussions. Z Geomorphol 38:105–119

Ibáñez JJ, De-Alba S, Bermúdez FF, García-Álvarez A (1995) Pedodiversity: concepts and measures. Catena 24:215–232. doi:10.1016/0341-8162(95)00028-Q

Ibáñez JJ, Vargas RJ, Vázquez-Hoehne A (2013) Pedodiversity state of the art and future challenges. In: Ibáñez JJ, Bockheim GJ (eds) Pedodiversity. CRC Press, Boca Raton, pp 1–28

IUSS Working Group WRB (2007) World reference base for soil resources 2006, first update 2007. World soil resources reports 103. FAO, Rome (pdf free available from www.fao.org/ag/agl/agll/wrb/doc/wrb2007_corr.pdf)

Jenny H (1941) Factors of soil formation. McGraw-Hill, New York

Loidi J, Fernández-González F (2012) Potential natural vegetation: reburying or reboring? J Veg Sci 23:596–604. doi:10.1111/j.1654-1103.2012.01387.x

Magurran AE (2003) Measuring biological diversity. Blackwell Publishing, Oxford

Michaud GA, Monger HC, Anderson DL (2013) Geomorphic-vegetation relationships using a geopedological classification system, northern Chihuahuan Desert, USA. J Arid Environ 90:45–54. doi:10.1016/j.jaridenv.2012.10.001, Free available from DigitalCommons@ University, Nebraska–Lincoln. Publications from USDA-ARS/UNL Faculty, USDA Agricultural Research Service (pdf free available from http://digitalcommons.unl.edu/cgi/viewcontent.cgi?article=2172&context=usdaarsfacpub)

Mirkin BM (1989) Plant taxonomy and syntaxonomy: a comparative analysis. Vegetatio 82:35–40. doi:10.1007/BF00217980

Petersen A, Gröngröft A, Miehlich G (2010) Methods to quantify the pedodiversity of 1 km^2 areas. Results from southern African drylands. Geoderma 155:140–146. doi:10.1016/j.geoderma.2009.07.009

Ricotta C (2005) Through the jungle of biological diversity. Acta Biotheor 53:29–38. doi:10.1007/s10441-005-7001-6

Rivas-Martínez S (2005) Notions on dynamic-catenal phytosociology as a basis of landscape science. Plant Biosyst 139:135–144, pdf free available from http://dx.doi.org/10.1080/11263500500193790

Saldaña A (2013) Pedodiversity and landscape ecology. In: Ibáñez JJ, Bockheim JG (eds) Pedodiversity. CRC Press, Boca Raton, pp 105–132

Stacey LW, Monger HC (2012) Banded vegetation-dune development during the Medieval Warm Period and 20th century, Chihuahuan Desert, New Mexico, USA. Ecosphere 3:art21 http://dx.doi.org/10.1890/ES11-00194.1 (pdf free available from http://www.esajournals.org/doi/abs/10.1890/ES11-00194.1

Toomanian N (2013) Pedodiversity and landforms. In: Ibáñez JJ, Bockheim JG (eds) Pedodiversity. CRC Press, Boca Raton, pp 133–152

Van Engelen V, Dijkshoom J (eds) (2013) Global and national soil and terrain databases (SOTER). Procedures manual version 2.0. ISRIC report 2013/04, Wageningen

Weber HE, Moravec J, Theurillat JP (2000) International code of phytosociological nomenclature. J Veg Sci 11:739–768. doi:10.2307/3236580

Westhoff V, van der Maarel E (1978) The Braun-Blanquet approach. In: Whittaker RH (ed) Classification of plant communities, 2nd edn. Dr W Junk, The Hague, pp 287–399

Williams BM, Houseman GR (2013) Experimental evidence that soil heterogeneity enhances plant diversity during community assembly. J Plant Ecol 7:461–469. doi:10.1093/jpe/rtt056

Zinck JA (1988) Physiography and soils, Lecture notes. International Institute for Aerospace Survey and Earth Sciences (ITC), Enschede

Zinck JA (2013) Geopedology. Elements of geomorphology for soil and geohazard studies, ITC special lecture note series. ITC, Enschede

Zonneveld IS (1989) The land unit: a fundamental concept in landscape ecology, and its applications. Landsc Ecol 3:67–86

Chapter 11
A New Soil-Landscape Approach to the Genesis and Distribution of Typic and Vertic Argiudolls in the Rolling Pampa of Argentina

H.J.M. Morrás and L.M. Moretti

Abstract The Rolling Pampa is one of the several subregions of the large Pampa plains in central Argentina. The most extensive and representative soils in the area are Typic Argiudolls, together with a smaller proportion of Vertic Argiudolls occurring mainly on relief tops and upper slope facets in a strip close to the Paraná – Río de la Plata fluvial axis. According to the traditional interpretation, the vertic soil properties are due to a combination of finer parent materials resulting from granulometric selection during eolian transport from the south-western Andean sources and intense smectite formation in a more humid eastern sector of the Pampa. A new sedimentological and geopedologic approach explains more accurately the development and spatial distribution of the main soils in the subregion. According to this, smectitic sediments coming from northern sources in the Paraná basin were deposited in the Rolling Pampa and later covered by illitic loess sediments from south-western Andean sources. In a subsequent humid period in the Holocene, the illitic sediments were eroded and the smectitic sediments were exposed on the upper parts of the undulating relief. As a consequence, Typic Argiudolls developed on the illitic and volcanoclastic Andean sediments, while vertic soils evolved in higher positions of the landscape on the smectitic sediments of older age and different origin.

Keywords Vertic Argiudolls • Typic Argiudolls • Pedogenesis • Parent material • Landscape

H.J.M. Morrás (✉) • L.M. Moretti
Soil Research Institute, INTA, Buenos Aires, Argentina
e-mail: hmorras@gmail.com; moretti.lucas@inta.gob.ar

© Springer International Publishing Switzerland 2016
J.A. Zinck et al. (eds.), *Geopedology*, DOI 10.1007/978-3-319-19159-1_11

193

11.1 Introduction

The Rolling Pampa is a subregion of the vast Pampa plains in Argentina that have
been subdivided on the basis of main ecological features (Pereyra 2003) (Fig. 11.1).
It is a strip of about 100 km wide, bordering with the Paraná and Río de la Plata
rivers to the north-east and the Salado river to the south-west. Several areas have
been recognized within the subregion due to specific combinations of geomorphology and soils (Scoppa and Vargas Gil 1969). According to Cappannini and
Dominguez (1961), the main subdivisions are the High Rolling Pampa and the Low
Rolling Pampa. The High Rolling Pampa extends from La Matanza river to the
north, occupying two thirds of the subregion and clearly showing the rolling morphology, with slopes of about 2 % and exceptionally about 4–5 %, associated with
a well-developed dendritic hydrographic network, whereas the Low Rolling Pampa
extends from La Matanza river to the south, being a transitional area to the Depressed
or Flooded Pampa. In the Low Rolling Pampa, the drainage network is less dense
and defined, and terrain undulations are less marked. The most extended and representative soils in the High Rolling Pampa are Typic Argiudolls (Soil Survey Staff

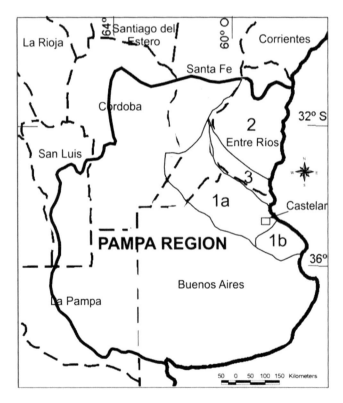

Fig. 11.1 Subregions of the Argentine Pampas. *1a*: High Rolling Pampa; *1b*: Low Rolling Pampa;
2: Mesopotamian Pampa; *3*: Delta of the Paraná River

2010), whereas those in the Low Rolling Pampa are Aquic Argiudolls. The dominant soils in the strip crossing both areas along the Paraná-Río de la Plata fluvial axis are Vertic Argiudolls (Etchevehere 1975). These vertic soils, sometimes associated with true Vertisols, usually appear on relief tops and upper slope facets (INTA 1989). The traditional interpretation of the genesis of vertic soils in this area based on smectitic clay neoformation is not satisfactory, particularly as to their occurrence on landscape summits and their relationship with other vertic soils in the neighboring Mesopotamian Pampa. Thus, detailed studies of sediments and soils in the frame of a different soil-landscape approach have been undertaken to obtain a better explanation of the genesis and distribution of soils in the Rolling Pampa.

11.2   The Parent Material of Pampean Soils: Two Contrasting Paradigms

11.2.1   Origin and Composition of Pampean Loess

The Pampean Region is a large sedimentary plain of primary and secondary (i.e. reworked) loess deposits in superposed mantles of varying thickness. In a simplified stratigraphic scheme for the north-eastern Pampa, the lower and thicker section deposited during the Early and Middle Pleistocene is named Lower Pampeano or Ensenada Formation. This is covered by a mantle of loessic sediments 6–7 m thick, deposited during the Late Pleistocene and known as the Upper Pampeano or Buenos Aires Formation. During the Holocene, eolian sediments named the Post Pampeano or La Postrera Formation were deposited in some areas of the plain (Nabel and Pereyra 2000; Zárate 2005).

Both the origin and the composition of these loessic sediments are still a rather controversial subject. There is consensus on that the main source areas for the bulk of Pampean loess and sand deposits are the andesitic and basaltic rocks and tuff deposits in the northern Patagonia and the Andes Cordillera (Teruggi 1957). Initially, it was considered that particles were directly windblown from these areas (Teruggi 1957; Sayago 1995). However, several authors proposed different alternatives considering a first stage of fluvial transport of sediments that were later on deflated from the floodplain deposits fringing the Pampa (Gonzalez Bonorino 1965; Zárate and Blasi 1993; Iriondo 1990; Iriondo and Kröhling 1996). In any case, all authors agree that eolian transport promoted granulometric sorting of the sediments in the Pampa, resulting in decreasing grain size from the south-west to the north-east. The addition of volcanic ash also played a remarkable role in the formation of Pampean sediments (Zárate and Blasi 1993; Zárate 2003).

Following the consensus on a main Andean sediment source, it has been generally considered that Pampean sediments are mineralogically and chemically homogeneous, except for variable volcanic glass contents (Fidalgo et al. 1975; Imbellone and Teruggi 1993; Sayago et al. 2001). It has been long accepted that the clay fraction of the most recent Pampean sediments is mineralogically homogeneous. Several authors have reported that the dominant clay mineral in the surficial sediments and soils of the Pampa is illite, and this would also be the case of the sediments and soils in the Chaco region (González Bonorino 1965, 1966; Scoppa 1976; Camilión 1993; Iriondo and Kröhling 1996).

Nevertheless, other studies on the mineralogy of the sand fraction from surface sediments and soils in the northern Pampa and southern Chaco have revealed, in addition to volcanoclastic components of Andean origin, evident contributions from the Pampean hill range of Córdoba, generally more abundant to the west of these regions. In contrast, sedimentary contributions from the Paraná river basin have been identified in the eastern sectors of the Pampa and the Chaco (González Bonorino 1965, 1966; Bertoldi de Pomar 1969; Morrás and Delaune 1985; Iriondo and Kröhling 1996, 2007; Morrás 2003; Etchichury and Tófalo 2004). Local inputs from the hilly systems of Ventania and Tandilia have also been identified in the sand fraction of surficial sediments in the southern Pampa region (Fidalgo et al. 1991; Blanco and Sánchez 1994; Pereyra and Ferrer 1997). Geochemical studies have allowed delineating several areas in the Pampa and southern Chaco plains on the basis of phosphorus and potassium contents in bulk samples as well as in specific granulometric fractions of soils and sediments. These results, correlated with sedimentological and mineralogical information, have permitted inferring that sediments from the Andean cordillera, Pampean hill ranges, and the Paraná river basin converge in these regions, in addition to local inputs from the Ventania and Tandilia hill ranges (Morrás 1996, 1999; Morrás and Cruzate 2002). Mineralogical studies of the clay fraction in soils from northern Pampa (Stephan et al. 1977) and southern Chaco (Morrás et al. 1980, 1982) have shown a progressive increase of smectites to the east that can be related to mineral contributions from the Paraná basin. It is important to mention that in the Mesopotamian Pampa, in the easternmost part of the region, dark deep Vertisols have developed on Early Quaternary lacustrine sediments rich in smectites (González Bonorino 1966; Iriondo 1994; Morrás et al. 1993; Durán et al. 2011). Fluvial sediments and present soils in the valley of the Paraná river (Morrás 1998) as well as in the Post Pampean estuarine sediments and soils in the coast of Río de la Plata (González Bonorino 1966; Imbellone et al. 2010) are also rich in smectites (Fig. 11.2).

11.2.2   Sediments and Soils in the Eastern Part of the Rolling Pampa

The Rolling Pampa presents a well expressed granulometric zonation. Iñiguez and Scoppa (1970) mentioned three eolian strips of Andean origin differing in granulometry. These sediments would have been deposited in three successive periods, the finer of which are found close to the Paraná river. Kröhling (1999) evaluated the silt content (2–50 µm) in surface sediments in the High Rolling Pampa from data

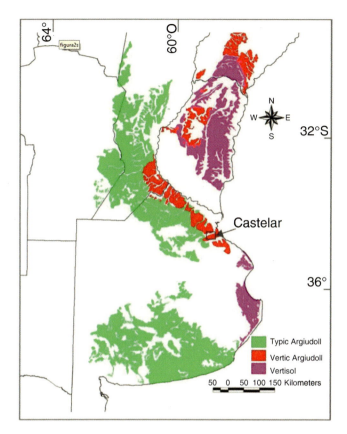

Fig. 11.2 Geographic distribution of Typic Argiudolls, Vertic Argiudolls, and Vertisols as main components of cartographic soil units in central Argentina

provided by soil maps. The silt content in these sediments considered of Andean origin progressively increased from SSW to NNE. The same author also indicated that ¨no other significant supply of sediments to the loess belt is evident from the maps, except for the smaller contributions from the Pampean ranges and occasional volcanic ash falls¨. More recently, Morrás and Cruzate (2000) evaluated the granulometry of the C horizon in about 1,400 soil profiles from the Rolling Pampa. Their classification according to the criteria proposed by Bidart (1992) for loessial sediments showed three main parallel strips from the Salado river to the NE: the first composed of sandy loess, the second composed of typical loess, and the third, the closest to the Paraná-Río de la Plata fluvial axis, characterized by the intercalation of typical and clayey loess.

Concerning the soils, small-scale maps show two parallel strips in the Rolling Pampa: one in the vicinity of the Paraná and Río de la Plata rivers where Vertic Argiudolls are dominant, and the other at some distance from the fluvial axis where Typic Argiudolls are the main soils (Etchevehere 1975) (Fig. 11.2). According to 1:50,000 scale soil maps, in the strip closer to the mentioned rivers, Vertic Argiudolls occur on interfluves and slope heads, while Typic Argiudolls appear on lower positions in the landscape.

With regard to soil mineralogy in the Rolling Pampa, González Bonorino (1966) considered that the mineral assemblages of the soils are wholly inherited from the parent materials and that illite is in many instances practically the only clay mineral in the soils in this area. In contrast, Iñiguez and Scoppa (1970) and Scoppa (1974) identified clay mineralogical variations in a west-east soil transect showing a progressive increase of smectite towards the Paraná river. As the authors assumed a common origin and a homogeneous composition of the parent material, they considered that mineralogical changes could only be explained by weathering processes, and that the higher smectite content eastwards, in vertic soils, would be the result of more intense neoformation in parent materials of finer grain size. No explanation was given about the processes that would drive the juxtaposition of Typic and Vertic Argiudolls in the landscape.

In spite of the pedogenic interpretation of the origin of smectitic clays in surface materials in the Rolling Pampa, some other studies on sediment-paleosol sequences carried out in excavations in the metropolitan areas of Buenos Aires and La Plata have revealed several sedimentary levels rich in smectite, particularly in the deep lower section of the Pampeano formation, whose source is in the Paraná basin (González Bonorino 1965; Riggi et al. 1986). Furthermore, a geotechnical study by Rimoldi (2001) in the city of Buenos Aires showed that not only deep sediments but also surficial loessic silts and present soils on higher topographic positions are expansive soils that may damage building foundations. Similarly, some paleosol sequences developed in the upper section of the Pampeano Formation that have been studied in different quarries near La Plata city revealed a high proportion or a predominance of smectitic clay in BC and C horizons (Teruggi and Imbellone 1987; Imbellone and Teruggi 1993; Blasi et al. 2001).

Thus, the traditional 'unicity and uniformity paradigm' about loess deposits, assumed as the foundation on which most investigations on Pampean soils are based, appear to be insufficient to explain some aspects of soil genesis and distribution in the Rolling Pampa, particularly in the framework of new studies revealing a heterogeneous mineralogical composition across the Chaco-Pampean plains (Morrás 1997, 2003). Consequently, more detailed investigations of sediments and soils have been carried out in a representative area of the Rolling Pampa.

11.3   Soil Parent Materials in the Castelar Area, South-Eastern High Rolling Pampa

11.3.1   Field and Laboratory Studies

Several studies have been carried out during the last years in the fields of the National Institute of Agricultural Technology (INTA) and its surroundings in Castelar, in the metropolitan area of Buenos Aires. This domain of about 650 ha is located in the proximity of the Reconquista river, i.e. at the boundary between High and Low Rolling Pampas. The landscape is dissected by waterways, and the sloping land surface is representative of the subregion (Fig. 11.3). A semi-detailed soil map shows Vertic Argiudolls on relief tops and slope heads and Typic Argiudolls on the lower slope segments and on the gently undulating to nearly level portions of the landscape at intermediate elevations (Gómez 1993).

A first mineralogical, geochemical, and magnetostratigraphic study was carried out on two main soil-sediment profiles named GAO and CAS (Nabel et al. 1999). The GAO profile is situated at the highest topographic position in the area at 22 m asl, in an excavation 6 m deep. A Vertic Argiudoll has developed on the surface of

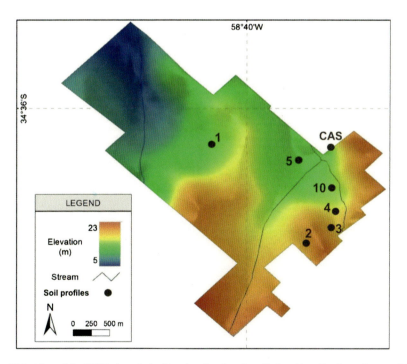

Fig. 11.3 Map of the INTA domain in Castelar showing the relief and localization of some of the profiles mentioned in the text. The GAO profile is outside the limits of the domain, at some distance to the south, at 22 m asl

the sedimentary sequence in a dull-orange, massive, silty-clayey sediment slightly pedogenized and with a high concentration of large calcareous nodules in its lower level. A laminar calcrete appears at 3.80 m depth, followed by a slightly structured sedimentary layer, coarser in texture than the surficial one. The CAS profile, at a mid-slope position at 17 m asl, shows a Typic Argiudoll followed by two laminar calcareous levels: one at 1.50 m depth is laminar and discontinuous, and has been correlated with the calcrete in the GAO profile; the second one appears at 4.10 m depth with a truncated paleosol atop (Fig. 11.4).

On the basis of particle size distribution, clay, silt, and sand mineralogy, magnetic susceptibility, and total Ti/Zr relationship, three different sedimentary units have been recognized in the studied profiles, which are here renamed I, S and LC. Briefly, in the GAO profile, the upper unit (S) is characterized by a high content of smectitic clay and a low content of fine quartzitic sand, while the lower unit (LC) is characterized by poorly crystallized 2:1 clay, and a high content of feldspar-rich sand. In the CAS profile, the upper part (Unit I) is characterized by a high content of illitic clay and small amounts of fine quartzitic sand, while the lower part of the profile is similar to the Unit LC in the GAO profile (Fig. 11.4).

Other analyses have confirmed the compositional differences between the surface sediments and the present soils in the GAO and CAS profiles. These studies found that the total contents of Fe, Mn, Ti, Cr, and Zn are higher in the soil at the GAO site (Unit S) than in the soil at the CAS site (Unit I), while the opposite has

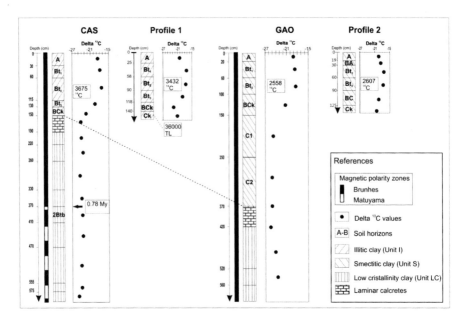

Fig. 11.4 Morphological and analytical characteristics of representative soil-sediment profiles in Castelar. The present soils in the CAS profile and profile 1 are Typic Argiudolls at intermediate topographic positions; the soils in the GAO profile and profile 2 are Vertic Argiudolls at the higher positions in the relief. For localization of profiles at the INTA field, refer to Fig. 11.3

been found for total Na (Morrás et al. 1998). Another mineralogical study of the coarse soil fractions has shown a higher content of volcanic glass shards and a higher proportion of feldspars relatively to quartz in the CAS profile than in the GAO profile (Liu et al. 2010). The content of free iron oxides and the magnetic properties of bulk materials and of lithogenic and pedogenic components differ in both profiles. For instance, the bulk magnetic susceptibility in the present CAS soil is three times higher than in the present GAO soil (Liu et al. 2010).

Additionally, soil mineralogical analyses have been carried out at numerous sites in the INTA fields. A clear relationship has been found between the soil clay composition and the cartographic and taxonomic soil units and their position in the landscape. In the upper positions of the higher interfluves, around 20 m asl, where Vertic Argiudolls are dominant, the BC and C horizons show a high proportion of smectites, together with a lower proportion of illite and traces of kaolinite. The Bt horizons the same composition, while the A horizons have mainly illite with some proportion of interstratified illite-smectite minerals. In the Typic Argiudolls found on lower slope facets and interfluves at intermediate elevations around 15 m asl, the clay fraction in the C horizon is composed of irregularly interstratified illite-smectite minerals, a similar proportion of illite, and traces of kaolinite. In the Bt horizons, the smectitic components increase slightly, while in the A horizon the predominant clay is illite (Morrás et al. 2002). Some soils with aquic and albic features in small local depressions at these mid-elevation positions show the same mineralogical composition as Typic Argiudolls. In other landscape positions, as well as

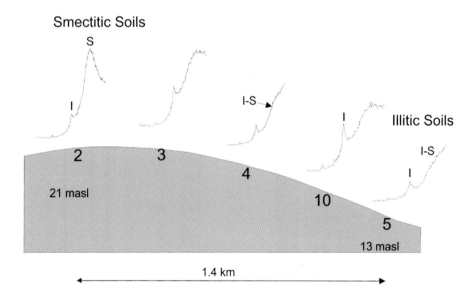

Fig. 11.5 Transect showing the relative position of selected soil profiles along the slope and the mineralogical composition of the clay fraction in the C horizons. Letters in the XRD diagrams stand for: *I* illite, *S* smectite, *I-S* irregularly interstratified illite-smectite. The localization of sites in the field is shown in Fig. 11.3

at intermediate slope facets or in the margins of small streams, the clay composition is generally intermediate between the smectitic and illitic types. Figure 11.5 shows the mineralogical change in the C horizon at five sites across a topographic transect 1.4 km long, from the top of the relief at 22 m asl where smectite dominates, to the footslope at 13 m asl where illite is the prevalent clay. Along the slope, the composition is heterogeneous, with variable proportions of smectite, illite, and interstratified I-S minerals.

Similarly, the soil magnetic susceptibility (MS) measured at a considerable number of sites in the INTA fields coupled with geostatistic data treatment has shown close spatial relationship between the position of the soils in the landscape and their magnetic values. Thus, in accordance with the first results obtained in the GAO and CAS sites, it has been observed that smectitic soils situated at the higher topographic positions, generally between 18 and 21 m asl, display the lowest MS values among the soils studied. Besides, MS values in the BC and C horizons of vertic soils are generally higher than in Bt and A horizons. In contrast, in the Typic Argiudolls on flat areas at intermediate elevations, generally between 13 and 15 m asl, MS values are higher than in vertic soils. In the BC and C horizons of these illitic soils, MS values are the same as or lower than in the corresponding Bt and A horizons. In other landscape positions and on slopes or in the vicinity of streams, MS values vary in space and with depth (Morrás et al. 2004a, b).

Also physical and chemical properties of soils representative of different taxonomic and cartographic units have been determined. Vertic Argiudolls have clearly higher CEC, COLE, and plasticity index values than Typic and Aquic Argiudolls, while several other properties such as water retention, water movement, porosity, and structural stability also vary with the mineralogical composition and landscape position of the soils (Castiglioni et al. 2005, 2006, 2007). Data on water retention capacity, soil bulk magnetic susceptibility, and clay mineralogical composition determined at a large number of sites in the INTA fields and GIS- processed showed to be spatially related (Morrás et al. 2004b). The highest moisture equivalent values were recorded in soils with high smectite content and low magnetic susceptibility located on higher topographic positions, while the opposite results were obtained for illitic soils on lower positions in the landscape.

11.3.2   Origin of Soil Parent Materials

There is evidence that the mineralogical differences between illitic and smectitic soil materials in the studied sector of the Rolling Pampa are not due to pedological processes as they were formerly interpreted. Firstly, the compositional differences in the clay fraction are clearly expressed in the C horizon of the present soils. Secondly, neat mineralogical differences are also found in the more stable sand and silt fractions, as well in the soil magnetic fraction all along the profiles. Thirdly, Typic and Vertic Argiudolls in INTA fields developed under the same bioclimatic conditions and on similar landscape positions although differing in elevation.

Finally, the relationship between soil mineralogy and topographic position is the reverse of the usual catenary sequence common in many environments, where smectitic clays develop downslope by neoformation (Bocquier 1973; Graham 2006; Morrás et al. 2009).

The results obtained show the existence of two surface sediments similar in lithology but clearly differing in mineralogy and physical and chemical properties. The illitic sediment rich in volcanic glass and with a high lithogenic magnetic signal should be originated from Andean sources, although there is evidence of some contribution from the Pampean ranges. In turn, the smectite-quartz-rich materials would be deflated from the neighboring alluvial plains of the Paraná river that carries sediments from an extensive basin to the north of Argentina (Morrás 1998). A palynofacies study at the GAO site revealed a high proportion of pollen from Myrtaceae in the upper smectitic section of the profile, which would reflect the input of wind-blown sediments from north-eastern Argentina and southern Brazil, an area where forests with these species develop (Grill and Morrás 2010). Zárate (2003) mentioned the Uruguayan shield as a potential source of sediments in the Rolling Pampa. Similarly to the proposal of Zárate and Blasi (1993) and Zárate (2003) for loessic sediments in the southern Pampa, a proximal source of smectitic sediments could be the paleo-floodplains of the Río de la Plata river that extend along the continental shelf as a consequence of marine regressions during glacial stages.

11.3.3 A Sedimentological and Landscape Evolution Model for the Rolling Pampa

A simplified reconstruction of the vertical and lateral relationship among the three sediments identified in the area of Castelar is represented in Fig. 11.6. Unit LC, characterized by loam or sandy loam texture, poorly crystallized clay, and high $CaCO_3$ content, occurs at the bottom of the studied profiles. Probably, this sediment crops out in lower landscape positions, in the floodplains of the small watercourses crossing the area. Following in the column, comes the loamy clay and smectite-rich Unit S that crops out at higher positions in the landscape and constitutes the parent material for vertic soils. Finally, the most recent sedimentary Unit I, silty clay loam and illitic, blankets most part of the terrain surface, including the backslopes of higher hills, and is the parent material for Typic Argiudolls.

Although the chronology of the deposits and the geomorphic processes are not yet well established, some dates are available that help frame the sequence of events. According to paleomagnetic data, the lowest part of Unit LC, i.e. the sediments with poorly crystallized clays below the paleosol found at 4 m depth in the CAS profile, belongs to the Matuyama Magnetic Polarity Zone (0.78–2.59 My) and is stratigraphically assigned to the Ensenada Formation (Nabel et al. 2005). In turn, the upper portion of Unit LC, also characterized by poorly crystallized clays, as well as the smectitic and illitic overlying sediments are in the Brunhes Magnetic Polarity

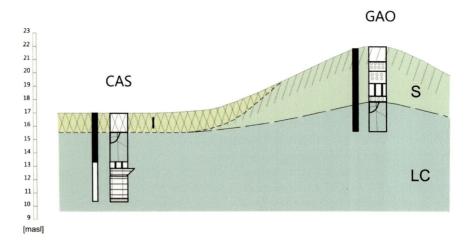

Fig. 11.6 Sketch showing the three sedimentary units (I, S, and LC) identified in the Rolling Pampa and their present vertical and horizontal relationships, as well as the relative position of the GAO and CAS profiles in the landscape. The *vertical black* and *white bars* represent magnetic polarity zones (See references in Fig. 11.4)

Zone (<0.78 My), deposited from the Middle Pleistocene onwards and stratigraphically assigned to the Buenos Aires Formation. TL dates of the C horizon in a Typic Argiudoll at 1.60 m depth gave an age of 36 ky BP (Zech W, written com.), which indicates that Unit I was deposited in the Late Pleistocene. Radiocarbon dates of Bt horizons in two Typic Argiudolls provided a mean age of 3,554 year BP and in two Vertic Argiudolls a mean age of 2583 year BP (AMS Laboratory, Arizona University, USA) (Fig. 11.4).

Based on the stratigraphic relationships between the three sedimentary units identified, a sequential geomorphic model is presented in Fig. 11.7. After the deposition of the smectitic loess sediments of Unit S, a climate change to humid conditions is assumed of having caused erosion and landscape incision, also cutting the calcretes in the uppermost part of Unit LC. Later, under dry climate in the Late Pleistocene, a loess mantle from western sources extends over all the landforms, giving rise to Unit I. Finally, in a new humid period in the Holocene, the highest crests and slopes would be eroded and smectite-rich Unit S exposed on the terrain surface. The ^{14}C age of the Bt horizons in the two Argiudolls seems to support this interpretation. Typic Argiudolls would be the first to develop on the illitic sediments blanketing the landscape, while Vertic Argiudolls would develop later, after the smectitic sediment was exhumed by erosion. The δ^{13}C values in the present soils (Fig. 11.4) correspond to a mix of C3 and C4 plants, indicating that all Argiudolls developed under an alternance of humid and dry periods (Morrás et al. 2007).

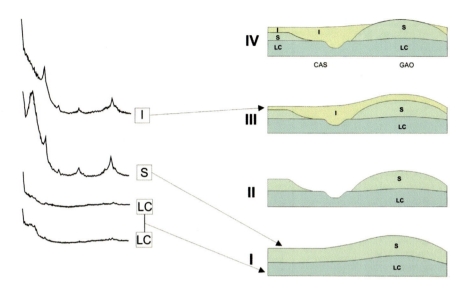

Fig. 11.7 Sequential model for loessial sedimentation and landscape formation in the Rolling Pampa from Early Pleistocene to present times. To the *left*, XRD diagrams representative of the clay fraction in the sedimentary units I, S, and LC. To the *right*, successive stages of deposition and erosion of sediments. **Stage I**: deposition of the smectitic sediment (Unit S) under dry climate during the Brunhes chron in the Late Pleistocene, above Unit LC of Early Pleistocene. **Stage II** : formation of a paleosurface by water erosion during a humid period. The incision in the landscape affects Units S and LC, including the calcretes at the top of Unit LC. **Stage III**: eolian deposition of Unit I blanketing the paleosurface, under dry climate during the Last Glacial. **Stage IV**: humid climate, water erosion and exhumation of Unit S at the higher topographic levels of the paleo-landscape in the Early-Middle Holocene. Finally, evolution up to the present includes pedogenesis and differentiation of Typic Argiudolls from the Vertic Argiudolls according to the mineralogical composition of the parent materials

11.4   Conclusions

During the last decades, studies of the Quaternary loess sediments covering the Pampa and Chaco regions have led to change the classic sedimentologic model to a more complex one, opening a wide and interesting panorama about the paleoenvi-ronmental evolution in these regions. Soil parent materials are more complex than originally thought, and this has led to modify the traditional pedogenic interpreta-tions, particularly the one concerning the development of vertic soils in the eastern sector of the Rolling Pampa. The study of soils in a representative area through a different geopedologic approach coupling soil-landscape evaluation with detailed mineralogical analysis offers a more accurate explanation about the development and spatial distribution of the main soils in the Rolling Pampa. The study carried out in Castelar may be considered as a pilot one to sustain detailed soil mapping and soil management in other catchments of the region. In a more general perspective and in accordance with recent considerations (Durán et al. 2011), the strip with dominant

Vertic Argiudolls and some Vertisols in the Rolling Pampa, instead of being interpreted as a pedogenic modification of south-western Andean materials in an extreme eastern position, may be visualized now as the western margin of an area including the Mesopotamian Pampa and a great part of Uruguay where vertic soils developed on smectitic clay sediments from several sources found in the Paraná basin.

References

Bertoldi de Pomar H (1969) Notas preliminares sobre la distribución de minerales edafógenos en la provincia de Santa Fe. Actas V Reunión Argentina de la Ciencia del Suelo, Santa Fe, pp 716–726

Bidart S (1992) Clasificación de los sedimentos eólicos del Pleistoceno Tardío-Holoceno del sur de la Provincia de Buenos Aires. Una propuesta. Actas Cuarta Reunión Argentina de Sedimentología II:159–166

Blanco M, Sánchez L (1994) Mineralogía de arenas en suelos loéssicos del sudoeste pampeano. Turrialba 4(3):147–159

Blasi A, Zarate M, Kemp R (2001) Sedimentación y pedogénesis cuaternaria en el noreste de la Pampa bonaerense: la localidad de Gorina como caso de estudio. Rev Asoc Argent Sedimentol 8(1):77–92

Bocquier G (1973) Genèse et évolution de deux toposéquences de sols tropicaux du Tchad. Mémoires ORSTOM 62, Paris

Camilion M (1993) Clay mineral composition of pampean loess (Argentina). Quat Int 17:27–31

Cappannini D, Dominguez O (1961) Los principales ambientes geoedafológicos de la Provincia de Buenos Aires. IDIA 163:33–39

Castiglioni M, Morrás H, Santanatoglia O, Altinier M (2005) Contracción de agregados de Argiudoles de la Pampa Ondulada diferenciados en su mineralogía de arcillas. Cienc del Suelo 23(1):13–22

Castiglioni M, Morrás H, Santanatoglia O (2006) Estudio de la porosidad de Argiudoles de la Pampa Ondulada mediante análisis digital de imágenes. Actas XX Congreso Argentino de la Ciencia del Suelo, Salta (edited in CD)

Castiglioni M, Morrás H, Santanatoglia O, Altinier M, Tessier D (2007) Movimiento del agua en Argiudoles de la Pampa Ondulada con diferente mineralogía de arcillas. Cienc del Suelo 25(2):109–121

Durán A, Morrás H, Studdert G, Xiaobing L (2011) Distribution, properties, land use and management of Mollisols in South America. Chin Geogr Sci 21(5):511–530

Etchevehere P (1975) Suelos. In: Relatorio Geología de la Provincia de Buenos Aires. VI Congreso Geológico Argentino, Bahía Blanca, pp 219–229

Etchichury M, Tofalo O (2004) Mineralogía de arenas y limos en suelos, sedimentos fluviales y eólicos actuales del sector austral de la cuenca Chaco-paranaense. Regionalización y áreas de aporte. Rev Asoc Geol Argent 59(2):317–329

Fidalgo F, de Francesco F, Pascual R (1975) Geología superficial de la Llanura Bonaerense. In: Relatorio Geología de la Provincia de Buenos Aires. VI Congreso Geológico Argentino, Bahía Blanca, pp 103–138

Fidalgo F, Riggi J, Gentile R, Correa H, Porro N (1991) Los "Sedimentos Pospampeanos" continentales en el ámbito sur bonaerense. Rev Asoc Geol Argent XLVI(3–4):239–256

Gómez L (1993) Carta básica semidetallada de suelos. Complejo de Investigaciones Castelar, INTA, Provincia de Buenos Aires. INTA-CIRN, Instituto de Suelos, Buenos Aires, p 114

González Bonorino F (1965) Mineralogía de las fracciones arcilla y limo del Pampeano en el área de la ciudad de Buenos Aires. Rev Asoc Geol Argent XX(1):67–148

Gonzalez Bonorino F (1966) Soil clay mineralogy of the Pampa plains, Argentina. J Sed Petrol 36(4):1026–1035

Graham R (2006) Factors of soil formation: topography. In: Certini G, Scalenghe R (eds) Soils: basic concepts and future challenges. Cambridge University Press, Cambridge, pp 151–163

Grill S, Morrás H (2010) Análisis palinofacial de sedimentos del Cenozoico Tardío en la Pampa Ondulada (Argentina): primeros resultados. Rev Brasileira Paleontol 13(3):221–232

Imbellone P, Teruggi M (1993) Paleosols in loess deposits of the Argentine Pampa. Quat Int 17:49–55

Imbellone P, Giménez J, Panigatti J (2010) Suelos de la Región Pampeana. Procesos de formación. Ediciones INTA, Buenos Aires, p 320

Iñiguez A, Scoppa C (1970) Los minerales de arcilla en los suelos zonales ubicados entre los ríos Paraná y Salado (Prov. de Buenos Aires). RIA Serie 3 VII (1):1–41

INTA (1989) Mapa de Suelos de la provincia de Buenos Aires, escala 1:500,000. Instituto Nacional de Tecnología Agropecuaria, Buenos Aires, p 533

Iriondo M (1990) Map of south American plains – its present state. Quat S Am Antartic Peninsula 6:297–308

Iriondo M (1994) Los climas cuaternarios de la Región Pampeana. Comunicaciones del Museo Provincial de Ciencias Naturales Florentino Ameghino, Santa Fe, 4(2):1–48

Iriondo M, Kröhling D (2007) Geomorfología y sedimentología de la cuenca superior del Río Salado (sur de Santa Fe y noroeste de Buenos Aires, Argentina). Lat Am J Sedimentol Basin Anal 14(1):1–23

Kröhling D (1999) Sedimentological maps of the typical loessic units in North Pampa, Argentina. Quat Int 62:49–55

Liu Q, Torrent L, Morrás H, Hong A, Jiang Z, Su Y (2010) Superparamagnetism of two modern soils from the northeastern Pampean region, Argentina and its paleoclimatic indications. Geophys J Int 183:695–705

Morrás H (1996) Diferenciación de los sedimentos superficiales de la región pampeana en base a los contenidos de fósforo y potasio. Actas VI Reunión Argentina de Sedimentología, Asociación Argentina de Sedimentología, Bahía Blanca, pp 37–42

Morrás H (1997) Origen y mineralogía del material parental de los suelos de la región pampeana. ¿Homogeneidad o heterogeneidad? 1° Taller de Sedimentología y Medio Ambiente. Asociación Argentina de Sedimentología, Buenos Aires, pp 19–20

Morrás H (1998) Mineralogía de arcillas de suelos de islas del Paraná Medio. Actas VII Reunion Argentina de Sedimentología, Salta, pp 194–202

Morrás H (1999) Geochemical differentiation of quaternary sediments from the pampean region based on soil phosphorous contents as detected in the early 20th century. Quat Int 62:57–67

Morrás H (2003) Distribución y origen de sedimentos superficiales de la Pampa Norte en base a la mineralogía de arenas. Resultados preliminares. Rev Asoc Argent Sedimentol 10(1):53–64

Morrás H, Cruzate G (2000) Clasificación textural y distribución espacial del material originario de los suelos de la Pampa Norte. Actas XVII Congreso Argentina de la Ciencia del Suelo, Mar del Plata (edited in CD)

Morrás H, Cruzate G (2002) Origen y distribución del potasio en suelos y sedimentos superficiales de la región Chaco-pampeana. In: Melgar R, Magen H, Lavado R (eds) El potasio en sistemas agrícolas argentinos INTA-IPI, Buenos Aires, pp 35–42

Morrás H, Delaune M (1985) Caracterización de áreas sedimentarias del norte de la provincia de Santa Fe en base a la composición mineralógica de la fracción arena. Cienc del Suelo 3(1–2):140–151

Morrás H, Postma J, Rapp M, Scoppa C (1980) Mineralogía de arcillas de algunos suelos del norte de la provincia de Santa Fe. Actas IX Reunión Argentina de la Ciencia del Suelo, Paraná, III: 1185–1191

Morrás H, Robert D, Bocquier G (1982) Caractérisation minéralogique de certains sols salsodiques et planosoliques du "Chaco Deprimido". Cah ORSTOM Sér Pédol XIX (2):151–169

Morrás H, Bayarski A, Benayas J, Vesco C (1993) Algunas características genéticas y litológicas de una toposecuencia de suelos vérticos en la Provincia de Entre Ríos (Argentina) In: Gallardo

J (ed) El estudio del suelo y de su degradación. Actas del XII Congreso Latinoamericano de la Ciencia del Suelo. Sociedad Española de la Ciencia del Suelo, Salamanca, II:1054–1061

Morrás H, Zech W, Nabel P (1998) Composición geoquímica de suelos y sedimentos loéssicos de un sector de la Pampa Ondulada. Actas Quintas Jornadas Geológicas y Geofísicas Bonaerenses, Mar del Plata, I:225–232

Morrás H, Altinier M, Castiglioni M, Grasticini G, Ciari G, Cruzate G (2002) Composición mineralógica y heterogeneidad espacial de sedimentos loéssicos superficiales en la Pampa Ondulada. XVIII Congreso Argentino de la Ciencia del Suelo, Pto. Madryn (edited in CD)

Morrás H, Altinier M, Castiglioni, M, Tessier D (2004a) Relación entre la mineralogía de arcillas y la susceptibilidad magnética en tres suelos del sur de la Pampa Ondulada. Actas XVIII Congreso Argentino de la Ciencia del Suelo, Paraná (edited in CD)

Morrás H, Ciari G, Grasticini C, Cruzate G, Altinier M, Castiglioni M (2004b) Variación espacial y relación entre la retención de humedad y la mineralogía magnética en suelos de la Pampa Ondulada. Actas XVIII Congreso Argentino de la Ciencia del Suelo, Paraná (edited in CD)

Morrás H, Moretti L, Hatté H, Zech W (2007) Perfiles de isótopos estables del Carbono en materiales Cenozoicos de las regiones Pampeana y Subtropical de la Argentina. In: Lázzari M, Videla C (eds) Isótopos estables en agroecosistemas. Ediuns, Universidad Nacional del Sur, Bahía Blanca, pp 157–162

Morrás H, Moretti L, Píccolo G, Zech W (2009) Genesis of subtropical soils with stony horizons in NE Argentina: autochthony and polygenesis. Quat Int 196:137–159

Nabel P, Pereyra F (2000) El paisaje natural bajo las calles de Buenos Aires. Museo Argentino de Ciencias Naturales Bernardino Rivadavia, Buenos Aires, p 124

Nabel P, Morrás H, Petersen G, Zech W (1999) Correlation of magnetic and lithologic features of soils and quaternary sediments from the Undulating Pampa. J S Am Earth Sci 12:311–323

Nabel P, Morrás H, Sapoznik M (2005) Magnetoestratigrafía de sedimentos cenozoicos en el oeste del gran Buenos Aires. Rev Asoc Geol Argent 60(2):383–388

Iriondo M, Kröhling D (1996) Los sedimentos eólicos del noroeste de la llanura pampeana. Actas XIII Congreso Geológico Argentino IV:27–48

Pereyra F (2003) Ecoregiones de la Argentina, vol 37. Anales SEGEMAR, Buenos Aires, p 191

Pereyra F, Ferrer J (1997) El material originario de los Molisoles de las Sierras Australes, Provincia de Buenos Aires, Argentina. Cienc del Suelo 15(2):87–94

Riggi J, Fidalgo F, Martínez O, Porro N (1986) Geología de los "sedimentos Pampeanos" en el Partido de La Plata. Rev Asoc Geol Argent XLI(3–4):316–333

Rimoldi V (2001) Carta geológico-geotécnica de la Ciudad de Buenos Aires. Servicio Geológico Minero Argentino (SEGEMAR), Buenos Aires (edited in CD)

Sayago J (1995) The Argentine neotropical loess: an overview. Quat Sci Rev 14:755–766

Sayago J, Collantes M, Karlson A, Sanabria J (2001) Genesis and distribution of the late Pleistocene and Holocene loess of Argentina: a regional approximation. Quat Int 76(77):247–257

Scoppa C (1974) The pedogenesis of a sequence of Mollisols in the Undulating Pampa (Argentine). Dr Sc thesis, Rijksuniversiteit Gent, Belgique, p 158

Scoppa C (1976) La mineralogía de los suelos de la llanura pampeana en la interpretación de su génesis y distribución. Actas VII Reunión Argentina de la Ciencia del Suelo, Bahía Blanca, IDIA, Suplemento 33, pp 659–673

Scoppa C, Vargas Gil J (1969) Delimitación de sub-zonas geomorfológicas en un sector de la región pampeana y sus relaciones edafogenéticas. Actas V Reunión Argentina de la Ciencia del Suelo, Santa Fé, pp 424–431

Soil Survey Staff (2010) Keys to soil taxonomy, 11th edn. Soil Survey Division USDA-Natural Resources Conservation Service, Washington, DC

Stephan S, De Petre A, de Orellana J, Priano L (1977) Brunizem soils of the central part of the Province of Santa Fe, Argentina. Pédologie XXVII(3):225–253

Teruggi M (1957) The nature and origin of Argentine loess. J Sed Petrol 27(3):322–332

Teruggi M, Imbellone P (1987) Palesosuelos loéssicos superpuestos en el Pleistoceno Superior-Holoceno de la región de La Plata, Provincia de Buenos Aires. Cienc del Suelo 5(2):175–188

Zárate M (2003) Loess of southern South America. Quat Sci Rev 22:1987–2006

Zárate M (2005) El Cenozoico Tardío continental de la Provincia de Buenos Aires. In: de Barrio R, Etcheverry R, Caballé M, Llambías E (eds) Relatorio del XVI Congreso Geológico Argentino, La Plata, pp 139–158

Zárate M, Blasi A (1993) Late Pleistocene-Holocene eolian deposits of the southern Buenos Aires Province, Argentina: a preliminary model. Quat Int 17:15–20

Chapter 12
Soil-Landform Relationships in the Arid Northern United Arab Emirates

C.F. Pain, M.A. Abdelfattah, S.A. Shahid, and C. Ditzler

Abstract The morphology and evolution of landforms, together with the materials of which they are composed, play a major role in the development and distribution of the soils in the northern United Arab Emirates, where landforms of aeolian origin in the west contrast with fluvial landforms in the east. These aeolian and fluvial landforms in turn contrast with a belt of coastal landforms along the Arabian Gulf and the Gulf of Oman. This chapter describes the landforms and soils of the Northern Emirates (NE), and shows how their form and evolution are closely related. At great group level (US Soil Taxonomy), the following soils were recognized: Torriorthents, Torripsamments, Haplocalcids, Haplocambids, Haplogypsids, Calcigypsids, Aquisalids, and Haplosalids. Twenty eight soil series were identified. Various combinations of these soil series were grouped into 42 map units, each consisting of two or more soil series and a number of minor soil types. At subgroup and family levels, these soils can be related to specific landform morphologies and processes.

Keywords United Arab Emirates • Aeolian soils • Alluvial soils • Arid soils • Soil-landform analysis

C.F. Pain (✉)
MED-Soil, Universidad de Sevilla, Sevilla, Spain
e-mail: colinpain@gmail.com

M.A. Abdelfattah
Faculty of Agriculture, Fayoum University, Fayoum, Egypt
e-mail: maa06@fayoum.edu.eg

S.A. Shahid
International Center for Biosaline Agriculture, Dubai, United Arab Emirates
e-mail: s.shahid@biosaline.org.ae

C. Ditzler
National Soil Survey Center, NRCS-USDA, Lincoln, NE, USA
e-mail: craig.ditzler@earthlink.net

© Springer International Publishing Switzerland 2016
J.A. Zinck et al. (eds.), *Geopedology*, DOI 10.1007/978-3-319-19159-1_12

12.1   Introduction

The development of landforms usually leads to the juxtaposition of different types of landforms and therefore different types of soils (Zinck 2013). This is a consequence of the influence of landforms on topography, soil parent material, and soil age. Geomorphology and geology, combined with time and climate, are the main factors that influence soil distribution. Understanding geomorphology is useful in understanding soil patterns. While the current arid climate suggests that wind erosion is the dominant factor shaping the geomorphology, this has not been always so, such as in wadis where water erosion has significant role in alluvial soil formation. At a detailed level there may be a close relationship between soil characteristics and position on a hillslope – the catena concept of Milne (1936). At a broader scale, different landform types will be formed of different materials, and be of different ages, and these factors will be reflected in the soil types present.

Soils in arid and semi-arid areas, especially those formed on depositional materials, tend to be very little modified from the original parent material (Dunkerley 2011). Nevertheless, some pedological alteration occurs. For example, dust falling on sand dunes contributes clay minerals to the material, and is found as thin clay coatings on sand grains. In other materials, especially in alluvium, calcium carbonate may form distinctive soil horizons where it cements the sediments. Near the coast, sea water intrusion introduces high salts in the soil leading to form marshlands and salt scalds locally called sabkha. The latter are devoid of any vegetation due to high salinity and near-surface water table of brine composition (Abdelfattah and Shahid 2007). Loose sandy material subject to aeolian movement creates various landforms: undulating, linear, transverse, and barchan sand dunes of different heights to over 200 m, as well as deflation plains. A number of dune formation periods probably occurred in the last 20,000–30,000 years, and older dunes now contain cores of sandstone. These aeolian processes of recent millennia have dominated the evolution of today's landscape. These and other processes tend to be controlled by materials that in turn are controlled by landforms.

The latter situation is the subject of this chapter, which introduces the Entisols and Aridisols that form the soils of the study area. It focuses on the landforms and soils of the Northern Emirates (NE), and describes the relationships between the two. It is based on data obtained during a soil survey of the area (EAD 2012).

12.2   Area and Methods

12.2.1   Regional Setting

The United Arab Emirates (UAE) is a federation of seven emirates in the south-east part of the Arabian Peninsula and adjacent to the Arabian Gulf (Fig. 12.1). It borders Oman and Saudi Arabia. The total area of the country is about 82,880 km². The NE

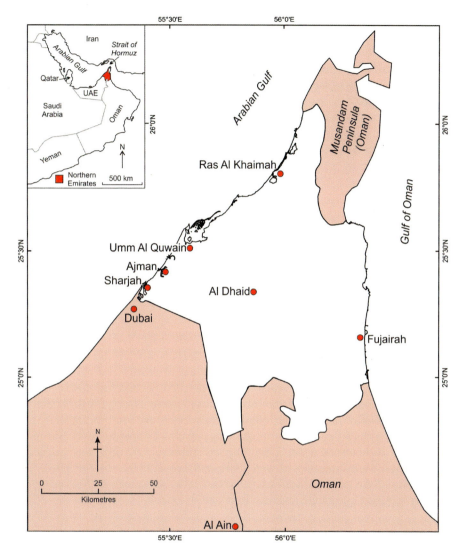

Fig. 12.1 The Northern Emirates and their location in the regional context of the Arabian Peninsula

consists of Sharjah, Ajman, Umm Al Quwain, Ras Al Khaimah, and Fujaira Emirates, and covers 6475 km², about 8.2 % of the country's surface area. The landscape is described in EAD (2012) and Pain and Abdelfattah (2015). It ranges from small areas of level coastal plains and sabkha to undulating desert sand plains, extensive areas of linear and transverse dunes, an alluvial plain up to 15 km wide, and mountainous rock outcrops along the Hajar Mountains. In the western part of the NE, linear dunes rise up to 100 m above the surrounding landscape, interspersed with small areas of almost level deflation plains and flats (Abdelfattah 2013a, b).

The UAE is in the Arabian Desert and is one of the hottest countries in the world. It has an arid climate with harsh dry summers, when temperatures regularly exceed 50 °C, and mild to warm winters with very little sporadic rainfall (80–160 mm in the NE) (Abdelfattah and Shahid 2007; Shahid and Abdelfattah 2008). The soil climate temperature regime in the NE is hyperthermic. There is a marked excess of evaporation over rainfall.

The oldest rocks in the NE are in the Hajar Mountains, where there are Permian to Cretaceous metamorphic, ophiolite, and sedimentary rocks including limestone (Styles et al. 2006). Surficial geology is dominated by Quaternary sediments, with aeolian dunes in the west and alluvial sediments on both sides and within wadis in the Hajar Mountains. The current shoreline of the NE consists of coastal lagoons, tidal flats, and marshes. The dunes and other sandy surficial materials are nowhere more than a few 10–100 s meters thick and overlie alluvial gravel inland and coastal and marine deposits near the coast (Fig. 12.2).

12.2.2 Data Collection

The study reported here was carried out during the Soil Survey of the NE (EAD 2012). Landforms were mapped from Google Earth images and a digital elevation model (DEM) derived from the Shuttle Radar Terrain Mission (SRTM), aided by reconnaissance fieldwork. Details of landform characteristics and materials were added during the soil survey, which involved detailed site and soil descriptions at 10,020 auger sites (2 m depth), 200 backhoe pits (2 m depth), and 150 drill observations (10 m depth). In sandy and silty materials augers and sand spears were used for routine observations, while in gravelly areas a Geoprobe corer was used (Geoprobe Systems, Kansas, USA http://geoprobe.com/). Soil and landscape descriptions were collected at every site and are available in the UAE Soil Information System at www.uaesis.ae (Abdelfattah and Kumar 2015). This information was used to compile final landform and soil maps.

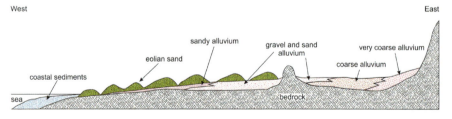

Fig. 12.2 Schematic cross section showing the main geological and geomorphic materials between the Hajar Mountains on the east and the coast on the west (not to scale)

12.3   Soils of the Study Area

12.3.1   Taxonomic Units

Twenty-eight soil series and one miscellaneous area (rock outcrop) were established according to the Soil Survey Manual (Soil Survey Staff 1993). They were allocated as components of 42 soil map units that make up the soil map of the Northern Emirates. The soil series are members of 2 soil orders, 6 suborders, 8 great groups, 13 subgroups, and 21 families as defined by the USDA Soil Taxonomy (Soil Survey Staff 1999). In addition to the Al Ain soil series, first identified in Abu Dhabi Emirate (EAD 2009; Shahid et al. 2013), 27 soil series were identified and described for the first time in this soil survey area. The soil orders are Aridisols and Entisols. The Aridisols are further divided into Calcids, Cambids, Gypsids, and Salids. The Entisols are divided into Orthents and Psamments (EAD 2012; Abdelfattah and Pain 2012; Abdelfattah 2013b). Most of the soils are either sandy or gravelly, but there is an important set of soils (Cambids) that are formed on fine alluvium in the area around Ras Al Khaimah. The collection of soil subgroups identified is shown in Fig. 12.3. Their relationships to landforms and parent materials are in Table 12.1.

12.3.2   Map Units

While the classification of soil profiles in the USDA soil classification is based on logical and hierarchical relationships between the different kinds of soils, map units reflect associations between soils in a landscape. A map unit will almost always include soil types that do not belong to the appropriate classification unit. These different soil types occur in areas that are too small to appear on the map; for example, soils on narrow floodplains in an area dominated by soils on aeolian sand dunes.

Soils were mapped at a scale of 1:50,000 (3rd order USDA level), with 42 map units being recognized. The name of a unit reflects the dominant soil or soils found within it, together with a general landscape characteristic that enables map units with similar soils to be separated on the basis of their landscape. Each unit typically consists of two or more soil series, or map unit components, together with a number of minor soil types. Each map unit description records the estimated proportion of each soil component and briefly summarizes the relationships between the components within that unit; the estimates were made from site observations, located between 500 and 1000 m apart. The individual map unit components consist of soil series described during the field survey.

Map unit descriptions were compiled on the basis of field observations of landscape patterns and an analysis of the soil and landscape classifications at sites within units described during the routine soil survey. The most common soil or soils were used to name the unit. Users of the information should be aware that each map unit will contain a wider range of soils than those described in the report, and individual

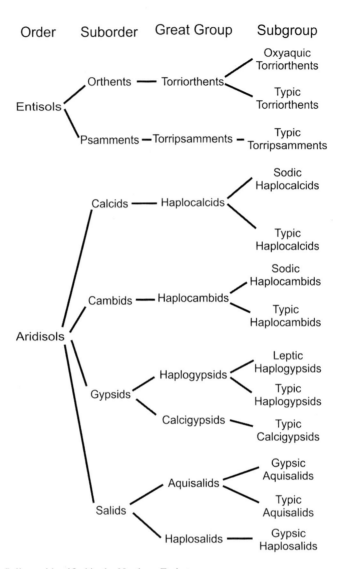

Fig. 12.3 Soil taxa identified in the Northern Emirates

delineations of the same unit, while having similar named soils, are likely to have a slightly different composition of minor and unreported soils.

Map unit characteristics, and interpretations made for different land uses for a unit, refer to the entire distribution of that unit unless specifically mentioned otherwise. Relative proportions of the named soil series may vary between delineations of individual map units, and minor soils may occur in all or only some of the unit delineations. Thus, the map and the definitions provide users with a guide to what they are likely to find in any particular part of the NE.

Table 12.1 Soil subgroups and their landforms and parent materials

Subgroups	Landforms	Parent materials
Oxyaquic Torriorthents	Coastal landforms	Sandy marine deposits with a thin eolian sand mantle
Typic Torriorthents	Floodplain, terrace fan	Alluvium (loamy sand to gravel, cobbles, and stones)
Typic Torripsamments	Sand dunes, floodplains within dunes	Eolian sands, alluvial sands
Sodic Haplocalcids	Floodplain, terrace fan	Loamy and sandy alluvium
Typic Haplocalcids	Floodplain, terrace fan	Gravelly alluvium
Sodic Haplocambids	Floodplain, terrace fan	Loamy alluvium
Typic Haplocambids	Floodplain, terrace fan	Loamy alluvium
Leptic Haplogypsids	Floodplain, terrace fan	Loamy, sandy and gravelly alluvium containing gypsum
Typic Calcigypsids	Floodplain, terrace fan	Loamy and gravelly alluvium containing gypsum as well as secondary calcium carbonate
Gypsic Aquisalids	Coastal landforms	Sandy, or sandy and loamy, marine deposits
Typic Aquisalids	Coastal landforms	Marine deposits over a lithified dune
Gypsic Haplosalids	Coastal landforms	Sandy, or sandy and loamy, marine deposits

12.4   Soil Forming Factors

The following sections describe the soil-forming factors that are related to geomorphology, in the context of the NE, and the processes that have contributed to the soil landscapes present today.

12.4.1   Parent Material

The nature of the parent material has a significant impact on the texture, mineralogy, and chemistry of the soils. Within the NE, parent material can be divided into three categories: aeolian sand, marine deposits in low-lying coastal areas, and alluvium derived from the various rock types of the Hajar Mountains. Each of these parent materials produces different soils depending on landscape position, climatic conditions, influence of plants and animals, and the amount of time these factors have had to alter the parent material.

12.4.1.1 Aeolian Sand

Pleistocene and Holocene aeolian sands occur throughout the western half of the NE in dunes and sand sheets. They consist of local coastal deposits, windblown sediments blown in from more distant areas, and older sediments derived from the then-exposed floor of the Arabian Gulf during drier glacial periods with lower sea levels. The aeolian sands of the NE are high in calcium carbonate equivalents (20 – >40 % by weight). The highest calcium carbonate contents are in the northern coastal areas, with a progressively higher proportion of silica sands and iron-oxides towards the mountains (White et al. 2001), giving the sands further inland a progressively redder color. Coastal areas are also influenced by additional windblown minerals, such as salt and gypsum, and often have an admixture of sea-shell pieces with the sand grains. The surface layers of the aeolian deposits are continuously being reworked, eroded, and re-deposited, and there is little opportunity for weathering and soil formation processes to occur, so horizons are only weakly developed. The Ajman Series (Typic Torriorthents) has formed in aeolian sands in a narrow band along the western coastal areas, while the slightly redder and coarser Sharjah Series (Typic Torripsamments) dominates the aeolian sands further inland (Fig. 12.4a).

12.4.1.2 Marine Deposits

On the coastal sabkha flats, the parent material consists mostly of recent sedimentary deposits of marine origin. Soil formation has been strongly influenced by the presence of near-surface saline groundwater and the accumulation of halite (sodium chloride salt), gypsum, and other soluble minerals that are moved upward through the soil profile by evaporation and then accumulate in the upper part of the profile. The Umm Al Quwain Series (Gypsic Aquisalids) is an example of a soil that formed in the coastal marine deposits parent material (Fig. 12.4b).

12.4.1.3 Alluvium

The level to gently undulating plains extending away from the Hajar Mountains are alluvial in origin, as evidenced by their stratified nature and inclusion of water-rounded pebbles. The size and amount of the pebbles are highest near the mountains (Fig. 12.4c) and decrease with distance from the mountains (Fig. 12.4d, e).

The coarsest soils formed in the gravelly and cobbly parent materials in the wadis within the mountain valleys (e.g. Bih Series – Typic Torriorthents). The plains between the mountain foothills and the edge of the aeolian sand dunes to the west also tend to be gravelly, but with fewer cobbles and smaller pebbles than in the mountain wadis (e.g. Al Dhaid Series – Typic Haplocalcids). Further east, the alluvial plain parent materials have been mostly covered by the younger aeolian sands of the dunes. In inter-dunal flats, the alluvial deposits are predominantly sandy with only a few pebbles mixed in, due to their greater distance from their mountain source.

Fig. 12.4 Examples of different parent materials in the NE. (**a**) thick aeolian sands of the Sharjah series (NE011); (**b**) marine deposits with shell fragments; (**c–e**) gravel deposits of the alluvial plains, getting finer with distance from the mountains; (**f**) fine-textured alluvium of the Ras Al Khaimah series (NE019); (**g**) fine alluvium in a wadi deposit

In the northern parts of the area, loamy alluvial sediments have been deposited in distal alluvial fans following transport from mountains upslope. These are some of the finest-textured parent materials in the NE (Fig. 12.4f) (e.g. Ras Al Khaimah Series – Typic Haplocambids). The present-day wadis extending away from the mountains and out into the desert are composed of thick, stratified, predominantly sandy parent materials of relatively recent alluvial origin. They are believed to be subject to rare flood events today (e.g. As Sirer Series – Typic Torriorthents) (Fig. 12.4g).

12.4.2   Climate

In the geopedologic context, climate contributes to soil formation in the NE mainly through its influence on vegetation, and by wind. The low vegetation density means that wind has direct access to the soil surface. Calcium carbonate tends to be present in dust that falls on the soils, and the limited precipitation has the effect of moving it downward with the wetting front, where it eventually precipitates in the soil, forming calcic and gypsic horizons (e.g. Al Kabkub Series – Typic Haplocalcids). In areas where a water table is present within 200 cm, dissolved minerals, such as sodium chloride salts and gypsum, are moved upward in the profile through evaporation. They accumulate in the soil to form gypsic and salic horizons. The Hisan Series (Gypsic Haplosalids) is an example of a soil with both a gypsic and a salic horizon formed in this way.

Sands from the soil surface at one location are blown off and deposited elsewhere. The result is minimal soil development due to the rapid loss or gain of soil material. On the gravely alluvial plains, wind removes the fine soil particles over time and leaves the heavier gravel behind, forming a pavement protecting the soil from further wind erosion (Fig. 12.5a).

12.4.3   Relief

Relief and topography affect soil development primarily be regulating the movement of water into and through the soil, and also by influencing the amount and intensity of sunlight that warms the soil. Convex landscape positions tend to shed water and limit infiltration, while concave positions tend to concentrate water flow and increase the potential for water infiltration. However, with the very low precipitation in the NE, these effects are limited. Slope steepness and aspect determine the amount and intensity of sunlight that hits the soil surface. South- and west-facing slopes tend to be warmer and drier than north- and east-facing slopes.

Fig. 12.5 (**a**) desert pavement after deflation, the removal of sand by wind; (**b**) salt crust near Umm Al Quwain

12.4.4 *Time*

Older soils tend to have more highly developed horizons relative to younger soils in similar environments. Soils in warm, arid environments, such as the NE, tend to develop horizons slowly compared to soils in other environments. Within the NE, the movement of windblown sands has not allowed any significant horizon development to occur in soils on the dunes and sand sheets, and these soils are the youngest and least developed in the area. More stable, older soils on the alluvial plains, especially those formed in loamy parent material, show more pronounced profile development, such as structure development, accumulations of salts, gypsum, and carbonates, and differences in the color of horizons.

12.5   Soil Forming Processes

Soil formation is the result of complex interactions of physical, chemical, and biological processes that occur in the soil over time. Despite their complexity, these
processes can be generalized into the four categories of additions, removals, transfers, and transformations (Simonson 1959). These are all related to a greater or
lesser extent to geomorphology. Examples of additions include the deposition of
sand and calcium carbonate-rich dust by wind and the accumulation of fine sediments on the surface of the soil as a result of periodic flooding and ponding in
wadis. Processes of removal include the deflation of some soil surfaces through the
action of wind, which removes the finer sand particles. This removal results in the
concentration of gravel on the soil surface and the formation of a desert pavement.
In the arid conditions of the NE, removal of soil materials from the soil profile by
leaching of water is uncommon and is generally restricted to areas under irrigation.
Transfer of materials in the soil can be seen through the dissolution of calcium carbonate (decalcification) and/or gypsum in the surface layer of the profile and their
downward movement with the wetting front and eventual precipitation and accumulation below in a calcic (calcification) or gypsic horizon (gypsification). Additionally,
transfer of materials in the soil is evidenced by the upward movement of saline
water driven by evaporation as moisture from a subsurface water table is drawn
upward and salts accumulate in the upper part of the profile. Transformation of soil
constituents is evidenced by the release of minerals to the soil as rock fragments
slowly weather in place. Also, in the soils of the coastal sabkha that have water
tables, iron has been chemically reduced and then oxidized to form reddish-colored
iron-accumulations in the soils.

12.5.1   Salinization

Salinization is the process responsible for the accumulation of soluble salts in the
soil profile. In the NE, it occurs due to the upward movement of solutes from an
underlying water table as water evaporates at the soil surface, and water rises
because of capillary suction. Accumulating salts result in the formation of a salic
horizon. This is a major process in soils of the coastal sabkha flats. In cases where
salinization is extreme, a salt crust a few centimeters thick covers the soil surface. A
polygonal pattern of soil cracking may develop in the crust as the salt crystals grow
and expand, causing surface heaving of a few centimeters in height (Fig. 12.5b).

12.5.2   Calcification and Gypsification

The processes responsible for the accumulation of calcium carbonate and/or gypsum in the soil are referred to as calcification and gypsification. Both calcium carbonate and gypsum are soluble in water and can therefore be relatively easily

dissolved, moved, and then re-precipitated within the soil (Fig. 12.6a). This most commonly occurs on landforms such as floodplains and terraces.

12.5.3   Aeolian Movement of Sand

Wind-blown sands blanket much of the landscape in the form of dunes and sand sheets. Some soils, such as the Al Madam Series (Typic Calcigypsids), are formed mostly in alluvial deposits, but have a cover of recently deposited aeolian sand a few tens of centimeters thick. Other soils, such as the Sharjah Series (Typic Torripsamments), are in areas of thick sand deposits on dunes and sand sheets and are formed entirely in aeolian sands (Fig. 12.6b). Still others, such as the Al Dhaid Series (Typic Haplocalcids), have had sand blown away from the surface, leaving a concentration of gravel armoring the surface and protecting it from further erosion by wind.

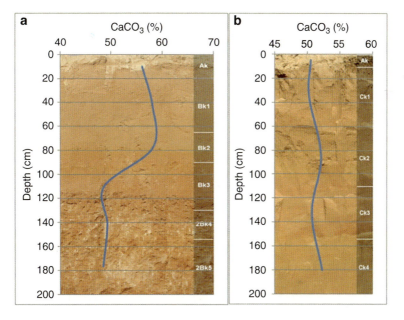

Fig. 12.6 (a) soil example (Al Kihef series, Typic Haplocalcids) with carbonate concentrating in the B horizon; (b) soil example formed on aeolian sand (Sharjah series, Typic Torripsamments) with minimal change in carbonate content with depth, reflecting little pedogenic alteration

12.6   Conclusions

Different landforms in the NE have different soil classes because of the influence of pedogenic processes such as salinization, decalcification and calcification, and gypsification. Although the rainfall is scanty in the desert environment, the recognition of calcic and gypsic horizons clearly demonstrates the operation of soil forming processes over a period of time. Coastal landscapes are dominated by Salids, the sand dunes and sand sheets by Psamments, and alluvial plains by Orthents and Calcids. The map units are based largely on combinations of different soil series that occur in specific geomorphic environments.

These conclusions demonstrate that a geopedology approach is just as important in arid and semiarid environments as it is in environments where soils are more strongly developed. The approach used here can be used in other arid areas where there is as yet a lack of detailed soil information. Such an approach is also important for land use planning because it provides a convenient and efficient way of obtaining soil and landform information.

References

Abdelfattah MA (2013a) Pedogenesis, land management and soil classification in hyper-arid environments: results and implications from a case study in the United Arab Emirates. Soil Use and Manage 29:279–294. doi:10.1111/sum.12031

Abdelfattah MA (2013b) Integrated suitability assessment: a way forward for land use planning and sustainable development in Abu Dhabi, United Arab Emirates. Arid Land Res Manage 27:41–64

Abdelfattah MA, Kumar AT (2015) A web based GIS enabled soil information system for the United Arab Emirates and its applicability in agricultural land use planning. Arab J Geosci 8:1813–1827. doi:10.1007/s12517-014-1289-y

Abdelfattah MA, Pain C (2012) Unifying regional soil maps at different scales to generate a national soil map for the United Arab Emirates applying digital soil mapping techniques. J Maps 8(4):392–405. doi:10.1080/17445647.2012.746744

Abdelfattah MA, Shahid SA (2007) A comparative characterization and classification of soils in Abu Dhabi coastal area in relation to arid and semi-arid conditions using USDA and FAO soil classification systems. Arid Land Res Manage 21:245–271. Available online: http://soils.usda. gov/technical/classification/taxonomy/. Last accessed 1 May 2015

Dunkerley DL (2011) Desert soils. In: Thomas DSG (ed) Arid zone geomorphology: process, form and change in drylands, 3rd edn. Wiley, New York, pp 101–129

EAD (2009) Soil survey of Abu Dhabi Emirate – extensive survey, vol 1. Environment Agency, Abu Dhabi

EAD (2012) Soil survey of the Northern Emirates. Three vol including soil maps. Environment Agency, Abu Dhabi

Milne G (1936) A provisional soil map of East Africa. East African Agriculture Research Station Amani Memoirs, Tanganyika Territory

Pain CF, Abdelfattah MA (2015) Landform evolution in the arid northern United Arab Emirates: impacts of tectonics, sea level changes and climate. Catena 134:14–29. doi:10.1016/j. catena.2014.09.011

Shahid SA, Abdelfattah MA (2008) Soils of Abu Dhabi Emirate. In: Perry RJ (ed) Terrestrial environment of Abu Dhabi Emirate. Environment Agency, Abu Dhabi, pp 71–91

Shahid SA, Abdelfattah MA, Othman Y, Kumar A, Taha FK, Kelley JA, Wilson MA (2013) Innovative thinking for sustainable use of terrestrial resources in Abu Dhabi Emirate through scientific soil inventory and policy development. In: Shahid SA, Taha FK, Abdelfattah MA (eds) Developments in soil classification, land use planning and policy implications: innovative thinking of soil inventory for land use planning and management of land resources. Springer, Berlin, pp 3–49

Simonson RW (1959) Outline of a generalized theory of soil genesis. Soil Sci Soc Am Proc 23:152–156

Soil Survey Staff (1993) Soil survey manual. Soil conservation service. US Department of Agriculture handbook 18. USDA National Soil Survey Centre, Lincoln, Nebraska

Soil Survey Staff (1999) Soil taxonomy: a basic system of soil classification for making and interpreting soil surveys, 2nd edn. US Department of Agriculture, Natural Resources Conservation Service. USDA handbook 436. USDA National Soil Survey Centre, Lincoln, Nebraska

Styles MT, Ellison RA, Arkley SLB, Crowley Q, Farrant A, Goodenough KM, McKervey JA, Pharaoh TC, Phillips ER, Schofield D, and Thomas RJ (2006) The geology and geophysics of the United Arab Emirates, vol 2: geology. British Geological Survey, Keyworth, Nottingham, and Ministry of Energy, UAE

White K, Goudie A, Parker A, Al-Farraj A (2001) Mapping the geochemistry of the northern Rub' Al Khali using multispectral remote sensing techniques. Earth Surf Proc Land 26:735–748

Zinck JA (2013) Geopedology. Elements of geomorphology for soil and geohazard studies, Special lecture notes series. ITC, Enschede

Chapter 13
Knowledge Is Power: Where Geopedologic Insights Are Necessary for Predictive Digital Soil Mapping

D.G. Rossiter

Abstract Much of current predictive digital soil mapping (PDSM) practice relies on terrain, climate, and remote sensing-derived covariates. These are easy to obtain and can serve as proxies to soil forming factors and from these to soil properties. However, mapping of soil bodies, not properties in isolation, is what gives insight into the soil landscape. A naïve attempt at correlating environmental covariates from current terrain, vegetation density, and surrogates for climate will not succeed in the presence of unmapped variations in parent material, soil bodies, and landforms inherited from past environments. Geopedology integrates an understanding of the geomorphic conditions under which soils evolve with field observations. Examples where simplistic DSM would fail but geopedology would succeed in mapping and, even better, explaining the soil distribution are shown: exhumed paleosols, low-relief depositional environments, and recent post-glacial landscapes.

Keywords Geomorphology • Predictive digital soil mapping • Soil-landscape relations • Pleistocene glaciation • Paleosols

13.1 Introduction

Digital Soil Mapping (DSM) is the term given to producing predictive soil maps by the use of mathematical models applied to field observations of soils and synoptic layers related to soil formation and distribution (McBratney et al. 2003); this was perhaps better named "predictive soil mapping" by Scull et al. (2003); the "digital" is a byproduct of current technology, and is meant to replace or extend the inductive reasoning of the expert soil mapper. The idea is to predict soil types or properties

D.G. Rossiter (✉)
Department of Crop and Soil Sciences, Cornell University, Ithaca, NY, USA
e-mail: dgr2@cornell.edu

© Springer International Publishing Switzerland 2016
J.A. Zinck et al. (eds.), *Geopedology*, DOI 10.1007/978-3-319-19159-1_13

over a landscape, based on some observations and a set of whole-field environmental covariates thought to be proxies for soil-forming factors. We restrict attention here to so-called 'scorpan'-based DSM, that is, where covariates are chosen to represent climate ('c'), organisms ('o'), relief ('r'), parent material ('p'), and time ('a'); known soils ('s') are used for calibration, and neighbourhood ('n') relations (i.e. local spatial correlation) may be used. In practice, 'r' (terrain) and 'o' as represented by vegetation indices or land use maps are the most widely-used covariates. The attraction of DSM is easy to understand: large areas can be covered with reduced field survey, the uncertainty shows the reliability of the map, and the models behind the predictions can be made explicit, often providing insight into soil geography. This is in contrast to previous approaches, which relied on the mapper's mental model of the soil landscape, spatialized by manual interpretation of aerial photographs (Farshad et al. 2013). The geopedologic approach of Zinck (2013) is the most theoretically-sound of these methods, because it is based on a systematic hierarchic soil-landscape analysis, not an ad hoc partitioning of the landscape based on perceived homogeneity.

Many digitally-produced soil maps are of single properties, notably soil organic C and particle-size distribution, sometimes showing the depth distribution (e.g. Liu et al. 2013) as specified by the GlobalSoilMap.net project (Arrouays et al. 2014). Geostatistical analysis based on point observations and correlation with spatially-complete covariates is well-suited for such mapping, although it provides no insight into soil geomorphology. By contrast, the geopedologic approach considers soil as a natural body with its own ecology and function. The actual soils form clusters in the very large potential space formed by each attribute taken separately, and the soil function can only be appreciated as a whole, much greater than the sum of its parts. Maps of these clusters, i.e. soil types, can then be interpreted for multiple uses, and in addition they form a sound basis for stratification in the mapping of single properties. Thus we restrict our attention to DSM efforts to map soil types. Some approaches start from existing maps, which implicitly contain rich geopedologic knowledge, and use digital methods to refine or update them (e.g. Kempen et al. 2009; Yang et al. 2011). We here consider the case where there is no existing soil-landscape map, only some point observations (usually purposive or opportunistic, not a probability sample) and a set of whole-field covariates.

DSM is the obvious soil mapping counterpart to similar data-driven approaches to knowledge in this computer age. Most current DSM models rely on terrain, climate, and vegetation intensity covariates. These are easy to obtain (see for example Hengl 2013) and can be used as proxies for soil forming factors; thus they are related via pedogenesis to many soil properties, and from the assemblage of properties to a soil type. An early statement of the hope of the digital soil mapper is from Zhu et al. (1996): "We assume that every soil series occurs under one or more typical environmental configurations or 'niches' and has a typical set of soil properties… can be characterized by a vector of environmental parameters in an m-dimensional parameter space". These authors reason by analogy in the so-called SoLIM (Soil-Landscape Inference Model) approach, which requires either a pre-existing map of soil types or expert knowledge of where each type occurs on the landscape. In some situations this hope has been justified, in others not.

Why is the 'scorpan' approach not always successful? Fundamentally, there is more to soil formation than the current environment. In particular, the soil forming factor 'time' is only approximately represented by landscape position, and the factor 'parent material' does not always have a close relation to topography. As early as 1935, Milne (1935) recognized that some east African toposequences (his 'catenas') developed on uniform parent rock, others on sequences of outcropping rocks. A direct correlation between soil types and slope positions was thus not possible. Variations in parent material (in the absence of a detailed surficial geology map) and the short time-scale of covariates compared with the time-scale of soil formation result in models that do not fully characterize the soil cover. In particular, soils may have inherited much of their current characteristics from previous climates and the associated vegetation, and indeed they may be the result of multiple cycles of soil formation. In younger landscapes, the topography may be relict from recent disruptions such as glaciation.

In this chapter, I give some examples where 'scorpan'-based DSM relying on the usual covariates will fail, but where geomorphic analysis results in successful landscape stratification, within which field observations can be placed, and will produce a reliable map. We consider three examples: exhumed paleosols, depositional low-relief environments, and young post-glacial landscapes. The last example is explained in detail. SoLIM approaches would also fail, except in the second example.

A separate issue is the complex and contingent nature of pedogenesis as evolution with continuously varying environmental conditions (Phillips 2001; Huggett 1998); this suggests that there is a chaotic, non-deterministic element to pedogenesis that cannot be inferred from observations of soils in similar niches. This is outside the scope of this chapter.

13.2 Analysis and Interpretation of Selected Examples

13.2.1 Exhumed Paleosols

Exhumed paleosols are soils, now at the surface or covered by a thin mantle of newer material that developed under a different climate than the present. They were then buried by new deposits, e.g. by a younger glacial till or loess, but then by landscape evolution (dissection, down wasting) exposed again at the surface. Their soil properties are largely controlled by conditions in the past, although of course now subject to current conditions for further evolution. A classic study is from Ruhe et al. (1967), who identified various glacial till, loess, and paleosol layers from four glacial and three interglacial stages in Iowa (USA). A detailed geomorphic investigation reveals, for example, relict fluvial surfaces (floodplain alluvium, slope fan alluvium) from the Sangamon interglacial which are now above the current base level where current fans and floodplains are located; further a relict pediment with stone line developed in Kansan till is mantled by a thin Wisconsin loess layer, and on the interfluves a modern soil developed in the loess but overlying a 'gumbotil' layer, i.e. very clayey weathered Kansan till. Some late Wisconsin-Recent slopes have cut back to interfluves, and on these erosional slopes Yarmouth-Sangamon paleosols outcrop, with younger soils above and below. These exhumed paleosols may also be truncated, so the paleo-B horizons are now at the surface.

How could DSM deal with this area? If a detailed soil map of an analogous area is available, this along with surveyor expert knowledge could be used in a SoLIM approach, but this is just a computer-assisted extension of knowledge-driven mapping. By contrast, the geomorphic analysis of Ruhe explains the soil distribution and provides a key for mapping. In geopedologic terms, the surfaces would be separated at the lithology level.

13.2.2 Depositional Low-Relief Environments

Soils in depositional low-relief environments such as fluvial systems with rapidly changing channels and variable infilling (e.g. the Rhine-Meuse delta of the Netherlands, see Berendsen 2005) cannot be mapped by interpolation, even with intensive boring campaigns, without geomorphic interpretation of the paleo-geography. Another example is the detailed study by Zinck (1987) of the alluvial and terrace soils associated with the Río Guarapiche in Monagas state, Venezuela. From the geomorphology, one can delineate various landscape components such as current and abandoned channels, backswamps, splay fans, and associate these with soil types. The relief is very subtle; vegetation differences can discover some of the differences, but only in areas where there has been no artificial drainage.

13.2.3 Young Post-glacial Landscapes

Large areas of northern North America and Europe are covered with soils developed in young post-glacial landscapes; smaller areas are from recent alpine glaciation. In these areas, the geomorphology and distribution of parent materials can only be understood by means of the detailed history of glaciation and deglaciation (e.g. proglacial lakes, outwash plains, sandurs) which have only an indirect relation with terrain variables. We illustrate this with an example from Tompkins and Tioga counties, New York State (USA).

Figure 13.1 is a fragment of the USGS 7.5′ 1:24,000 topographic map West Danby and Willseyville (NY) sheets. An analyst following the geopedologic approach would use stereo-pairs of remote-sensed images, e.g. airphotos, but even without stereo view the map clearly shows features that are immediately recogniz-

Fig. 13.1 Fragment of the USGS 7.5′ 1:24,000 topographic map West Danby and Willseyville (NY) sheets. Annotations are geomorphic features (*black numbers*) and sites where soils are discussed (*red letters*); see text

able to a trained analyst familiar with the Pleistocene history of the region (von Engeln 1961): (1) a terminal moraine of the Valley Heads stage, behind which are (2) hummocks and kettles from stagnating ice; (3) pro-moraine outwash terraces, breached on the E and NE margin by (4) post-retreat outflow channels which formed (5) outwash terraces transecting the end moraine; (6) truncated spurs and post-glacial incisions; (7) in the NE edge a high-level terrace formed above the moraine when it was blocking outflow; (8) high-level outflows from the main glacial tongue, when it was pressed up against the E margin; (9) post glacial fans from upland erosion; (10) a large kettle, now a shallow lake and swamp, in front of the centre of the moraine, corresponding to a large block of ice separated from the glacier.

Figure 13.2 shows the detailed soil survey of the same area, provided by the NRCS (USA) Web Soil Survey (http://websoilsurvey.sc.egov.usda.gov), here

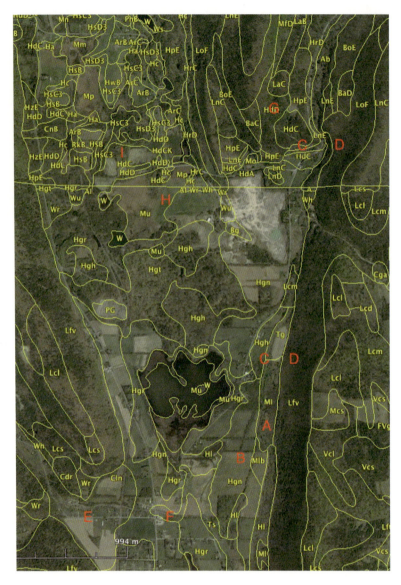

Fig. 13.2 Detailed soil survey of the area shown in Fig. 13.1, provided by the NRCS (USA) Web Soil Survey (http://websoilsurvey.sc.egov.usda.gov), displayed on a Google Earth background by the SoilWeb app (http://www.gelib.com/soilweb.htm). Annotations as in Fig. 13.1 See SoilWeb for map unit codes and descriptions

Table 13.1 Tentative geopedologic legend for the area shown in Figs. 13.1 and 13.2

Landscape	Relief type	Lithology	Landform	Soil series
Dissected plateau	Truncated ridge	Thin till from Devonian shales and mudstones	Convex summit	Lordstown
			Concave backslope	Volusia
		The same, plus outcropping bedrock	Straight, very steep front slope	Lordstown, Arnot
		Deep till	Side slopes	Langford
	Side valleys	Recent poorly-sorted alluvium from upland material	Narrow valley, moderate gradient	Chenango, coarser
			Alluvial fan	Chenango, finer
	Terraces	Glacial outwash	Dissected	Howard
			Flat	Howard
	Ice-margin complex	Dissected thin till and outcropping bedrock	Overflow channel (upland)	Valois
		Glacial outwash	Overflow channel (upland margin)	Howard, Valois
Through valley	End moraine complex	Wisconsonian poorly-sorted pushed material	End moraine	Howard, Palmyra
		The same, plus recent organic sediment and sorted fine sand	Hummocks and kettles	Howard, Arkport, saprists, water
		The same, plus recent organic sediment	Post-moraine lake and marsh	Saprists
	Outwash plain	Glacial outwash from end moraine material	Plain	Howard
		Ice-block inclusions	Pro-moraine kettle	Saprists, water
	Recent overflow channels	Alluvium	Flat-bottomed channel	Tioga, Middlebury

displayed on a Google Earth background by the SoilWeb application (Beaudette and O'Geen 2009; http://www.gelib.com/soilweb.htm).

Table 13.1 shows a tentative geopedologic legend for this area.

Referring to this figure, we identify several situations where a DSM approach using the usual covariates will not work, but where geomorphic knowledge results in an easy landscape interpretation:

- Positions A and B have identical slopes (flat), differ in elevation by less than one meter, are the same distance from streams, have almost the same wetness index, both are agricultural fields, yet the soils are quite different. A is mapped as the somewhat poorly-drained Middlebury (coarse-loamy Fluvaquentic Eutrochrepts) and well-drained Tioga series (coarse-loamy Dystric Fluvent Eutrochrepts), aggrading alluvial soils in silty and sandy alluvium from the present-day outlet of Michigan Creek, while B is mapped as the Howard series (loamy-skeletal Glossoboric Hapludalfs), a well-drained well-developed (considering the approximately 12 k years since the retreat of the glacier) gravelly loam from pro-glacial outwash, with about 30 % rock fragments, mostly rounded cobbles of mixed origin.

- Positions C and D (two examples) have identical very steep slopes and slope shapes (straight), both well vegetated with native hardwoods, yet the soils are radically different. C is again the Howard series, but truncated by the modern outlet of Michigan Creek to expose an outcrop of gravelly outwash, while D is mapped as the Lordstown series (coarse-loamy Typic Dystochrepts), a channery silt loam with about 20 % large to medium rock fragments from Devonian shale and mudstone; on the steepest slopes the soils are probably in the Arnot series (loamy-skeletal Lithic Dystochrepts).
- Positions E and F are adjacent, with similar terrain parameters, elevation and land use, but are easily recognized as a modern alluvial fan (9, E) and glacial outwash (10, F). Again, F is the Howard series; here E is mapped as Chenango (loamy-skeletal Typic Dystrudepts), a younger soil with periodic flash floods (e.g. due to hurricanes) resulting in additions of subrounded poorly sorted gravels (mudstone and sandstone) from the surrounding uplands.
- Position G is especially interesting. It is at a high elevation, has moderately steep slopes, is in native forest vegetation, yet is also mapped as the Howard series, i.e. it is glacial outwash, not soil in residuum as is the case of the surrounding Lordstown soils with the same topography and vegetation. The geomorphic clue here is outside the figure: Michigan Hollow (seen entering on the NE) is a through valley where the original drainage divide, about 5 km N, was removed by the glacier; subsequently as that tongue melted, a large amount of outwash was deposited in what was then a lake behind the terminal moraine (1). Apparently there were two levels; the higher one (G) was subsequently easily eroded by upland runoff; the lower terrace (between G and C) remains almost flat. The incision at (8) is also explained by a period where the ice filled the valley (NW in the figure) so that meltwater had to follow this channel to produce some of the outwash (5). The W margin of this hill shows the same phenomenon but from when the ice had melted enough to allow water to flow along its margins at the base of the truncated spur.
- Positions H and I differ by only 30 m elevation, are both flat, both with dense vegetation; yet while I is again mapped as Howard (glacial outwash), H is mapped as Typic and Terric Medisaprists, i.e. an organic soil. Geomorphically this is easy to understand: both positions are part of the kettle moraine (2). Some similar positions to I are mapped as Arkport (coarse-loamy Psammentic Hapludalfs); these are further behind the end moraine where meltwater was sandier.

Although 'scorpan'-based DSM would not be able to find these differences, some other approaches might have some success. To do so, they would have to emulate the geopedologic interpretation. For example, it might be possible to identify post-glacial alluvial fans by their relative landscape position: where narrow steep side valleys emerge onto outwash plains. Also, their shape is diagnostic: narrow at the proximal (upstream) end, widening at the distal end. These might be revealed by a segmentation, which then considered adjacency and oriented (proximal-distal) shape relations. However, the boundary between the fan and the outwash which it overlays (E vs. F) is quite subtle. Although visible to the geopedologist, it seems difficult to delineate automatically. The difference between C and D might be revealed by total slope length and position on the slope.

Another covariate that might have some success is hyperspectral remote sensing, which might allow vegetation communities to be distinguished (e.g. between positions H and I).

13.3 Discussion: What Is the Place of the Geopedologic Approach in DSM?

DSM methods are important additions to the soil mapper's toolkit, especially when large areas need to be mapped, and when estimates of uncertainty are needed. In simple landscapes with close correlation between topographic parameters, land use, and soil type, it has shown good success. However as the above examples show, there are situations where a geomorphic understanding is necessary to identify locations where each soil type is expected.

Object-oriented image segmentation applied to stacks of terrain parameters (Dragut et al. 2009) offers a digital approach to discovering landscape units, which can perhaps be interpreted and correlated to soil types. However several segments may have similar landscape parameters, yet be of contrasting origin, for example, alluvial terraces vs. glacial outwash terraces.

This begs the question as to whether geomorphology, as opposed to geomorphometry, can be digitally mapped (Bishop et al. 2012). If so, the digital geomorphic map could be used as a powerful covariate for digital soil mapping; perhaps geopedology would not be necessary. The most promising method so far is object-oriented analysis, followed by geomorphometric characterization (Hengl and Reuter 2008), leading, it is hoped, to interpretable terrain units. However Bishop et al. (2012) are clear on the limitations: "Although this scale-dependent approach is conceptually pleasing, it is nonetheless fundamentally a cartographic approach to mapping that does not formally address issues of processes, internal and external forcing factors, feedback mechanisms and systems, or spatio-temporal dynamics." In other words, geomorphology, and hence geopedology, is not simply terrain analysis, no matter how sophisticated. Evans (2012) has a similarly pessimistic view of the prospects for automated geomorphic mapping.

13.4   Conclusion

There are situations where neither DSM nor geopedology will be successful, and where intensive systematic field observation is the only way to map important soil differences. An example is given by Toomanian (2013) of a playa in the Zayandeh-rud valley, Iran, where an aeolian mantle creates a uniform surface; this mantle covers a wide diversity of aeolian, lagoonal, and alluvial layers deposited during the Quaternary and Tertiary. The geomorphometry is uniform, the soil surface reflectance and vegetative cover as well. Although surface salinization can be detected, this is not related to important subsurface differences. There is no solution but to grid sample and interpolate. But for many soil landscapes, the integration of geomorphic understanding and its relation to soil genesis allows successful mapping, where simple environmental correlation using 'scorpan' covariates as presumed proxies for soil-forming factors is not successful.

References

Arrouays D, Grundy MG, Hartemink AE et al (2014) Global soil map. Adv Agron 125:93–134

Beaudette DE, O'Geen AT (2009) Soil-web: an online soil survey for California, Arizona, and Nevada. Comput Geosci 35:2119–2128. doi:10.1016/j.cageo.2008.10.016

Berendsen HJA (2005) Landschappelijk Nederland: de fysisch-geografische regio's, 3rd edn. Uitgeverij Van Gorcum. Assen, The Netherlands

Bishop MP, James LA, Shroder JF Jr, Walsh SJ (2012) Geospatial technologies and digital geomorphological mapping: concepts, issues and research. Geomorphology 137:5–26. doi:10.1016/j.geomorph.2011.06.027

Dragut L, Schauppenlehner T, Muhar A et al (2009) Optimization of scale and parametrization for terrain segmentation: an application to soil-landscape modeling. Comput Geosci 35:1875–1883

Evans IS (2012) Geomorphometry and landform mapping: what is a landform? Geomorphology 137:94–106. doi:10.1016/j.geomorph.2010.09.029

Farshad A, Shrestha DP, Moonjun R (2013) Do the emerging methods of digital soil mapping have anything to learn from the geopedologic approach to soil mapping and vice versa? In: Shahid SA, Taha FK, Abdelfattah MA (eds) Developments in soil classification, land use planning and policy implications. Springer, Dordrecht, pp 109–131

Hengl T (2013) WorldGrids. http://worldgrids.org/doku.php. Accessed 28 Nov 2014

Hengl T, Reuter HI (eds) (2008) Geomorphometry: concepts, software, applications, vol 33, Developments in Soil Science. Elsevier, Amsterdam

Huggett R (1998) Soil chronosequences, soil development, and soil evolution: a critical review. Catena 32(3–4):155–172

Kempen B, Brus DJ, Heuvelink GBM, Stoorvogel JJ (2009) Updating the 1:50,000 Dutch soil map using legacy soil data: a multinomial logistic regression approach. Geoderma 151:311–326

Liu F, Zhang GL, Sun YJ et al (2013) Mapping the three-dimensional distribution of soil organic matter across a subtropical hilly landscape. Soil Sci Soc Am J 77:1241–1253. doi:10.2136/sssaj2012.0317

McBratney AB, Mendonça Santos ML, Minasny B (2003) On digital soil mapping. Geoderma 117:3–52

Milne G (1935) Some suggested units of classification and mapping, particularly for east African soils. Soil Res Suppl Proc Int Soc Soil Sci 4:183–198

Phillips JD (2001) Contingency and generalization in pedology, as exemplified by texture-contrast soils. Geoderma 102(3–4):347–370

Ruhe RV, Daniels RB, Cady JG (1967) Landscape evolution and soil formation in southwestern Iowa, Technical bulletin. US Department of Agriculture, Washington, DC

Scull P, Franklin J, Chadwick O, McArthur D (2003) Predictive soil mapping: a review. Prog Phys Geogr 27:171–197

Toomanian N (2013) Fundamental steps for regional and country level soil surveys. In: Shahid SA, Taha FK, Abdelfattah MA (eds) Developments in soil classification, land use planning and policy implications. Springer, Dordrecht, pp 203–227

von Engeln OD (1961) The finger lakes region: its origin and nature. Cornell University Press, Ithaca

Yang L, Jiao Y, Fahmy S et al (2011) Updating conventional soil maps through digital soil mapping. Soil Sci Soc Am J 75:1044–1053. doi:10.2136/sssaj2010.0002

Zhu AX, Band LE, Dutton B, Nimlos T (1996) Automated soil inference under fuzzy logic. Ecol Model 90:123–145

Zinck JA (1987) Aplicación de la geomorfología al levantamiento de suelos en zonas aluviales y definición del ambiente geomorfológico con fines de descripción de suelos. Instituto Geográfico "Augustín Codazzi", Bogotá

Zinck JA (2013) Geopedology. Elements of geomorphology for soil and geohazard studies, ITC special lecture notes series. Faculty of Geo-information Sciences and Earth Observtion, University of Twente, Enschede

Chapter 14
Geopedology, a Tool for Soil-Geoform Pattern Analysis

A. Saldaña

Abstract The soilscape is the pedologic portion of the landscape. Soil scientists have examined it mainly within the field of soil landscape analysis, which traditionally regards a quantitative characterization of the spatial pattern and complexity of soil landscapes. Landscape ecology emphasizes the interaction between spatial pattern and ecological process, that is, the causes and consequences of spatial heterogeneity across a range of scales. The spatial component of the environment is crucial in any environmental analysis, and new approaches to soil patterns are necessary for appropriate landscape planning, management, and conservation. Therefore, the integration of landscape ecology fundamentals together with soil science principles can be helpful in this regard. This contribution deals with the principles of soilscape-pattern analysis complemented with the application of landscape ecology metrics. An example of the application of this type of analysis to the Jarama-Henares interfluve, central Spain, is presented.

Keywords Soilscape • Geopedologic units • Landscape metrics • Diversity • Fractal dimension

14.1 Introduction

The soil is an essential component of ecosystems (Yaalon 2000) and has been considered as a background of the ecosystem (Klink et al. 2002). Soil is a very dynamic system which performs many functions and delivers services vital to human activities and ecosystems survival. Good-quality information from soil and land resource survey is necessary for wise natural resource management (McKenzie et al. 2000; Sanchez et al. 2009). However, more attention has been devoted to the biotic part than the abiotic component of ecosystems. Thus, new approaches to soil spatial patterns are necessary for appropriate landscape planning, management, and conservation.

A. Saldaña (✉)
Departamento de Ciencias de la Vida – UD Ecología, Facultad de Biología, Ciencias Ambientales y Química, Universidad de Alcalá, Alcalá de Henares, Madrid, Spain
e-mail: asuncion.saldana@uah.es

© Springer International Publishing Switzerland 2016   239
J.A. Zinck et al. (eds.), *Geopedology*, DOI 10.1007/978-3-319-19159-1_14

Several definitions of landscape are available in the literature. From an ecological point of view, landscape has been defined as a combination of elements that vary in size, shape, and arrangement, and are under the continuous influence of natural and anthropogenic events (Forman and Godron 1986; Krummel et al. 1987; Turner 1990; Turner and Gardner 1991). It has also been defined as an area that is spatially heterogeneous in at least one factor of interest (Turner et al. 2001, 2005). From a pedological point of view, the soil landscape (soilscape) is the pedologic portion of the landscape (Buol et al. 1973; Fridland 1974; Hole 1978). It is made up of multi-polypedonic units delineated at different scales, and, at whatever scale of general ization, may be characterized by its internal make-up and relationship to surrounding soilscapes (Hole 1978). Soilscapes are among the most complex and intricate of all physical landscapes and have been examined mainly within the field of soil landscape analysis (Hupy et al. 2004, and references therein). Traditionally, the latter has involved a quantitative characterization of the pattern and complexity of soil landscapes (Fridland 1974, 1976; Hole 1978; Hole and Campbell 1985).

Geomorphology and pedology have evolved as separate disciplines until the 1960–1970s. Modern research has shown the close relationship between soils and landforms (Huggett 1975; Birkeland 1984; Gerrard 1992; Zhu et al. 2001; Scull et al. 2005; Ziadat 2005). In some instances soils are considered to help interpret the evolution of landscape elements (e.g. Birkeland 1984); in other cases, the point is how geomorphology can help understand the genesis, evolution, distribution, and mapping of soils. Huggett (1975) proposed the concept of the soil-landscape system, i.e. any landscape unit in which landforms and soils, and the geomorphic and pedologic processes which create them, are seen as a whole. This concept was designed to link soil and geomorphic processes in a landscape, a topic pursued by pedologists with a geomorphic orientation and geomorphologists interested in soils (Huggett 1995, and references therein). In 1992 Hudson proposed the so-called soil-landscape paradigm in which soil-landscape units are narrower defined than landforms; they can be thought of as a landform further modified by soil-forming factors. Although the concept was adopted by many scientists, Hudson acknowledged a major weakness, i.e. an important dependence on tacit knowledge. A geopedologic approach could help compensate this weakness for the application of the soil-landscape paradigm (Zinck 2013).

The present contribution deals with the quantification of soil-landscape relationships. It is structured in three parts: the first one focuses on the soil cover pattern; the second one deals with soil mapping and the qualitative and quantitative analysis of soilscape, within an ecological landscape approach; and the third part presents an example of the integration of landscape ecology fundamentals together with soil science principles to analyze the complexity of the soilscapes in the Jarama-Henares interfluve, central Spain.

14.2   The Soil Cover Pattern

Soil cover is a more or less regular spatial arrangement of different kinds of soil bodies and associated bodies of not-soil (Hole and Campbell 1985). Soils genetically linked to various degrees produce a definite pattern (*struktura* in Fridland's terminology) in the soil mantle (Fridland 1965, 1974, 1976). Nevertheless, the soil cover pattern is distinguished from the zonal or regional soil pattern that is expressed by a gradual change in soil over large areas resulting from climatic gradients (Fridland 1976).

The pedological structure or pattern of the landscape regards the size, shape, and arrangement of component soil bodies (Hole and Campbell 1985). The idea of a structured soil mantle was developed by several Russian pedologists such as Sibirtsev, Dokuchaev, Ivanova, Gerasimov, and particularly Fridland (1976).

According to Fridland, an elementary soil area (ESA) is the simplest soil cover element. It is a soil formation unit free from any internal pedogeographic boundary and with variable size. The ESAs can be: (1) homogeneous (i.e. soils belonging to one classification unit of the lowest rank, occupying a space that is bounded on all sides by other ESAs or not-soil formations); or (2) heterogeneous (i.e. formed by contrasting soil pedons), either as sporadically or as regular-cyclic soil ESAs. ESAs assemble to constitute soil combinations, which are formed by spatially and genetically related ESAs. According to the character of genetic links between the components and the degree of contrast, Fridland defined the following six classes of soil combinations: complexes, patches, catenas, variations, mosaics, and tachets (Fridland 1974). ESAs may be described according to their content, geometry, place in soil combinations, and ecology. Soil combinations are characterized by their composition, genetic and geometric forms, differentiating factors, history of development, degree of stability, complexity and contrast (Fridland 1974, 1976). In a similar fashion, Hole and Campbell (1985) distinguished entities such as elementary soil body, simple combinational soil body, complex combinational soil body, and very complex combinational soil body. They address several items in soil landscape analysis: setting, scale factor, principal kinds of patterns on a plan view, origin of the soil cover pattern, and measurements for soil bodies and soil cover description.

14.3   Quantifying Soil-Landscape Relationships

14.3.1   Soil Mapping and the Geopedologic Approach

Detailed soil information is indispensable for land resources management and environmental modelling (Gobin et al. 2001; Ziadat 2005; Sanchez et al. 2009). New geographic information technologies (GPS, GIS, remote sensing, etc.) have created opportunities in soil mapping by providing critical spatial information and new methods to analyze data. The change of paradigm in soil mapping is leading to: (1) an increase in the types of spatial information available as base maps; (2) the replacement of base maps by GPS as the primary source of positional referencing; (3) the decoupling of the relationships between different aspects of map scale (Miller and Schaetzl 2014); and (4) the emergence of digital soil mapping (McBratney et al. 2003; Boettinger et al. 2010).

Although conventional soil survey remains as the main source of soil spatial information (Zhu et al. 2001; Scull et al. 2003; Sanchez et al. 2009), it has been criticized on the following grounds: it relies on a qualitative analysis of landscapes where expert knowledge is essential; polygons often portray soil variation as being discontinuous; the format and detail of conventional soil maps are not compatible with other data derived from detailed digital terrain analyses and remote sensing techniques, among others (Hudson 1992; McBratney et al. 1992; McKenzie et al. 2000; Zhu et al. 2001; Scull et al. 2005; Ziadat 2005; Sanchez et al. 2009; Esfandiarpoor et al. 2010). Nevertheless, digital soil mapping methods, which are based on soil-landscape relationships using geomorphometry and land cover as predictors, perform poorly in low relief areas such as alluvial and coastal plains (Zhao et al. 2014, and references therein). According to Zinck (2013), geopedology aims at supporting soil survey, combining pedologic and geomorphic criteria to establish soil map units and analyze soil distribution on the landscape. Geomorphology provides the contours of the map units (i.e. the container), while pedology provides the taxonomic components of the map units (i.e. the content). Therefore, the geopedologic map units are more than the conventional soil map units, since they also contain information on the geomorphic context in which soils are found and have developed. Esfandiarpoor et al. (2009) analyzed the effect of survey density on the results of applying the geopedologic approach to soil mapping and concluded that this approach works satisfactorily in reconnaissance or exploratory surveys. The same authors, using statistical and geostatistical methods, concluded that the geopedologic soil mapping approach is not completely satisfactory for detailed mapping scales (Esfandiarpoor et al. 2010). In both cases, to increase the accuracy of the geopedologic results at large scales, the authors suggested to add the category of landform phase, which is already included in the geopedologic approach (Zinck 1988, 2013). Rossiter (2000) considers that the geopedologic approach is adequate for semi-detailed studies (scales 1:35,000–1:100,000).

14.3.2 Soil Pattern Analysis

Soilscape analysis provides an important link between pedology and geomorphology. It is typically performed using soil maps, which are usually the primary data source. The analysis assumes that the soil mapper has accurately represented the pedogenic pattern in a map (Schaetzl 1986). A variety of metrics to quantitatively describe and evaluate the soilscape have been used related to: (1) density (e.g. mean density of soil bodies, the count of the mean number of soil boundaries intersected by a transect of unit length, etc.); (2) composition (e.g. the count of soil bodies, the proportionate extents of components, etc.); (3) diversity (e.g. the number of soil map legend units per unit area, the index of heterogeneity, etc.); (4) size and shape (e.g. the soil body shape index, the coefficient of dissection, etc.). It should be noticed that after a promising initial period, little research has been done on soils pattern analysis in the last decades (e.g. Schaetzl 1986; Saldaña 1997; Hupy et al. 2004; Saldaña et al. 2011). More effort has been posed on the pedometrics and diversity of soil landscapes (Hupy et al. 2004, and references therein; see also Chap. 10 by Ibáñez and Pérez in this book).

14.3.3 An Ecological Approach to Spatial Pattern Analysis

Landscape ecology is a subdiscipline of ecology that emphasizes the interaction between spatial pattern (i.e. the amount and configuration of something within an area) and ecological processes (e.g. disturbance, fragmentation, connectivity, etc.), that is, the causes and consequences of spatial heterogeneity across a range of scales. Landscape ecology focuses on four aspects of landscape systems: (1) the evolution and dynamics of spatial heterogeneity, i.e. how the landscape mosaic is created and how it changes; (2) the interactions between and exchanges of materials across heterogeneous landscapes, i.e. how materials and organisms move from one patch to another; (3) the influence that the spatial heterogeneity of the landscape mosaic has upon biotic and abiotic processes in the landscape; and (4) the management of spatial heterogeneity (Turner 1989, 2005; Turner et al. 1989; Wu and Hobbs 2002; Wu 2013). As expected, landscape ecology increasingly relies on remote sensing data and GIS allowing the integration of data obtained at different spatial scales.

Landscape pattern or structure denotes spatial heterogeneity. The latter has two components: the amount of different possible entities and their spatial arrangements (i.e. composition and configuration in terms of landscape ecology) (Fahring 2005). Landscape pattern indices are common tools of landscape ecologists, affording comparisons of different study areas or the same study area at different times (Corry 2005). Many features and measures (patch size and shape, heterogeneity, diversity, neighborhood, and interaction) have already been implemented in commercial or free-of-charge software. Nevertheless, three issues deserve some attention: (1) several authors (e.g. Wu and Hobbs 2002; Haines-Young 2005; Turner 2005) warned that contemporary work on pattern has mainly focused on the analysis or description of spatial geometry and has failed to explore relationships between pattern and processes; (2) Riitters et al. (1995) and Li et al. (2005) showed that none of the available indices is appropriate for all aspects of a landscape pattern; and (3) landscape ecologists deal with processes occurring at a wide range of spatial and temporal scales. So it is important to carefully define both spatial and temporal scales because they influence the conclusions drawn by an observer, and whether the results can be extrapolated to other times and locations (Turner et al. 2001).

Therefore, there is no "best" scale at which to study the environment but the appropriate scale depends on the research question at hand (Wiens 1989; Levin 1992; Noss 1992). Thus, an exploration of a broad spectrum of spatial and temporal scales is recommended (Wiens et al. 1986; Wu 2004).

14.4 An Example of Soilscape Analysis at Different Scales in Central Spain

A quantitative analysis of the soil-geoform patterns in the Jarama-Henares interfluve, central Spain, can be found in Saldaña (1997) and Saldaña et al. (2011). The general aim of this research was to analyze the structure and evolution of the soilscape in the Jarama-Henares interfluve and Henares River valley during the Plio-Quaternary. Several techniques, including classical statistics, numerical classification, fuzzy sets, geostatistics, and soilscape pattern analysis were used (Saldaña 1997). Some results derived from the quantification of the soil-geoform patterns in the area by means of soilscape and landscape metrics are offered hereafter.

Three main physiographic units have been identified in the area: (1) a calcareous plateau of Plio-Villafranchian age, (2) a piedmont lying at the foot of the Paleozoic range of Ayllón and Alto Rey that consists of relatively flat water-divide surfaces and Raña surfaces, and (3) the valleys of the Jarama and Henares rivers and two tributaries, the Torote and Camarmilla rivers. The Henares River valley is asymmetric, with 20 topographic benches along its right bank and a series of incised glacis-terraces on its left bank. This area has attracted the attention of many researchers for a long time, with main focus on geology, geomorphology, and vegetation. Soils were addressed from the 1950s onwards as a mean to understand the landscape evolution (for more details, see Medina (1977) and Saldaña (1997), and references therein).

Geomorphic units of the area were delineated either by photo-interpretation or extracted from existing maps using the geopedologic approach proposed by Zinck (1988). Soil characterization of the geomorphic units resulted in the establishment and mapping of the geopedologic units forming the soilscape. In this way, two geopedologic maps were prepared at regional (1:50,000) and local (1:18,000) scales, respectively.

The first step of the soil pattern analysis regarded a qualitative description of the patterns following Fridland (1976) and Hole and Campbell (1985) (i.e. material patterns, form patterns, and microclimate patterns), and the landscape model by Forman and Godron (1986) (i.e. patches, corridors, and matrix). Then, a quantification of the geopedologic combinations present in the maps, at the two scales, was done using 21 indices including soil pattern and landscape metrics, selected from the literature. The latter were grouped into four main sets:

(a) Density measures, regarding the density of the geopedologic units (mean density of soil bodies, index of heterogeneity, and mean density of soil map units).
(b) Diversity indices, which take into account the frequency of geopedologic units within a given area (landscape richness, spatial richness, landscape diversity, spatial diversity, dominance, fragmentation, mean number of nodes, and degree of chronological uniformity).
(c) Neighborhood and interaction indices, focusing on the relationships between adjacent map units (isolation of a soil map unit, interaction among patches, juxtaposition, binary comparison matrix, number of different classes, and center versus neighbors).
(d) Size and shape indices resulting from geomorphic and pedologic processes (patch size, soil map unit shape index, fractal dimension, and mean soil boundary length).

Diversity indices at the regional scale yielded the highest values for the oldest valleys in the area. This can be understood because they account for the largest number of units derived from the river deposition and later dissection. In addition, in the Henares River valley, diversity index values decreased from young to old landscapes. Thus, these indices can be used as indicators of soilscape evolution.

Size and shape indices provided information about geopedologic units. For example, at the regional scale, the highest values of fractal dimension corresponded to the Raña surfaces, i.e. large and flat surfaces of upper Pliocene where Palexeralfs and Palexerults were described. They have been strongly dissected and show contorted shapes. Thus, the intensity of dissection counteracts the normal trend of soil evolution towards homogeneization of the soil mantle. At the local scale, the highest fractal dimension values were for the most dissected landscape, i.e. the piedmont made up of a series of incised glacis-terrace levels on the left bank of the Henares River. These results suggest that size and shape indices are good indicators of terrain stability and relief dissection.

The contribution of the neighborhood and interaction indices to the understanding of soil and landscape evolution was limited given that it was very difficult to establish a trend because the values were evenly distributed at both scales. All results were scale-dependent, because the values of the indices were higher at the local scale. For example, the maximum value of spatial diversity was 0.278 for a low terrace of the Henares River, at the regional scale, while a value of 0.299 was obtained for the vales located on the left bank of the Henares River, at the local scale. The maximum value of the fractal dimension at the regional scale was 1.462 for the terrace scarp of the Henares River, while it was 1.595 for the vales on the left bank of the Henares River, at the local scale.

Saldaña (1997) looked for correlations among pattern indices, at both regional and local scales, to identify redundancies and select the most efficient ones (Table 14.1). At the regional scale, nine indices were efficient (i.e. landscape richness, spatial diversity, dominance, fragmentation, isolation of a soil map unit, interaction among patches, center versus neighbor, patch size, and the fractal dimension), while at the local scale seven indices were found to be efficient (i.e. spatial diversity,

Table 14.1 Correlation matrix of metrics at the regional scale

Metrics	1	2	3	4	5	6	7	8	9	10	11	12	13
1	1.00	0.31	0.06	0.42	-0.11	0.19	0.45	0.23	**1.00**	0.00	-0.24	0.20	**0.58**
2	0.31	1.00	**-0.53**	**0.73**	-0.04	0.19	**0.78**	**0.95**	0.31	0.14	-0.11	0.03	0.19
3	0.06	**-0.53**	1.00	0.19	0.06	-0.03	**-0.68**	**-0.76**	0.06	0.02	0.08	0.19	0.18
4	0.42	**0.73**	0.19	1.00	0.00	0.20	0.36	0.49	0.42	0.18	-0.06	0.18	0.36
5	-0.11	-0.04	0.06	0.00	1.00	-0.44	0.00	-0.06	-0.11	0.19	0.08	-0.07	-0.13
6	0.19	0.19	-0.03	0.20	-0.44	1.00	0.07	0.16	0.19	-0.14	0.00	0.10	0.20
7	0.45	**0.78**	**-0.68**	0.36	0.00	0.07	1.00	**0.84**	0.45	0.09	-0.17	-0.26	-0.02
8	0.23	**0.95**	**-0.76**	0.49	-0.06	0.16	**0.84**	1.00	0.23	0.10	-0.12	-0.05	0.08
9	**1.00**	0.31	0.06	0.42	-0.11	0.19	0.45	0.23	1.00	0.00	-0.24	0.20	**0.58**
10	0.00	0.14	0.02	0.18	0.19	-0.14	0.09	0.10	0.00	1.00	-0.01	0.03	0.05
11	-0.24	-0.11	0.08	-0.06	0.08	0.00	-0.17	-0.12	-0.24	-0.01	1.00	0.40	0.16
12	0.20	0.03	0.19	0.18	-0.07	0.10	-0.26	-0.05	0.20	0.03	0.40	1.00	**0.87**
13	**0.58**	0.19	0.18	0.36	-0.13	0.20	-0.02	0.08	**0.58**	0.05	0.16	**0.87**	1.00

Correlation >0.5 and significant at 95 % probability in bold

Extracted from Saldaña (1997)

1 spatial richness; 2 spatial diversity; 3 dominance; 4 fragmentation; 5 isolation; 6 interaction; 7 juxtaposition; 8 binary comparison matrix (BCM); 9 number of different classes (NDC); 10 centre versus neighbour (CVN); 11 patch size; 12 map unit shape index; 13 fractal dimension

dominance, fragmentation, isolation of a soil map unit, interaction among patches, patch size, and fractal dimension).

14.5 Discussion and Conclusions

Soil mapping is an important part of soil science. The two approaches to soil mapping, traditional soil survey and digital soil mapping, have advantages but also some drawbacks. Although they have often been presented as incompatible in the literature, Hengl and Rossiter (2003) and Zinck (2013) have shown that the geopedologic approach could be useful also in digital soil mapping.

The qualitative description of the soil patterns (Fridland 1976; Hole and Campbell 1985) provides a soil-geoform classification according to parent material, geomorphic or anthropic origin, relief, composition of soil combinations, and shape of the units. This can be used as a general framework for ecological studies.

The quantitative analysis of the geopedologic patterns showed that some metrics are useful to identify trends in the soilscape evolution (Saldaña 1997). In addition, the correlation among some indices implies that not all are necessary to quantitatively characterize the soilscape of the Jarama-Henares interfluve. The most appropriate ones are: spatial diversity, dominance, fragmentation, isolation of a soil map unit, interaction, patch size, and fractal dimension. The spatial diversity and the diversity indices can be used as indicators of soil evolution in the landscape, while shape indices, in particular the fractal dimension, are useful indicators of terrain stability and relief dissection. Other authors have also shown that numerous correlations occur among landscape pattern indices (Riitters et al. 1995; Cain et al. 1997).

The influence of scale on spatial patterns has been acknowledged and is a critical issue in many sciences, mainly those that study phenomena embedded in space and time. This holds now when geographic technologies are widely used for environmental analysis (Goodchild 2011; Miller and Schaetzl 2014). The example presented here (Saldaña 1997) shows that all metrics were scale-dependent, with higher values obtained at the local scale. Besides, the number of indices required to describe appropriately the soilscape patterns was smaller at the local than at the regional scale. This can be explained because the soilscape is richer in geopedologic units at the local than at the regional scale given that more detail is depicted. Other researchers (e.g. Turner et al. 1989) have already reported that spatial patterns found on a large scale map may even vanish at small scales. Hupy et al. (2004) quantified the amount of additional soil information that can potentially be gained by mapping at larger scales and concluded that time and money to map at a larger scale should be dedicated to those soilscapes that would show the greatest increase of information.

Acknowledgements This work was funded by projects CGL2010-18312 from CICYT of Spain and S2009AMB-1783-REMEDINAL-3 from the Madrid Autonomous Government. I am grateful to the reviewer for his useful suggestions that improved an earlier version of this contribution.

References

Birkeland PW (1984) Geomorphology and soils. Oxford University Press, Oxford

Boettinger JL, Howell DW, Moore AC, Hartemink AE, Kienast-Brown S (2010) Digital soil mapping. Bridging research, environmental application, and operation. Springer, Dordrecht

Buol SW, Hole FD, McCracken RJ (1973) Soil genesis and classification. Iowa State University Press, Ames

Cain DH, Riitters K, Orvis K (1997) A multi-scale analysis of landscape statistics. Landsc Ecol 12:199–212

Corry RC (2005) Characterizing fine-scale patterns of alternative agricultural landscapes with landscape pattern indices. Landsc Ecol 20:591–608

Esfandiarpoor I, Salehi MH, Toomanian N, Mohammadi J, Poch RM (2009) The effect of survey density on the results of geopedological approach in soil mapping: a case study in the Borujen region, Central Iran. Catena 79:18–26

Esfandiarpoor I, Mohammadi J, Salehi MH, Toomanian N, Poch RM (2010) Assessing geopedological soil mapping approach by statistical and geostatistical methods: a case study in the Borujen region, Central Iran. Catena 82:1–14

Fahring L (2005) When is a landscape perspective important? In: Wiens JA, Moss M (eds) Issues and perspectives in landscape ecology. Cambridge University Press, Cambridge, pp 3–10

Forman RTT, Godron M (1986) Landscape ecology. Wiley, New York

Fridland VM (1965) Makeup of the soil cover. Sov Soil Sci 4:343–354

Fridland VM (1974) Structure of the soil mantle. Geoderma 12:35–41

Fridland VM (1976) Pattern of the soil cover (*Struktura pochvennogo pokrova*, originally in Russian, 1972). Geographical Institute of the Academy of Sciences of the USSR. Israel Program for Scientific Translation, Jerusalem

Gerrard J (1992) Soil geomorphology. An integration of pedology and geomorphology. Champman and Hall, London

Gobin A, Campling P, Feyen J (2001) Soil-landscape modelling to quantify spatial variability of soil texture. Phys Chem Earth Part B: Hydrol Oceans Atmos 26:41–45

Goodchild MF (2011) Scale in GIS: an overview. Geomorphology 130:5–9

Haines-Young R (2005) Landscape pattern: context and process. In: Wiens JA, Moss MR (eds) Issues and perspectives in landscape ecology. Cambridge University Press, Cambridge, pp 103–111

Hengl T, Rossiter DG (2003) Supervised landform classification to enhance and replace photo-interpretation in semi-detailed soil survey. Soil Sci Soc Am J 67:1810–1822

Hole FD (1978) An approach to landscape analysis with emphasis on soils. Geoderma 21:1–23

Hole FD, Campbell JB (1985) Soil landscape analysis. Rowman and Allanheld Publishers, Totowa

Hudson B (1992) The soil survey as paradigm-based science. Soil Sci Soc Am J 56:836–841

Huggett RJ (1975) Soil landscape systems: a model of soil genesis. Geoderma 13:1–22

Huggett RJ (1995) Geoecology. An evolutionary approach. Routledge, London

Hupy CM, Schaetzl RJ, Messina JP, Hupy JP, Delamater P, Enander H, Hughey BD, Boehm R, Mitroka MJ, Fashoway MT (2004) Modeling the complexity of different, recently deglaciated soil landscapes as a function of map scale. Geoderma 123:115–130

Klink HJ, Potschin M, Tress B, Tress G, Volk M, Steinhardt U (2002) Landscape and landscape ecology. In: Bastian O, Steinhardt U (eds) Development and perspectives of landscape ecology. Kluwer Academic Publishers, Dordrecht, pp 1–47

Krummel JR, Gardner RH, Sugihara G, O'Neill RV, Coleman PR (1987) Landscape patterns in a disturbed environment. Oikos 48:321–324

Levin SA (1992) The problem of pattern and scale in ecology. Ecology 73:1943–1967

Li XZ, He HS, Bu RC, Wen QC, Chang Y (2005) The adequacy of different landscape metrics for various landscape patterns. Pattern Recognit 38:2626–2638

McBratney AB, De Gruijter JJ, Brus DJ (1992) Spacial prediction and mapping of continuous soil classes. Geoderma 54:39–64

McBratney AB, Santos MLM, Minasny B (2003) On digital soil mapping. Geoderma 117:3–52

McKenzie NJ, Gessler PE, Ryan PJ, O'Connell DA (2000) The role of terrain analysis in soil mapping. In: Wilson JP, Gallant JC (eds) Terrain analysis: principles and applications. Wiley, New York, pp 245–265

Medina A (1977) Evolución de los suelos en el valle del Henares. PhD thesis, Complutense University of Madrid, Madrid

Miller BA, Schaetzl RJ (2014) The historical role of base maps in soil geography. Geoderma 230–231:329–339

Noss RF (1992) Issues of scale in conservation biology. In: Fieldler PL, Jain SK (eds) Conservation biology: the theory and practice of nature conservation, preservation, and management. Chapman & Hall, New York, pp 239–250

Riitters K, O'Neill RV, Hunsaker CT, Wickham JD, Yankee DH (1995) A factor analysis of landscape pattern and structure metrics. Landsc Ecol 10:23–39

Rossiter DG (2000) Methodology for soil resource inventories, Lecture notes. International Institute for Aerospace Survey and Earth Sciences (ITC), Enschede

Saldaña A (1997) Complexity of soils and soilscape patterns on the southern slopes of the Ayllón range, central Spain. A GIS-assisted modelling approach. International Institute for Aerospace Survey and Earth Sciences (ITC), Enschede

Saldaña A, Ibáñez JJ, Zinck JA (2011) Soilscape analysis at different scales using pattern indices in the Jarama–Henares interfluve and Henares River valley, central Spain. Geomorphology 135:284–294

Sanchez PA, Ahamed S, Carré F, Hartemink AE, Hempel J, Huising J, Lagacherie P, McBratney AB, McKenzie NJ, Mendonça-Santos ML, Minasny B, Montanarella L, Okoth P, Palm CA, Sachs JD, Shepherd KD, Vågen TG, Vanlauwe B, Walsh MG, Winowiecki LA, Zhang G (2009) Digital soil map of the world. Science 325:680–681

Schaetzl RJ (1986) Soilscape analysis of contrasting glacial terrains in Wisconsin. Ann Assoc Am Geogr 76:414–425

Scull P, Franklin J, Chadwick OA, McArthur D (2003) Predictive soil mapping: a review. Prog Phys Geogr 27:171–197

Scull P, Franklin J, Chadwick OA (2005) The application of classification tree analysis to soil type prediction in a desert landscape. Ecol Model 181:1–15

Turner MG (1989) Landscape ecology: the effect of pattern on process. Annu Rev Ecol Syst 20:171–197

Turner MG (1990) Spatial and temporal analysis of landscape pattern. Landsc Ecol 4:21–30

Turner MG (2005) Landscape ecology: what is the state of the science? Annu Rev Ecol Evol Syst 36:319–344

Turner MG, Gardner RH (1991) Quantitative methods in landscape ecology: an introduction. In: Turner MG, Gardner RH (eds) Quantitative methods in landscape ecology. Springer, New York, pp 3–14

Turner MG, O'Neill RV, Gardner RH, Milne BT (1989) Effects of changing spatial scale on the analysis of landscape pattern. Landsc Ecol 3:153–162

Turner MG, Gardner RH, O'Neill RV (2001) Landscape ecology in theory and practice. Pattern and process. Springer, New York

Wiens JA (1989) Spatial scaling in ecology. Funct Ecol 3:385–397

Wiens JA, Addicott JF, Case TJ, Diamond J (1986) The importance of spatial and temporal scale in ecological investigations. In: Diamond J, Case TJ (eds) Community ecology. Harper & Row, New York, pp 145–153

Wu JG (2004) Effects of changing scale on landscape pattern analysis: scaling relations. Landsc Ecol 19:125–138

Wu JG (2013) Key concepts and research topics in landscape ecology revisited: 30 years after the Allerton Park workshop. Landsc Ecol 28:1–11

Wu JG, Hobbs R (2002) Key issues and research priorities in landscape ecology: an idiosyncratic synthesis. Landsc Ecol 17:355–365

Yaalon DH (2000) Down to earth. Nature 407:301

Zhao M, Rossiter DG, Li D, Zhao Y, Liu F (2014) Mapping soil organic matter in low-relief areas based on land surface diurnal temperature difference and a vegetation index. Ecol Indic 39:120–133

Zhu AX, Hudson B, Burt J, Lubich K, Simonson D (2001) Soil mapping using GIS, expert knowledge, and fuzzy logic. Soil Sci Soc Am J 65:1463–1472

Ziadat FM (2005) Analyzing digital terrain attributes to predict soil attributes for a relatively large area. Soil Sci Soc Am J 69:1590–1599

Zinck JA (1988) Physiography and soils, Lecture notes. International Institute for Aerospace Survey and Earth Sciences (ITC), Enschede

Zinck JA (2013) Geopedology. Elements of geomorphology for soil and geohazard studies, ITC special lecture notes series. International Institute for Aerospace Survey and Earth Sciences (ITC), Enschede

Chapter 15
Use of Soil Maps and Surveys to Interpret Soil-Landform Assemblages and Soil-Landscape Evolution

R.J. Schaetzl and B.A. Miller

Abstract Soils form in unconsolidated parent materials, which make them a key link to the geologic system that originally deposited the parent material. In young soils, i.e. those that post-date the last glaciation, parent materials can often be easily identified as to type and depositional system. In a GIS, soil map units can then be geospatially tied to parent materials, enabling the user to create maps of surficial geology. We suggest that maps of this kind have a wide variety of applications in the Earth Sciences, and to that end provide five examples from temperate climate soil-landscapes.

Keywords Soil surveys • Soil maps • Soil parent materials • Soil geomorphology • Soil landscapes • Lithologic discontinuities

15.1 Introduction

Soils form from (and in) unconsolidated parent materials. Parent material is one of the five main soil-forming factors (Jenny 1941), and thus pre-conditions soil development and the pedogenic system from the inception of soil formation. For example, soils forming in dune sand will never be clayey, and are likely to always be highly permeable. Similarly, soils forming in lacustrine clays will never be sandy. Glacial till parent materials are lacking in areas that have never been glaciated, and marine clays do not exist in interior, continental locations. By extension, proper interpretation of soils, as they exist today, can provide key links between them, the

R.J. Schaetzl (✉)
Department of Geography, Michigan State University, East Lansing, MI, USA
e-mail: soils@msu.edu

B.A. Miller
Department of Agronomy, Iowa State University, Ames, Iowa, USA

© Springer International Publishing Switzerland 2016   251
J.A. Zinck et al. (eds.), *Geopedology*, DOI 10.1007/978-3-319-19159-1_15

soil landscape, and the geologic or geomorphic processes that emplaced the soil parent material at some time in the past (Ehrlich et al. 1955; Gile 1975; Schaetzl 1998). That is, soils can provide key information about past sedimentologic or geologic processes and systems, by virtue of their parent materials (e.g. Schaetzl et al. 2000).

Some parent materials overlie a previously formed soil, i.e. a buried paleosol (Follmer 1982; Schaetzl and Sorenson 1987). If the overlying parent material is thin, pedogenesis may "weld" the soil formed at the surface to the paleosol below (Ruhe and Olson 1980), which complicates both parent material interpretations as well as pedogenesis in the surface soil (Wilson et al. 2010). We provide this example only to note that, in this chapter, we will focus on the more common and straightforward situations, in which soils form in fresh and permeable parent material. These kinds of soils provide the best opportunity for establishing the linkages between soil type and character with the parent material type and the processes that emplaced that parent material.

Such examples abound. Many landscapes, especially those that have recently undergone recent glaciation, are rich in parent materials that are relatively unaltered and "fresh" at the time that pedogenesis began. Examples of such parent materials include dune sand, till, volcanic ash, and flood deposits. In most cases, this material is easily identified by excavating deeply, i.e. below the solum and into the C horizon. All of the materials mentioned above are unconsolidated, porous, and permeable. Hence, pedogenesis, largely driven by percolating water, can operate freely in such materials, and can begin immediately after time$_{zero}$. Thus, a clear and often indisputable link can be made between the soil and some form of past geologic/sedimentologic process.

Although much can also be gained from the proper and careful interpretation of soil parent materials on old, stable sites in continental interiors (Brown et al. 2003; Eze and Meadows 2014), most applications involving soil parent materials are found on younger landscapes. Young soils, e.g. Entisols and Inceptisols, resemble their parent materials most directly, because pedogenesis has had little time to operate and alter these materials. In these and other soils that are minimally weathered, soil parent materials can often be readily identified as to type. In older soils, however, especially highly weathered Oxisols and Ultisols, determining the type of parent material present at time$_{zero}$ can be more difficult, mainly because many of the primary minerals in such soils have been altered or destroyed by weathering. Also, erosion may have removed some of the material or brought in other materials from upslope or from upwind. Textures may have been changed by pedogenesis.

With this introduction in mind, we observe that the study of parent materials in soils has much to offer the geoscience, geomorphology, and even the landscape ecology community. Our focus will be on providing examples of studies or situations where careful examination of uniform parent material type and distribution can provide important information about the geomorphic attributes and history of the landscape.

We also provide one important caveat: many soils have formed in "stacked" parent materials, in which a thin layer of one parent material lies immediately atop a distinct but different parent material. The two parent materials are separated by a lithologic discontinuity (Schaetzl 1998). Although this situation sometimes makes parent material interpretations more difficult, it also often provides even greater opportunities for paleoenvironmental interpretation, because such soils can enlighten us about a depositional process or system (the lower material) that then changed to another type of system, i.e. twice the amount of information is potentially available. Examples follow in the text below.

15.2   Methodological Approach

The approach we present can be operationalized with a soil map and a digital elevation model (DEM). Both must be in digital form, so they can be manipulated in a GIS. Soil maps focus on surficial materials and are usually more detailed than available geologic maps due to investments in agricultural development and land valuation. Normally, we overlay the soil information on a hillshade DEM product, so that the soil information can also be matched to topography. For sites in the USA, digital soil data is provided by the Natural Resources Conservation Service (NRCS) web site via the Geospatial Data Gateway (http://datagateway.nrcs.usda.gov). Downloadable files from this site can be added to a GIS.

A key additional step is incorporating supplementary information into the GIS file. We normally code as many of the soil series as possible to parent material by using a two-step process. First, for each soil series we look up its official description on the NRCS web site (https://soilseries.sc.egov.usda.gov/osdname.asp), if we do not already know it. From the official series description, we note the parent material and code it into the GIS as one of several parent material classes, e.g. till, outwash, glacio-fluvial sediment, loess, lacustrine sediment, dune sand, residuum, and a few other, minor categories (Miller et al. 2008). For soils with loess and underlying sediment listed as the parent material, e.g. loess over till, the loess thickness and the type of underlying sediment can also be noted in separate fields.

It should be noted that the NRCS soil maps in the USA are very detailed, often produced at a scale of 1:15,840, resulting in maps that regularly subdivide parent material areas by changes in other soil forming factors. Therefore, interpreting these detailed maps for parent material generally results in an aggregation of map units. Although the relationship between soils and their parent materials is ubiquitous, the scale and purpose of the soil map could potentially deemphasize the parent material-related information available in the map.

The approach described here enables the user to display maps of parent material (and possibly loess thickness) in a GIS, and the data are matched nicely to topography. We have also added additional fields to the GIS attribute table, centering around soil texture, e.g. texture of the surface mineral (usually A) horizon, as well as its parent material (lowest horizon). We have also noted when the texture modifier on the lowest horizon contained the words "gravelly," "cobbly," or "stony," allowing us to compile a data layer for soils that contain significant amounts of coarse fragments in their parent materials. The result is a digital map of surficial geology attributes with greater detail and coverage than is typically available from other sources.

15.3 Results: Analysis and Interpretation of Selected Examples

15.3.1 A Detailed Surficial Geology Map of Iowa, USA

This example illustrates how the methodological approach described above can efficiently convert soil survey information into a format customized for investigations of soil-landform assemblages and soil-landscape evolution over large areas.

In Iowa, surficial geology maps with a high level of detail are only available for a fraction of the state. In contrast, detailed soil maps are available for the entire state. Although the relationship between those producing the respective maps is strong and information is freely shared between the two groups (geologists and soil scientists), differences in disciplinary practices have left a gap in available map products. Notably, geologists here often use NRCS soil maps as base maps, but verify and enhance the information with consideration of deeper bore holes and interpretation for more specifics, e.g. age, about the respective geologic formation, stratigraphy, etc. These investigations require additional time and resources, which help explain the limited coverage of the surficial geologic maps produced in this way. Benefiting from the investment in land use and management information over the past century, detailed soil maps fully cover the state. However, they focus on the top 2 m, and only include a brief attribution of the parent material to the geology, as understood at the time of map production.

Using the methodological approach we described, Miller and Burras (2015) constructed a relational database for each of the soil series mapped in Iowa. Although the NRCS soil database does contain a parent material attribute field, it does not contain as much information as could be found in the geomorphic setting of the official soil series descriptions. However, even the official soil series descriptions often do not directly link the soil series to the recognized geologic formation and geomorphic landform; some interpretations are required. For example, the Clarion soil is described as having calcareous till as parent material, occurring on convex slopes of gently undulating to rolling Late Wisconsin till plain, and with loam to clay loam textures. These characteristics, combined with the geographic extent of the soil series, clearly match what geologists would recognize as the Dows Formation. Additional geomorphic information is gained from soils mapped in the same catena. The Webster soil is generally mapped in the swales below Clarion delineations and is described as being formed in glacial till or in local alluvium derived from till. Thus the spatial juxtaposition of these two and similar soils indicates the pattern of hillslope erosion and basin fill processes along with landform structure (Fig. 15.1b).

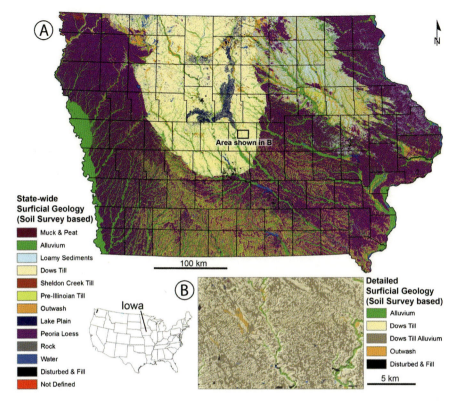

Fig. 15.1 Surficial geology maps for Iowa, USA, based on digital soil survey maps and interpretation of official soil series descriptions. After Miller and Burras (2015). (**a**) Although the same soil series in different counties are technically different soil map units, they are still constrained by definitions set in the official series description. This relationship allows for several county-scale maps to be efficiently translated to desired attribute classes. (**b**) The attribute scale can be customized by the user to include as much or as little detail as needed for the map's purpose. At this larger cartographic scale, it is useful to distinguish soils formed in till of the Dows Formation versus soils formed in the slopewash alluvium derived from that till. Patterns of parent material at this scale are complemented by the elevation hillshade that makes landscape structure more visible

Creating this translation between information recorded in the soil survey to terms useful for geomorphic purposes requires knowledge of the local geology and sometimes careful consideration of context. Nonetheless, after evaluating 863 soil series across Iowa, Miller and Burras (2015) leveraged the soil maps to efficiently and accurately create a detailed surficial geology map covering 145,700 km^2 (Fig. 15.1a). Although the resulting map does not contain as much attribute information as the maps produced by geologists, 67–99 % of the pixels in it are in agreement, and the map provides considerably more spatial information and detail than the geology maps. This level of information is often vital to environmental and geomorphic research.

15.3.2   The Loess-Covered Landscapes of Western Wisconsin, USA

This example illustrates how detailed soil surveys can help determine loess thicknesses across a landscape. Loess covers most upland sites in western Wisconsin (Hole 1976). Most of this loess originated from the Mississippi River, which was a major conduit for silt-rich glacial meltwater and which forms the western boundary of the state (Scull and Schaetzl 2011). In most cases, the loess overlies bedrock or bedrock residuum, as this part of the state has never been glaciated.

Here, soil map units in county-scale soil maps are described with a typical loess thickness and thus the maps can provide detailed information about loess thickness and distribution (Fig. 15.2). Some soil series are formed in "thick" loess, i.e. thicker than the typical 60-in. profile description, and in these cases, loess thicknesses provided by the soil maps represent only a minimum value. Most soils, however, are formed in <60 in. of loess over another parent material, e.g. residuum, bedrock, colluvium, or alluvium. For example, the official description for the Dubuque series states that it "consists of moderately deep, well drained soils formed in 46–91 cm (18–36 in.) of loess and a thin layer of residuum from limestone bedrock or reddish paleosol…" Another common soil in the area, Norden, is "formed in loess and in the underlying loamy residuum weathered from glauconitic sandstone." Note that the parent material description for Norden soils does not include loess thickness. In this case, one must examine the official profile description to determine the typical loess thickness. Norden soils have the following typical horizonation: Ap 0–8 in., Bt1 8–11 in., Bt2 11–20 in., 2Bt3 20–25 in., 2Bt4 25–33 in., 2Bt5 33–37 in., and 2Cr 37–60 in.. All horizons above the lithologic discontinuity at 20 in. are silt loam in texture, as is typical for loess. Thus, where mapped, Norden soils can be assumed to have formed in approximately 20 in. of loess.

This type of procedure can be adopted for all soils in the region, and after the loess thicknesses have been entered into the GIS, detailed maps of loess thickness can be readily created. Figure 15.2 illustrates this approach at a variety of scales. This approach has been successful in a number of loess studies performed in the upper Midwest, USA (Jacobs et al. 2011; Luehmann et al. 2013; Schaetzl and Attig 2013; Schaetzl et al. 2014). Such data are extremely valuable for determining the source areas for loess, which is usually thickest near its source. These types of maps are also useful for guiding land management decisions.

15.3.3   The Recently Deglaciated Landscape of North-Eastern Lower Michigan, USA

This example illustrates how detailed soil surveys can help interpret the geomorphic history of a recently glaciated landscape. Northeastern Lower Michigan was deglaciated roughly 12,300 cal years ago (Larson et al. 1994). At that time, ice associated

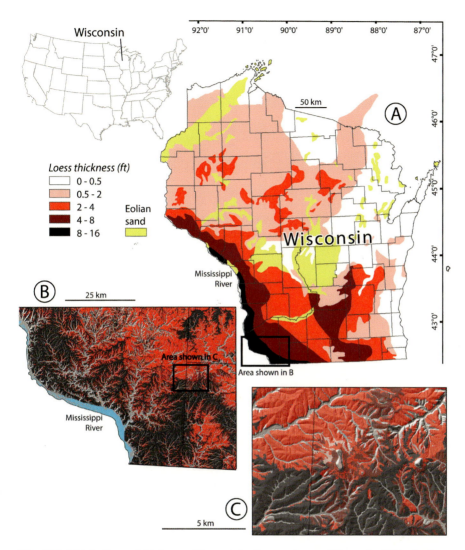

Fig. 15.2 Distribution and thickness of loess and eolian sand across Wisconsin, USA; the loess thickness color legend is similar for all three maps. (**a**) Regional loess thickness, and legend data for loess thicknesses. After Hole (1950) and Thorp and Smith (1952). (**b**) Loess thickness for south-western Wisconsin, as determined in a GIS by using soil series descriptions. (**c**) A more detailed map of loess thickness, created using similar methods but shown at a larger cartographic scale

with the Greatlakean advance of the Laurentide ice sheet had moved rapidly into the region from the northwest, out of the Lake Michigan basin. The ice then stagnated and is assumed to have melted in place (Schaetzl 2001). Associated with the Greatlakean advance and the stagnant ice margin were several shallow, short-lived, proglacial lakes, or at least this has frequently been assumed. The Greatlakean advance left no conspicuous end moraine, and thus the exact location of the outer limit of the ice advance is not known, and has been the subject of considerable debate (Melhorn 1954; Burgis 1977; Schaetzl 2001). Thus, it is conceivable that soil data (maps) may be able to help resolve the extent of this ice advance, as it has been shown to do elsewhere (Millar 2004).

Fortunately, detailed (1:15,840) soil maps and 10-m DEMs exist for this area (Knapp 1993). These maps can be used to help interpret the most recent sedimentary systems that were operational during deglaciation, because post-glacial modifications to these materials have been minimal. Topographic data are not particularly insightful for determining the limit of the Greatlakean ice in this area, because the glacier left no end moraine. However, because water presumably ponded in front of the ice, the northernmost limits of clayey glacio-lacustrine sediment can suggest a likely glacial margin (Fig. 15.3). Indeed, Schaetzl (1991) used this type of data as well as some others, gleaned from soil parent material descriptions, to infer an ice margin just to the north of large areas of glaciolacustrine sediment. Similar sediment behind (north of) this inferred margin is associated with a later, high-level paleolake and is thus clearly not associated with Greatlakean ice (Fig. 15.3).

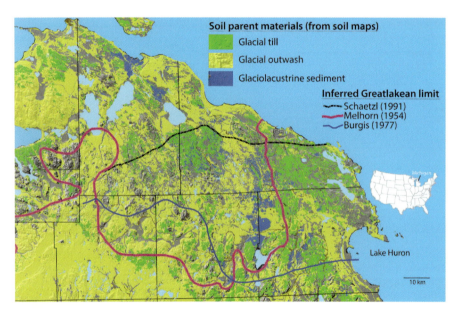

Fig. 15.3 Soil parent materials in north-eastern Lower Michigan, as determined from soil maps and the official soil series descriptions, in a GIS. Also shown are the inferred limits of the Greatlakean ice advance, ca 12,300 cal years ago

15.3.4 An Enigmatic Soil Parent Material on the Outwash Plains of Southwestern Michigan, USA

This example illustrates how field and laboratory data can help determine the parent materials for soil series that have only been described "generically", and how soils with a lithologic discontinuity can potentially provide excellent information about past changes in depositional systems.

Many soils on the low relief outwash plains of southwestern Michigan have loamy upper profiles, despite (as expected) being underlain by coarse, sandy outwash. The origin of this upper material has long been an enigma to soil scientists and geologists alike. It was too thin to be a separate layer of glacial till, and too fine-textured to be glacial outwash.

The main soils that occur on these outwash surfaces are in the Kalamazoo and Schoolcraft series. Kalamazoo soils are described as having formed in "loamy outwash overlying sand, loamy sand, or sand and gravel outwash on outwash plains", whereas Schoolcraft soils have "formed in loamy material over sand or gravelly sand on outwash plains." Typically, this generically described "loamy material" is 40–90 cm thick, and has a diffuse lower boundary. For lack of a better term, we refer to this layer as a loamy mantle.

Soil textural data, as determined by laser diffraction, from two representative pedons (Fig. 15.4) illustrate that the outwash at depth is dominated by sand, whereas the loamy mantle is either silty (Fig. 15.4a) or has a distinctly bimodal particle size distribution – with both sand and silt peaks (Fig. 15.4b). Textural data for the loamy mantle (not shown here) are almost always bimodal, and the sand peak aligns with the same peak in the outwash below. These data suggest that the loamy mantle is a mixed sediment – sand from the outwash mixed with a silty sediment above, but of unknown origin.

In a recent study, Luehmann et al. (2016) sampled and determined the textural distributions of 167 locations across the outwash plains of southwestern Michigan. The loamy mantle in almost all of these soils had a bimodal particle size distribution. Using a "filtering" method first reported in Luehmann et al. (2013), they were able to isolate the textural pattern of the original, silty sediment, and map its characteristics across the region. Spatial patterns for the loamy mantle were easily interpretable, illustrating that the silty material is silt-rich loess, and that it has been subsequently mixed with sand from below by pedoturbation. The mantle is thickest near a large meltwater valley that existed during deglaciation (Fig. 15.5), suggesting that it was the main loess source. Textural data of various sorts (not shown here) also confirmed that the loess that comprises the loamy mantle gets finer-textured and better sorted to the east, away from this channel. This type of spatial pattern is typical for loess.

This work showed that the heretofore enigmatic mantle on the outwash plains of southwestern Michigan is silt-rich loess that was derived from the Niles-Thornapple Spillway and its major tributary channels. The Spillway was active for approximately 500 years, between ca 17,300 and 16,800 cal years ago, carrying silt-rich meltwater. This study highlights the fact that not all soil parent materials are

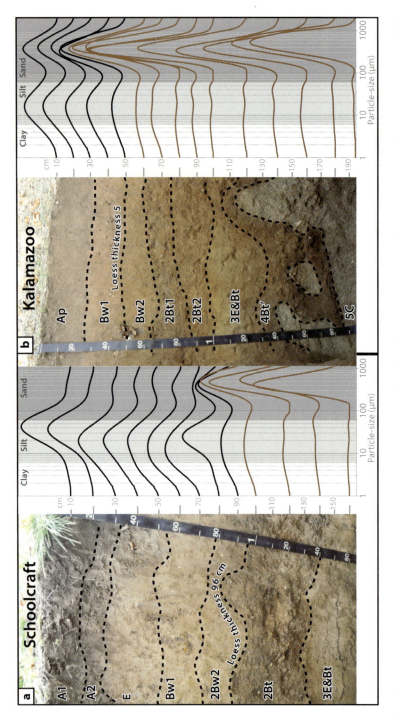

Fig. 15.4 Photos, with horizon boundaries marked, and textural curves for a (**a**) Schoolcraft and a (**b**) Kalamazoo soil profile (After Luehmann et al. 2016)

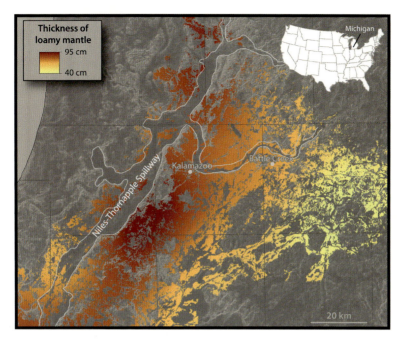

Fig. 15.5 Interpolated map, using ordinary kriging, of the thickness of the upper sediment, which is interpreted as loess, on the outwash plains of southwestern Michigan. Interpolated data are shown only in areas where outwash soils with a loamy mantle are mapped (After Luehmann et al. 2016)

"obvious" or stated in their official series descriptions, but with some work the genetic origin of the sediment can often be determined.

15.3.5   A Watershed with a Complex Geology in the Western Grand-Duchy of Luxembourg

This example illustrates how the use of detailed soil surveys for interpreting soil-landform assemblages can also be applied to non-glaciated landscapes, and thus can provide key information for other scientific inquires. In particular, relationships between bedrock parent materials, soil morphology, and indicative vegetation patterns can provide important information for hydrological modelling.

The available geologic map (1:25,000) for the Huewelerbach experimental catchment in western Luxembourg shows the locations of several geologic formations in the watershed, including units of sandstone, limestone, and claystone. Some of these formations have alternating layers of marl. The catchment also contains a colluvial-alluvial complex at the bases of many hillslopes. Complicating the spatial distribution of these formations is a fault that is believed to run mostly northwest of, and parallel to, the main trunk stream. Because of this fault, the hydrologic characteristics of the opposing hillslopes are not identical. Parent materials yielding soils with B and/or C horizons consisting of heavy clay lead to an environment dominated by overland flow. In contrast, parent materials yielding thicker soils with sandy to silty-sandy textures facilitate better infiltration and deeper percolation, and hence more lateral subsurface flow and less surface runoff.

Juilleret et al. (2012) conducted a soil survey of the catchment, classifying 6 soil map units with 70 hand auger drillings to a depth of 110 cm. They subsequently verified the relationships between the properties of the soil profile with the parent material, using a mechanized coring machine to sample a maximum depth of 400 cm at 12 locations along two transects. Using the World Reference Base (IUSS 2006) to classify the soils, they found Calcisols corresponded with geologic formations containing units of marl, Podzols with a sandstone formation that lacks marl layers, and Colluvisols with the colluvial-alluvial complex. The Podzols correspond with the occurrence of conifers, whereas the other soils occur under deciduous vegetation. Under grasslands, Pelosols and Brunisols were identified. In the Bw and C horizons of these soils, a distinctive sequence of a red clayey layer and a grey sandy layer helped reveal the presence of an additional geological formation recognized in the area, but not previously depicted on the existing geologic map. For this catchment of soils formed in a variety of sedimentary rocks, standard soil survey methods were able to improve upon the information available from the standard geologic map. This information was valuable for improving the mapping of geologic formations and for providing key information for modelling hillslope hydrology.

15.4   Summary and Conclusions

The relationship between soils and the material in which they form connects soil survey maps and geological maps. Different information collected for, and portrayed on, the respective maps – due to differences in purpose, focus, or resources – can assist investigations in other disciplines. This multiple utility is especially true for studying soil-landform assemblages and soil-landscape evolution.

Although the pedogenic pathway of a soil is constrained by the parent material, interpretation of soil properties to infer parent material origins needs to carefully consider the potential for complicating factors. For example, other factors of soil formation can alter the material, especially over long periods of time. Also, buried paleosols within the modern soil profile can result in new horizons with properties that are influenced by the interaction of modern pedogenesis with the properties of the old horizons.

Because of the interconnection between soils and geology, one should beware of the potential for circular reinforcement of information. The reason soil maps often provide more spatial information than available geologic maps is because of the greater spatial density of field sampling and greater availability of easily-observed covariates for spatial prediction. However, soil mappers also use geologic maps as one of the base maps for their soil maps (Miller and Schaetzl 2014). Therefore, the potential exists for an error on one type of map to become circularly reinforced. Only field investigation is capable of catching these problems and better informing all maps.

In many cases, soil maps and surveys –together or singly –provide information that is not available from any other source, particularly with regard to spatial detail and characteristics of the top meter of unconsolidated material. Therefore, these maps often represent an untapped potential for improving our geomorphic understanding of landscapes (Brevik and Miller 2015).

References

Brevik EC, Miller BA (2015) The use of soil surveys to aid in geologic mapping with an emphasis on the Eastern and Midwestern United States. Soil Horiz 56(4).

Brown DJ, Helmke PA, Clayton MK (2003) Robust geochemical indices for redox and weathering on a granitic landscape in central Uganda. Geochim Cosmochim Acta 67:2711–2723

Burgis WA (1977) Late-Wisconsinan history of northeastern lower Michigan. PhD dissertation, University of Michigan

Ehrlich WA, Rice HM, Ellis JH (1955) Influence of the composition of parent materials on soil formation in Manitoba. Can J Agric Sci 35:407–421

Eze PN, Meadows ME (2014) Texture contrast profile with stonelayer in the Cape Peninsula, South Africa: autochthony and polygenesis. Catena 118:103–114

Follmer LR (1982) The geomorphology of the Sangamon surface: its spatial and temporal attributes. In: Thorn C (ed) Space and time in geomorphology. Allen and Unwin, Boston, pp 117–146

Gile LH (1975) Causes of soil boundaries in an arid region: I. Age and parent materials. Soil Sci Soc Am Proc 39:316–323

Hole FD (1950) (reprinted, 1968) Aeolian sand and silt deposits of Wisconsin. Wisconsin Geological and Natural History Survey map, Madison

Hole FD (1976) Soils of Wisconsin. University of Wisconsin Press, Madison

IUSS (2006) World reference base for soil resources, vol 103, 2nd edn, IUSS working group, world soil resources report. FAO, Rome

Jacobs PM, Mason JA, Hanson PR (2011) Mississippi Valley regional source of loess on the southern Green Bay Lobe land surface, Wisconsin. Quat Res 75:574–583

Jenny H (1941) Factors of soil formation. McGraw-Hill, New York

Juilleret J, Iffly JF, Hoffmann L, Hissler C (2012) The potential of soil survey as a tool for surface geological mapping: a case study in a hydrological experimental catchment (Huewelerbach, Grand-Duchy of Luxembourg). Geol Belg 15(1–2):36–41

Knapp BD (1993) Soil survey of Presque Isle County, Michigan. USDA Soil Conservation Service, US Government Printing Office, Washington, DC

Larson GJ, Lowell TV, Ostrom NE (1994) Evidence for the Two Creeks interstade in the Lake Huron basin. Can J Earth Sci 31:793–797

Luehmann MD, Schaetzl RJ, Miller BA, Bigsby M (2013) Thin, pedoturbated and locally sourced loess in the western Upper Peninsula of Michigan. Aeolian Res 8:85–100

Luehmann MD, Peter B, Connallon CB, Schaetzl RJ, Smidt SJ, Liu W, Kincare K, Walkowiak TA, Thorlund E, Holler MS (2016) Loamy, two-storied soils on the outwash plains of southwestern lower Michigan: pedoturbation of loess with the underlying sand. Ann Assoc Am Geogr 105 (in press)

Melhorn WN (1954) Valders glaciation of the southern peninsula of Michigan. PhD dissertation, University of Michigan

Millar SWS (2004) Identification of mapped ice-margin positions in western New York from digital terrain-analysis and soil databases. Phys Geogr 25:347–359

Miller BA, Burras CL (2015) Comparison of surficial geology maps based on soil survey and in depth geological survey. Soil Horiz 56

Miller BA, Schaetzl RJ (2014) The historical role of base maps in soil geography. Geoderma 230–231:329–339

Miller BA, Burras CL, Crumpton WG (2008) Using soil surveys to map Quaternary parent materials and landforms across the Des Moines lobe of Iowa and Minnesota. Soil Surv Horiz 49:91–95

Ruhe RV, Olson CG (1980) Soil welding. Soil Sci 130:132–139

Schaetzl RJ (1991) A lithosequence of soils in extremely gravelly, dolomitic parent materials, Bois Blanc Island, Lake Huron. Geoderma 48:305–320

Schaetzl RJ (1998) Lithologic discontinuities in some soils on drumlins: theory, detection, and application. Soil Sci 163:570–590

Schaetzl RJ (2001) Late Pleistocene ice flow directions and the age of glacial landscapes in northern lower Michigan. Phys Geogr 22:28–41

Schaetzl RJ, Attig JW (2013) The loess cover of northeastern Wisconsin. Quat Res 79:199–214

Schaetzl RJ, Sorenson CJ (1987) The concept of "buried" vs "isolated" paleosols: examples from northeastern Kansas. Soil Sci 143:426–435

Schaetzl RJ, Krist F, Rindfleisch P, Liebens J, Williams T (2000) Postglacial landscape evolution of northeastern lower Michigan, interpreted from soils and sediments. Ann Assoc Am Geogr 90:443–466

Schaetzl RJ, Forman SL, Attig JW (2014) Optical ages on loess derived from outwash surfaces constrain the advance of the Laurentide ice from the Lake Superior Basin, Wisconsin, USA. Quat Res 81:318–329

Scull P, Schaetzl RJ (2011) Using PCA to characterize and differentiate the character of loess deposits in Wisconsin and Upper Michigan, USA. Geomorphology 127:143–155

Thorp J, Smith HTU (1952) Pleistocene eolian deposits of the United States, Alaska, and parts of Canada. Map 1:2,500,000. Geological Society of America, New York

Wilson MA, Indorante SJ, Lee BD, Follmer L, Williams DR, Fitch BC, McCauley WM, Bathgate JD, Grimley DA, Kleinschmidt K (2010) Location and expression of fragic soil properties in a loess-covered landscape, Southern Illinois, USA. Geoderma 154:529–543

Chapter 16
Applying a Geopedologic Approach for Mapping Tropical Forest Soils and Related Soil Fertility in Northern Thailand

M. Yemefack and W. Siderius

Abstract This chapter describes the application of a geopedologic approach for delineating and characterizing soil units and related soil fertility in tropical forest highlands of northern Thailand. The study area contains four types of landscape including mountain, hilland, piedmont, and valley, and five types of parent material including igneous, metamorphic and sedimentary rocks. Soils are distributed over three slope positions, i.e. summit, backslope, and footslope. Soil variability is high, including Oxisols, Ultisols, Alfisols, Inceptisols, Mollisols, and Entisols, found either in consociation or association. A mathematical approach for analysing relations between individual soil bodies was applied to study the soil fertility variation as related to the categorial levels of a hierarchic geoform classification system. This relationship was displayed by means of numerical values of the Similarity Index (SI) and the Fertility Distance (FD), computed by integrating eight soil properties (pH, C, N, K, CEC-soil, CEC-clay, clay, and base saturation) assumed to influence soil fertility. The study showed that the geopedologic approach for characterizing soils of this complex area was suitable and allowed the results obtained in sample areas to be extrapolated to similar areas. It has the advantage not only to be based on strong integration of geomorphology and pedology, but also to take into account the parent material at lower categorial levels of the system.

Keywords Geopedology • Soil mapping • Soil fertility variation • Similarity index • Fertility distance

M. Yemefack (✉)
International Institute of Tropical Agriculture (IITA) and Institute of Agricultural Research for Development (IRAD), Yaounde, Cameroon
e-mail: Myemefack@cgiar.org

W. Siderius
Faculty of Geo-Information Science and Earth Observation (ITC), University of Twente, Enschede, The Netherlands
e-mail: wr.siderius@hetnet.nl

© Springer International Publishing Switzerland 2016   265
J.A. Zinck et al. (eds.), *Geopedology*, DOI 10.1007/978-3-319-19159-1_16

16.1   Introduction

Soil productivity can be assessed using fertility indicators that limit plant growth. Such limiting factors may also be related to soil types which depend themselves on various soil forming factors. Several approaches are used to evaluate soil fertility. The most widespread ones are based on soil testing and simple fertilizer trials. They are site-specific and situation-specific (Sanchez 1976, 1977; Sanchez et al. 1983). Therefore, their results and recommendations are difficult to extrapolate, especially in areas with short distance soil variability. In landscapes dominated by repeating landform types, the topographic position and the nature of the parent material may help predict the fertility of the soil units, the relation with plant growth, and the vulnerability to degradation under alternative land uses.

Soil fertility studies using the Fertility Capability Soil Classification System (FCC) have shown that soil individuals in one FCC unit may belong to different taxonomic classes in the USDA soil taxonomy or other natural taxonomic systems (Sanchez et al. 1982, 2003). Research efforts have been made to interpret soil maps in terms of the productivity potential of the soil. The FAO land evaluation system (FAO 1976) evaluates very well the land for a particular type of utilization but does not give enough information on the overall capability of the soil or its fertility status. Parametric methods have also been used in land evaluation with the advantage of eliminating subjectivity because curves or functional relations are indexing and compounding (Driessen and Konijn 1992); they reveal orders of magnitude or trends in components of land-use systems. They are developed and tested for application in a particular area, and may therefore not be applicable equally well elsewhere. The fertility capability soil classification system (FCC) is one of the first attempts to link soil distribution and soil fertility (Sanchez et al. 1982, 2003), with the advantage to group soils with similar fertility limitations using quantitative limits. It can be used at any scale, but soils of the same FCC class may not produce exactly the same yield from the field under the same management practice. For site differentiation at larger scales, we thus hypothesized that a combination of a geomorphic soil survey approach with parametric methods may produce soil information with better link to soil fertility.

The aim of this study was to show how a combination between a geopedologic approach to collect soil data and a mathematical approach for analysing relations between individual soil bodies can be used to display the relationship between soil distribution and soil fertility in a natural forest area in northern Thailand. The inventory and characterization of the soil resource patterns paid special attention to soil fertility in relation to landform and parent material using geometric models that integrate soil properties assumed to influence soil fertility.

16.2 Materials and Methods

16.2.1 The Study Area

The study area of approximately 10,000 ha is located in northern Thailand, Mae Taeng District, Chiang-Mai Province. The area lies west of the Ping river valley, between latitudes 19°01′–19°08′ N and longitudes 98°48′–98°56′ E. Elevation ranges from 320 masl in the lowlands to about 1200 masl in the western part of the district. The mean annual rainfall is approximately 1200 mm with six rainy months. Mean monthly temperature varies from 21 °C to about 29 °C, and the mean monthly relative humidity between 60 % (March–April) and 85 % (August–September). The area is characterized by a large diversity of rock substrata described in the geological map of Amphoe Mae Taeng at the scale of 1:50,000 (Tansathien et al. 1984), including Precambrian igneous rocks (granite) and metamorphic rocks (gneiss, micaschist, quartzite, marble), Tertiary Fe-Mn conglomerate and breccia, and Quaternary sediments (alluvial deposits, gravelly clay, lateritic gravel). The geomorphology is strongly related to the geology and lithology. Four main landscape types were distinguished, including valley (320–340 m), piedmont (330–450 m), hilland (380–700 m), and mountain (380–1200 m) (elevations in masl).

16.2.2 Geopedologic Survey

Topographic and geologic maps at 1:50,000 scale and aerial photographs at 1:18,000 scale from 1977 were interpreted. From the preliminary interpretation map, sample areas were selected based on geology and landscape. Using the hierarchic system of geoform classification of the geopedologic approach (Zinck 1988), geoforms were differentiated according to four categorial levels from landscape to landform.

Fieldwork was concerned with the general survey, identification of transects, soil profile description, and soil sampling. Soil survey was carried out in samples areas and knowledge from sample areas was extrapolated throughout the whole study area. Survey methods used were taken from Soil Survey Manual (Soil Survey Division Staff 1993), while soil profiles were described based on the FAO guidelines for soil profile description (FAO 1990). For the study of soil fertility, several transects were preselected from the aerial photographs taking into account the geological setting. During the reconnaissance survey, five transects were finally selected from mountain (T1, T2, T5) and hilland (T3, T4) landscapes based on accessibility, lithology, and landscape type (Figs. 16.1, 16.2, 16.3, 16.4, 16.5, and 16.6). Along each transect, one observation point of approximately 60 cm

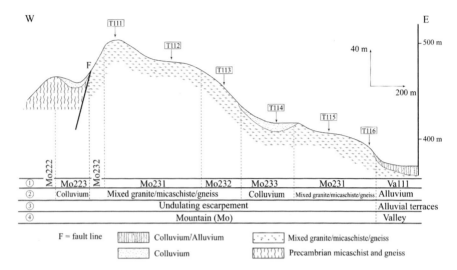

Fig. 16.1 Transect T1 in mountain landscape (Map units in legend of Table 16.2)

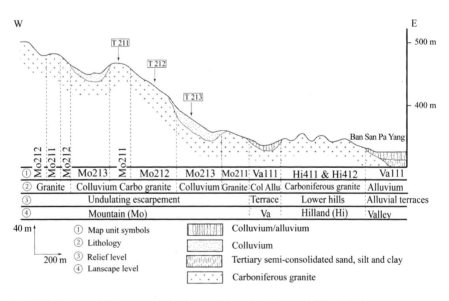

Fig. 16.2 Transect T2 in mountain landscape (Map units in legend of Table 16.2)

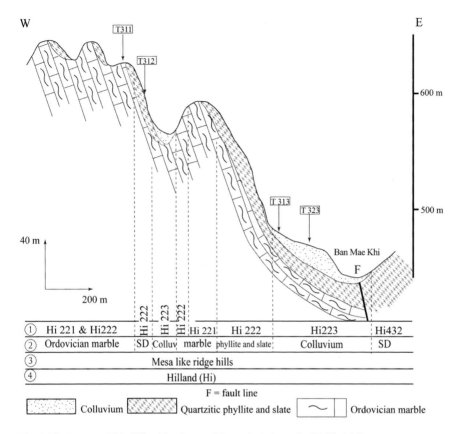

Fig. 16.3 Transect T3 in hilland landscape (Map units in legend of Table 16.2)

depth was described in each landform unit (i.e. summit, backslope, footslope), and samples were taken at three constant depths: composite A horizon, 20–25 cm, and 40–50 cm depth.

Soils were classified down to the family level according to the USDA Soil Taxonomy (Soil Survey Staff 1998). Maps compiled from aerial photo-interpretation at scale of 1:20,000 (Yemefack et al. 1994) were digitized and handled using the ILWIS software (ITC 1993). Soil samples were analysed at ISRIC soil laboratory in Wageningen using procedures and methods described by Van Reeuwijk (1993) for the following determinations: pH water and KCl, organic carbon, total nitrogen, cation exchange capacity (CEC), exchangeable bases (Mg, Ca, K, Na), free $CaCO_3$, and particle size distribution.

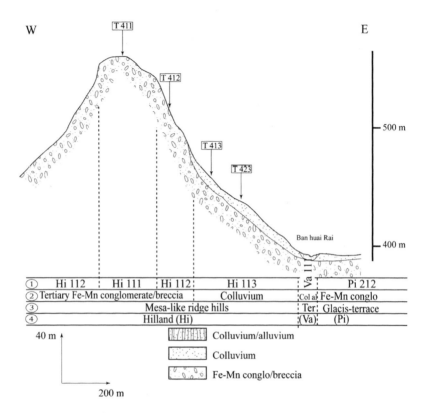

Fig. 16.4 Transect T4 in hilland landscape (Map units in legend of Table 16.2)

16.2.3   Assessing the Relationship between Soil Fertility and Soil Map Units

Soil fertility variation as related to the categorial levels of the geoform classification system was measured by a geometric approach and displayed by means of numerical values of the Similarity Index (SI) and the Fertility Distance (FD). Values were computed by integrating eight soil properties assumed to influence soil fertility: pH water, organic carbon (OC), total nitrogen (N), potassium (K), soil cation exchange capacity (CEC soil), clay cation exchange capacity (CEC clay), base saturation, and clay content. SI and FD values were calculated with samples from three soil depths for a 0–50 cm slice.

16.2.3.1   Similarity Index

The similarity index (SI) uses a geometric model to formalize the presumed relationship between individuals, as described in Webster and Oliver (1990) and Webster (2000). Assuming a single variable, the distance separating two individuals of this variable is the measure of their relation. The closer they are, the more alike

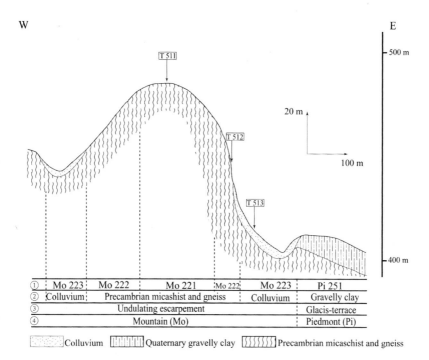

W   E

①	Mo 223	Mo 222	Mo 221	Mo 222	Mo 223	Pi 251
②	Colluvium	Precambrian micashist and gneiss			Colluvium	Gravelly clay
③		Undulating escarpement				Glacis-terrace
④		Mountain (Mo)				Piedmont (Pi)

Colluvium   Quaternary gravelly clay   Precambrian micaschist and gneiss

Fig. 16.5 Transect T5 in mountain landscape (Map units in legend of Table 16.2)

they are, and vice versa. This distance can be calculated by Pythagoras' theorem as follows. If the coordinates of two points i and j are x_{i1}, x_{i2}, and x_{j1}, x_{j2}, then the distance Δ_{ij} between them is given by: $\Delta_{ij} = \sqrt{\left[\left(x_{i1} - x_{j1}\right)^2 + \left(x_{i2} - x_{j2}\right)^2\right]}$. The prin ciples hold for any number of dimensions or number of properties of each point. If there are p variables (properties), then a p-dimensions character space can be postulated, and the above equation can be generalized as follows: $\Delta_{ij} = \sqrt{\sum \left(x_{ik} - x_{jk}\right)^2}$.

The distance Δ is known as Pythagorean distance or Euclidian distance between the individuals. It increases with the number of characters p involved in the comparison. It can be divided by p or by the square root of p to give an 'average' distance d_{ij}. The distance d_{ij} measures the dissimilarity between the individuals i and j. The d_{ij} can be scaled or normalized so that it lies in the range from zero for maximum similarity to one for maximum dissimilarity. To convert d_{ij} to a measure of similarity, the index S_{ij} is calculated as $S_{ij} = 1 - d_{ij}$. S_{ij} is known as similarity coefficient or similarity index. The similarity index SI gives the level of resemblance between two units. When SI is 1, there is total similarity; while the index zero (0) connotes full dissimilarity.

Prior to calculation, data standardization is essential to scale the measurement of properties so that the interpretation of results can be done more readily. The values of a variable were divided by the standard deviation of that variable in the sample. Every scale then had a standard deviation of 1. Table 16.1 shows some laboratory data and their corresponding standardized values used in this study.

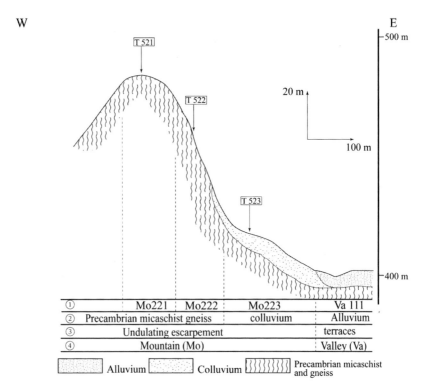

Fig. 16.6 Transect T5 under cultivation in mountain landscape (Map units in legend of Table 16.2)

Table 16.1 Original laboratory data and standardized data of selected soil properties

Soil properties (0–10 cm)	OC%	N%	K	CECs	CECcl	BS%	Clay%	pH
Original values (as from the laboratory)	3.99	0.25	1.1	27.1	59.2	85	45.7	6.7
Standardized values (by the standard deviation)	3.91	3.57	2.2	2.51	1.08	5.48	3.84	13.4

K, CEC, and CEC-clay in cmol(+)/kg; pH-water; *OC%* organic carbon content, *N%* total nitrogen content, *K* potassium, *CECs* soil cation exchange capacity, *CECcl* clay cation exchange capacity, *Clay%* clay content, *BS%* base saturation percentage

16.2.3.2 Fertility Distance

The fertility distance (FD) method was developed to estimate the fertility status of each map unit and rank the soil units according to their fertility status. It was hypothesized that soil fertility has a starting point where all the properties contributing to its quality have a value zero. The fertility distance was then supposed to be a vector F whose origin is a point at the intersection of the soil forming factors influencing the fertility status of the soil. So, the fertility distance for each individual (or soil sample) is calculated as: $FD = \sqrt{\sum (x_{ik})^2}$, where x_{ik} is the standardized values of each soil property k in point i. FD increases also with the number of properties k. FD computed for each soil permitted a relative classification of individuals according to their FD values. The higher the FD value, the better the fertility of a soil.

16.3   Results and Discussion

16.3.1   Characteristics of the Geopedologic Map Units

The application of the geopedologic approach to aerial photo-interpretation and field verification yielded a map legend with 45 classes of soil map units corresponding to different landform types distributed over four landscapes (Fig. 16.7 and Table 16.2). Main characteristics of the map units at landscape level are summarized hereafter.

The *valley (Va)* landscape is an extensive nearly level surface, extending along the Ping river and its tributaries. It is composed of high, middle, and lower terraces

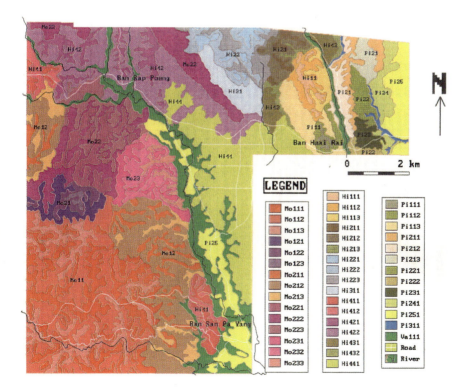

Fig. 16.7 Geopedologic soil map of the study area (Map units in legend of Table 16.2)

Table 16.2 Legend of the geopedologic soil map of the study area (map in Fig. 16.7)

Landscape	Relief/molding	Lithology/parent material	Landform	Symbol	Slope (%)	Main and associated soils	Area (ha)
Mountain (Mo)	Undulating stepped plateau surface	Granite and derived colluvium.	Summit/shoulder	Mo111	0–10	Kandiustalfic Eutrustox	645
			Backslope	Mo112	20–100	Kandiustalfic Eutrustox, Humic Rhodic Eutrustox	915
			Footslope	Mo113	2–30	Fluventic Ustropepts, Typic Kandiustults, Ustic Kandihumults	241
		Micaschist, gneiss and derived colluvium	Summit/shoulder	Mo121	0–10	Kandiustalfic Eutrustox	135
			Backslope	Mo122	20–100	Kandiustalfic Eutrustox, Kanhaplic Rhodustalfs	66
			Footslope	Mo123	2–30	Typic Kandiustalfs, Kanhaplic Haplustalfs	13
	Undulating escarpment	granite and derived colluvium. (T2)	Summit/shoulder	Mo211	0–10	Typic Kandiustults	254
			Backslope	Mo212	20–100	Typic Kandiustults	671
			Footslope	Mo213	2–30	Ustic Kandihumults	188
		Mica-schist, gneiss and derived colluvium (T5)	Summit/shoulder	Mo221	0–10	Kandiustalfic Eutrustox	182
			Backslope	Mo222	10–100	Kanhaplic Rhodustalfs, Kandiustalfic Eutrustox	639
			Footslope	Mo223	2–30	Kanhaplic Haplustalfs	174
		Mixed granite and micaschist/gneiss and derived colluvium (T1)	Summit/shoulder	Mo231	0–18	Kandiustalfic Eutrustox	144
			Backslope	Mo232	10–70	Kandiustalfic Eutrustox	172
			Footslope	Mo233	2–30	Udic Kandiustalfs	50

Hilland (Hi)	Mesa-like ridge high hills	Fe-Mn conglomerate / breccia and derived colluvium (T4)	Summit/shoulder	Hi111	0–5	Typic Kanhaplustults	38
			Backslope	Hi112	20–100	Kanhaplic Haplustults	93
			Footslope	Hi113	5–25	Typic Kandiustults	61
	Rounded ridge high hills	Quartzitic phyllite, slate (light to dark grey) and derived colluvium	Summit/shoulder	Hi211	0–10	Typic Kandiustults	52
			Backslope	Hi212	10–100	Typic Kandiustults, Typic Kandiustalfs	91
			Footslope	Hi213	5–30	Typic Kandiustalfs	16
		Marble and derived colluvium (T3)	Summit/shoulder	Hi221	0–10	Entic Haplustolls	70
			Backslope	Hi222	10–100	Typic Argiustolls, Entic Haplustolls, T. Haplustox	169
			Footslope	Hi223	5–30	Typic Argiustolls, Typic Kanhaplustults	126
	Glacis	Marble and derived colluvium (T3)	Hummocky complex slope	Hi311	0–30	Typic Argiustolls, Typic Kanhaplustults	154
	Lower hills	Granite	Summit/shoulder	Hi411	0–8	Typic Kandiustalfs, Typic Ustropepts	128
			Back/footslope	Hi412	10–20	Typic Kandiustalfs	131
		Mica-schist, gneiss	Summit/shoulder	Hi421	0–8	Typic Kandiustalfs	224
			Back/footslope	Hi422	10–20	Typic Kandiustalfs, Kanhaplic Haplustalfs	681
		Quartzitic phyllite and slate	Summit/shoulder	Hi431	0–8	Typic Kandiustalfs, Typic Kandiustults	99
			Back/footslope	Hi432	8–16	Ustoxic Dystropepts	39
		Semi-consolidated sand, silt and clay	Erosional glacis	Hi441	0–25	Entic Haplustolls, Paleustults, Vertic Argiustolls	877

(Continued)

Table 16.2 (Continued)

Landscape	Relief/molding	Lithology/parent material	Landform	Symbol	Slope (%)	Main and associated soils	Area (ha)
Piedmont (Pi)	High glacis-terrace	Stone and gravel	Summit/shoulder	Pi111	0–5	Lithic Ustorthents	16
			Backslope	Pi112	10–35	Lithic Ustorthents	55
			Footslope	Pi113	5–10	Typic Kandiustults	152
	Lower glacis-terrace	Quartzitic Fe-Mn concretionary conglomerate/ breccia	Slightly dissected summit/shoulder	Pi211	0–4	Typic Kandiustults	97
			Slightly dissected, gentle back/ footslope	Pi212	5–10	Rhodic Kandiustults, Paleustults	93
			Dissected back/ footslope	Pi213	10–30	Typic Kandiustults	72
		Stone, gravel and clay	Dissected slope facet	Pi221	10–35	Plinthic Kandiustults	81
			Slightly dissected slope facet	Pi222	10–20	Plinthic Kandiustults	30
		Sand, greenish clay	Strongly dissected badlands	Pi231	35–40	Badlands	61
		Gravelly clay	Dissected slope facet	Pi241	20–40	Typic Kandiustults	132
		Gravelly clay	Slightly dissected slope facet	Pi251	0–10	Typic Kandiustults	491
	Vale	Colluvium/alluvium	Vale bottom	Pi311	0–4	Typic Ustropepts	37
Valley (Va)	Alluvial terraces	Alluvium	Tread/riser, floodplain complex	Va111	0–4	Vertic Epiaqualfs	741

that were not separable and put into one map unit (Va111). The *piedmont (Pi)* landscape is a gently inclined land surface along the foot of hillands or mountains, forming a sharp boundary with the valley. It was subdivided into 11 map units. The *hilland (Hi)* landscape is a rugged terrain characterized by the repetition of hills with uneven summit heights, separated by a moderate to coarse drainage pattern. Seventeen map units were delineated from three hills (Hi1, Hi2, and Hi3) distinguished according to their internal relief. The *mountain (Mo)* landscape is a rugged, strongly dissected landscape characterized by higher summits than the surroundings and an important internal dissection with elevation differences of 80–150 m. This landscape was divided into 16 map units in two main relief forms (Mo1, Mo2).

16.3.2   Soil Variability as Highlighted by the Geopedologic Survey

Main soil classes occurring in the study area are Oxisols, Ultisols, Alfisols, Mollisols, Inceptisols, and Entisols, with considerable variations in soil properties. Most of these soils are deep to very deep, well-developed and friable, predicting a relatively good quality as to their productivity. They occur usually associated with various landscapes. On the same landscape, parent material plays an important role in controlling soil patterns at landform level. This is in accordance with Hendricks' conclusion (Hendricks 1981) that the most important factor for forest soil formation in northern Thailand is parent material. Several well-defined soil-parent-material relationships were established, including:

– Oxisols-Ultisols-Inceptisols sequences on granite;
– Oxisols-Alfisols-Inceptisols sequences on gneiss;
– Mollisols-Alfisols sequences on marble;
– Oxisols-Ultisols/Alfisols sequences on quartzitic phyllite and slate;
– Ultisols-Inceptisols-Entisols associations on well-drained sedimentary depositions.

At landform level (slope facet), the slope form and steepness seem to have less influence on soil formation in the area. On the same parent material, summit and backslope soils show usually the same characteristics; while footslope soils are different. The concept of slope-complex defined by the Department of Land Development of Thailand (DLD 1984) may have led to a genetic bias in grouping all slope soils in one single map unit at semi-detailed survey level (scale 1:50,000). The soils are better grouped by parent material types at this scale. The geopedologic soil survey approach appears to be particularly useful in these conditions. It also allows the results from sample areas to be easily extrapolated to similar situations outside the sample areas.

16.3.3 Soil Fertility Variation

Although many soil properties important to plant growth are used at several lower levels in Soil Taxonomy (Soil Survey Staff 1998), a simple characterization and classification of soils might not take into account much about their potentiality and limitation levels to agricultural use (Buol and Couto 1981; Sanchez et al. 2003). In this study, soil fertility variation was analysed and assessed as induced by two major soil-forming factors: relief as slope facet units and lithology/parent material at site level. Soil fertility is taken in a comprehensive sense covering the physical and chemical soil properties that contribute to plant growth and soil productivity (Demolon 1952; Fairbrige and Finkl 1972). Soil fertility has thus to do principally with plant nutrient elements and soil conditions (Kauffman et al. 1998; Sanchez 1976).

SI as single variable was used to measure the relationship between two individual soil units, with SI=1 for total similarity and SI=0 for total dissimilarity. SI values comparing pairs of soil units varied between 0.1 and 0.94 in our study area (Tables 16.3 and 16.4). FD was computed to estimate and rank the soil units at landscape and relief type levels according to their fertility status. The higher is the FD, the better the fertility of the soil. In our study area, soil fertility differed significantly (p=0.000) from one soil map unit to another. The map units were ranked as follows based on their FD class: FD 16–18: Hi221 and Hi222; FD 14–16: Mo221, Mo231, Mo232 and Mo233; FD 12–14: Mo223, Mo211, Mo212 and Mo213; FD 10–12: Hi113, Hi112 and Hi111.

Table 16.3 Similarity index (SI) and fertility distance (FD) values (in 0–50 cm soil depth) used for landform and transect comparisons

Landform pairs	Method	Transects				
		T1	T2	T3	T4	T5
Summit-backslope	SI	0.84	0.84	0.80	0.73	0.81
	FD	14.5–13.5	12.3–12.3	16.6–17.4	10.5–11.8	14.7–15.5
Summit-footslope	SI	0.82	0.70	0.29	0.75	0.72
	FD	14.5–14.4	12.3–12.9	16.6–11.9	10.5–12.0	14.7–13.7
Backslope-footslope	SI	0.75	0.74	0.20	0.93	0.69
	FD	13.5–14.4	12.3–12.9	17.4–11.9	11.8–12.0	15.5–13.8

Table 16.4 Matrices of similarity index of site couples, compared by landforms (in 0–50 cm soil depth)

	Summit						Backslope						Footslope				
	T1	T2	T3	T4	T5		T1	T2	T3	T4	T5		T1	T2	T3	T4	T5
T1	1	.57	.41	.39	.94		1	.59	.22	.70	.67		1	.76	.55	.63	.79
T2		1	.14	.61	.53			1	.16	.81	.42			1	.78	.81	.78
T3			1	.10	.39				1	.16	.30				1	.80	.61
T4				1	.35					1	.48					1	.66
T5					1						1						1

SI values of more than 0.6 are shown in shaded cases

16.3.3.1 Soil Fertility Variation in Relation to Landform Types

Three landforms were considered, namely summit, backslope, and footslope, to anal-yse the fertility variations along a toposequence. The SI and FD indices were used to assess fertility variation from one landform to another comparing landform pairs (Table 16.3). Soil fertility varies between slope facets, showing major differences between the footslope and other slope positions (i.e. summit and backslope). Footslope soils being at a "receiving" position are expected to be more fertile than upslope soils. The results of this study do not fully confirm this hypothesis (Table 16.3). It is only true in the case of lower slope facets that are related with sum-mits genetically very infertile, as shown by the examples of T4 on Fe-Mn conglom-erate and T2 on granite. The opposite occurs on lower slope facets that are associated with summits and backslopes originally more fertile such as T3 on marble, T5 on gneiss/micaschist, and T1 on mixed gneiss/micaschist/granite. Thus, footslope soils are just fairly or moderately fertile. Summit and backslope soils are relatively similar in their fertility status, but vary considerably in the area due to parent material.

16.3.3.2 Soil Fertility Variation in Relation to Lithology

Parent rocks strongly interpose in the variation of soil fertility. Their effect was greatest on summit and backslope positions where the country rock has a strong influence on soil parent material (Table 16.4). Soil fertility in these two landforms is thus strongly related to the nature of the underlying bedrock. This is shown by high SI (more than 70 %) in all transects. Footslope soils showed fairly moderate fertility with less variation within the five transects. Nine of the ten options compar-ing footslope soils from the five transects showed a SI of more than 0.6, while only a few had an SI greater than 0.6 on summit and backslope respectively (Table 16.4).

The marble substratum (T3) produces the most fertile soils of the area (with FD of 16–18). Indeed, the dissolution of this rock provides more bases to the soil, lead-ing to high pH values (6.5–7) and high base saturation percentage (around 90 %). Soil texture is loamy. The schisto-gneissic rocks produce also fertile soils (T5 with FD of 14–16) owing their richness in phyllosilicate minerals.

Granitic rocks, dominated by tectosilicates, contribute less to soil fertility (FD of 10–12). The iron-manganese concretionary conglomerate/breccia produces the least fertile soils of the group (FD of 10–12). In fact, this material is composed of quartzitic rock fragments bedded in iron-manganese concretionary cement. The re-dissolution and redistribution of the iron produces a soil dominated by sesquioxides with low adsorption surface soil minerals, unfavourable to soil fertility.

On footslopes, bedrocks lose their influence because they are buried under col-luvial debris originating from upslope erosion. These soils tend to have the same brownish matrix and similar textures ranging from fine loamy to fine clayey through-out the area. One explanation of these changes at footslope positions may involve the reduction or chelation of iron hydroxides in the presence of organic matter in the surface horizons, and the subsequent transportation and deposition of discoloured topsoil as colluvium downslope (van Wambeke 1992; Kauffman et al. 1998). The colluvia acting as soil parent material on footslopes tend to produce finer textured soils, with low activity clay and presence of organic matter. These processes generate footslope soils of moderate fertility in our study area.

16.4   Conclusion

Soil information collected using the geopedologic approach at four hicrarchic levels (i.e. landscape, relief type, lithology, and basic landform) provided a useful basis to implement geometric models for soil fertility assessment. The parent material (i.e lithology) appeared to be the most important factor influencing soil distribution patterns and soil fertility variation, while the landform level (slope facets) interfered at a lower degree. The study area is complex and comprises a variety of soil types which could be mapped at detailed or semi-detailed level using the geopedologic approach. Also extrapolation of the information from sample areas to similar areas was made easier.

Summit and backslope soils on the same lithology have similar properties that are strongly influenced by the underlying parent rock. Meanwhile, soils on footslopes appeared to be less controlled by the parent country rocks and to depend more on colluvial debris with different properties originating from upslope erosion. The fertility distance method was useful for ranking individuals, relative soil classification, and site selection. It is flexible because the FD can be calculated from any fixed point and with any number of variables. Detailed research involving crop production as a function of FD may help scale the method, making it a powerful tool for land evaluation and land-use planning.

References

Buol SW, Couto W (1981) Soil fertility capability assessment for use in the humid tropics. In: Greenland DJ (ed) Characterization of soils in relation to their classification and management for crop production: examples from some areas of the humid tropics. Clarendon Press, Oxford

Demolon A (1952) Principes d'agronomie. Tome I: Dynamique du sol, 5th edn. Editions Dunod, Lorient

DLD (1984) Semi-detailed soil map of Amphoe Mae Taeng (1:50,000 scale), sheet 4747 II. Department of Land Development, Bankok

Driessen PM, Konijn NT (1992) Land-use systems analysis. Department of Soil Science and Geology, Malang. INRES, Den Haag

Fairbridge RW, Finkl CW Jr (1972) Encyclopedia of soil science, Part 1 physics, chemistry, biology, fertility, and technology. Dowden, Hutchinson and Ross, Stroudsburg

FAO (1976) A framework for land evaluation, FAO soils bulletin 32. FAO, Rome

FAO (1990) Guidelines for soil description. Soil Resources, Management and Conservation Service; Land and Water Development Division, FAO, Rome

Hendricks CA (1981) Soil-vegetation relations in the North Continental Highland region of Thailand: a preliminary investigation of soil-vegetation correlation, Technical bulletin 32. Soil Survey Division, Ministry of Agriculture and Cooperatives, Bankok

ITC (1993) The Integrated Land and Water Information System ILWIS 1.4. User's manual. International Institute for Aerospace Survey and Earth Sciences, Enschede

Kauffman S, Sombroek W, Mantel S (1998) Soils of rainforests: characterization and major constraints of dominant forest soils in the humid tropics. In: Schute A, Ruhiyat D (eds) Soils of tropical forest ecosystems: characteristics, ecology and management. Springer, Berlin, pp 9–20

Sanchez PA (1976) Properties and management of soils in the tropics, A Wiley-interscience publication. Wiley, New York

Sanchez PA (1977) Soil management under shifting cultivation. In: Sanchez PA (ed) A review of soils research in tropical Latin America. North Carolina State University, Raleigh, pp 46–60

Sanchez PA, Couto W, Buol SW (1982) The fertility capability soil classification system: interpretation, applicability and modification. Geoderma 27:283–309

Sanchez PA, Vilachia JH, Bandy DE (1983) Soil fertility dynamics after clearing a tropical rainforest in Peru. Soil Sci Soc Am J 47:1171–1178

Sanchez PA, Palm CA, Buol SW (2003) Fertility capability soil classification: a tool to help assess soil quality in the tropics. Geoderma 114:157–185

Soil Survey Division Staff (1993) Soil survey manual, Soil conservation service, US Department of Agriculture handbook 18. US Government Printing Office, Washington, DC

Soil Survey Staff (1998) Keys to soil taxonomy, 5th edn. SMSS technical monograph 19. USDA, Blacksburg, Virginia, USA

Tansathien W, Sripongpan P, Dhamdusdi Y, Paksamut N (1984) Geological map of Thailand, sheet of Amphoe Mae Taeng (1:50,000 scale). Geological Survey Division, Department of Mineral Resources, Bangkok

Van Reeuwijk LP (1993) Procedures for soil analysis. Technical paper, 4th edn. ISRIC, Wageningen

van Wambeke A (1992) Soils of the tropics: properties appraisal. McGraw Hill, New York

Webster R (2000) Statistics to support soil research and their presentation. Eur J Soil Sci 52:331–340

Webster R, Oliver MA (1990) Statistical methods in soil and land resource survey. Spatial Information Systems. Oxford University Press, Oxford

Yemefack M, Ndyeshumba P, Rashid MS, et al. (1994) Semi-detailed soil map of Mae Taeng, Chiang-Mai, Thailand. Soil survey group I, Sol3/2 students 1993/94. ITC, Enschede

Zinck JA (1988) Physiography and soils, Lecture notes. ITC, Enschede

Part III
Methods and Techniques Applied to Pattern Recognition and Mapping

Chapter 17
Soil Mapping Based on Landscape Classification in the Semiarid Chaco, Argentina

C. Angueira, G. Cruzate, E.M. Zamora, G.F. Olmedo, J.M. Sayago, and I. Castillejo González

Abstract The semiarid Chaco is an ecosystem shared by Argentina, Paraguay, Bolivia, and Brazil where land use changes from forest to commercial agriculture and social conflicts have been intensive during the last decade. These changes and the lack of reliable soil information at suitable scales are threatening the sustainable development of the region. In Santiago del Estero province, Argentina, a soil survey was carried out with the objective of reducing the knowledge gap. Due to the large area, geomorphology diversity, limited funding, and high demand of information, a geopedologic survey using remote sensing and GIS was considered an appropriate approach. Map units were determined based on the integration of geoforms and soils, knowledge of landscape and soil forming factors, field observations, and laboratory determinations. Three main landscape units were recognized: (1) a fluvio-eolian Chaco plain including a megafan with Haplustolls and Torripsamments,

C. Angueira (✉)
Instituto Nacional de Tecnología Agropecuaria (INTA), Estación Experimental Agropecuaria Santiago del Estero (EEASE), Santiago del Estero, Argentina
e-mail: cristina.angueira@gmail.com

G. Cruzate
INTA, Centro de Investigación en Recursos Naturales (CIRN), Buenos Aires, Argentina
e-mail: gcruzate@gmail.com

E.M. Zamora
INTA, EEA Manfredi, Córdoba, Argentina
e-mail: eduardomaxizamora@hotmail.com

G.F. Olmedo
INTA Agricultural Experimental Station Mendoza, Luján de Cuyo, Argentina
e-mail: olmedo.guillermo@inta.gob.ar

J.M. Sayago
Instituto de Geociencias y Medio Ambiente, Universidad Nacional de Tucumán, S. M. de Tucumán, Argentina
e-mail: jmsayago@arnet.com.ar

I. Castillejo González
Departamento de Ingeniería Gráfica y Geomática, Universidad de Córdoba, Córdoba, España
e-mail: ilcasti@uco.es

© Springer International Publishing Switzerland 2016   285
J.A. Zinck et al. (eds.), *Geopedology*, DOI 10.1007/978-3-319-19159-1_17

(2) the Rio Dulce valley with Torripsamments, and (3) the alluvial migratory plain of Río Salado with Torripsamments, Ustifluvents, and Natraqualfs. The used approach helped speed up the soil information collection at appropriate scale for land use planning.

Keywords Geopedology • Soil mapping • Remote sensing • GIS • Landscape classification

17.1 Introduction

The semiarid Chaco ecosystem (Cabrera 1976; Vargas Gil 1988; Sebastián et al. 2006) is a flat mixed woodland-grassland landscape shared by Argentina, Paraguay, Bolivia, and Brazil. Temperature rises from south to north and rainfall from west to east. Using unrestricted forest clearing and fire, traditional land use has changed to commercial agriculture in the last decade (Morello et al. 2006). This process, together with the lack of adequate soil information for land use planning, is threatening the sustainable development of the region.

Mitigation of the impacts caused by agricultural expansion on fragile natural resources and land use sustainability require balanced environmental performance, better understanding of the physical and anthropogenic factors affecting land use, and systematic organization of regional data and information (Angueira 1994).

The knowledge of soils and their geographic distribution is basic information for (a) agricultural research and modelling (Jhorar et al. 2003; Walter et al. 2007), (b) land capability assessment (Bouma and Bregt 1989; Angueira and Zamora 2003); (c) sustainable land use planning at regional, local, and farm scale (McRae and Burnham 1981; Rossiter and van Wambeke 1991), and (d) integration of relevant information in geographic information systems (Zinck 1994; Burrough and McDonnell 1998; Angueira et al. 2007).

Earlier traditional soil inventory methods were expensive and time consuming due to the high cost of remote sensing documents, difficulties to plan fieldwork properly, and limited application of integrated soil-landscape survey. Nowadays, these restrictions have been overcome by the development of tools, methods, and systems in remote sensing, geostatistics, GIS, and data processing (McBratney et al. 2003), provided by progress in computer technology and informatics.

The geomorphic approach helps understand the spatial distribution of soils on the landscape, and geomorphic processes influence soil formation and features (Bockheim et al. 2005). The synergism between pedology and geomorphology is important for predicting areal distribution of soils on the landscape, and it is the basis of geopedology, a transdisciplinary soil survey approach (Zinck 1994, 2012, 2013).

Satellite images offer the possibility of segmenting the landscape into units whose soil composition can be determined by conventional or advanced methods, and their use supports extending soil survey to inaccessible areas by reducing time and fieldwork (Mulder et al. 2011). Landsat satellite data have been used for physiographic soil mapping (Sayago 1982), geological mapping (Moore et al. 2007), and surface features mapping (Metternicht and Zinck 2003). Moreover, the combination of multi-source geographic datasets (Krol et al. 2007), digital elevation model (DEM), and spectral satellite data may improve landform classification in complex landscapes (Dobos et al. 2000).

In Argentina, a national soil map at scales of 1:500,000 and 1:1,000,000 shows cartographic units labeled on the basis of soil-landscape relationships (INTA 1990). A soil survey of the Pampa region carried out at scale 1:50,000 provides map units in terms of soil associations and complexes (Echevehere 1976). The semiarid Chaco is covered by detailed soil surveys only for small parts (Peña and Salazar 1978; Angueira and Vargas Gil 1993; Angueira and Zamora 2003).

This chapter presents the geopedologic survey of an area covering the southwestern semiarid Chaco region at 1:500,000 scale. The combination of geopedology and modern geomatic techniques was considered appropriate for soil inventory in the semiarid Chaco because of the lack of soil studies at suitable scales, large area extent, diversity of geomorphology, soil types and land uses, high demand of information, and limited funding and trained staff.

17.2   Materials and Methods

17.2.1   Study Area

The semiarid Chaco ecosystem is a sedimentary plain fractured and dislocated in depth. A depression with subsequent accumulation of sediments formed during the Tertiary, followed by an uplift accompanied by strong parallel and transverse folds and faults (Abitbol 1997; Martin 1999; Peri and Rosello 2010).

The present landscape, the fluvial network, and the Mar Chiquita depression were formed in the late Pleistocene and covered by aeolian sediments (Sayago 1995; Carignano et al. 2014). These physiographic processes determined changes in the Salado and Dulce river systems. The Rio Dulce formed complex alluvial fans (Martin 1994; Barbeito and Ambrosino 2007) and its main channel is assumed to have shifted southward to the Salinas Ambargasta Sumampa or together with Rio Salado to Mar Chiquita, until the current position (Martin 1999).

The study area of 8,800 km^2 is located in the south-west of the semiarid Chaco region (Vargas Gil 1988), in Santiago del Estero province, between 27°30′–28°35′S and 63°45′–64°35′W (Fig. 17.1). The climate according to the Thornthwaite classification is DB'4da' semiarid, without or with little excess of water, mesothermal, with rainfall concentrated during the summer months.

The landscape is gently sloping from the west to north-east, east and south-east, and shows a set of landforms resulting from exogenous and endogenous processes.

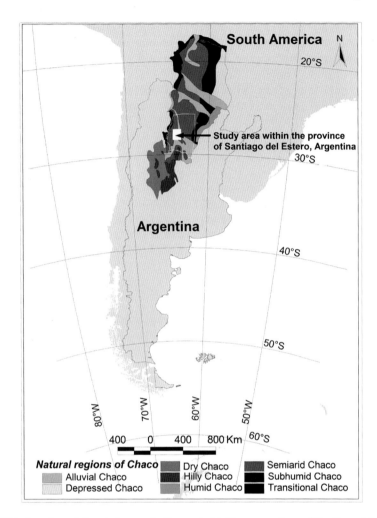

Fig. 17.1 Location of the study area in the semiarid Chaco, province of Santiago del Estero, Argentina (Vargas Gil 1988)

It comprises part of a fluvio-eolian plain and the alluvial plains of Rio Dulce and Rio Salado (Angueira et al. 2007) with their ancient and present meanders characteristic of rivers in areas with low slope or energy.

17.2.2 Materials

Data were obtained from maps and reports, remote sensing documents, field observations, and laboratory determinations of soil properties. The material used included the geomorphology map of Santiago del Estero (Angueira et al. 2007), the soil maps of Santiago del Estero (INTA 1990; Angueira et al. 2007) and the right bank of Rio Dulce (Angueira and Zamora 2003), and topographic data derived from a DEM 90-m (CGIAR-CSI 2004) with altitude values with 3 arc sec interval.

Landsat satellite images (NASA-USGS 1972) were selected on the basis of adequate spatial resolution for the scale of work and availability of long term records, even from the decade of the 1970s with unchanged native vegetation. Selected scenes included MSS 246-79 (02/75); TM 230-079 and 229-079 in dry and wet seasons from 1984 to 2011; and 8 OLI (05/14). Complementary data were obtained from CBERS (INPE 2008) and SAC-C (CONAE 2000) images.

ArcGis 9.3 (ESRI), Imagine 9.3.1 (Leica), and SAGA (Böhner et al. 2006) software were used for mapping, digital processing of satellite images, calculating derivative morphographic and morphometric attributes, and displaying the results.

17.2.3 Methods

17.2.3.1 General Methodology for the Geopedologic Survey

The work started with a review of available information and an overview of the area to identify major geomorphic features, while maps were displayed in a GIS environment for easy and efficient handling.

Using an iterative and exhaustive visual interpretation of topographic data and satellite imagery, preliminary map units were delineated, a draft hierarchic geoform legend as cartographic frame was established to define soil-landscape relationships, and sites for describing soils and checking boundaries in the preliminary physiographic units were identified. Physical and chemical soil properties were determined, and soils were classified according to Soil Survey Staff (2010). The interpretative map was converted into geopedologic map after confirming boundaries, legend, characterization of soils and their spatial distribution patterns.

17.2.3.2 Geospatial and Soil Analysis Tools and Methods

Morphographic and morphometric attribute maps were established from DEM 90-m to describe, identify, and classify by visual interpretation the geoforms at different levels of the taxonomic system (Zinck 2013). The following attributes were calculated with the ArcGis software: slope, hillshade, profile curvature, viewshed, wetness index, flow direction, flow accumulation, flow length, stream link, stream network, stream order, drainage network, watershed basin, aspect, cross-sectional curvature, longitudinal curvature, convergence index, and closed depressions. To improve visualization of the drainage network, the ratio of surface area drained by each outlet more appropriate to the scale of work was selected.

For automated classification of geoforms, the attributes of analytical hillshading, slope, aspect, cross-sectional curvature, longitudinal curvature, convergence index, and closed depressions were combined using the isodata cluster analysis by SAGA software for an isoform map.

Systematic visual interpretation of Landsat, SAC, and CBERS satellite images was carried out after preprocessing, radiometric correction, and georeferencing on the Gauss-Krüger projection IGN (2000). Criteria and procedures of visual interpretation, analysis of elements, landscape patterns, and physiographic methods (Goosen 1967) were applied. Tone, texture, color, pattern, shape, size, height, elevation, location, and their association with other objects, were all elements considered to characterize the physiographic system that has controlled the formation of the area.

DEM-derived maps were combined with Landsat images to improve visual interpretability and understanding of the relationships between the landscape elements (Shepande 2010). Images from dry and wet seasons were analyzed with different band composition to identify water bodies, sediments in water, drainage networks, vegetation types, texture, soil moisture, terrain features, and soil conditions.

The SAC-C scene was used to identify patterns of waterlogging following the extremely heavy rainfall of 30.03.2006. CBERS high-resolution (CCD) scenes were used to improve the delineation of fluvial landforms, meanders, and spill areas with accumulation of material.

A set of 176 georeferenced soil profiles and observations, approximately 1/50 km^2, was described according to the guidelines of Echevehere (1976). Laboratory determinations on dried soil samples included soil texture by Bouyoucos, pH in soil paste, EC in saturated soil paste extract, organic carbon by Walkley-Black, CEC by ammonium acetate 1N pH 7, and percentage of $CaCO_3$ by the Scheibler method.

17.3   Results

17.3.1   Geospatial Analysis

Visual interpretation of a DEM 90-m contour map with 1 m vertical intervals (Fig. 17.2) allowed distinguishing relief characteristics such as faults (1) and slope changes, while a 5 m interval map was useful to identify runoff ways and separate high- and low-lying areas.

The drainage network map shows a watershed flowing to the north and east of the Rio Salado, an alluvial fan to the south-east, and a main course to the south of Río Dulce. From the isoform map (Fig. 17.3) were recognized the following features: a fan with its apex in the west and a divergent gently sloping area to north-east and south-east (1), a main and a secondary fault (2), north-south parallel valleys at the foot of the main fault (3), and a shallow sag pond (4) at the foot of the secondary fault.

Water bodies, main rivers, streams, meanders, soil wetness, vegetation type changes were identified by distinguishing color patterns, tones, and texture, analyzed

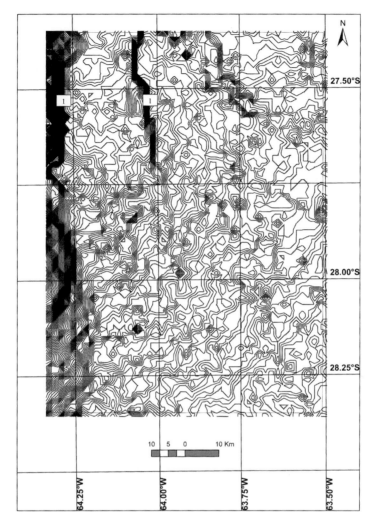

Fig. 17.2 Contour map with 1 m vertical intervals, (1) main fault in the west and secondary in the east

in sequences of different band composition of satellite images. Visual analysis revealed that bands (2,3,4), (3,4,5), (3,5,7), and in gray colors, provided optimal contrast.

17.3.2 Soil-Landscape Relationships

Iterative analysis of the soil-landscape relationships, fieldwork, and laboratory data allowed classifying the soil subgroups recognized in each landform. Mean values of selected soil properties are shown in Table 17.1.

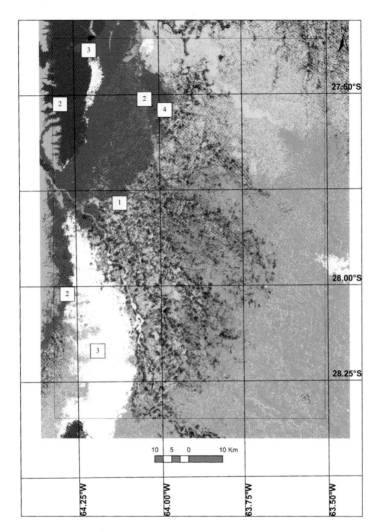

Fig. 17.3 Isoform map showing *1* fan, *2* secondary fault, *3* valleys, *4* shallow sag pond at the foot of the secondary fault

Geoforms and soils influence each other, being one of them the dominant factor according to circumstances, natural conditions, and landscape types. The main characteristic in the study area is that the sedimentary processes control soil distribution and properties, the type of pedogenesis, and the degree of soil development in all landscapes.

In the loess-covered proximal megafan dominate Torriorthentic Haplustolls without cambic horizon, together with Aquic Haplustolls in blowout depressions that have excess water and aquic conditions in some periods of most years.

In the distal megafan, characterized by a radial drainage network and interfluvial plains, Torriorthentic Haplustolls associated with Aridic Haplustolls are the main

Table 17.1 Mean values of relevant soil properties per landscape unit and subgroup

Hor	Taxon	Text	Clay %	Silt %	Sand %	OM %	CO₃ %	pH	EC dS/m	CEC cmol kg⁻¹	ESP %
Fluvio-eolian Chaco plain – proximal megafan (Río Sali-Dulce) – 1 loess cover – 1P											
A	Torriortentic Haplustoll	SiL	11	60	29	2.1	0.0	0.0	0.3	15.7	0.0
AC		SiL	8	60	32	0.9	0.0	7.2	0.2	13.1	4.2
Ck		SiL	6	58	36	0.4	1.5	7.7	1.0	11.6	18.7
Fluvio-eolian Chaco plain – proximal megafan (Río Sali-Dulce) – 2 blowout depression – 2P											
A	Aquic Haplustoll	L	10	60	29	1.6	0.0	5.7	0.2	2.0	2.3
AC		L	8	64	28	1.0	0.0	7.2	0.3	15.3	2.8
C		L	8	65	27	0.7	0.0	7.7	0.3	14.9	3.2
Fluvio-eolian Chaco plain – distal megafan (Río Sali-Dulce) – 3 interfluvial plain – 3P											
A	Aridic Haplustoll	SiL	20	54.9	25.3	2.2	0.9	7.5	2.2	18.0	10.7
Bw		SiL	19	53.5	27.5	0.8	1.4	7.8	3.1	16.1	19.1
Ck		SiL	20	52.6	27.7	0.6	2.2	7.8	3.8	16.2	19.9
A	Torriortentic Haplustoll	SiL	14	53	34	2.0	0.3	6.8	3.5	14.4	6.3
AC		SiL	13	54	33	0.8	0.7	7.3	2.7	12.6	6.7
Ck		SiL	12	52	36	0.4	2.6	7.6	4.1	10.7	17.3
A	Aridic Argiustoll	SiL	21	51	28	2.8	1.5	7.2	1.0	19.7	4.0
Bt		CL	28	46	27	1.3	0.0	7.6	0.6	17.7	8.0
BC		SiL	20	51	22	0.7	1.4	7.5	0.9	17.2	25.0
Ck		L	23	44	33	0.2	1.5	7.6	1.0	16.1	22.5
Fluvio-eolian Chaco plain – distal megafan (Río Sali-Dulce) – 4 infilled channel – 4P											
A	Ustic Torripsamment	LSa	3	30	67	0.8	0.0	6.9	13.5	7.4	19.0
C		LSa	1	20	79	0.3	2.1	7.4	17.3	7.0	21.0
A	Ustic Torriorthent	SiL	10	38	52	1.5	0.0	8.1	1.0	10.2	2.5
AC		SiL	8	43	49	0.6	0.0	7.9	0.5	9.8	2.1
Cl		SiL	7	45	48	0.3	3.0	8.1	0.6	9.4	5.8

(continued)

Table 17.1 (continued)

Hor	Taxon	Text	Clay %	Silt %	Sand %	OM %	CO₃ %	pH	EC dS/m	CEC cmol kg⁻¹	ESP %
Fluvio-eolian Chaco plain – old alluvial overland flow (Río Sali-Dulce) – 5 overflowed depression – 5P											
A	Aridic Haplustalf	L	24	48	28	1.6	0.0	6.9	5.8	21.0	3.0
Bt		CL	37	43	20	1.7	0.0	6.4	17.0	20.1	3.0
BC		L	26	43	31	1.0	1.7	6.6	37.5	15.8	22.0
Fluvio-eolian Chaco plain – old alluvial overland flow (Río Sali-Dulce) – 6 alluvial overflow levee – 6P											
A	Ustic Torriorthent	L	13	43	44	2.0	0.9	7.3	18.7	14.7	21.0
AC		L	15	48	37	0.7	1.4	7.6	53.0	13.6	31.0
C		L	13	47	40	0.4	6.1	7.6	66.5	12.8	33.0
A	Ustic Torrifluvent	SiL	8	70	22	2.0	1.3	6.8	2.5	13.6	6.0
AC		SiL	6	70	24	0.4	1.2	7.0	12.4	10.0	15.0
II		SiC	40	44	16	1.0	0.0	6.8	17.5	18.8	15.0
Valley (Rio Dulce) – middle terrace – 7 Levee and overflows (mt) – 7D											
A	Aridic Haplustoll	SiL	19	63	18	4.3	0.0	6.1	0.0	15.4	16.0
Bw		SiL	16	68	16	1.2	0.0	6.4	0.0	13.8	18.0
Ck		SiL	19	61	20	1.4	1.9	6.7	0.0	11.0	23.0
Valley (Rio Dulce) – low terrace – 8 levee and overflow (lt) – 8D											
A	Typic Haplustoll	SaL	6	44	50	3.5	1.1	6.8	6.1	13.1	6.0
C		SaL	3	36	61	0.1	3.0	8.2	1.3	7.8	27.0
Valley (Rio Dulce) – active floodplain – 9 river – 9D											
A	Ustic Torripsamment	Sa	1	3	96	0.2	0.0	7.3	0.1	3.2	0.2
C		Sa	1	5	94	0.6	0.0	6.3	0.2	2.9	6.9
Alluvial migratory plain (Río Salado) – active fluvial valley- alluvial overflow plain 10S											
A	Typic Natraqualf	L	18	40	42	3.1	0.0	6.3	0.9	15.8	5.0
Bn		L	35	43	22	0.6	0.0	6.1	10.6	13.8	15.0
BC		L	26	48	26	0.3	0.0	6.3	36.0	12.6	20.0
Ck		L	22	46	32	0.1	3.0	7.4	16.3	11.6	16.0

Alluvial migratory plain (Rio Salado) – active fluvial valley – levee 11S

A	Ustic Torripsamment	SiL	14	52	34	1.4	1.0	7.7	1.8	12.4	13.0
AC		L	10	50	40	0.4	2.2	7.4	49.0	10.6	36.0
Csa		L	14	44	42	0.1	2.7	7.5	65.0	11.0	36.0

Alluvial migratory plain (Río Salado) – active floodplain – 12 alluvial overflow swamp – 12S

I	Ustic Torrifluvent	SiL	11	63	26	3.5	3.0	7.9	1.9	21.2	11.7
II		SiL	9	65	26	0.6	2.5	8.2	4.3	12.2	42.0
III		SiL	9	35	56	0.4	0.9	7.9	12.0	9.1	28.0

Alluvial migratory plain (Río Salado) – Fluvio-eolian terrace remnant – 13 alluvial flat – 13S

A	Aridic Haplustoll	L	16	48	36	2.2	0.0	7.2	4.0	14.8	5.0
Bw		L	15	47	38	1.2	0.0	7.2	10.8	12.7	16.0
BC		L	16	48	36	0.6	0.0	7.4	18.5	12.6	22.0
Ck		L	14	50	36	0.4	2.0	7.7	19.4	12.5	24.0

Alluvial migratory plain (Río Salado) – Fluvio-eolian terrace remnant – 14 alluvial channel – 14S

A	Ustic Torripsamment	SaL	9	34	57	2.4	0.0	6.4	0.7	9.0	6.0
AC		SaL	7	34	59	0.8	0.0	6.8	12.9	9.0	22.0
C		SaL	4	32	64	0.5	0.0	7.6	17.7	7.8	36.0

soils, while Aridic Argiustolls occur in micro-depressions. Ustic Torriorthents are common on sandy alluvial overflow levees and Aridic Haplustalfs in overflowed depressions.

In the Río Dulce valley, soils of different levels of development and contrasting textures occur, with poorly developed Entisols on modern floodplain deposits and more developed Aridic and Typic Haplustolls on terrace levels. In the alluvial migratory plain of Río Salado, Typic Natraqualfs developed on the overflow plain with high water table, and Ustic Torripsamments on the sandy levees.

In the active floodplain of the streams draining the Huyamampa depression occur Ustic Torrifluvents together with alluvial saline-sodic soils. On the fluvio-eolian terrace remnant, Aridic Haplustolls are the dominant soils, with Ustic Torripsamments in elongated and irregular depressions.

17.3.3   Geopedologic Map and Legend

The hierarchic classification of the geopedologic map units in landscapes, moldings, and landforms together with their soil components at various scales was established from the integrated analysis of the relationships and interactions between geoforms and soils. The spatial distribution of soils is related to landforms at all scales. The map units are soil associations and complexes, consisting of two or more soil taxa geographically associated in a landform.

The study area comprises 3 landscape types, 9 molding types, and 14 landform types. They are described including name, symbol, and soil composition, and shown in the geopedologic map (Fig. 17.4) and legend (Fig. 17.5).

17.3.3.1   Fluvio-eolian Chaco Plain

The fluvio-eolian Chaco plain in the west and center of the study area, a slightly convex landscape with 0.5–1 % slope, presents three moldings belonging to the Rio Sali-Dulce system and including a proximal megafan, a distal megafan, and an old alluvial overland flow area, with their corresponding landform units and soil components.

(a) The *proximal megafan* is a gently sloping terrain covered by a loess mantle including at the level of landform: a *loess cover unit* (*1P*), a gently sloping surface with Torriorthentic Haplustolls, dissected by *blowout depressions* (*2P*), which are wide and elongated shallow areas occasionally acting as runoff paths with Aquic Haplustolls.

(b) The *distal megafan* is a cone-shaped deposit of sand and finer materials formed in the area where the river slows down and spreads into a flatter plain at the exit of the Huyamampa N-S fault. It is composed of two landforms: an *interfluvial plain* (*3P*), flat or gently sloping areas, with dominant Aridic Haplustolls and Torriothentic Haplustolls, and subordinate Aridic Argiustolls, and an *infilled*

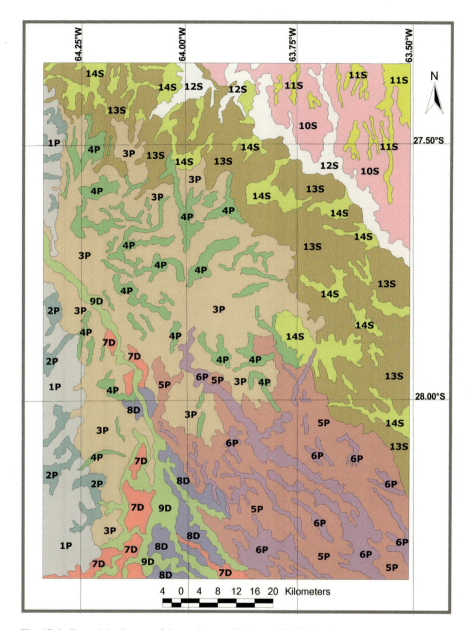

Fig. 17.4 Geopedologic map of the study area (See legend in Fig. 17.5)

channel (*4P*), flat irregular, elongated or curvilinear shallow depressions, back-filled with sediments, located within the interfluvial plain, with Ustic Torripsamments as dominant and Ustic Torriorthents as subordinate soils.

(c) The *old alluvial overland flow area* with an *overflowed depression* (*5P*), a relatively flat, nearly closed fan-shaped accumulation of sand and finer sediments

LANDSCAPE	MOLDING	FACIES	LANDFORM	CODE	SOILS
Fluvio-eolian Chaco plain (Rio Sali-Dulce)	Proximal megafan	Eolian	1 Loess cover	1P	Torriorthentic Haplustolls
			2 Blowout depression	2P	Aquic Haplustolls
	Distal megafan	Alluvial	3 Interfluvial plain	3P	Aridic Haplustolls Torriorthentic Haplustolls Aridic Argiustolls
			4 Infilled channel	4P	Ustic Torripsamments Ustic Torriorthents
	Old alluvial overland flow	Alluvial	5 Overflowed depression	5P	Aridic Haplustalfs
			6 Alluvial overflow levee	6P	Ustic Torriorthents Ustic Torrifluvents
Valley (Rio Dulce)	Middle terrace	Alluvial	7 Levee and overflows (mt)	7D	Aridic Haplustolls
	Low terrace		8 Levee and overflows (lt)	8D	Typic Haplustolls
	Active floodplain		9 River	9D	Ustic Torripsamments
Alluvial migratory plain (Rio Salado)	Active fluvial valley	Alluvial	10 Alluvial overflow plain	10S	Typic Natraqualfs
			11 Levee	11S	Ustic Torripsamments
	Active floodplain	Alluvial	12 Alluvial overflow swamp	12S	Ustic Torrifluvents
	Fluvio-eolian terrace remnant	Eolian over alluvial	13 Alluvial flat	13S	Aridic Haplustolls
			14 Alluvial channel	14S	Ustic Torripsamments

Fig. 17.5 Legend of the geopedologic map of the study area (See map in Fig. 17.4)

that formed where the currents slow down and dissipate, with Aridic Haplustalfs as dominant soils. It is crossed by many channels and levees, elongated parallel areas oriented NW-SE crossing the slightly lower floodplain, named *alluvial overflow levees* (*6P*), with Ustic Thorriothents as dominant and Ustic Torrifluvents as subordinate soils.

17.3.3.2 Rio Dulce Valley

The Rio Dulce Valley is characterized by watercourses and three molding types, i.e. a middle terrace, a low terrace, and an active floodplain.

(a) The *middle terrace* is the higher terrain area formed by the river on its right side with levee and former watercourse included in the landform labelled as *levee and overflows* (*mt*) (*7D*), with Aridic Haplustolls as dominant soils.

(b) The *low terrace* was identified at the left side of the river. The main landform labelled as *levee and overflows* (*lt*) (*8D*) comprises flat surfaces above the floodplain that are formed by the deposition of alluvium adjacent to the river exposed to periodic overflows, with Typic Haplustolls as dominant soils.

(c) The *active floodplain* includes the landform named *river* (*9D*), formed by the main course of the Rio Dulce and lower order courses generally dry, with floodwaters spilling out of the riverbed. Ustic Torripsamments are the dominant soils with subordinate Ustic Torrifluvents.

17.3.3.3 Alluvial Migratory Plain of Rio Salado

The alluvial migratory plain of Rio Salado is an area with 0.5–1 % slope to the south-east, composed by three moldings including an active fluvial valley, an active floodplain, and a fluvio-eolian terrace remnant.

(a) The *active fluvial valley* comprises two landforms: an *alluvial overflow plain* (*10S*), an extensive, depressed area between natural levees and terraces with Typic Natraqualfs as dominant and Ustic Torrifluvents as subordinate soils, and *levees* (*11S*), elongated high areas, almost parallel in north-south direction, distributed throughout the alluvial overflow plain, with dominant Ustic Torripsamments and subordinate Aridic Natrustalfs.

(b) The *active floodplain* consists of an *alluvial overflow swamp* (*12S*), streams and a low-lying saturated ground, intermittently covered with water and vegetated by shrubs and trees, with Ustic Torrifluvents as dominant soils.

(c) The *fluvio-eolian terrace remnant* is formed by two landforms: an *alluvial flat* (*13S*), a large gently sloping area, nearly level, erosional remnant of an alluvial plain without drainage network, with dominant Aridic Haplustolls and subordinate Ustic Torriorthents, and an *alluvial channel* (*14S*), a concave shallow microrelief through which runoff is drained in periods of high water, with Ustic Torripsamments as dominant soils.

17.4 Discussion

The geopedologic survey provided information highlighting the complex and intricate soil-geoform patterns of a representative semiarid Chaco area. The integration between geomorphology and pedology along the survey process is reflected in the legend structure, the geomorphic units providing the cartographic frames for the soil types.

Because soils have formed from loess in the west and from alluvial material in the center and north-east, the textures in the C horizons vary: they are mainly silty loam in the megafan landforms (1P and 2P); in alluvial landscapes the sand content increases and is variable according to the position on the relief with sandy loam in levees (9D, 12S and 14S), loam (3P, 4P, 5P, 8D, 10S) and silty loam (6P, 7D, 11S, 13S) in wide flat positions, and variable in slightly concave low positions (Fig. 17.6).

The versatility of multi-spectral satellite images was demonstrated to study soil-landscape features in different seasons and scales. During the dry season (June–September), characterized by the scarcity of vegetation, the visibility of the terrain surface was better, consistent with observations by Shepande (2010).

Visual image interpretation, despite being time consuming and subjective, allows the surveyor to use his/her own knowledge and experience to improve the delineation of map units, as stated by Sarapaka and Netopi (2010) and Trotter (1991) quoted by Shepande (2010). Maps generated through visual interpretation have de advantage of being relatively simple and inexpensive (Manchanda et al. 2002).

The use of DEM and Landsat imagery in a GIS framework proved to be an improved method for mapping soil patterns in the Chaco flat study area, as well as in hilly terrain (Aksoy et al. 2009). DEM was important to derive morphographic and morphometric attributes that are used in soil-landscape characterization at regional scale (Dobos et al. 2000) and to reduce the disadvantage caused by the absence of stereovision in visual interpretation of images.

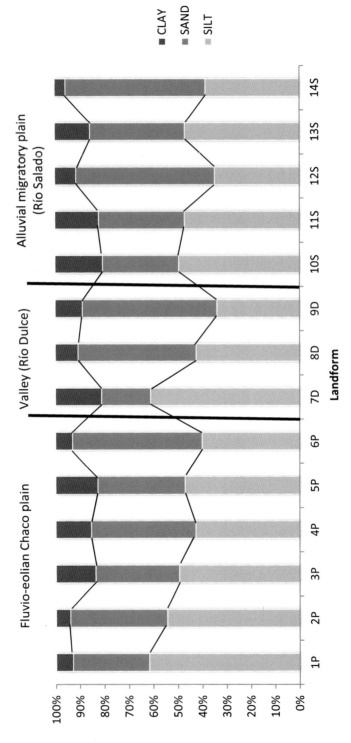

Fig. 17.6 Clay, silt, and sand contents of the C horizons in landscape and landform map units. (*1P* loess cover, *2P* blowout depression, *3P* interfluvial plain, *4P* infilled channel, *5P* overflowed depression, *6Pc* alluvial overflow levee, *7D* levee and overflows (mt), *8D* levee and overflows (lt), *9D* river, *10S* alluvial overflow plain, *11S* levee, *12S* alluvial overflow swamp, *13S* alluvial flat, *14S* alluvial channel)

Geopedology improves the perspective of soil studies at regional scale, together with digital soil mapping to improve and complete the spatial and thematic coverage of regional soil-landscape relationships.

17.5 Conclusions

The geopedologic survey proved to be useful for mapping large and complex geomorphic areas, with a variety of landscapes, very sparsely populated, and lacking sufficient all-weather roads and basic infrastructure.

The use of DEM map derivatives, multiple spectral, temporal and spatial resolution satellite images, and visual interpretation techniques were useful to enhance the ability to identify and classify landscapes and soils.

The application of the geopedologic approach based on remote sensing data, use of modern survey techniques, knowledge of landscape and soil forming factors, and fieldwork contributed to soil mapping at appropriate scales in areas of agricultural expansion for land evaluation and planning.

Acknowledgement This study was funded by the National Institute for Agricultural Technology (INTA), Argentina. The support of Sanchez de la Orden M. from the Universidad de Córdoba, Spain is appreciated. We express our gratitude to our colleagues Lorenz G, Boetto M, Martin A, and Barbeito O for sharing information and knowledge about this region and to López J and Barraza G for their contribution to computer aspects.

References

Abitbol AE (1997) Programa de desarrollo de pequeñas comunidades. Informe Técnico Santiago del Estero. CFI, Buenos Aires

Aksoy E, Ozsoy G, Sabri Dirim M (2009) Soil mapping approach in GIS using Landsat satellite imagery and DEM data. Afr J Agric Res 4(11):1295–1302

Angueira C (1994) Evaluación de tierra, esquema FAO: Lavalle-Tapso-Frias. INTA, Santiago del Estero

Angueira C, Vargas Gil JR (1993) Carta de suelos de Lavalle-Tapso-Frias. INTA, Santiago del Estero

Angueira C, Zamora E (2003) Oeste del área de riego del Río Dulce, Santiago del Estero, Argentina. Carta de Suelos. INTA, Santiago del Estero

Angueira C, Prieto D, López J, Barraza G (2007) Sistema de información geográfica de Santiago del Estero, 2.0. CD ROM and http://sigse.inta.gov.ar. INTA, Santiago del Estero

Barbeito O, Ambrosino S (2007) Estudio de áreas de inundación extraordinaria del curso inferior del Río Dulce; análisis geomorfológico. Hidroeléctrica Río Hondo SA

Bockheim JG, Gennadiyev AN, Hammer RD, Tandarich JP (2005) Historical development of key concepts in pedology. Geoderma 124:23–36

Böhner J, McCloy KR, Strobl J (2006) SAGA: analysis and modeling applications, vol 115, Göttingen Geographische Abhandlungen. Verlag Erich Goltze GmbH, Göttingen

Bouma J, Bregt AK (1989) Land qualities in space and time. In: Proceedings of a symposium organized by ISSS. Pudoc, Wageningen, The Netherlands

Burrough PA, McDonnell RA (1998) Principles of geographical information systems. Oxford University Press, Oxford

Cabrera AL (1976) Regiones fitogeográficas argentinas. Enciclopedia Argentina de Agricultura y Jardinería. Tomo II Fs 1 Acme, Buenos Aires, pp 1–85

Carignano C, Kröhling D, Degiovanni S, Cioccale M (2014) Geomorfología. Relatorio XIX Congreso Geológico Argentino, Córdoba

CGIAR-CSI (2004) The CGIAR Consortium Spatial Information. http://srtm.csi.cgiar.org

CONAE (2000) Mission SAC-C. http://www.conae.gov.ar/satelites/sac-c.html

Dobos E, Michelli E, Baumgardner MF, Biehl L, Helt T (2000) Use of combined digital elevation model and satellite radiometric data for regional soil mapping. Geoderma 97:376–391

Echevehere PH (1976) Normas de reconocimiento de suelos. INTA-CIRN, Pub 152 2da edn, Buenos Aires

Goosen D (1967) Aerial photo interpretation in soil survey, Soils Bull 6. FAO, Rome

IGN (2000) http://www.ign.gob.ar/NuestrasActividades/ProduccionCartografica

INPE-CBERS (2008) http://www.cbers.inpe.br

INTA (1990) Atlas de suelos de la República Argentina. Tomo II. Proyecto SAG-INTA PNUD Argentina 85/010

Jhorar RK, Rossiter W, Siderius W, Feddes RH (2003) Calibration of effective soil hydraulic parameters of heterogeneous soil profiles. J Hydrol 28:233–247

Krol B, Rossiter DG, Siderius W (2007) Ontology-based multi-source data integration for digital soil mapping. In: Lagacherie P, McBratney AB, Voltz M (eds) Digital soil mapping. An introductory perspective, vol 31, Developments in soil science. Elsevier, Amsterdam, pp 119–133

Manchanda MI, Kudrat M, Tiwari AK (2002) Soil survey and mapping using remote sensing. Trop Ecol 43(1):61–74

Martín AP (1994) Hidrogeología del abanico aluvial del Río Dulce en las ciudades de Santiago del Estero y La Banda XXIV Congreso Iberoamericano de Ingeniería Sanitaria. AIDIS. Tomo II Diagua 44:22. Buenos Aires, Argentina

Martín AP (1999) Hidrogeología de la Provincia de Santiago del Estero. Edición del Rectorado de la Universidad Nacional de Tucumán, Tucumán, Argentina

McBratney AB, Mendonça Santos ML, Minasny B (2003) On digital soil mapping. Geoderma 117:3–52

McRae CP, Burnham SG (1981) Land evaluation. Clarendon, Oxford

Metternicht GI, Zinck JA (2003) Remote sensing of soil salinity: potentials and constraints. Remote Sens Environ 85:1–20

Moore CA, Hoffmann GA, Glenn NF (2007) Quantifying basalt rock outcrops in NRCS soil map units using Landsat-5 data. SSSA Madison Soil Surv Horiz 48:59–62

Morello J, Pengue W, Rodríguez A (2006) Etapas de uso de los recursos naturales y desmantelamiento de la biota del Chaco. In: Brown A, Martinez Ortiz U, Acerbi M, Corcuera J (eds) La situación ambiental argentina 2005. Fundación Vida Silvestre Argentina, Buenos Aires, pp 83–90

Mulder VL, de Bruin S, Schaepman ME, Mayr TR (2011) The use of remote sensing in soil and terrain mapping. Geoderma 162(1–2):1–19

NASA-USGS (1972) Landsat mission. http://landsat.usgs.gob

Peña Zubiate C, Salazar Lea Plaza J (1978) Carta de suelos de los Departamentos Belgrano y Gral. Taboada. INTA, Buenos Aires

Peri VG, Rosello EA (2010) Anomalías morfoestructurales del drenaje del Río Salado sobre las Lomadas de Otumpa (Santiago del Estero y Chaco) detectadas por procesamiento digital. Rev Asoc Geol Argent 66(4):634–645

Rossiter DG, van Wambeke AR (1991) Automated land evaluation system (ALES). Cornell University, Ithaca

Sarapaka B, Netopi P (2010) Erosion processes on intensively farmed land in the Czech Republic: comparison of alternative research methods. 19th World Congress of Soil Science, Soil Solutions for a Changing World, Brisbane

Sayago JM (1982) Interpretability of Landsat images for physiography and soil mapping in the sub-humid region of the Northeast of Argentina. In: Proceedings of the international symposium on remote sensing of environment, Buenos Aires, pp 977–987

Sayago JM (1995) The Argentine neotropical loess: an overview. Q Sci Rev 14:755–766

Sebastián A, Torrella SA, Adámoli J (2006) Situación ambiental de la ecorregión del Chaco Seco. In: Brown A, Martinez Ortiz U, Acerbi M, Corcuera J (eds) La situación ambiental argentina. Fundación Vida Silvestre Argentina, Buenos Aires, pp 75–82

Shepande C (2010) Development of geospatial analysis tools for inventory and mapping of soils of the Chongwe Region of Zambia. PhD dissertation, Minnesota University USA

Soil Survey Staff (2010) Keys to soil taxonomy, 11th edn. USDA-NRCS, Washington, DC

Trotter G (1991) Remotely-sensed data as an information source for geographical information systems in natural resource management: a review. Int J Geogr Inform Syst 5(2):225–239

Vargas Gil JR (1988) Chaco Sudamericano. Regiones Naturales en X Reunión Grupo Técnico Regional en Forrajeras. FAO, Córdoba

Walter C, Lagacherie P, Follain S (2007) Integrating pedological knowledge into digital soil mapping. In: Lagacherie P, McBratney AB, Voltz M (eds) Digital soil mapping. An introductory perspective, vol 31, Developments in soil science. Elsevier, Amsterdam, pp 281–300

Zinck JA (1994) Soil survey: perspectives and strategies for the 21th century. ITC publication 21. Enschede

Zinck JA (2012) Geopedología. Elementos de geomorfología para estudios de suelos y de riesgos naturales, ITC special lecture notes series. ITC, Enschede

Zinck JA (2013) Geopedology. Elements of geomorphology for soil and geohazard studies, ITC special lectures notes series. ITC, Enschede

Chapter 18
Updating a Physiography-Based Soil Map Using Digital Soil Mapping Techniques

D.J. Bedendo, G.A. Schulz, G.F. Olmedo, D.M. Rodríguez, and M.E. Angelini

Abstract Research work carried out in Entre-Rios province (Argentina) for mixed land use planning and management in relation to suitable soil conditions required high-resolution soil information at farm level. Basic information was provided by a 1:20,000 scale soil map made using physiographic analysis with intensive aerial photo-interpretation of soil-landscape relationships and landscape-oriented field survey. Continuous productivity-index (PI) classes were predicted from a number of environmental covariates, mostly DEM derivatives, using regression and geostatistical techniques. The PI land classification was used to adjust the soil-landscape/soil-series interpretation of the existing choropleth soil map by means of correlating discrete PI values obtained from a conventional mapping procedure with continuous PI values obtained by soil digital mapping procedures.

Keywords Conventional soil map • DEM-derived environmental covariates • Regression analysis • Kriging analysis • Continuous soil productivity index

18.1   Introduction

Land evaluation is an essential tool in the land use planning process. It allows assessing soil suitability, predicts soil behavior under current or future use, provides a basis for monitoring control measures of degradation processes, and contributes to formulate land use strategies by comprehensively considering all soil functions within the ecosystem.

D.J. Bedendo (✉) • G.A. Schulz
INTA Agricultural Experimental Station Paraná, Oro Verde, Argentina
e-mail: bedendo.dante@inta.gob.ar; schulz.guillermo@inta.gob.ar

G.F. Olmedo
INTA Agricultural Experimental Station Mendoza, Luján de Cuyo, Argentina
e-mail: olmedo.guillermo@inta.gob.ar

D.M. Rodríguez • M.E. Angelini
INTA Soil Research Institute Castelar, Buenos Aires, Argentina
e-mail: rodriguez.dario@inta.gob.ar; angelini.marcos@inta.gob.ar

© Springer International Publishing Switzerland 2016
J.A. Zinck et al. (eds.), *Geopedology*, DOI 10.1007/978-3-319-19159-1_18

Productivity index (PI) is one of the many indicators used for land classification among conventional land evaluation procedures. PI determination aims at establishing a numeric valuation of land production capability in a given region. The index gives a value proportional to the maximum potential yield attainable by common crops ecologically adapted under given management practices (Tasi and Schulz 2008).

Although time-consuming and costly, detailed soil maps made by conventional survey method are usually used to derive PI mean-value maps at farm level. However, such maps only reflect soil variability within discrete polygons based on the surveyors' mental model of the soil-landscape relationships. Conventional soil survey maps provide knowledge of the soil-landscape relationships acquired by soil mapping in a given survey area. This knowledge is not explicitly documented in the polygon data model, which simplifies the complex, continuous distribution of soil types across the landscape into discrete polygons with definite boundaries within which the spatial soil variation is not captured. Continuous spatial variation of soil characteristics, in response to changes in environmental conditions that are noticed during field mapping is nearly impossible to depict on conventional soil maps. Such variability of soil properties within polygon boundaries is difficult to quantify by using the vector data model (Zhu et al. 2003).

Therefore, thematic (soil productivity) maps derived from a soil survey map might not be the best guide for subsequent land-use planning as required by local land-use planners, who would rather prefer a parcel-to-parcel determination of soil productivity. The addition of digital soil mapping (DSM) technologies may improve the overall mapping process and make it more quantitative (Giasson et al. 2011). DSM outputs are soil properties or soil classes derived using a spatial inference system. The usual procedure is based on a number of predictive approaches that involve prior soil information in point and map form (McBratney et al. 2003). Field and laboratory observational methods are coupled with spatial and non-spatial soil-inference systems (Lagacherie and McBratney 2006) in order to allow for the prediction of soil properties or classes by using soil information and environmental covariates. DSM techniques and procedures have many advantages, such as cost, consistency, and documentation, as well as the ease of updating when new data become available. Their key component is the capability of deriving uncertainties for predicted outcomes, thus allowing tracking error propagation through the whole process (Carré et al. 2007).

The research approach of the present work aims at (a) predicting continuous productivity-index (PI) classes from a number of environmental covariates, mostly DEM derivatives, using DSM regression techniques and processes; (b) determining the correlation between PI "discrete" values as obtained from the conventional mapping procedure and "continuous" PI values as obtained by soil digital mapping procedures; and (c) analyzing the average differences found among the correlation values, in comparison to the local expert knowledge of the expected variability of productivity-related soil properties within each map unit, as a basis for adjusting the soil-landscape/soil-series interpretation of an existing detailed choropleth soil map.

18.2   Materials and Methods

18.2.1   Study Area

From an ecological point of view, the Entre-Ríos province belongs to the northeastern fringe of the Pampas region, a transitional territory between the country's fertile central plains and the Mesopotamian region of Argentina, where an unprecedented expansion of cropland into marginal areas during the past 50 years has affected soil resources and overall soil quality. Dominant soils in this region are good fertility Vertisols, mostly Hapluderts that require careful management to minimize the impact of monoculture and lack of biodiversity on soil degradation through water erosion, organic matter depletion, and nutrient loss (Wingeyer et al. 2015).

Since 2009, INTA has been conducting soil research in a 5300 ha farm ("Santa Inés de las Estacas") located in the north of Entre-Ríos province. This farm has been selected as a high-management level sample area where a mixed type of agriculture combined with livestock production is currently being planned in relation to suitable soil conditions. Most of the livestock activity is based on grazing of natural grassland and/or cultivated forage under natural woodland on Uderts and Aqualfs. Patches of annual crops including soybean with sorghum, corn, and/or wheat under rotation with permanent grassland, are extensive on the more fertile Udolls.

18.2.1.1   Main Geological and Geomorphic Features

Last Andean epeirogenic movements (Miocene/Pliocene) uplifted and faulted the older block deposits in the area and subsequently triggered a cycle of erosion processes that have dominated up to present times the development of the Entre-Ríos "peneplain" landscape (Plan Mapa de Suelos 1990), being accentuated or attenuated according to hydrographic base-level changes during the Quaternary marine transgressions/regressions. During the lower-middle Pleistocene, thick loess-like sediments extensively covered the faulted blocks and underwent posterior transformations into secondary loess-like clayey materials (palustrine sediments locally known as "limos calcáreos") according to climatic changes related to glacial/periglacial alternating periods.

Landscape development continued during the late Pleistocene and Holocene with further erosion/redeposition of colluvial loess sediments downslope onto the broad excavated valleys. The latest events include the infilling of colluvial depressions with recent alluvial deposits and younger materials that in some places directly overlie Pliocene sandy and sandy-clayey sediments previously exposed by Pleistocene dissection.

Two contrasting geomorphic environments (Table 18.1) are identified: the peneplain and the valley landscapes. The peneplain is divided into (1) highlands with stable, flat or very gently undulating upper slopes covered by clayey "calcareous limo" as parent material of the local Alfisols and Vertisols showing poor surface

Table 18.1 Soil map units and physiographic legend

Landscape	Relief features	Soil parent materials	Landform/slope facets	Soil subgroups	Soil series and phases	Map symbol	PI[a] mean value
Peneplain	High, wide slope tops without gilgai microrelief	Very thin layer of loess-like mantle (Fm S Guillermo) over Fm Hernan-darias	Flat/very gently undulating levels	Vertic Ochraqualfs	Saucesito	Sau	33
			Closed, concave depressions within slope tops		Saucesito, frequently ponded	Sau.e1	28
			Flat/very gently undulating slope tops		Colonia Trece	CT	28
	Middle slopes with linear gilgai microrelief	Very thick deposits of calcareous "limo" (Fm Hernandarias)	(idem, with gilgai)	Argiudollic Pelluderts	Ramblones, flat	Ra.p0	37
			Flat, uneroded upper slopes		Ramblones	Ra	40
			Slightly-to-severely eroded upper gently-undulating slopes		Ramblones, slightly eroded	Ra.h1	40
					Ramblones, slight to moderately eroced	Ra.h1/h2	35
					Ramblones, modera-tely eroded	Ra.h2	26
					Ramblones, severely eroded (+ gullies)	Ra.h3 + C	13
	Lower slopes without gilgai microrelief	Undifferentiated colluvial/retrans-ported loess	Very gently undulating lower slopes	Vertic Argiudolls	Banderas	Ba	57
					Banderas, slightly eroded	Ba.h1	57
					Banderas, modera-tely eroded	Ba.h2	38
			Slightly convex footslopes		Banderas, cumulic	Ba.x	60
			Footslopes/sparse, old terraces (partly over Fm Ituzaingó)	Aquic Argiudolls	Tacuaras	Tc	86
Valley	River/creeks plains	Undifferentiated alluvial deposits	Flat/concave levee/basin flood-plain complex	Haplaquepts, Fluvents Psamments	Arroyo Estacas Complex	Co.A°Est	6
			Narrow-stream small vales	Haplaquents	Alluvial plain (ur.d.)	Ap	6

[a]PI productivity index

drainage, and (2) a gently undulating "peneplain" showing linear-gilgai features of partly eroded Vertisols and footslopes where colluvial loess infilled internal depressions and waterways with Vertic Argiudolls.

Valleys include (1) narrow alluvial plains with a complex of alluvial materials ranging from clayey depressions and ancient meanders to more sandy areas in levees and terraces, and (2) dendritic networks of parallel, peneplain-dissecting streams (small temporary creeks) flowing in a straight southeast-northwest direction, as main sources of sedimentation in the valley plains.

18.2.1.2   Soil-Parent Material Relationships

Main soil parent materials are a represented by the thick package of Quaternary sediments including extensive Middle Pleistocene expansive clay deposits (Hernandarias Fm) overlying Pliocene sandy, fluvial sediments (Ituzaingó Fm). Holocene deposits of La Picada Fm and the recent thin, topsoil aeolian mantle of the San Guillermo Fm are much less extensive.

Uderts and Aqualfs developed in the Hernandarias Fm, while the very thin mantle of the San Guillermo Fm constitutes the mollic epipedon of some of the local Vertisols. Locally, the more suitable soil series for agricultural use are developed in the loess-like colluvial deposits (undifferentiated).

18.2.2   Materials

A detailed soil map (Fig. 18.1 and Table 18.1) covers the farm area (Walter 2007). This map was made using a physiographic analysis approach with intensive aerial photo-interpretation of soil-landscape relationships from 1964 to 1966 aerial photographs and photo-mosaics at scale of 1:20,000.

A preliminary photo-interpretation map was used as framework for the landscape-oriented auger-pit field survey. Subsequent correlation was made to determine soil type/landscape map units according to the corresponding soil landscape/soil series relationships established on the general soil map of the Departmento La Paz (Plan Mapa de Suelos 1990). Although it was published at a small soil-reconnaissance scale, this general map shows associations of soil series so that the features of those series can be used for developing interpretations at larger scales as stated in the Soil Survey Manual (Soil Survey Division Staff 1993).

A discrete productivity-index map obtained after digitalization of the basic soil map and its attribute data base in a geographical information system (GIS) was available for this research as well.

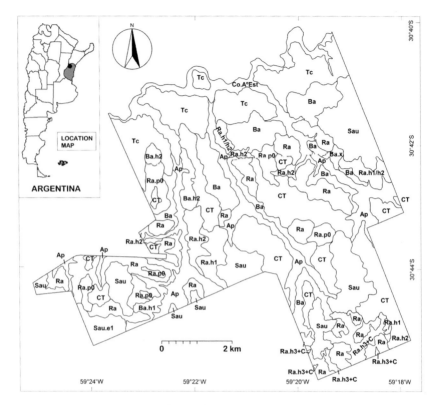

Fig. 18.1 Physiographic soil map of the study area

18.2.3   Methods

18.2.3.1   Morphometric Indexes and Sampling Scheme

To apply the digital soil mapping models, Shuttle Radar Topographic Mission data (SRTM) were used as complementary information. These data were originally produced at 30 m resolution using C- and X-bands and preprocessed by the Jet Propulsion Laboratory (JPL). The error characteristics for interferometric SAR have been summarized by Rodriguez et al. (2006). The principal errors concern uncompensated Shuttle mast motion, which produces a striping effect, and random phase noise proper of radar signal (Walker et al. 2007). For our study area, Rodriguez et al. (2006) mention that the relative height error is 5.5 m, with a random error of about 2 m. Another work shows that the mean error for Entre-Ríos province is 2.36 m (IGN 2014). The latter was not calculated for the SRTM30 but for a product derived from it. The striping effect was not observed in the study area.

The following morphometric indexes were calculated from the SRTM30 DEM by using the SAGA software (SAGA Development Team 2008): slope (Slp), slope length factor (LSF), altitude above channel network (AACN), channel network base level (CNBL), topographic wetness index (TWI), profile curvature (PC), convergence index (CI), slope height (SlH), standardized height (StdH), mid-slope position (Pos), and valley depth (VD).

The resulting indexes together with the physiography-based 1:20,000 soil map were input into a Conditioned Latin Hypercube (cLHS) sampling scheme method (Minasny and McBratney 2006) to define a soil sampling scheme for subsequent fieldwork during which an Eijkelkamp soil column (1 m long, 9 cm diameter) cylinder auger was used to obtain 48 undisturbed soil profile samples (2 replicates per site) (Schulz et al. 2010; Wilson et al. 2010). The first set of samples was used to identify and describe soils series in the laboratory according to Etchevehere (1976) and Schoeneberger (2002) guidelines, while the remaining replicates were used for complete physical and chemical laboratory analyses.

18.2.3.2 Spatial Prediction of Soil Variables

The maps of soil variables for each PI factor were interpolated to obtain the continuous PI map. Continuous variables included clay content, sand and silt content of surface and subsurface horizons, organic matter content, and cation exchange capacity of surface and subsurface horizons. They were interpolated using a regression-kriging model (Hengl 2009; Olmedo et al. 2012). Discrete variables (e.g. clay expansivity of surface and subsurface horizons, drainage degree, current and potential erosion) were interpolated using a decision-tree model. All models were processed with the R software (R Core Team 2014).

The regression-kriging model consisted in the generation of a multiple linear regression between the 48 variable values measured at field sites and the morphometric indexes. The selection of prediction variables among the morphometric indexes was made using a stepwise method, the choice criterium being based on the AIC (Akaike Information Criteria) value. The variable value for each pixel was calculated from the morphometric indexes. Regression residuals were subsequently calculated as the difference between the 48-point mean values and the regression model output map (data not shown) by ordinary kriging interpolation. A map of the maximum likelihood value for each soil variable was calculated by adding the map obtained by the regression model to the ordinary kriging interpolation values. Finally, to verify the model adjustment, maps were validated by means of a leave-one-out cross validation scheme whereby the mean prediction error, the mean standardized-prediction error and the mean square-normalized error together with the correlation between observed and predicted data (OBSvsPRE) as well as the correlation between predicted and residual data (PREvsRES) were also calculated (Table 18.2).

On the other hand, maps of clay, silt, and sand contents were used as input for a pixel-by-pixel textural class calculation (according to the USDA textural triangle) by running the package "soiltexture" in R (Moeys and Shangguan 2014).

The decision-tree model was built using the R version of the Quinlan C5.0 algorithm (Kuhn et al. 2014) to predict the variable values by using the morphometric

Table 18.2 Regression-kriging model results

Variables	Covariates	R^2	ME	MSPE	MSNE	COP	CPR
Surface clay content	AACN, CNBL	0.2162	0.0327	15.75	1.037	0.4453	−0.0959
Surface silt content	CNBL, SlH, StdH, VD	0.2026	−0.0880	14.30	1.084	0.3296	−0.2146
Surface sand content	AACN, CNBL, TWI	0.3826	0.0022	1.06	0.784	0.8709	−0.0089
Surface OM	PC, SlpH, StdH, TWI	0.1642	−0.0023	3.54	1.756	0.0771	−0.4189
Subsurface clay content	AACN, CNBL, LS, Slp, DEM, VD, TWI	0.3487	0.0177	36.03	3.452	0.2032	−0.4518
Subsurface silt content	CNBL, LS, SlpH, Slp, StdH	0.2094	0.0044	31.35	3.084	0.2002	−0.3853
Subsurface sand content	AACN, CNBL	0.2372	−0.0727	10.34	2.708	−0.2418	0.1468
Subsurface cation exchange capacity	AACN, Slp	0.1441	0.1306	126.65	1.040	0.1949	−0.2303

R^2 coefficient of determination, *ME* mean error, *MSPE* mean standardized prediction error, *MSNE* mean square normalized error, *COP* [Corr(obs,pred)] correlation observed and predicted, *CPR* [Corr(pred,res)] correlation predicted and residual

index values as the predictor elements. The algorithm selected 1–3 predictor variables by considering their capability to generate homogeneous groups. After the decision trees were generated, both the predictors sets used and the wrongly-classified data were then also analyzed (Table 18.3). The decision trees were subsequently used to calculate the pixel-by-pixel variable values based on the morphometric index values.

18.2.3.3 Continuous PI Determination

To perform the continuous PI calculation, the PI equation was adjusted to the Entre-Ríos soil conditions. The present and potential erosion factor was split into two new factors: 'present erosion' and 'potential erosion' to allow for possible combinations not considered in the original publication of INTA-SAGyP (1987). The relationship with the factor value for existing combinations was however maintained.

$$PIc = H \times D \times Def \times Ta \times Tb \times Sa \times Na \times OM \times T \times Epre \times Epot \qquad (18.1)$$

Where: PI_c = productivity index; H = climatic condition; D = drainage; P_e = effective depth; T_a = surface horizon texture; T_b = subsurface horizon texture; Sa = salinity; Na = alkalinity; OM = organic matter; T = cation exchange capacity; E_{pre} = present water erosion; E_{pot} = potential water erosion.

The values for each PI factor were normalized (Table 18.4, Fig. 18.2) to a 0–1 scale as defined in the INTA-SAGyP (1987) reference document.

Finally, both a continuous PI mean value and a standard deviation value were calculated for each polygon and later compared with the respective PI values obtained by the conventional method (Table 18.4, Figs. 18.3 and 18.4).

Table 18.3 Decision-trees model results

Variables	Covariates	Size	Errors
Surface swelling clay	DEM, PC, TWI	4	0.083
Subsurf swelling clay	CNBL	1	0.146
Drainage class	CNBl, VD, NH, DEM, TWI, Pos	6	0.042
Present water erosion	VD, PC, CI	8	0.125
Potential water erosion	VD, CNBL, DEM, TWI, PC, SlpH, AACN	4	0.188

Table 18.4 Continuous PI equation factors normalization

Factor	Class	Value
Drainage	Moderately well drained	0.9
	Imperfectly drained	0.8
Surface horizon texture	SiCl	0.7
	SiClLo and swelling clay	0.7
	SiClLo and no swelling clay	0.9
	SiLo	0.9
Subsurface horizon texture	Cl and swelling clay	0.7
	Cl and no swelling clay	0.8
	SiCl and swelling clay	0.8
	SiCl and no swelling clay	0.9
	SiClLo and swelling clay	0.9
	SiClLo and no swelling clay	1
	SiLo	1
Organic matter	<2 %	0.7
	2–4 %	0.85
	>4 %	1
Cation exchange capacity	<10 cmol(+)/kg	0.6
	10–20 cmol(+)/kg	0.8
	>20 cmol(+)/kg	1
Present erosion	No erosion	1
	Accumulation	0.3
	Slight erosion	0.9
	Moderate erosion	0.7
	Severe erosion	0.3
Potential erosion	Without erosion hazard	1
	Slight erosion hazard	0.9
	Moderate erosion hazard	0.7
	Severe erosion hazard	0.5
	Possibility of accumulation	0.3

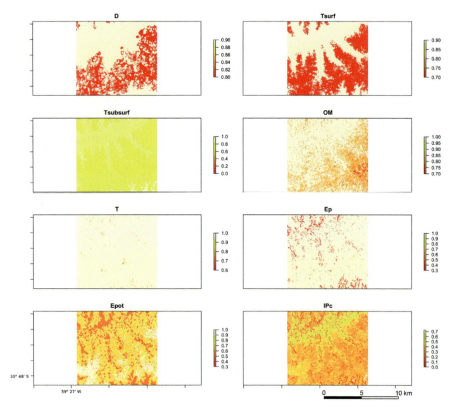

Fig. 18.2 Continuous soil productivity index factors *D* drainage, *Tsurf* surface horizon texture, *Tsubsurf* subsurface horizon texture, *OM* organic matter, *T* cation exchange capacity, *Ep* present water erosion, *Epot* potential water erosion, *IPc* continuous productivity index

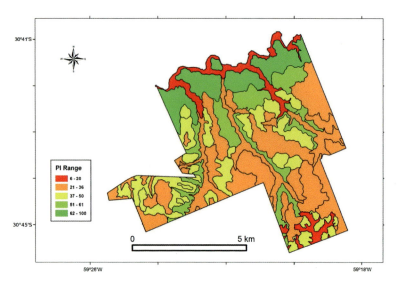

Fig. 18.3 Discrete productivity index map

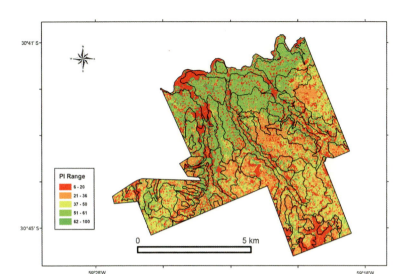

Fig. 18.4 Continuous productivity index map

18.3   Results and Discussion

Several maps of continuous soil properties related to the productivity-index calcula-
tions (climate condition, drainage, effective depth, surface horizon texture, subsurface horizon texture, salinity, alkalinity, organic matter content, cation exchange
capacity, present and potential water erosion) were produced (Fig. 18.2).

18.3.1   Climatic Condition, Effective Depth, Salinity, and Alkalinity

The factors of climatic condition, effective depth, salinity, and alkalinity were val-
ued as 1 according to the following criteria. The study area is located in the Southern
Chaco-Pampean climatic region that is considered optimal for most crop development and production, without wetness or temperature limitations in the critical
yield-defining period (INTA-SAGyP 1987). The effective depth in all 48 sampled
sites was more than 100 cm; clayey horizons did not hinder root development
throughout the solum. In all samples, EC values within 75 cm depth were lower than
4 dS/m (non-saline). Alkalinity values (ESP) were lower than 2 % at 0–20 cm depth
and lower than 15 % at 50–100 cm depth.

18.3.2 Drainage, Texture, Organic Matter, Cation Exchange Capacity, and Erosion

Soils are moderately well drained and imperfectly drained. Moderately well drained soils (0.9) occur in the lower areas of the landscape, close to creeks, with high sand content that facilitates the internal water movement. In contrast, landscape upslope soils are imperfectly drained (0.8) because of high clay content that hinders soil drainage.

Particle size distribution maps (i.e. sand, silt, and clay) were obtained by regression-kriging. Clay values were adjusted by using a clay-expansivity factor to account for its influence on soil water movement that directly affects crop yield and consequently the final PI value. Surface horizon texture was rated in two classes (i.e. clay-loam and silty-clay-loam) which coincided with the spatial distribution of Alfisols and some high-clay-content Vertisols. The dominant textures in subsurface horizons were clay-loam and silty-clay-loam, with clay and silty-loam textures to a lesser extent.

Organic matter content in surface horizons varied from less than 1 % up to 10 %. The Tacuaras soil series had mean values higher than 5 % in spite of having been deforested and put into agricultural production 5 years ago. It is located on a very gently sloping landscape (less than 1.5 % slope) and benefits from a system of excess-water evacuation terraces which minimizes land degradation. The Colonia Trece and Suacesito soil series have values higher than 4 % organic matter, some-what higher than the modal profile values, which could be influenced by the present land use for livestock production under natural woodland. The Ramblones soil series shows mean values of about 4 % organic matter, somewhat lower than the modal profile (5.4 %) values, which could be related to the fact that some samples have been taken in sectors affected by water erosion causing topsoil loss. Lower organic matter contents (2 %) were found in Entisols near natural waterways.

18.3.3 Comparison of Conventional and Continuous Productivity Index Values

The average difference between the continuous PI mean value and the PI value calculated by the conventional method is 2.83 PI units (Table 18.5). This difference becomes much higher if only the Banderas soil series (with differences between 11.06 and 29.71 values) or the Tacuaras soil series map units are considered.

The mean standard deviation for the continuous PI value for every soil unit is 12.25. Some units of the Banderas soil series reach a mean standard deviation value ranging between 14.96 and 19.27.

Table 18.5 Average differences between conventional PI and continuous cPI values

Map unit	PI	Mean cPI	S.D. cPI	DIFF
Banderas	57.00	45.94	13.09	11.06
Banderas	57.00	44.94	11.26	12.06
Banderas	57.00	38.54	19.27	18.46
Banderas, slightly eroded	55.00	34.84	17.24	20.16
Banderas	57.00	35.77	16.71	21.23
Banderas, cumulic	60.00	34.54	17.10	25.46
Banderas	57.00	27.28	14.96	29.72
Tacuaras	86.00	50.56	4.07	35.44
Tacuaras	86.00	48.27	11.14	37.73
Tacuaras	86.00	47.83	10.56	38.17
Tacuaras	86.00	45.02	15.55	40.98
Tacuaras	86.00	42.77	16.55	43.23

18.4 Conclusions

The generation of a continuous-PI variability map by DSM methods could over-come a main limitation of PI maps derived from conventional soil maps by showing the variability of productivity-related soil properties within each map unit.

However, current DSM methods have some limitations. One of these is the difficulty to predict a large number of soil properties simultaneously while preserving the relationships among them. Another important limitation is the inability to include pedological knowledge in the prediction models. The conventional soil map is, therefore, the only model available for soil productivity analysis that provides a set of soil-landscape relationships for every map unit.

Because it is not possible to derive such a relationship model from any map of independent soil properties, cartographic boundaries of conventional maps should be preserved as the basic structure underlying any research in soil productivity variability.

Future research should include 'continuous indices' of all the soil variables considered in the PI equation which would be obtained from linear adjustment procedures, in replacement of 'discrete indices' of the same parameters such as used in this work.

Acknowledgements This research was carried out within the framework of both the Soil Cartography (PNSUELO-1134032) and the Land Evaluation (PNSUELO-1134033) projects of the INTA National Soil Program. We acknowledge our project colleagues Lucas Martín Moretti, Julieta Irigoin, and Leonardo Mauricio Tenti Vuegen for providing data and expert knowledge as well as assistance during the field survey operations.

References

Carré F, McBratney AB, Mayr T, Montanarella L (2007) Digital soil assessments: beyond DSM. Geoderma 142(1–2):69–79. doi:10.1016/j.geoderma. 2007.08.015, http://linkinghub.elsevier.com/retrieve/pii/S0016706107002261

Etchevehere P (1976) Normas de reconocimiento de suelos, 2nd edn, INTA-CIRN Suelos, publicacion 52. Castelar, Buenos Aires

Giasson E, Sarmento EC, Weber E, Flores CA, Hasenack H (2011) Decision trees for digital soil mapping on subtropical basaltic steeplands. Sci Agric 68:167–174

Hengl T (2009) A practical guide to geostatistical mapping, 2nd edn. University of Amsterdam, Amsterdam

IGN (2014) Modelo digital de elevaciones de la República Argentina. Instituto Geográfico Nacional, Buenos Aires, http://www.ign.gob.ar/archivos/InformeMDE

INTA-SAGyP (1987) Índices de productividad. Estudios para la implementación de la reforma impositiva agropecuaria, Proyecto PNUD Argentina 85/019. Área Edafológica, Buenos Aires

Kuhn M, Weston S, Coulter N (2014) C50: C5.0 decision trees and rule-based models. http://CRAN.R-project.org/package=C50, R package version 0.1.0–21

Lagacherie P, McBratney A (2006) Spatial soil information systems and spatial soil inference systems: perspectives for digital soil mapping. In: Lagacherie P, McBratney AB, Voltz M (eds) Digital soil mapping: an introductory perspective. Developments in soil science, vol 31. Elsevier, pp 3–22. doi:http://dx.doi.org/10.1016/S0166-2481(06)31001-X, http://www.sciencedirect.com/science/article/pii/S016624810631001X

McBratney AB, Mendonca Santos ML, Minasny B (2003) On digital soil mapping. Geoderma 117(1):3–52

Minasny B, McBratney AB (2006) A conditioned Latin hypercube method for sampling in the presence of ancillary information. Comput Geosci 32(9):1378–1388. doi:10.1016/j.cageo.2005.12.009, http://www.sciencedirect.com/science/article/pii/S009830040500292X00149

Moeys J, Shangguan W (2014) Soiltexture: functions for soil texture plot, classification and transformation. http://CRAN.R-project.org/package=soiltexture. R package version 1.2.19

Olmedo GF, Angelini ME, Vallone RC, Moretti L (2012) Estimación de variables dáficas en el Oasis Productivo De Tupungato, Mendoza. XIX Congreso Latinoamericano de la Ciencia del Suelo. XXIII Congreso Argentino de la Ciencia del Suelo. Mar del Plata

Plan Mapa de Suelos de la Provincia de Entre Ríos (1990) Carta de suelos de la República Argentina. Departamento La Paz. Convenio INTA – Gobierno de Entre Ríos. EEA Paraná, Serie Relevamiento de Recursos Naturales N° 7

R Core Team (2014) R: a language and environment for statistical computing. R Foundation for Statistical Computing, Vienna, http://www.R-project.org/

Rodriguez E, Morris CS, Belz JE (2006) A global assessment of the SRTM performance. Photogramm Eng Remote Sens 72(3):249–260, http://essential.metapress.com/index/GP76H362U7L66153.pdf, 00327

SAGA Development Team (2008) System for automated geoscientific analyses (SAGA GIS). Germany. http://www.saga-gis.org/

Schoeneberger PJ (2002) Field book for describing and sampling soils, v 3.0. Government Printing Office, URL http://www.nrcs.usda.gov/Internet/FSE_DOCUMENTS/nrcs142p2_052523.pdf

Schulz G, Bedendo D, Wilson M, Oszust J, Pausich G (2010) Muestreador columnar de suelos. Alternativas de uso con fines edafológicos. 2 Relevamiento expeditivo de suelos (prospección rápida). In: Actas del XXII Congreso Argentino de la Ciencia del Suelo, Rosario, Argentina (expanded-summary PDF file on CD-ROM, 4 pages)

Soil Survey Division Staff (1993) Soil survey manual. Soil Conservation Service. US Department of Agriculture Handbook 18

Tasi H, Schulz G (2008) Índices de productividad específicos para el cultivo de arándanos en el Departamento Concordia, Provincia de Entre Ríos. In: Resúmenes XXI Congreso Argentino de la Ciencia del Suelo. San Luis (summary of oral presentation)

Walker WS, Kellndorfer JM, Pierce LE (2007) Quality assessment of SRTM C- and X-band interferometric data: implications for the retrieval of vegetation canopy height. Remote Sens Environ 106(4):428–448, DOI http://dx.doi.org/10.1016/j.rse.2006.09.007, URL http://www.sciencedirect.com/science/article/pii/S003442570600349X

Walter R (2007) Soil survey technical report of "Santa Inés de las Estacas" sample area, La Paz, Entre Ríos. Unpublished raw data

Wilson M, Oszust J, Sasal MC, Schulz G, Gvozdenovich J, Pioto AC (2010) Muestreador columnar de suelos. Alternativas de uso con fines edafológicos. 1 Densidad aparente y agua útil. In: Actas del XXII Congreso Argentino de la Ciencia del Suelo. Rosario (expanded-summary PDF file on CD-ROM, 3 pages)

Wingeyer AB, Amado TJC, Pérez-Bidegain M, Studdert GA, Varela CHP, Garcia FO, Karlen DL (2015) Soil quality impacts of current southamerican agricultural practices. Sustainability 7(2):2213–2242. doi:10.3390/su7022213, http://www.mdpi.com/2071-1050/7/2/2213

Zhu AX, Burt JE, Moore AC, Smith MP, Liu J, Qi F (2003) SoLIM: a new technology for soil mapping using GIS, expert knowledge and fuzzy logic. Dept Geography, Univ Wisconsin-Madison-National Resources Conswervation Service, U.S. Dept. of Agriculture. http://solim.geography.wisc.edu/pubs/Overview2007-02-16.pdf

Chapter 19
Contribution of Open Access Global SAR Mosaics to Soil Survey Programs at Regional Level: A Case Study in North-Eastern Patagonia

H.F. Del Valle, P.D. Blanco, L.A. Hardtke, G. Metternicht,
P.J. Bouza, A. Bisigato, and C.M. Rostagno

Abstract The Japan Aerospace Agency (JAXA) recently released multi-temporal global SAR mosaics derived from a 4-year data acquisition project (2007–2010) of the Advanced Land Observing Satellite (ALOS) PALSAR, L-band at 25 m spatial resolution. These open access data sets could assist traditional soil surveys and/or digital soil mapping programs undertaken at regional and subregional scales. Through improving mapping accuracy and reducing fieldwork time, together with digital identification and classification of landscape types and geomorphic features, soil survey programs could be completed over extensive areas currently lacking reliable soil information. Argentina is a country that needs to establish operational digital soil mapping (DSM) initiatives to address challenges and potential solutions of soil surveys at detailed and semi-detailed scales. These efforts could provide useful soil information to complement or update existing soil survey data, and document methods and results. Although remote sensing has been recognized as an efficient technology to support data gathering and information generation for soil and terrain mapping, the Argentine national knowledge of how to operationalize these techniques is still incomplete. Limited research has been carried out on the potential of microwave remote sensing data for spatial estimation of different topsoil properties, excepting soil moisture. This chapter intends to narrow down this knowledge gap by assessing the potential of ALOS PALSAR image mosaics for identifying and mapping land covers, as soil cartographic base, or as a value-added layer for

H.F. Del Valle (✉) • P.D. Blanco • L.A. Hardtke • P.J. Bouza • A. Bisigato • C.M. Rostagno
Consejo Nacional de Investigaciones Científicas y Técnicas (CONICET), Centro Nacional Patagónico (CENPAT), Instituto Patagónico para el Estudio de los Ecosistemas Continentales (IPEEC), Puerto Madryn, Chubut, Argentina
e-mail: delvalle@cenpat-conicet.gob.ar; blanco@cenpat-conicet.gob.ar; hardtke@cenpat-conicet.gob.ar; bouza@cenpat-conicet.edu.ar; bisigato@cenpat-conicet.gob.ar; rostagno@cenpat-conicet.gob.ar

G. Metternicht
Institute of Environmental Studies, University of New South Wales, Sydney, NSW, Australia
e-mail: g.metternicht@unsw.edu.au

© Springer International Publishing Switzerland 2016
J.A. Zinck et al. (eds.), *Geopedology*, DOI 10.1007/978-3-319-19159-1_19

integration in thematic soil mapping. The chapter also analyses changes in L-band backscatter overtime, and their relation to land degradation processes. To this end, a test area covering the north-eastern Patagonia region was chosen for its diversity of geology, geomorphology, soil, and land use, as well as for the existing soil expertise and an ongoing regional soil-mapping project.

Keywords L-band SAR • Dual polarization • Surface roughness • Image classification • Soil and terrain mapping • Argentina

19.1   Introduction

In Argentina, soil surveys begun as a need for intensifying and expanding the production of food crops. In the 1980s, the Soil Atlas of Argentina, at scales of 1:500,000 and 1:1,000,000, was produced with financial support of the United Nations. This project integrated existing information from data-rich regions of significant agricultural productivity with others of low agricultural suitability, which lacked good soil cartography (SAGyP-INTA 1990). Results of this project show discrepancies with other studies and disagreements in soil interpretation, particularly between administrative (i.e. provincial) boundaries as pointed out by del Valle (1998) who argued that precision was lost because soil units were re-categorized into more general taxonomic classes. The methodology applied in the identification and classification of the soil map units (SMUs) precluded their use for gathering geopedologic information (see definition in Part I of this book), as more attention should have been focused on soil diversity (e.g. facies, etc.). Therefore, extensive zones still lack comprehensive soil information at suitable scales and survey levels. This in turn hinders good land or territorial planning strategies, since it is difficult to generate relevant knowledge when information does not exist, or it is incomplete and very fragmented (del Valle 2008).

19.1.1   Digital Soil Mapping Supported by Radar Remote Sensing

The success of digital soil mapping (DSM) is probably a convergence of many factors: the availability of spatial digital data (DEM, satellite images), the accessibility of computational power (hardware and software), the development of data mining tools and GIS, and geostatistics applications (McBratney et al. 2003; Lagacherie et al. 2007).

According to Mulder (2013), existing remote and proximal sensing methods can support three main components of DSM: (1) Remote sensing data may help in segmenting the landscape into homogeneous soil-landscape units in which soil composition can be assessed; (2) Remote and proximal sensing methods allow for inference of soil properties using physically based and empirical methods; and (3) Remote sensing data assist spatial interpolation of sparsely sampled soil property data, as primary or secondary data source.

Different methodologies based on passive and active remote sensors have been proposed for the estimation of soil parameters (Zribi et al. 2011), accounting for the significant differences between optical and microwave wavelengths in the mechanics of imaging, and in the way that characteristics of a target are measured. Microwave signals depend mainly on the dielectric constant of the target, that is, a measure of how well electromagnetic waves interact with a given type of material. Low soil moisture increases the influence of surface soil roughness, soil-vegetation interaction (double scattering), and soil volume scattering due mainly to scatterers within the topsoil (Fung and Chen 2010).

A returned radar signal varies considerably in response to variations in terrain morphology, topography, and surface cover (Ridley et al. 1996; Levin et al. 2008; del Valle et al. 2010). To explain these variations, it is necessary to understand the nature of the interaction between active microwave radiation and surface properties. Radar scattering from an air-soil interface is dominated by surface scattering, unless significant penetration occurs in the top surface layer, in which case volume scattering may become appreciable. In arid landscapes, vegetation cover is generally sparse and the terrain surfaces are usually erodible lands; consequently, surface scattering is heavily dependent on surface roughness, which is in turn a dynamic geomorphic property (del Valle et al. 2013).

19.1.2 Addressing a Knowledge Gap

Although remote sensing has been recognized as an efficient technology for soil surveys, our knowledge of how to apply these techniques to soil and terrain mapping is still incomplete. This knowledge gap is not limited to Argentina; it extends to other regions of the world as well. Mulder et al. (2011) conclude that in general, there is no coherent methodology established in which approaches of spatial segmentation, measurements of soil properties and interpolation using remotely sensed data are integrated in a holistic way to achieve complete area coverage. Furthermore, limited research has been carried out showing the potential of microwave remote sensing data for spatial estimation of various soil properties, with exception of soil moisture (Saha 2011).

In November 2014, the Japan Aerospace Exploration Agency (JAXA) released a 25-m spatial resolution global mosaic prepared using 4 years (2007–2010) from Phased Array type L-band Synthetic Aperture Radar (PALSAR) on board the Advanced Land Observing Satellite (ALOS) (Shimada et al. 2014). The Agency invited soil scientists to assess the potential of this product for specific and/or localized applications, such as mapping and monitoring of soil properties (soil moisture, roughness) and land degradation processes. Such information could be used alone, or in combination with open access optical data (e.g. free LANDSAT data) to derive thematic maps and advance DSM initiatives.

Therefore, this chapter aims to assess the potential of the L-band ALOS-PALSAR image mosaics to assist in identification and mapping of soil covers at regional level, as well as a soil cartographic base, or value-added layer for integration in thematic soil mapping. The chapter also analyses changes in L-band backscatter overtime, and their relation to land degradation processes. To this end, we selected a test area covering the north-eastern Patagonia region of Argentina; this area is suitable for the aforementioned assessments given its diversity in geology, geomorphology, soil, and land use, the existing soil expertise and an ongoing regional soil map.

19.2 Materials and Methods

19.2.1 Study Area

Our initial study area is located in the north-eastern part of the Chubut province stretching over the southern portion of the Monte Desert biome (Abraham et al. 2009) and the north-eastern Patagonian Steppe (Fig. 19.1). It comprises approximately 27,293 km^2, centered at 43°39′15″S and 65°07′52″W. Wildfires and domestic grazing are the main disturbances in this region (Villagra et al. 2009). Our work

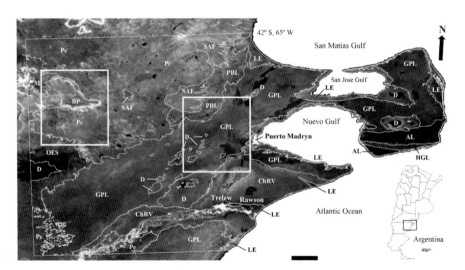

Fig. 19.1 Multitemporal ALOS PALSAR mosaic between June and October during 2007, 2009 and 2010, RGB (LHH$_{2007}$LHV$_{2009}$LHH$_{2010}$). *Gray-scale* Image mode. The wide-ranging Geomorphic Landscape: Gravel Plain Levels (GPL), Chubut River Valley (ChRV), Older Erosion Surfaces (OES), Exhumed and Covered Peneplain (Pe), Low Mountains (M), Secondary Alluvial Fans (SAF), Pediments and Bottomlands (PBL), Major and minor Depressions (D), Littoral Environment (LE) and Aeolian Landforms (AL). *White boxes* represent the test sites studied: El Moro (*westwards*) and San Luis (*eastwards*), respectively. *Bar* length=20 km

focused on two representative test sites within this large area, covering a surface area of 2,730 km^2 each one, named El Moro (westwards) and San Luis (eastwards), respectively.

A simplified landscape classification was developed and tested for the study area. Ancillary data and expert knowledge were used to compile the landscape dataset, which was later complemented with SRTM 1 arcsec resolution (http://earthexplorer. usgs.gov/) and terrain derivatives as well (Fig. 19.2). Table 19.1 summarizes the principal soil environmental features of the test sites used as auxiliary data source (del Valle et al. 1997; Bouza and del Valle 2014).

El Moro test site is characterized by landscapes of low mountains, exhumed and covered peneplains, and volcanic plains. This test site represents the southern portion of the *Somun Cura* Massif or Northern Patagonian Massif (Ramos 1999). The structure is characterized by large basement blocks with inclined grabens, affected by the Andean orogeny. The climate is arid, average annual temperature is 12.5 °C and the mean annual precipitation is 187 mm. Winds are mainly from the west or south-west (3.2 m/s). The dominant physiognomy is typical of the Monte, with shrubs (20–45 %). As elevation increases, herbaceous elements are mixed with shrubs (León et al. 1998).

The San Luis test site presents different geomorphic systems related to successive episodes of aggradation and erosion, with a complex drainage ancient alluvial fan. Its successive developments are linked to relict landforms such as the proto Chubut river course and the Simpson paleo-valley (Gonzalez Díaz and Di Tommaso

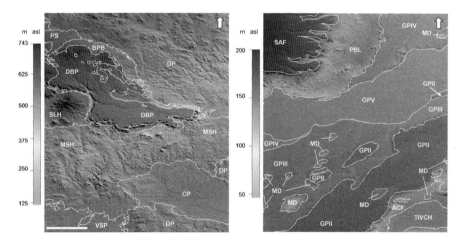

Fig. 19.2 Landscape unit delineations with the SRTM-based procedure 1 arcsec, meters above sea level. *Left*: El Moro landscapes. Basaltic Plateau Border (BPB), Covered Peneplain (CP), Dissected Basalt Plain (DBP), Dissected Peneplain (DP), Medium Slope Hill (MSH), Piedmont Slope (PS), Steep Slope Hill (SLH), Valley Slopes and Plains (VSP). *Right*: San Luis landscapes. Alluvial Colluvial Fans (ACF), Gravel Plain II (GPII), Gravel Plain III (GPIII), Gravel Plain IV (GPIV), Gravel Plain V (GPV), Secondary Alluvial Fans (SAF), Pediments and Bottomlands (PBL), Terrace IV Chubut River (TIVCH). *Bar* length = 10 km

Table 19.1 Soil environmental characteristics of the test sites studied

Class	Landscape	Dominant slope %	Surface Rockiness (1)	Stoniness (2)	Texture dominant Topsoil	Subsoil	Erosion	Degree of erosion	Dominant soils IUSS WRB (2014)	Vegetation types
El Moro										
VSP	Valley slopes and plains	2–5		Many	sl	sl, sc	Mass movement, gully/wind erosion, salt deposition	Moderate/severe	Haplic solonetz Haplic calcisols	Barren, scrubs/shrubs
DP	Dissected peneplain	5–10	Dominant			Rock		Severe	Lithic leptosols Eutric regosols	
CP	Covered peneplain	2–5	Many	Common		sl, ls, sc		Moderate/severe	Haplic calcisols Calcar c regosols	Shrubs/scrubs
MSH	Medium slope hill					sl	Mass movement		Haplic solonetz Haplic calcisols	
SLH	Steep slope hill	15–30	Dominant		sl, l	Rock	gully erosion	Very severe	Eutric regosols	Barren, trees/shrubs in gallery
PS	Piedmont slope	2–5	Many	Common	sl	sl, l	Water/wind erosion	moderate/severe	Haplic calcisols	Tall shrubs, grasses
DBP	Dissected basalt plain	0.5–2			scl, ls	rock	Gully/tunnel erosion, wind erosion		Mollic leptosols Lithic leptosols	Grasses, scrubs, barren
BPB	Basaltic plateau border	5–15	Abundant		sl	scl, ls	Mass movement, gully/tunnel erosion, wind erosion		Lithic leptosols Eutric regosols	Barren, tall shrubs, shrubs/scrubs, grasses

San Luis

Code	Landform	Stoniness (%)	Rockiness		Texture	Erosion	Erosion severity	Soil classification	Vegetation
GPII	Gravel plain II	0.5–2	Many	sl	sl, sc	Sheet/rill erosion, wind erosion	Slight/moderate	Haplic calcisols; Eutric cambisols	Shrubs with grasses
GPIII	Gravel plain III							Calcaric regosols	
GPIV	Gravel plain IV								
GPV	Gravel plain V								
TIVCH	Terrace IV Chubut river								
ACF	Alluvial and colluvial fans	2–5			sl, l	Water/wind erosion	Moderate	Haplic calcisols	
SAF	Secondary alluvial fans	2–5	Abundant		sl, sc	Gully/wind erosion, salt deposition	Moderate/severe	Calcaric regosols	Shrubs/scrubs with grasses
PBL	Pediments/bottomlands				l, scl	Gully/wind erosion, salt deposition		Haplic calcisols; Calcaric regosols	Barren, shrubs/scrubs
MD	Minor depressions		Common		sl, scl			Gleyic solonchaks	Barren, scrubs, shrubs with grasses

Surface rockiness: many 15–40 %, abundant 40–80 %, and dominant 80 %
Surface stoniness (coarse fragments >2 mm), completely or partly at the surface: common 5–15 %, many 15–40 %, abundant 40–80 %
USDA texture classification: *l* loam, *ls* loamy sand, *sc* sandy clay, *scl* sandy clay loam, *sl* sandy loam

2011). The alluvial fan dates back to an uncertain period between the late Pleistocene and late Sangamon interglacial at the end of the last glaciation. Average annual temperature is 13 °C and the mean annual precipitation is 236 mm, evenly distributed along the year. Winds are mainly from the west or south-west (4 m/s). Evergreen shrubs are the dominant vegetation type, although deciduous shrubs and grasses are common on slightly grazed areas. Plant cover varies from 15 to 60 %, as a function of grazing disturbance (Bisigato et al. 2005).

19.2.2 Multitemporal ALOS PALSAR Data

For the generation of the study area subsets, 25-m global mosaics from ALOS PALSAR L_{HH} and L_{HV} data acquired between June and October (winter-early spring) during 2007, 2008, 2009 and 2010 were used (Shimada et al. 2014). Table 19.2 presents the characteristics of the tiles used to generate the local mosaics.

The environmental conditions prevailing at the time of the SAR image acquisition have implications for mapping and monitoring the extent of land covers. Meteorological records were consulted to assess mainly the potential impact of precipitation on radar backscatter. Long-term climatic records for the season of image acquisition show that the study area receives less than 50 % of the annual rainfall at this time of the year. On average, this applied for the length of this study (2006–2010), with some exceptional years and individual locations (Harris et al. 2014). Years 2006 and 2010 experienced above average rainfall, while 2008 and 2009 were dry years, well below average. Among these extremes, 2007 and 2010 show uneven patterns of precipitation. Year 2010 may be considered an "average" year for the region, with exception of the southernmost area, and 2007 does not show a definite spatial trend.

Table 19.2 Characteristics of the tiles used to generate the local ALOS PALSAR mosaics

Characteristic	Description
Reference location	Latitude and longitude of north-west corner
Coordinate system	Latitude-longitude coordinates
Spacing	0.8 arcsec unit providing spacing of 25 m
Resolution of SAR image	36 m (azimuth) × 20 m (range)
Number of pixels	4,500 columns × 4,500 rows
Mode	Fine Beam Dual (FBD)
Polarization	HH – HV
Local incidence angle	34.3°
Orbit	Ascending
Contents	Amplitude data in HH and HV, mask information (ocean flag, effective area, void area, layover, shadowing), local incidence angle, total dates from the ALOS launch
No. of images per year	2007:22; 2008:20; 2009:23; 2010:24

19.2.3   Methodology

Land cover maps were produced applying the sequence of steps illustrated in Fig. 19.3. The processing steps are discussed briefly below.

19.2.3.1   Pre-processing

A new open-source software tool from the Sentinels Application Platform (SNAP), i.e. SENTINEL-1 Toolbox v1.1.0 (https://sentinel.esa.int/web/sentinel/toolboxes/sentinel-1), was used for image pre-processing; radiometric distortions of the sea were masked out for each subset using the DEM derived from the SRTM 1 arcsec resolution imagery (http://earthexplorer.usgs.gov/). The SRTM model was evaluated in different sectors of the test sites with a digital elevation model of higher resolution (25 m) performed by del Valle et al. (2002). The SRTM received an overall accuracy of 7.3 m and 4.5 m RMSE for El Moro and San Luis test sites, respectively.

19.2.3.2   Conversion to Radar Cross Section

The ALOS-PALSAR data are distributed as amplitude data, and therefore digital numbers (DN) need to be converted to normalized radar cross section (NRCS) in decibels (dB), applying a sensor specific Calibration Factor (CF) according to the following equation (Rosenqvist et al. 2007):

$$\sigma^0 = 10 * \log_{10}[DN^2] + CF$$

Where $CF = -83.0$ dB for both HH and HV polarizations.

The magnitude of σ^0 is a function of the physical and electrical properties of the target, the wavelength and polarization of the ALOS-PALSAR Fine Beam Dual (FBD), and the incident angle (34.3°) as modified by the local slope.

Speckle reduction was a minor issue, since the PALSAR data is multi-looked (four looks) by JAXA (Shimada et al. 2014).

19.2.3.3   Surface Roughness

Surface roughness refers to the unevenness of the earth's surface due to natural processes and/or human activities (Smith 2014). The root mean square (rms) height, the correlation length, and an autocorrelation function statistically describe it (Ulaby and Long 2014). Microscale and mesoscale roughness was described, associated respectively with image brightness (tone) and image texture (Henderson and Lewis 1998). In addition, we used complementary spatial context, understanding of backscattering characteristics, and expert knowledge.

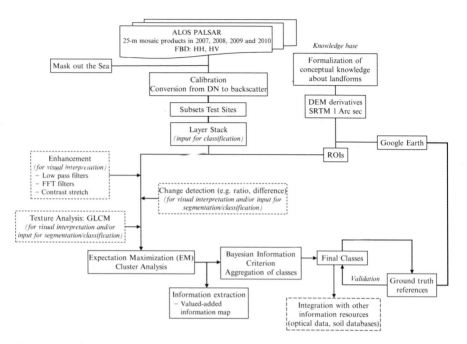

Fig. 19.3 Visual modeling of multitemporal mosaic processing and analysis workflow. *Dash rectangles* represent at this stage a partial analysis

Microscale roughness refers to the scale of small components (targets) within an individual pixel such as rocks, stones, or leaves and branches of shrubs. Microwave radiations are differently scattered by targets as a function of various target characteristics, among others their roughness (Table 19.3). A breakpoint between smooth and rough surfaces and the corresponding dominant radar scattering mechanisms are inferred by the empirical Rayleigh criterion. The Rayleigh criterion modified by Peake and Oliver (1971) provides a good estimate of the range to be considered to interpret the surface roughness influence (Deroin et al. 2014). Therefore, the sensitivity of L-band PALSAR data to roughness ranges theoretically from 1 to 7 cm. These dimensions are in the size range of rock fragments and stones that cover the eroded soils in the area. In our case, most places show small root mean square (rms) surface height values and the maximum height can be used to roughly define smooth, moderate, and rough surfaces.

Mesoscale surface roughness is related to image texture (see Sect. 19.2.3.4) and is a function of the characteristics of numerous pixels covering a target. On the other hand, the topographic slope and the aspect of the terrain influence macroscale surface roughness.

Table 19.3 Surface scattering as a function of surface roughness

Levels of radar backscatter	Surface scattering	Roughness	Rayleigh criterion ALOS PALSAR $\lambda = 23.6$ cm, $\theta = 34.3°$ Peake and Oliver (1971)	Image tone
Very high backscatter (above −5 dB)	Lambertian diffuse	Very rough surface	rms $\leq \lambda/4.4$ $\cos \theta = 6.5$ cm	Bright
High backscatter (−10 dB to 0 dB)		Rough surface		
Moderate backscatter (−20 dB to −10 dB)	Non Lambertian diffuse	Moderately rough surface	rms $\leq \lambda/8 \cos$ $\theta = 3.6$ cm	Intermediate
Low backscatter (below −20 dB)	Specular	Smooth surface	rms $\leq \lambda/25$ $\cos \theta = 1.1$ cm	Dark

19.2.3.4 Image Change Detection

Algebraic operations between bands (L_{HH} and L_{HV}) and texture analysis in this phase were used to improve visual and digital image interpretation. Image ratio and difference of the polarimetric bands served to emphasize particular signal-target interactions and, hence, interpret land targets of interest (geologic materials, landforms, soil erosion, vegetation, drainage patterns, etc.). By expressing differences through changes in color hue, it is possible to visually enhance surface conditions of the test sites (Shimada et al. 2010).

Previous research (Kux and Henebry 1994; Kandaswamy et al. 2005; Wang and Yong 2008) shows that textural features based on gray-tone spatial dependencies have a general applicability in SAR image classification. Therefore, this research used the Gray level co-occurrence matrix (GLCM) texture extraction feature. This traditional method contains some improvements in the GLCM operator from SNAP (e.g. the automatic calculation of the GLCM's mean, variance, and correlation). A scheme with 64 gray levels was used to preserve detailed information while avoiding high computational load. To keep the size of textural feature space manageable, only one window size was tested (5×5). Moreover, inter-pixel distance was fixed at one (d = 1) since most relevant correlation exists between adjacent pixels, and omnidirectional features were obtained by averaging out the results in the four directions of angle $(\theta)=0$, 45, 90, and 135°. Seven second-order GLCM-based texture measures, including contrast, dissimilarity, homogeneity, angular second moment, energy, maximum probability, and entropy, were extracted from every quantized layer of HH and HV backscattering coefficients for each one of the years considered.

In addition, a preliminary object-based segmentation method was applied on the color composite of the GLCM mean of 2007 (Red), 2009 (Green) and 2010 (Blue), but were not compared with the pixel-based result (Sect. 19.2.3.5). We used the mean shift segmentation algorithm of the open source Monteverdi2 v0.8 software (http://sourceforge.net/projects/orfeo-toolbox/files/Monteverdi2-0.8.0).

19.2.3.5   Feature Extraction and Classification

Unsupervised classification is an important technique for the automatic analysis of SAR data. Several unsupervised classification approaches for polarimetric SAR data have been proposed (Ferro-Famil et al. 2001; Liu et al. 2002; Bruzzone et al. 2004; Ince 2010). One approach to unsupervised classification is based on statistical clustering, which has the advantage of identifying classes that do not perfectly align with pure or isolated physical scattering mechanisms. Instead, objects with an arbitrary but similar backscattering are grouped. To this end, the very popular Expectation Maximization (EM) algorithm was used, where each pixel is assigned with different degrees of class membership to all possible classes in a way that maximizes the posterior probability of the assignment with respect to a mixture model (Dempster et al. 1977).

In an EM classification algorithm, one attempts to assign N pixels to M different classes, while the optimal set of class centers $\Sigma^{(0)}$ remains to be found by the algorithm. For this purpose, one defines the log-likelihood function, the joint sample likelihood conditioned upon a set of class centers:

$$\sum_j^{(0)} = \frac{1}{N_j} \sum_{i \in \omega j} \mathbf{C}_i$$

N_j = number of pixels in ω_j, $\mathbf{C}_i$ = observed covariance matrices

Initial seed regions were determined by a random assignment of pixels to each one of the M classes. Subsequently, in the so-called expectation step (*ES*), the *a posteriori* probabilities for each pixel and class were estimated, i.e. the probability that a pixel belongs to class j, given its covariance $\mathbf{C}_i$ and a set of class centers $\Sigma^{(0)}$ (per σ^0_{HH}, σ^0_{HV} and year). The probability describes both the classes of interest and the non-identified clusters, which are in the image (Davidson et al. 2002).

The maximization step (*MS*) computes parameters maximizing the expected log-likelihood found on the *ES*. All cluster centers and covariance matrixes were recalculated from the updated posteriors, so that the resulting data likelihood function is maximized. When the iteration was completed, each pixel was assigned to the cluster where the posterior probability was maximal.

A layer stack of the derived backscatter coefficients of all the SAR layers together was used as the predictor variables for the unsupervised classification. The EM algorithm implemented in SNAP included the following parameters for both test sites: number of clusters (25), number of iterations (60), random seed (used to generate initial clusters, the default was 31415), ROI-mask (used as to restrict the cluster analysis to a specific area of interest) and addition of probability.

The Bayesian Information Criterion (BIC), also known as Schwarz's Bayesian criterion (SBC), was used to determine how many final clusters (classes) are present in each test site. It gives an indication of which classes are closer statistically and how classes split and merge. This generic function was implemented in the R statistic packages (v3.1.3), and it calculates for one or several fitted model objects for which a log-likelihood value can be obtained, according to the formula $-2 * \log-likelihood + npar * \log(nobs)$, where *npar* represents the number of parameters and *nobs* the number of observations in the fitted model. Cluster labeling was the final step undertaken with the assistance of available ground truth data.

19.2.3.6 Validation

Classification accuracy was assessed using independent information gathered from remotely sensed data of higher accuracy existing for the same study area (see Bisigato et al. 2013; Bouza and del Valle 2014; del Valle et al. 1997, 2013; Hardtke et al. 2011). A stratified-random selection of validation sites was used to construct the error matrix ensuring at least 25 samples per class in each test site. This information was then compiled in a contingency table so that the accuracy of each class can be determined. In addition, the classified image maps were displayed in image windows and linked to Google Earth (export view as kmz) for the visual evaluation of the classes.

19.3 Results and Discussion

To consider in an inclusive way the advantages and disadvantages of ALOS PALSAR mosaics acquired in different years, we begun assessing the surface roughness to identify landscape composition and soil processes, as well as assessing the band operations and texture analysis for terrain characterization. Then, we classified the different land covers of the study area; in this step, the accuracy of the resultant classification was established by digital integration and comparison to validation information derived from ground truth.

19.3.1 Landscape Roughness

Figures 19.4 and 19.5 show the basic statistics of backscattering coefficients at different polarizations per year and landscape for El Moro and San Luis test sites, respectively.

The level of L_{HH} backscatter was higher as compared to the L_{HV} band, being in agreement with previous investigations reported by del Valle et al. (2010, 2013). It appears that co-polarized data (HH) are more responsive to surface roughness or

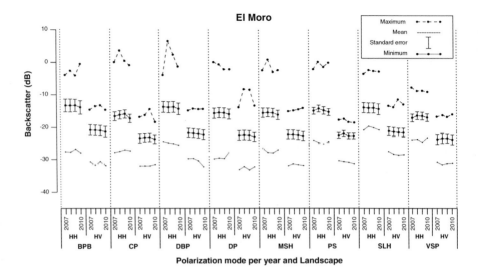

Fig. 19.4 Backscatter coefficients (σ^0) from El Moro site per polarization mode, year, and land-scape. Basaltic Plateau Border (BPB), Covered Peneplain (CP), Dissected Basalt Plain (DBP), Dissected Peneplain (DP), Medium Slope Hill (MSH), Piedmont Slope (PS), Steep Slope Hill (SLH), Valley Slopes and Plains (VSP)

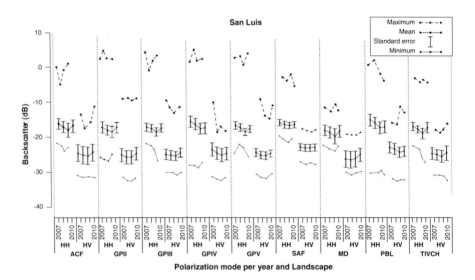

Fig. 19.5 Backscatter coefficients (σ^0) from San Luis site per polarization mode, year, and land-scape. Alluvial Colluvial Fans (ACF), Gravel Plain II (GPII), Gravel Plain III (GPIII), Gravel Plain IV (GPIV), Gravel Plain V (GPV), Secondary Alluvial Fans (SAF), Pediments and Bottomlands (PBL), Terrace IV Chubut River (TIVCH)

volume scatterers on the scale of L-band while cross-polarized data (HV) are most responsive to the geometry or texture of surface or volume scatterers (Ulaby and Long 2014).

The averaged backscatter differences for L_{HH} and L_{HV} can be attributed to changes in surface scattering of the landscape types, and are therefore largely determined by surface roughness. El Moro site presents mainly a decreasing tendency over time, while San Luis site shows a decrease in the first 3 years and increases in 2010 (wet conditions). The decreasing tendency observed at El Moro site is likely due to the complex interactions between different factors (e.g. topography, rocky surface, soil texture, and vegetation cover) that influence the soil moisture content (Tromp-van Meerveld and McDonnell 2006). Backscatter of terrain is modulated by the surface geometry of plains, hills, and low mountains. This modulation is a function of slope steepness, slope orientation, and the scattering mechanism of the terrain.

Almost all landscapes have places with relatively heterogeneous surfaces that correspond to direct backscattering of more or less rough surfaces. Landscape roughness changes are noticeable when mobile roughness elements such as flexible vegetation are present. Dependence of the backscattered intensities might be due to natural disturbances (drought, lightning-caused wildfires) and human impacts (wildfires, overgrazing) that affected mainly the San Luis site (2007–2009), causing a decrease in vegetation cover (high backscattering) or showing dry vegetation where the low backscattering is related to low values of the dielectric property. When fire has affected vegetation, there is a remarkable change in radar backscatter and/or image texture (del Valle et al. 2013).

The generalized order of landscape roughness observed was as follows:

- BPB > DBP $\cong$ SLH > PS > MSH $\cong$ DP > CP $\cong$ VSP (El Moro site; for acronyms see Fig. 19.4).
- GSOP $\cong$ PBL > ACF $\cong$ GPIV $\cong$ GPV $\cong$ GPIII $\cong$ GPII $\cong$ TIVCH > MD (San Luis site; for acronyms see Fig. 19.5).

The maximum backscatter values were higher for L_{HH} than for L_{HV}. On the other hand, the minimum backscatter values also showed the same tendency for both polarizations. The lower noise level detectable on the PALSAR image coincided with the estimated values obtained by Shimada et al. (2009).

The capability of the L-band to penetrate mixed sandy loam with low content of rock fragments and pebbles (<1 m thick) and return information about geologic and geomorphic buried features is showed in Fig. 19.6. An ancient drainage pattern cuts in the bedrock, where most of times it appears as a dark-gray network in L_{HH} (buried paleochannels). The signal of radar images waves from smoother channel filling and their attenuation by the same material causes the dark-gray aspect of the buried paleodrainage contrasting with the brightness of the hard substratum (Paillou et al. 2010). On the other hand, the white lines in L_{HV} in the same sector (Fig. 19.6, ref. *1*) correspond to hard rock layers, covered by a thin sandy loam layer, abundant rock fragments and pebbles (<0.40 m thick). This is an example of volume scattering, as the radar signal penetrated the dry cover, with backscattering when it met hidden rocks.

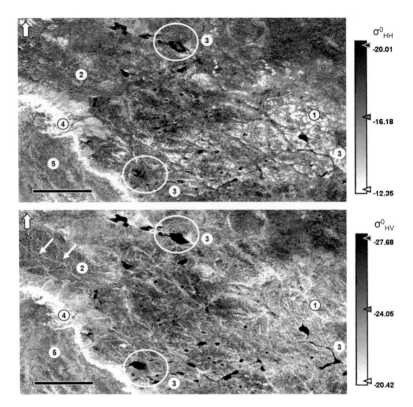

Fig. 19.6 Backscatter coefficients (σ^0) comparison between L_{HH} and L_{HV} (2010). El Moro site *1* Paleochannels buried (*dark-gray lines* in L_{HH}). Hard rock layers covered by a thin layer of mixed sandy loam, abundant rock fragments and pebbles (*white lines* in L_{HV}); *2* Paleomicrorelief, subsoil erosion (argillic horizons) showing signs of buried channels erosion in L_{HV} (*white arrows*); *3 White circles* shows the spatial differences in playa-lake sediments (enhanced aquifer in L_{HV}); *4* Basaltic plateau border; *5* Differential erosion of the basaltic 'cap rocks' more evident in L_{HV}. Bar length = 5 km

Figure 19.6 (ref. *2*) indicates the existence of water erosion in argillic horizons showing buried paleochannels in L_{HV}. Súnico et al. (1996) studied this paleomicro-relief near the study area. The argillic horizons are discontinuous, appearing and disappearing over short distances showing signs of erosion and burial. In Patagonia, the argillic horizon thickness or distribution is closely related to lateral and vertical facies variations, and to different soil moisture conditions in the past (del Valle 1998).

The playa lakes (saline lakes) appear dark in the image (below −20 dB), but with spatial differences in both polarizations (Fig. 19.6, ref. *3*). L_{HV} enhances the aquifer with braided discharge distributaries (white lines).

The influence of topography was evident on the BPB landscape (Fig. 19.6, ref. *4*). The surface roughness is controlled by the weathering of the bedrock and soil erosion or the reworking of surficial deposits (mass movement, alluvial sorting, wind erosion, etc.). This landscape exhibits a stronger influence on the radar signal given by bright returns and shadows (front/back slopes), rock outcrops, basaltic rock aquifers, and vegetation. Vegetation changes on the plateau (DBP, Fig. 19.6, ref. 5) from top to base, where semi-arboreal vegetation (tall shrubs) is nearly 2 m high. When the surface is no longer horizontal (outstanding targets), preferential backscattering occurs from edges and corners. The radar shadows exist and are only a few pixels in content, and do not cause any major classification errors.

In general, the relatively flat topography (prevailing in San Luis site) and low elevation variations are controlled mainly by the microrelief features (Fig. 19.7). Microrelief consists of mounds associated with shrubs, and intermounds where desert pavements and vesicular layers have developed. The smoother areas correspond to recently burnt places, playa lakes, roads, very dry terrain (deflation areas without ridges), and paleo drainage channels (Fig. 19.7, refs. 1 to 3). They appear dark in the image.

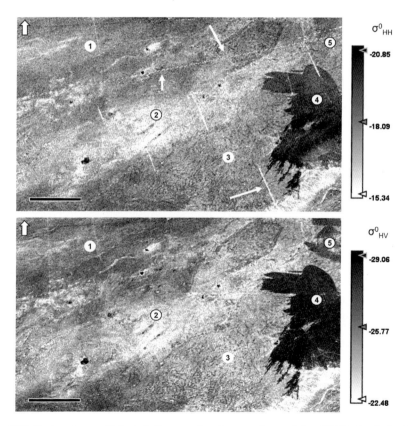

Fig. 19.7 Backscatter coefficients (σ^0) comparison between L_{HH} and L_{HV} (2008). San Luis site. Identification of paleo drainage channels with similar patterns in both polarizations mode (1–3). Note the observed channel pathways from the different gravel plains. 4–5 Burnt areas. *White arrows* indicate man-made-structures. *Bar* length=5 km

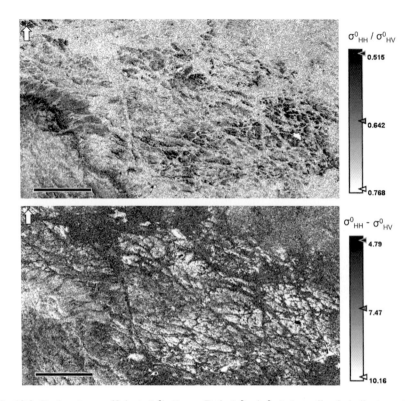

Fig. 19.8 Backscatter coefficients (σ^0). *Upper*: Ratio (σ^0_{HH} / σ^0_{HV}). A small ratio indicates a strong cross-polarized response (volume scattering). The opposite is where a low cross-polarized response and a higher ratio occurs (surface scattering). *Lower*: Difference (σ^0_{HH} − σ^0_{HV}). *Bar* length = 5 km

Distinction between burnt and unburnt areas is remarkable. However, the backscattering intensity of burnt areas after bushfire clearly increases compared to that before the fire (Fig. 19.7, refs. *4* and *5*). Reduced volume scattering, bare dry soil and a decreasing dielectric constant result after wildfire in a low backscatter. In fact, the scrublands were severely burned, leaving only the trunks of the shrubs (del Valle et al. 2013).

19.3.2 Polarimetry Band Ratio, Difference and Texture Analysis

We applied band ratio, difference, and texture analysis for the separation and identification of terrain features. Results show that polarization shift inherent to the scattering process provides benefit to statistical and band analysis of SAR for terrain characterization.

Figure 19.8 (upper) is an example of the ALOS PALSAR ratio (σ^0_{HH} /σ^0_{HV}). The major linear features and variations of the vegetation cover are evident. A small ratio value (dark tone) indicated a strong cross-polarized (HV) response. In addition, cross-polarized signal seems to provide better discrimination between specular and diffuse signal return. The opposite is where a low cross-polarized response and a higher ratio occurs (bright tone). This information is consistent with the one obtained in the landscape roughness analysis.

The difference ($\sigma^0_{HH} - \sigma^0_{HV}$) is shown in Fig. 19.8 (lower). This difference can be considered also as a measurement of surface roughness. A scene is expected to respond to radar illumination in the same manner regardless of the polarimetry parameters involved, producing in our case study a difference of non-uniform intensity. The uniformity decreases as the number of different features increases, raising the general brightness of the difference as well as its spatial variability. Therefore, it can be assumed that pixels in the different polarizations that are close to each other in backscatter intensity denote mainly a dominant double bounce, and rough surfaces.

Figures 19.9 and 19.10 (left) show the Gray Level Co-occurrence Matrix (GLCM) mean in RGB for both test sites. The results reproduce the co-occurrence mean features that were appropriate for characterizing different landscape structures in 2007, 2009 and 2010. Color information improves the results of the gray-scale textural features. Figures 19.9 and 19.10 (right) show different classes that could be separated very well through the GLCM mean RGB segmentation. This shows the efficiency of the object-based classification and the performance of Monteverdi2 software in precise segmentation. This technique could be used for feature extraction and image classification.

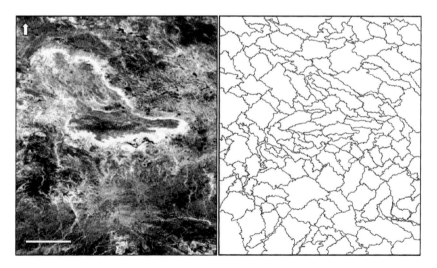

Fig. 19.9 El Moro site. *Left*: Gray Level Co-occurrence Matrix (GLCM) mean, RGB (LHH$_{2007}$LHV$_{2009}$LHH$_{2010}$). Gray-scale image mode. *Right*: Result of the texture segmentation process. *Bar* length = 10 km

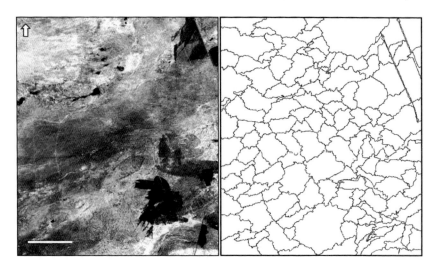

Fig. 19.10 San Luis site. *Left*: Gray Level Co-occurrence Matrix (GLCM) mean, RGB (LHH$_{2007}$LHV$_{2009}$LHH$_{2010}$). Gray-scale image mode. *Right*: Result of the texture segmentation process. *Bar* length = 10 km

19.3.3 Mapping Approach

The optimum number of clusters was computed from the so-called BIC. Consequently, together with the ground truth available, the 25 initial classes were aggregated into 6 and 7 classes for El Moro and San Luis sites, respectively as shown in Table 19.4 and Fig. 19.11. Over the study sites, radar signal backscattering mechanisms can be simplified into four major categories: surface scattering, volume scattering, double-bounced scattering, and specular scattering.

El Moro test site (Fig. 19.11, left) shows the effect of different scatterers in the land cover classes. The effects of narrow slumps (classes 1 and 2, very rough surfaces) causes preferential backscattering from their edges and corners (double-bounced scattering). This is known as edge effect and corner reflectors (Fung and Chen 2010). Orientation of the slumps corresponds to that of the ALOS PALSAR ascending orbit. Therefore, the side looking is perpendicular to the structure. These classes particularly present intensive linear erosion along the edges mainly of the basalt plateau. Fractures in the basalt rocks and subsequent water percolation enable the development of mass movements. Gullies where streams are incised were associated with slump and flow erosion (class 2). The soils have low fertility supporting a grass cover with scrubs. Class 3 (bright and intermediate rough surface) shows a dependence of surface scattering (tall shrubs on gullies, alluvial fans, and basalt flows). Vegetation on basalts lives on shallow soils, as streams readily disappear into the porous fractured lava flows from which moderate water volumes emerge (plateau border). The dielectric constant (moisture content) of the plateau border governs the strength of the backscatter. Multiple volume-surface scattering was observed mainly in class 4 (moderately rough and rough surfaces). The presence of

Table 19.4 Cluster centre of the aggregation classes using the Bayesian Information Criterion (BIC) and ground truth

Year/σ^0/surface	El Moro[a]						San Luis[b]						
	1	2	3	4	5	6	1	2	3	4	5	6	7
2007													
σ^0_{HH}	-9.6	-11.7	-13.7	-15.8	-17.6	-20.9	-12.3	-14.5	-15.9	-17.0	-18.1	-20.2	-20.9
σ^0_{HV}	-18.5	-20.2	-21.6	-23.1	-24.8	-28.5	-21.5	-22.0	-23.2	-24.8	-26.3	-27.8	-28.5
2008													
σ^0_{HH}	-9.6	-11.7	-13.4	-15.6	-17.4	-20.8	-12.9	-15.4	-16.6	-17.4	-18.5	-20.8	-20.9
σ^0_{HV}	-16.6	-20.2	-21.6	-22.9	-24.7	-28.5	-22.0	-22.7	-23.6	-25.2	-26.6	-28.5	-28.8
2009													
σ^0_{HH}	-9.6	-11.8	-13.8	-15.6	-17.3	-20.7	-13.7	-16.4	-17.3	-18.5	-19.4	-20.6	-21.9
σ^0_{HV}	-18.5	-20.3	-21.8	-23.1	-24.8	-28.5	-22.6	-23.3	-23.9	-25.5	-26.7	-28.0	-28.7
2010													
σ^0_{HH}	-10.0	-12.2	-14.3	-16.4	-18.0	-20.4	-13.4	-16.1	-16.7	-17.4	-18.3	-20.5	-20.8
σ^0_{HV}	-19.0	-20.6	-22.1	-23.6	-25.1	-28.3	-22.2	-23.0	-23.4	-24.6	-25.6	-27.7	-28.7
Surface (%)	1.2	3.5	19.2	44.4	31.3	0.5	0.7	16.4	33.0	26.2	21.0	1.9	0.8

[a] 1 Rock talus (slumps), 2 Rock talus (slumps and gullies), 3 Tall shrubs on gullies, alluvial fans and basalt flows, 4 Erosional landforms (paleosurfaces), 5 Mixed barren land (rock outcrops/sandy loam/sandy clay), 6 Playa lake (saline lake). Class 1 to Class 6 decrease the roughness surface

[b] 1 Man-made structures, 2 Pediments and bottomlands, and gravel plains, 3 Secondary alluvial fans and gravel plains, 4 Gavel plains and alluvial flats with burned areas, 5 Gravel plains, burned areas, 6 Bare soils, burned areas, 7 Playa lake (saline lake). Class 1 to Class 7 decrease the roughness surface

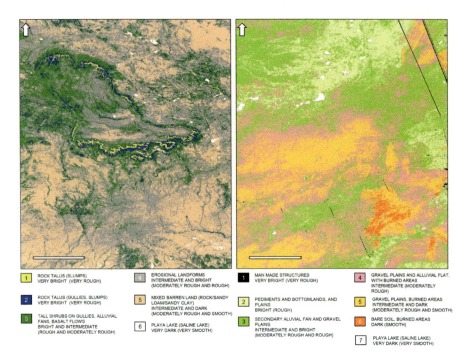

Fig. 19.11 Spatial distribution of the land cover classes with their tone and their surface roughness, using the Expectation Maximization (EM) algorithm and the Bayesian Information Criterion (BIC). *Left*: El Moro land cover classes. Class *1* to Class *6* decrease the roughness surface. *Right*: San Luis land cover classes. Class *1* to Class *7* decrease the roughness surface. *Bar* length = 10 km

paleosurfaces in this class shows an ancient landscape that was eroded by the incision of waterways and subsequently buried (Súnico et al. 1996). Class 5 (moderately rough and smooth surfaces) represent mixed barren land (rock/sandy loam/sandy clay), areas of bedrock, desert pavement, scarps, talus, and volcanic material. Generally, vegetation accounts for less than 35 % of total cover. Dry salt flats (playa or saline lake) occurring on the bottoms of interior desert basins were included in class 6 (very smooth surface, specular scattering).

The San Luis test site (Fig. 19.11, right) comprises different physical features (stony surface, gravel contents in soil depth, soils of contrasting textures, shrub structure, etc.) very interesting to evaluate with SAR remote sensing. The microrelief consists of mounds associated with shrubs and intermounds where mainly desert pavements and vesicular layers have developed (interaction surface-volume scattering). The vegetation is patchy and consists of grasses under bushes or shrublike groups separated in some classes by vast bare soil areas (Bisigato et al. 2013). Access roads, boundary fences, farmlands, power lines, etc. were recognized in class 1 (very rough, surface and double-bounced scattering). Pediments, bottomlands, alluvial fans and plains (class 2, rough surfaces) have the second highest mean σ^0 values. We attributed this to the double-bounced backscattering due to geologic erosion, sedimentation, and soil erosion, which are severe in this class (Gonzalez Díaz and Di Tommaso 2011). On shrublands canopies, SAR penetrates through the vegetation to reach the ground surface. Then, the multiple scatterings between vegetation and the ground surface attenuate the incoming radar signal, and the energy returning to the radar is reduced (class 3, moderately rough and rough surfaces). Rough surface is related mainly to soil erosion (rill, stony surface) that reduces the transmission of the radar wave and enhances the backscattering return. Classes 4 (moderately rough surface) and 5 (moderately rough and smooth surfaces) seem to be associated to moderate-severe overgrazing and the moisture conditions of recent and old burned areas, i.e. becoming detectable when there is an increase in roughness and moisture content in the soil (del Valle et al. 2013). The lower value of the backscattering (class 6, smooth surface) relates to the fact that signal return is dominated by the soil contribution. Class 7 (very smooth surface, specular scattering) presents the same characteristics as those of the El Moro site.

The overall classification accuracy obtained for the different classes (El Moro-San Luis sites, respectively) is 91.2–90.7 % and the KHAT statistic value is 90.3–89.7 %.

19.4   Conclusions and Outlook

The results presented in this chapter suggest that it is feasible using the new PALSAR mosaic data as input in soil survey programs. Furthermore, active microwave remote sensing can represent an appropriate support for DSM. It is important to recognize that the radar image represents physical processes and that it is interpretable based on the understanding of these processes.

A significant knowledge base exists in the microwave remote sensing literature that focuses on the development and refinement of methods to estimate soil physical properties. However, more case studies in Argentina's soil surveys will contribute to an improved understanding of the applicability of radar analysis techniques for traditional mapping as well as for DSM. Further research should be conducted to compare classifiers' performance, including hierarchical and regression tree approaches, extension from basic land covers to more complex land cover classification schemes, and image fusion with optical data.

If DSM supported by active remote sensing is selected as a tool for a typical soil survey programme with high production goals, it requires to budget time for staff to learn its use, and to explore how to apply the outputs for improved performance. Coordinating the long-term implementation of DSM methods should be seen as a way to more accurate and easier soil surveys.

Integration of DSM methods into existing Argentinean soil survey protocols is a challenging task. Our goal with this and future operational initiatives is to formalize the application of microwave remote sensing methods, and to provide an operational framework within which DSM can grow in Argentina.

For operational soil monitoring systems (larger) time series of dual-polarization data should continue. In this regard, the forthcoming SAOCOM (Spanish for Argentine Microwaves Observation Satellite) mission is also of particular interest. In addition to L-band data, current COSMO SKYMED (X-band) and SENTINEL-1 (C-band) missions of easy data availability for Argentina may be also utilized to increase accuracy of landscape discrimination and characterization.

Acknowledgement This research was carried out within the framework of the forthcoming SAOCOM mission, supported by the National Commission on Space Activities (CONAE) jointly with the Department of Science, Technology and Productive Innovation (MINCyT) from Argentina. The SAOCOM project was established for the definition and development of L-band SAR products oriented towards applications for risk and emergency management. This work was also partially supported by CONICET (PIP-11220080101238 and PIP-11220110100220).

References

Abraham EM, del Valle HF, Roig FA, Torres L, Ares JO, Coronato FR, Godagnone R (2009) An overview of the geography of the Monte Desert biome (Argentina). In: Cavagnaro JB, Villagra PE, Bertiller MB, Bisigato AJ (eds) Deserts of the World Part III: The Monte Desert – The Monte. J Arid Environ 73(2):144–153

Bisigato AJ, Bertiller MB, Ares JO, Pazos GE (2005) Effect of grazing on plant patterns in arid ecosystems of Patagonian Monte. Ecography 28:561–572

Bisigato AJ, Hardtke LA, del Valle HF (2013) Soil as a capacitor: considering soil water content improves temporal models of productivity. J Arid Environ 98:88–92

Bouza PJ, del Valle HF (2014) Propiedades y génesis de acumulaciones de carbonatos en Aridisoles del centro-este del Chubut. Capítulo VIII. In: Imbellone P (ed) Suelos con acumulaciones calcáreas y yesíferas de Argentina. INTA- AACS Editions, Buenos Aires, pp 199–219

Bruzzone L, Wegmüller U, Wiesmann A (2004) An advanced system for the automatic classification of multitemporal SAR images. IEEE Trans Geosci Remote Sens 42:1321–1334

Chen CT, Chen KS, Lee JS (2003) The use of fully polarimetric information for the fuzzy neural classification of SAR images. IEEE Trans Geosci Remote Sens 41:2089–2100

Davidson G, Ouchi K, Saito G, Ishitsuka N, Mohri N, Uratsuka S (2002) Polarimetric classification using expectation methods. In: Polarimetric and interferometric SAR workshop, Communications Research Laboratory, Tokyo

del Valle HF (1998) Patagonian soils: a regional synthesis. Ecol Aust 8:103–123

del Valle HF (2008). Controversias y tendencias de la modelación cartográfica ambiental. In: Cantú MP, Becker AR, Bedano JC (eds) Evaluación de la sustentabilidad ambiental en sistemas agropecuarios. Editorial Fundación Universidad Nacional de Río Cuarto (EFUNARC), pp 89–96

del Valle HF, Frulla L, Gagliardini DA (1997) Segmentation of textures in ERS-1/SAR images applied to evaluate land degradation of rangelands (Central Patagonia, Argentina). International seminar on the use and applications of ERS in Latin America. ESA (European Space Agency). ESA SP-405, pp 177–184

del Valle HF, Buck A, Mehl H (2002) Digital elevation models as tools in soil research in northeastern Patagonia. XVIII Congreso Argentino de la Ciencia del Suelo, Puerto Madryn (Chubut), CD-ROM, 11 pp

del Valle HF, Blanco PD, Metternicht GI, Zinck JA (2010) Radar remote sensing of wind-driven land degradation processes in northeastern Patagonia. J Environ Qual 39:62–75 del Valle HF, Hardtke LA, Blanco PD, Sione W (2013) Assessment of land degradation using Shannon Entropy approach on PolSAR images in Patagonian coastal deserts. Geofocus 13(2):84–111

Dempster AP, Laird NM, Rubin DB (1977) Maximum likelihood from incomplete data via the EM algorithm. J R Stat Soc Ser B 39:1–38

Deroin JP, Djemai S, Bendaoud A, Brahmi B, Ouzegane K, Kienast JR (2014) Integrating geologic and satellite radar data for mapping dome-and-basin patterns in the In Ouzzal Terrane, Western Hoggar, Algeria. J Afr Earth Sci 99:652–665

Ferro-Famil L, Pottier E, Lee JS (2001) Unsupervised classification of multi-frequency and fully polarimetric SAR images based on the H/A/alpha-Wishart classifier. IEEE Trans Geosci Remote Sens 39(11):2332–2342

Fung AK, Chen KS (2010) Microwave scattering and emission, models for users. Artech House, Boston, 427 pp

González Díaz EF, Di Tommaso I (2011) Evolución geomorfológica y cronológica relativa de los niveles aterrazados del área adyacente a la desembocadura del río Chubut al Atlántico. Rev Asoc Geol Argent 68(4):507–525

Hardtke LA, del Valle HF, Sione W (2011) Spatial distribution of wildfire risk in the Monte biome (Patagonia, Argentina). J Maps 2011:588–599. doi:10.4113/jom.2011.1184

Harris I, Jones PD, Osborn TJ, Lister DH (2014) Updated high-resolution grids of monthly climatic observations – the CRU TS3.10 Dataset. Int J Climatol 34(3):623–642

Henderson FM, Lewis AJ (1998) Radar fundamentals: the geoscience perspective. In: Ryerson RA (ed) Principles & applications of imaging radar, vol 2, 3rd edn, Manual of remote sensing. Wiley, New York, pp 131–181

Ince T (2010) Unsupervised classification of polarimetric SAR image with dynamic clustering: an image processing approach. Adv Eng Softw 41:636–646

IUSS Working Group WRB (2014) World reference base for soil resources 2014. International soil classification system for naming soils and creating legends for soil maps. World soil resources reports no. 106. FAO, Rome

Kandaswamy U, Adjeroh DA, Lee MC (2005) Efficient texture analysis of SAR image. IEEE Trans Geosci Remote Sens 43(9):2075–2083

Kux HJH, Henebry G (1994) Multi-scale texture in SAR imagery: landscape dynamics of the Pantanal, Brazil. Geoscience and remote sensing symposium 1994. IGARSS '94. Surface and atmospheric remote sensing: technologies, data analysis and interpretation, vol 2, pp 1069–1071

Lagacherie P, McBratney AB, Voltz M (2007) Digital soil mapping: an introductory perspective, 1st edn. Elsevier, Amsterdam

León RJC, Bran D, Collantes M, Paruelo JM, Soriano A (1998) Grandes unidades de vegetación de la Patagonia extrandina. Ecol Aust 8:275–308

Levin N, Ben-Dor E, Kidron GJ, Yaakov Y (2008) Estimation of surface roughness (Zo) over a stabilizing coastal dune field based on vegetation and topography. Earth Surf Process Landf 33:1520–1541

Liu X, Skidmore AK, Osten HV (2002) Integration of classification methods for improvement of land-cover map accuracy. ISPRS J Photogramm Remote Sens 56:257–268

McBratney AB, Mendonça Santos ML, Minasny B (2003) On digital soil mapping. Geoderma 117(1–2):3–52

Mulder VL (2013) Spectroscopy-supported digital soil mapping. Wageningen University, Wageningen

Mulder VL, de Bruin S, Schaepman ME, Mayr TR (2011) The use of remote sensing in soil and terrain mapping: a review. Geoderma 162:1–19

Paillou P, Lopez S, Farr T, Rosenqvist A (2010) Mapping subsurface geology in Sahara using L-band SAR: first results from the ALOS/PALSAR Imaging Radar. IEEE J Sel Top Earth Obs Remote Sens 3(4):632–636

Peake WH, Oliver TL (1971) The response of terrestrial surfaces at microwaves frequencies. Technical report 2770-7. Ohio State University, Columbus

Ramos V (1999) Las provincias geológicas del territorio argentino. In: SEGEMAR (ed) Geología Argentina. Instituto de Geología y Recursos Minerales, Buenos Aires. Anales 29:341–396

Ridley J, Strawbridge F, Card R, Phillips H (1996) Radar backscatter characteristics of a desert surface. Remote Sens Environ 57:63–78

Rosenqvist A, Shimada M, Ito N, Watanabe M (2007) ALOS PALSAR: a pathfinder mission for global-scale monitoring of the environment. IEEE Trans Geosci Remote Sens 45:3307–3316

SAGyP-INTA (1990) Atlas de Suelos de la República Argentina. Proyecto PNUD Arg-85/019, Buenos Aires

Saha SK (2011) Microwave remote sensing in soil quality assessment. Int Arch Photogramm Remote Sens Spat Inf Sci XXXVIII-8/W20:34–39

Shimada M, Isoguchi O, Tadono T, Isono K (2009) PALSAR radiometric and geometric calibration. IEEE Trans Geosci Remote Sens 47(2):3915–3932

Shimada M, Tadono T, Rosenqvist A (2010) Advanced land observing satellite (ALOS) and monitoring global environmental change. Proc IEEE 98(5):780–799

Shimada M, Itoh T, Motooka T, Watanabe M, Shiraishi T, Thapa R, Lucas R (2014) New global forest/non-forest maps from ALOS PALSAR data (2007–2010). Remote Sens Environ 155:13–31

Smith MW (2014) Roughness in the Earth Sciences. Earth Sci Rev 136(2014):202–225

Súnico A, Bouza PJ, del Valle HF (1996) Erosion of subsurface horizons in northeastern Patagonia, Argentina. Arid Soil Res Rehabil 10:359–378

Tromp-van Meerveld HJ, McDonnell JJ (2006) Threshold relations in subsurface stormflow: 1. A 147-storm analysis of the Panola hillslope. Water Resour Res 42:W02410. doi:10.1029/2004WR003778

Ulaby FT, Long DG (2014) Microwave radar and radiometric remote sensing. University of Michigan Press, Ann Arbor

Villagra P, Defosse G, del Valle H, Tabeni S, Rostagno C, Cesca E, Abraham E (2009) Land use and disturbance effects on the dynamics of natural ecosystems of the Monte Desert: implications for their management. J Arid Environ 73:202–211

Wang ZZ, Yong JH (2008) Texture analysis and classification with linear regression model base on wavelet transform. IEEE Trans Image Process 17(8):1421–1430

Zribi M, Baghdadi N, Nolin M (2011) Remote sensing of soil. Appl Environ Soil Sci 2011, Article ID 904561, 2 p, doi:10.1155/2011/904561

Chapter 20
Geopedology Promotes Precision and Efficiency in Soil Mapping. Photo-Interpretation Application in the Henares River Valley, Spain

A. Farshad, J.A. Zinck, and D.P. Shrestha

Abstract Two approaches to prepare photo-interpretation maps that guide the location of field observations and serve as frames for soil cartography are compared. The physiographic approach is mainly descriptive and aims at separating relief units on the basis of their physiognomic appearance. The geopedologic approach highlights relationships between soils and geoforms and aims at predicting patterns of soil distribution prior to soil survey. Both approaches have been applied in the Henares river valley (Spain). The two interpretation maps are compared in terms of soil pattern and density of delineations.

Keywords Photo-interpretation • Geopedologic approach • Physiographic analysis • Soil survey • Henares valley • Spain

20.1 Introduction

Soil as a natural three-dimensional body is mainly an underground entity that is conventionally accessed by means of point observations. Adjacent soil bodies form a continuum on the landscape whose spatial variability between observation points is inferred by using surface indicators. Soil mapping aims at segmenting the soil continuum into delineations as homogeneous as possible. However, it is "almost never feasible to delineate accurately on a map the area that soils of one taxonomic

A. Farshad (✉) • D.P. Shrestha
Faculty of Geo-Information Science and Earth Observation (ITC), University of Twente, Enschede, The Netherlands
e-mail: abbasfarshad@gmail.com; d.b.p.shrestha@utwente.nl

J.A. Zinck
Faculty of Geo-Information Science and Earth Observation (ITC), University of Twente, Enschede, The Netherlands

Institute of Environmental Studies, University of New South Wales, Sydney, NSW, Australia
e-mail: alfredzinck@gmail.com

© Springer International Publishing Switzerland 2016
J.A. Zinck et al. (eds.), *Geopedology*, DOI 10.1007/978-3-319-19159-1_20

class occupy in the field" (SMSS 1986). The use of geomorphology can contribute to overcome this issue, as geomorphic features help recognize and explain the systematic variations in soil patterns (Wilding and Drees 1983). The aim of this chapter is to show the role that geomorphology, as compared to physiography, can play in air photo-interpretation to increase accuracy and efficiency in soil survey.

20.2   Trends in Soil Mapping and Pattern Analysis

Soil delineation is a polygon of a soil map that is relatively homogeneous at the mapping scale considered. However, it contains usually more than one single taxonomic class. As most of the soil properties are hidden below the terrain surface, it is not possible to follow on the ground their actual boundaries. Soil surveyors have to rely on surrogate indicators such as topography, vegetation, surface color, and other terrain features to delineate soil units on a map. These external criteria can be identified and mapped through interpretation of remote-sensed documents.

The process of inventory/mapping includes collecting spatial and point data and storing them in a database. Combined use of image analysis techniques and improved locational accuracy by GPS facilitates field data collection. Usually, remote-sensed data, mostly covering the visible and near-infrared portion of the electromagnetic spectrum, only provide information on the land surface. On the other hand, the advancement in the domain of data base management systems (DBMS), both spatial and non-spatial, in a GIS environment allows storing and retrieving data in point, vector, and/or raster format as required.

New digital technologies for field data acquisition and management are taking over conventional survey approaches, sometimes competing with each other, sometimes advantageously combined (Farshad et al. 2013). Soil mapping has effectively benefited from constant progress in the domains of GIS, digital terrain modeling, pattern analysis, expert system, decision support system, among others (Saldaña 1997; Moran and Bui 2002; Bui 2003; Hengl and Reuter 2009). Remote sensing including electromagnetic, hyperspectral, and penetrating sensors, has been less used, often at research level.

Soil pattern is a relevant concept in soil cartography. It refers to the spatial arrangement of soil bodies described in terms of configuration and taxonomic composition. Fridland (1976) and Hole and Campbell (1985) have defined and classified soil patterns in terms of elementary soil body, combinational soil body, and soil cover pattern. Soil distribution in patterns helps address and quantify the spatial complexity and variability of soilscapes both lateral and vertical. For this purpose, heterogeneity indices such as mean density of soil bodies, index of heterogeneity, and mean density of soil map units, and size and shape indices such as fractal dimension and soil body or soil map unit shape index, have been applied (Saldaña 1997; Hansakdi 1998; Saldaña et al. 2011).

Size, shape, and contour irregularity of the delineations control the geometry of the units at any given level (Fridland 1976; Hole and Campbell 1985). The irregularity of many natural boundaries obeys fractal laws (Mandelbrot 1982). According to the fractal theory, the morphological outline of any polygon can be related to the value of the fractal dimension D. The D value provides a combined measure of irregularity and fragmentation across all the spatial scales considered. It increases with increasing fragmentation, and helps describe the structure of soil patterns, especially dissection patterns.

20.3 The Relevance and Role of Photo-Interpretation in Soil Survey

20.3.1 Retrospect

Before the advent and development of digital soil mapping, the approaches, methods, and techniques applied to soil inventory have evolved over time from intensive field survey to increasing use of remote-sensed documents. Earlier soil mapping was based on grid survey. Surveying instruments, including theodolite, were used to plot the observation points that were subsequently clustered to form soil map units. This time demanding procedure was substantially improved when in the 1950s aerial photo-interpretation was introduced in soil survey using physiographic and later geomorphic criteria for segmenting the landscape and delineating units. The Manual of Photographic Interpretation (ASP 1960) was a cornerstone reference promoting the use of aerial photography in surveys. However, the role of geomorphology in photo-interpretation for pattern recognition in soil survey was not yet properly addressed (Frost 1960). Buringh (1960) introduced the procedure of element analysis in aerial photo-interpretation for soil survey. Layers of information extracted from aerial photographs were overlaid. This procedure was very useful for teaching aerial photo-interpretation but time consuming. Simultaneously, two approaches called respectively physiognomic analysis and physiographic analysis were developed. The physiognomic analysis consisted of a mental combination of a number of elements that was used together with the physiographic analysis. The latter was developed in the 1950–1960s at ITC (International Training Centre for Aerial Survey) in the Netherlands by Buringh (1960), followed up by Vink, Goosen, and Bennema, as well as at CSIRO in Australia (Stewart 1968) and elsewhere. At this stage, the role of geomorphology in soil survey was clearly recognized (Goosen 1967). In the late 1980s, the geopedologic approach, based on the integration of geomorphology and pedology and soil-geoform relations, was introduced at ITC as a teaching subject matter for training specialists in soil survey (Zinck 1988).

20.3.2 Approaches to Air Photo-Interpretation

Aerial photographs have been for several decades the main remote-sensed source of information to support soil survey. At large scale, an air photo is still one of the most reliable document to extract information on natural resources in general and the soil cover in particular. Stereoscopic vision added by basic knowledge in geomorphology allows reading and segmenting the physical landscape in photo-interpretation units to support soil survey.

Although controlled by geomorphic criteria, photo-interpretation is somewhat subjective and reflects the formation and experience of the interpreter. Bie and Beckett (1973) showed to what extent soil maps prepared by different surveyors could be different. Four skilled soil surveyors from three countries were asked independently to map soils in an area of 19 km² in Cyprus, using air-photo interpretation. Comparing the four maps demonstrated that the four interpreters used different approaches to legend construction and boundary delineation for mapping the same soilscape. They concluded that the quality of soil surveys by air-photo interpretation is more sensitive to the surveyor's choice of soil classes and mapping units than to his/her skill in locating boundaries. Whether a larger scale would have helped is questioned, as the photo scale in this case was 1:10,000.

Hereafter, two different modalities to carry out photo-interpretation are addressed and further applied in a case study.

20.3.2.1 Physiographic Analysis

Physiography comprises the study and understanding of the features that determine the outlook and characteristics of a landscape. Besides the geomorphology and geology of the study area, other factors such as hydrology, vegetation, and land use play a role (Goosen 1967; Bennema and Gelens 1969). Geomorphic processes responsible for the formation of the landforms are identified according to their appearance on aerial photographs. This guides the delineation of physiographic units as a basis for analyzing soil patterns. The photo-interpretation units are characterized using geomorphic terms together with descriptive (physiognomic) terms that refer to vegetation cover, land use, grey tone, etc. (Bie and Beckett 1973). Usually three levels are used to stratify the terrain features, namely landtype, sub-landtype, and map unit.

20.3.2.2 Geopedologic Analysis

Geomorphology forms the backbone of the geopedologic approach using a taxonomic system that comprises six categorial levels including from high to low: geo-structure, morphogenic environment, landscape, relief/molding, lithology/facies, and landform. The first two levels are appropriate for very small-scale mapping. Regional surveys start usually at the level of landscape, followed by relief/molding, lithology/facies, and landform. Seven major landscape types are recognized: mountain, plateau, hilland, piedmont, peneplain, plain, and valley. The middle and lower categories of the system contain an increasingly larger number of classes downwards (Zinck 2013). Each landscape can be composed of a number of relief types (e.g. hill, mesa, and glacis) that may occur on different substratum materials (lithology/facies). Any relief type may comprise a set of landforms. An advantage of the geopedologic analysis in photo or image interpretation is the stepwise procedure following the hierarchic consecutive levels of the geoform system for feature identification. In this way, a structured legend can be built up step by step following a logical sequence of interpretation activities that is described hereafter.

(1) Photo-lecture: visual exploration of the image, using for instance a photo-pair under the stereoscope to identify the main landscape types present in the study area;
(2) Sketching: tracing master-lines to separate the main landscape units distinguished by visual exploration, as a first step for segmentation of the study area;
(3) Selecting cross sections that traverse the study area in appropriate directions, usually perpendicular to outstanding landscape features;
(4) Pattern recognition: identifying the different geoform classes that can be recognized along the selected cross sections;
(5) Delineation: extrapolating the geoform classes from the segmented cross sections to the rest of the area to delineate preliminary photo-interpretation map units;
(6) Composition of legend: structuring the legend in columns, one for each hierarchic level of the geoform classification system, as above mentioned, and labelling the identified geoform types;
(7) Interpretation of landscapes and their elements, applying physiographic logic for which a thorough knowledge of applied geomorphology is required;
(8) Fieldwork preparation: elaborating a predictive soil mapping frame on the basis of the remote-sensed geoform characteristics (observed surface features, assumed parent materials); designing the observation and sampling scheme on the basis of the nature and spatial distribution of the geomorphic photo-interpretation units.

20.3.3 Comparative Example of Air Photo-Interpretation

20.3.3.1 The Study Area

The study area is located in the Henares river valley about 40 km NE of Madrid in Spain, on the southern slope of the Ayllón mountain range (Fig. 20.1). Geologically, it belongs to the Madrid Basin, a Tertiary depression falling within the Tajo river drainage basin. The current climate has been prevailing since the upper Pliocene, when climate changed from wet to more arid. Both tectonic movements and climate change have contributed to the present regional physiography, comprising landscapes of plateau, piedmont, and valley. The Henares valley is entrenched in a limestone plateau and shows a system of alluvial terraces and colluvio-alluvial glacis formed during the Quaternary. Soils include Entisols, Inceptisols, Alfisols, and Ultisols, occurring approximately in this order from recent to old depositional units of the Quaternary (Saldaña 1997).

Fig. 20.1 Location map depicting plateau and piedmont in the SE corner, and the terraces banking the Henares river; *red* areas are irrigated fields on Landsat false color composite

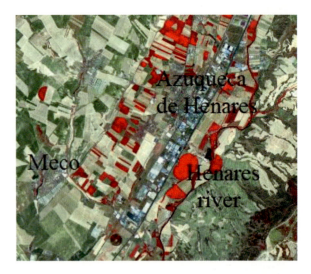

Fig. 20.2 Aerial photograph that was visually interpreted to produce the maps of Figs. 20.3 and 20.4

20.3.3.2   Method and Materials

The two methods above described, namely the physiographic and the geopedologic approaches, were implemented to prepare photo-interpretation maps using aerial photos at 1:35,000 scale (Fig. 20.2). The maps were compared visually and by means of pattern indices including the number of soil map units and soil delineations (polygons) per type of map unit. The soil map of the area prepared by Saldaña (1997) on the basis of geopedologic principles was used as ground truth to control the results.

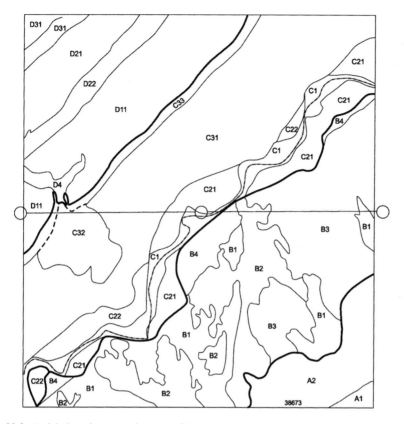

Fig. 20.3 Aerial photo-interpretation map of the area covered by the photograph in Fig. 20.2 using physiographic analysis (Bennema and Gelens 1969)

20.3.3.3   Results

(a) Physiographic analysis

The photo-interpretation map of Fig. 20.3 and its legend (Table 20.1) were obtained applying the physiographic analysis approach according to Bennema and Gelens (1969) to the aerial photo shown in Fig. 20.2. The four major map units called landtypes are plateau, dissected footslope, river plain with younger terraces, and older terraces, eroded and higher. The landtypes are divided into sublandtypes. Further subdivision into map units is done only for some selected landtypes. Units are described using descriptive, physiognomic terms (Table 20.1). The sequence landtype-sublandtype-map-unit helps identify some basic organization of the landscape. The description of the photo-interpretation units highlights relevant features of the regional physiography. However, the hierarchic structure of the landscape is not formally addressed and laid out. The inference of geopedologic relationships concerning soil formation and distribution is

Table 20.1 Legend of the photo-interpretation map shown in Fig. 20.3

Major map unit (landtpye)	Sublandtype	Map unit
(A) Plateau	(A1) Undulating plateau top, cultivated	
	(A2) Dissected (gullied) steep escarpment and adjoining land	
(B) Dissected footslope of A	(B1) Steep small hills (forest and grass, scattered parceling)	
	(B2) Broad valleys and long gentle slopes (cultivated)	
	(B3) Complex of steep small hills and gentle slopes, both about half of area (mostly under grass)	
	(B4) Pediment slope	
(C) River plain with younger terraces	(C1) Flood plain and streamed, not cultivated	
	(C2) Lower terrace	(C21) Lower terrace under cultivation
		(C22) Lower terrace, grassland
		(C23) Lower terrace slightly higher than C22 locally grading into C3
	C3 Higher terrace	(C31) Terrace
		(C32) Alluvial fan
		(C33) Colluvial zone along upward scarp
(D) Older terraces, more eroded, much higher than C3	(D1) First terrace	(D11) First terrace (plain)
	(D2) Second terrace	(D21) Second terrace (plain)
		(D22) Eroded terrace scarp
	(D3) Third terrace	(D31) Third terrace- (plain)
		(D32) Third terrace scarp
	(D4) Tertiary terrace (plain)	(D41) Ending on the level of C31

Bennema and Gelens (1969)

weakly approached, hampering the process of converting the photo-interpretation map into a soil map.

(b) Geopedologic analysis

The same area covered by the aerial photograph shown in Fig. 20.2 was interpreted applying geopedologic analysis (Zinck 1988). Photo-interpretation prior to geopedologic field survey is based on heavy input of geomorphology and the implementation of the hierarchic geoform classification system. Stereoscopic vision provides preliminary identification and delineation of the geoforms on the basis of their external characteristics, while the classification system allows organizing the photo-interpretation information in a hierarchically structured legend at four categorial levels (Fig. 20.4 and Table 20.2).

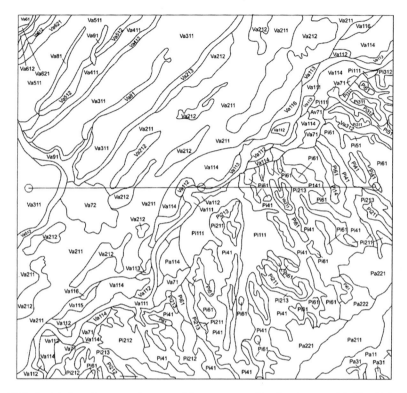

Fig. 20.4   Aerial photo-interpretation map of the area covered by the photograph in Fig. 20.2 using geopedologic analysis (Zinck 1988)

Figure 20.4 displays the asymmetric configuration of the Henares river valley with elongated stepped alluvial terraces on the right bank and dissected colluvio-alluvial piedmont glacis and glacis-terraces on the left bank. The stratified photo-interpretation information will guide the field survey and contribute substantially to the final soil map.

20.4   Discussion

The two photo-interpretation approaches use different criteria (physiographic vs geomorphologic) to extract information from the photos and therefore the results are not directly comparable. The physiographic photo-interpretation is exclusively based on the external, topographic appearance of the landscape elements. It provides information somewhat similar to the morphographic features one can extract from digital elevation models. In contrast, the geopedologic photo-interpretation is an inference exercise that pursues not only the recognition and delimitation of the geoforms, but also intends to infer their origin and predict relationships with soil

Table 20.2 Legend of the photo-interpretation map shown in Fig. 20.4

Landscape (level 4)	Relief/molding (level 3)	Lithology/facies (level 2)	Landform (level 1)
Pa Plateau	Pa1 Mesa	Pa11 Sedimentary rocks	
	Pa2 Escarpment	Pa21 Sedimentary rocks	Pa211 Scarp
		Pa22 Sediment rocks + colluv	Pa221 Dissected talus
		Pu22 Sed. rocks + colluvium	Pa222 Badlands (gullies)
	Pa3 Vale	Pa31 Colluvium	
Pi Piedmont	Pi1 Low glacis-terrace	Pi11 Alluvium	Pi111 Tread
			Pi112 Riser (not mapped)
	Pi2 Middle glacis-terrace	Pi21 Alluvium (torrential)	Pi211 Flat tread
			Pi212 Undulating tread
			Pi213 Scarp
	Pi3 High glacis-terrace	Pi31 Alluvium (torrential)	Pi311 Tread
			Pi312 Scarp
	Pi4 Glacis	Pi41 Colluvium + solifluction	
	Pi5 Vale	Pi51 Colluvium	
	Pi6 Hill	Pi61 Alluvium	
Va Valley	Va1 Floodplain	Va11 Alluvium	Va111 Active channel banks
			Va112 Non-active point bars
			Va113 Levee
			Va114 Overflow mantle
			Va115 Basin
			Va116 Erosion level
	Va2 Low terrace	Va21 Alluvium	Va211 Levee/overflow mantle
			Va212 Basin complex
			Va213 Riser (not mapped)
	Va3 Lower middle terrace	Va31 Alluvium	Va311 Tread
			Va312 Riser (scarp + talus)
	Va4 Upper middle terrace	Va41 Alluvium	Va411 Tread
			Va412 Riser (scarp + talus)
	Va5 High terrace	Va51 Alluvium	Va511 Tread
			Va512 Riser (scarp + talus)
	Va6 Very high terrace	Va61 Alluvium	Va611 Tread
			Va612 Scarp
		Va62 Colluvium	Va621 Talus
	Va7 Fan	Va71 Younger alluvium	
		Va72 Older alluvium	
	Va8 Swale	Va81 Colluvium	
	Va9 Vale	Va91 Colluvio-alluvium	

Zinck (1988)

formation and distribution. The conceptual differences between the two approaches, one more descriptive, the other more explanatory, led to outcomes that are different in several aspects.

Both approaches are subjective interpretation exercises in which the level of reference of the interpreter plays a role. However, in the case of the geopedologic approach, interpretation is guided stepwise by the pre-established frame of the legend in four consecutive levels. The interpreter follows a zooming-in perception of the landscape allowing stratification and segmentation of landscape elements. In contrast, the physiographic approach is more flexible in structuring the information and leads to ad hoc applications.

The labelling of the photo-interpretation legend levels is based on explicit geomorphic terms in the geopedologic approach versus non-explicit land terms in the physiographic approach. As a result, the correlation between landscape and landtype, relief/molding and sublandtype, landform and map unit is rather loose. Furthermore, the physiographic approach does not include a level to recognize or infer the lithology of the bedrocks or the facies of the surface formations. For these reasons, map overlay was not intended.

The geopedologic approach generates more photo-interpretation units, especially at the lower levels, than the physiographic approach because geomorphic features recognition is more detailed than land features recognition (Tables 20.3 and 20.4).

The total number of single photo-interpretation units labelled on the maps is 37 vs 20. In posterior field survey, some of the photo-interpretation units were correlated on the basis of similar soil contents, reducing the number of soil map units to 26 as shown in Fig. 20.5 (Saldaña 1997).

Table 20.3 Number of photo-interpretation map units and map delineations

Landscape	Physiography-based (Fig. 20.3)	Geopedology-based (Fig. 20.4)	Soil map of the area (Fig. 20.5)
Plateau (Pa/A)	2	5	2
Piedmont/footslope (Pi/B)	4	10	8
Valley/Terraces (Va/C&D)	14	22	16
Total single API units labelled on the maps	20	37	26
Total map delineations (polygons)	41	143	102

A Plateau, *B* Footslope, *C* Young terraces, *D* Old terraces from Fig. 20.3
Pa plateau, *Pi* piedmont, *Va* valley from Fig. 20.4

Table 20.4 Number of photo-interpretation map units at individual legend levels

Legend level	Physiographic analysis	Geopedologic analysis	Difference in number of units
Landtype/landscape	4	3	−1
Sublandtype/relief-type	13	18	+5
Map unit/landform	13	28	+15

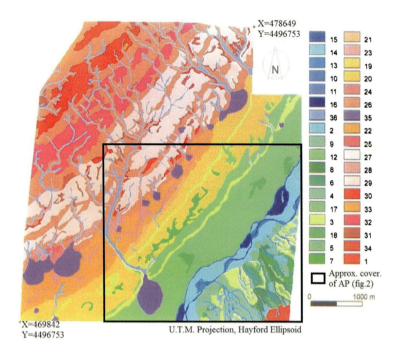

Fig. 20.5 Geopedologic map of a sector of the Henares river valley comprising 36 soil units, with unit 1 in plateau, units 2–12 in piedmont, and units 13–36 in valley (Saldaña 1997). The photo-interpretation area delineated by the frame contains 26 of these map units

The shape of the photo-interpretation map units is different. For being more comprehensive, the physiographic analysis generates polygons of relatively simple configuration (Fig. 20.3), as compared to the more complex shapes shown by the geopedologic-based delineations, especially in the piedmont landscape (Fig. 20.4).

Mapping density is different. The use of geomorphic criteria to stratify and segment the landscape generates a number of delineations (i.e. polygons) substantially higher than the one obtained from physiographic analysis, with 143 vs 41 (Table 20.3). Thus the map of Fig. 20.4 is an efficient introduction to field survey, as it was proven later by Saldaña (1997) when applying the geopedologic approach to produce the soil map in Fig. 20.5. Efficiency is not only in terms of time and energy that a surveyor spends on mapping the soilscape following a scientifically-backed model, but also in terms of opportunities to interpret the resulted soil map for multi-purpose applications (Farshad 2013; Shrestha et al. Chap. 28, in this book).

20.5 Conclusion

The introduction of air photo-interpretation has improved and accelerated soil survey. However, analyzing and quantifying the spatial and temporal variability remains an issue. The geopedologic approach to soil survey contributes to disentangle systematic soil variability and promotes cartographic precision in an efficient way. It also helps when interpreting the results for studies in landscape ecology, land degradation, land evaluation. The application of the physiographic approach is hampered by the fact that it does not rest on a formalized hierarchic system of physiographic units in contrast to the geoform taxonomy implemented in the geopedologic approach.

References

ASP (1960) Manuel of photographic interpretation. American Society of Photogrammetry, Washington, DC

Bennema J, Gelens HF (1969) Aerial photo-interpretation for soil surveys, ITC Lecture Notes. ITC, Enschede, The Netherlands

Bie SW, Beckett PHT (1973) Comparison of four independent soil surveys by air-photo interpretation, Paphos area (Cyprus). In: van der Weele AJ, Steiner D (eds) Photogrammetria, vol 29. Elsevier, Amsterdam

Bui EN (2003) Soil survey as knowledge system. Geoderma 120:17–26

Buringh P (1960) The applications of aerial photographs in soil surveys. Manual of photographic interpretation. American Society of Photogrammetry, Washington, DC, pp 633–666

Farshad A (2013) Geopedology reports historical changes in climate and agroecology: a case study from Northwestern Iran. ECOPERSIA 1(2):145–159

Farshad A, Moonjun R, Shrestha DP (2013) Do the emerging methods of digital soil mapping have anything to learn from the conventional geopedologic approach to soil mapping? In: Shabbir AS, Faisal KT, Abdelfattah MA (eds) Developments in soil classification, land use planning and policy implications. Springer, Dordrecht

Fridland VM (1976) Pattern of the soil cover. Geographical Institute of the Academy of Sciences of the USSR, Moscow 1972. Israel Program for Scientific Translation

Frost RE (1960) Photo-interpretation of soils. Manual of photographic interpretation. American Society of Photogrammetry, Washington, DC, pp 343–397

Goosen D (1967) Aerial photo-interpretation in soil survey, vol 6, Soil Bull. FAO, Rome

Hansakdi E (1998) Soil pattern analysis and the effect of soil variability on land use in the Upper Pasak area, Petchabun, Thailand. Unpublished MSc thesis. ITC, Enschede

Hengl T, Reuter HI (2009) Geomorphometry. Concepts, software, applications, vol 33, Developments in soil science. Elsevier, Amsterdam

Hole FD, Campbell JB (1985) Soil landscape analysis. Rowman and Allanheld, Totowa

Mandelbrot B (1982) The fractal geometry of nature. WH Feeman, New York

Moran CJ, Bui EN (2002) Spatial data mining for enhanced soil map modeling. Int J Geogr Inf Sci 16(6):533–549

Saldaña A (1997) Complexity of soils and soilscape patterns on the southern slopes of the Ayllon range, central Spain. A GIS-assisted modeling approach. ITC, Enschede, The Netherlands, 49

Saldaña A, Ibáñez JJ, Zinck JA (2011) Soilscape analysis at different scales using pattern indices in the Jarama-Henares interfluve and Henares river valley, Central Spain. Geomorphology 135:284–294

SMSS (1986) Guidelines for using soil taxonomy in the names of soil map units. Technical monograph 10. USDA and Department of Agronomy of Cornell University, New York

Stewart GA (ed) (1968) Land evaluation. CSIRO symposium Canbera. McMillan of Australia, Victoria

Wilding LP, Drees LR (1983) Spatial variability and pedology. In: Wilding LP, Smeck NE, Hall GF (eds) Pedogenesis and soil taxonomy. I Concepts and interactions. Elsevier, Amsterdam, pp 83–116
Zinck JA (1988) Physiography and soils, ITC lecture notes. ITC, Enschede, The Netherlands
Zinck JA (2013) Geopedology. Elements of geomorphology for soil and geohazard studies, ITC special lecture notes series. ITC, Enschede, The Netherlands

Chapter 21
Geomorphometric Landscape Analysis of Agricultural Areas and Rangelands of Western Australia

B. Klingseisen, G. Metternicht, G. Paulus, and D. Wilson

Abstract Several techniques exist for generating landform units and these differ in terms of their categorical structure. The geopedologic approach to landform classification is based on a strong integration of geomorphology and pedology using geomorphology as a tool to improve and speed up soil mapping. Likewise, the Australian classification of landforms proposes a two-level descriptive procedure for a systematic, parametric description of landforms into landform patterns and landform elements. This chapter examines geopedology in the context of soil-landscape studies in Australia, and discusses two case studies from Western Australia, where GIS-based geomorphometric tools were used for semi-automated classification of landform elements, based on topographic attributes like slope, curvature or elevation percentile. The case studies illustrate how results of the geomorphic classification add value to management decisions related to rangelands, precision agriculture, spatial analysis, and modelling of land degradation, and other spatial modelling applications where landscape morphometry is an influential factor in the processes under study.

Keywords Geomorphometry • Semi-automated landform classification • Soil-landscape studies • Western Australia • Landscape analysis

B. Klingseisen (✉) • D. Wilson
Department of Spatial Sciences, Curtin University, Perth, Australia
e-mail: bernhard.klingseisen@gmail.com; deanna.wilson@postgrad.curtin.edu.au

G. Metternicht
Institute of Environmental Studies, University of New South Wales, Sydney, NSW, Australia
e-mail: g.metternicht@unsw.edu.au

G. Paulus
School of Geoinformation, Carinthia University of Applied Sciences, Villach, Austria
e-mail: g.paulus@fh-kaernten.at

21.1   Introduction

The shape of the landscape is influential for many land surface processes, including
the flow of surface water, transport of sediment and pollutants, climate on both local
and regional scales, as well as the distribution of habitats for plant and animal species (Blaszczynski 1997). The ability to computationally analyse and quantify the
form of the land surface efficiently and objectively has become essential for many
application areas including geomorphology, soil science, engineering, and ecology.
In the following sections we briefly introduce a systematic classification of landforms used in Australia for soil landscape studies and software available for the
extraction of landform elements and/or patterns, as well as topographic position.

21.1.1   The Australian Landform Classification System

There are several approaches for deriving landform units and these differ in terms of
categorical structure (Moore et al. 1993; Gessler et al. 1995; McKenzie and Ryan
1999). For instance, Speight (1974, 1990) developed the Australian landform classification system using a two-level descriptive procedure for a systematic and parametric description of landforms, into landform patterns and landform elements.
This system views a landform as a hierarchical mosaic of tiles whereby the larger
tiles form *landform patterns* with an average radius of 300 m; the *patterns* consist
of smaller tiles, or landform elements, with a usual radius in the order of 20 m
(Speight 1990). About 40 types of *landform patterns* (e.g. floodplain, dunefield, and
hills), and over 70 types of *landform elements* (e.g. cliff, footslope, and valley flats)
are included in Speight's classification system. Relief type and stream occurrence
describe landform patterns; while landform elements may be described by five attributes namely slope, morphological type (i.e. topographic position), dimensions,
mode of geomorphological activity and geomorphological agent.

Speight (1990) distinguished ten types of topographic positions in which landform elements can be clustered (see Table 21.1); a full description of these morphological types is provided in Speight (1990) and Fig. 21.1 presents examples of
terrain profiles divided into morphological types of landform elements.

Speight's (1990) description of landforms is a key component that contributes
towards the systematic description of the landscape and recording of field observations in Australian soil and land surveys, and as such many existing survey records
consist of Speight's (1990) landform descriptions. Furthermore, landform information is used by field botanists, ecologists, and other natural resource scientists and
managers. For this reason, approaches for geomorphic analysis of Australian landscapes benefit from adopting the main concepts of this framework.

Table 21.1 Morphological type (topographic position) classes by Speight (1990)

Name	Definitions of Speight (1990)
Crest	Area high in the landscape, having positive plan and/or profile curvature
Depression (open/ closed)	Area low in the landscape, having negative plan and/or profile curvature, closed: local elevation minimum; open: extends at same or lower elevation
Flat	Areas having a slope <3 %
Slope	Planar element with an average slope >1 %, subclassified by relative position
Simple slope	Adjacent below a crest or flat and adjacent above a flat or depression
Upper slope	Adjacent below a crest or flat but not adjacent above a flat or depression
Mid slope	Not adjacent below a crest or flat and not adjacent above a flat or depression
Lower slope	Not adjacent below a crest or flat but adjacent above a flat or depression
Hillock	Compound element where short slope elements meet at a narrow crest <40 m
Ridge	Compound element where short slope elements meet at a narrow crest >40 m

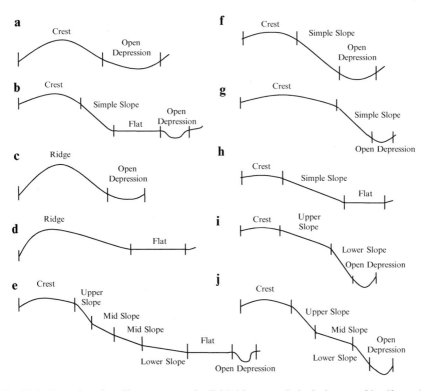

Fig. 21.1 Examples of profiles across terrain divided into morphological types of landform elements (Adapted from Speight 1990)

21.1.2   Land Surface Parameters and Geomorphometry Software for Landform Classification

Landform elements such as those described above may be distinguished by their shape, size, orientation, relief, and contextual position, and can be derived automatically from a digital elevation model (DEM) using topographic attributes or land surface parameters. Land surface parameters can be divided into local geometric parameters describing the shape of the land surface (e.g. slope, aspect, plan and profile curvature) and regional statistical parameters describing the relative position of a point within its surroundings (e.g. local relief, deviation from mean elevation) (Blaszczynski 1997; Gallant and Wilson 2000; MacMillan and Shary 2009).

Coops et al. (1998) introduced a set of techniques enabling to determine topographic positions from a DEM, where the classes are equivalent to Speight's (1990) morphological types. Coop's approach uses thresholds to key land surface parameters, dimension, as well as relative position within a toposequence to extract landform elements. Several free or commercial software products are currently available that are similarly capable of deriving land surface parameters for the determination of landform elements and topographic position. Table 21.2 provides an overview of software with a focus of the applicability to the Australian Classification of Speight. All software enables deriving land surface parameters such as average slope, profile curvature, plan curvature, and relative elevation. Fewer, however can recognise landforms (e.g. crest, depression, flat). A comprehensive summary of land surface parameters and geomorphometry software, with many practical examples can be found in Hengl and Reuter (2009).

One important aspect that Table 21.2 summarises is the ability for the user to customise the software application and adjust the classification parameters taking

Table 21.2 Software with geomorphometric functionalities, and capability to derive landform elements

Software	Scriptable API	Land surface parameters				Landform classification					Custom
		SLP	PRC	PLC	RELEL	CR	D	F	SS	US MS LS	Parameters
ArcGIS	+/+	+	+	+	+	(+)	(+)	(+)	(+)	(+)	(+)
GeoMedia	+/+	+	+	+	+	(+)	(+)	(+)	(+)	(+)	(+)
GRASS	+/+	+	+	+	+	+	+	+			+
Landserf	+/+	+	+	+	+	+	+	+			+
SAGA	/+	+	+	+	+	+	+	+	+	+	
TAS GIS	/	+	+	+	+			+		+	
ILWIS	+/	+	+	+	+	+	+	+			

+ … Natively supported; (+) … supported after customisation or by using existing scripts
SLP Average Slope, *PROFC* Profile Curvature, *PLANC* Plan Curvature, *RELEL* Relative Elevation (not referring to a specific parameter), *CR* Crest, *D* Depression, *F* Flat, *SS* Simple Slope, *US* Upper Slope, *MS* Mid Slope, *LS* Lower Slope; *US, MS* and *LS* are representative for the ability to determine relative slope position

into account the regional, physiographic context as well as dominant geomorphic processes. Klingseisen et al. (2008) for example implemented a semi-automated GIS tool using GeoMedia Grid to generate topographic attributes for landform classification in an agricultural landscape; this information can assist decision making related to precision agriculture, as shown in the next section.

21.2 Case Study: Geomorphometric Landscape Analysis in the Western Australian Wheatbelt Using a Customised GIS Application

21.2.1 Introduction

Site-specific crop mnagement (SSCM) aims at optimising resource application (seed, fertilizer, pesticide, water) to increase farm returns and minimise chemical input and environmental hazards. The spatial framework for SSCM are so called land management units (LMUs), which are homogeneous zones within a field that can be used by farmers in a similar way due to their similar physical characteristics. Soil maps of Western Australia derived using traditional survey methods are available at scales 1:100,000 and 1:500,000 at its best, where a farm of 2000 ha may be covered by only two soil map units at that cartographic scale, insufficient for management decisions related to SSCM. LMUs based on terrain attributes derived from a DEM can provide a more detailed subdivision of the landscape. Thus the need arises for an application that provides the functionality to carry out a semi-automated landform classification in order to analyse the relationship between the derived topographic information and soil properties.

Topographic attributes and landforms have been recognised as important input into the definition of LMUs due to their relationship with soil properties, surface water flow, erosion and sedimentation (Warren et al. 2006). In traditional methods of soil survey, experienced soil surveyors define soil boundaries using stereo aerial photo-interpretation methods based on soil formation models and functional models of prediction which exist in the minds of the surveyor (McKenzie et al. 2000). Hence, soil maps across an area may be of varying quality, depending on the surveyor's experience and knowledge. Research into automated, repeatable processes providing quantitative instead of cognitive expressions of relationships between soil properties and terrain attributes has been conducted since the early 2000s (Ventura and Irvin 2000); this research has developed automated and semi-automated landform classifications for extracting topographic information (Klingseisen et al. 2008).

21.2.2 Study Area and Data

The project was carried out at the Muresk Institute of Agriculture Farm, 100 km north-east of Perth, in the Western Australian wheatbelt. The Muresk Farm covers an area of 1720 ha used for cropping, sheep farming, and cattle production. The elevation of the flat to slightly rolling terrain, with a mean slope of 5 %, ranges from about 154 to 274 m above sea level. Height data with a vertical resolution of 0.01 m were derived from stereo aerial photography at 1:40,000 scale, on a 10 m grid; an iterative adaptive filter was used to remove small discontinuities (Caccetta 2000).

21.2.3 Methodology

A three-tier semi-automated methodology inspired by previous work of Skidmore (1990) and Coops et al. (1998) was adopted for this case study, and implemented as a custom GIS application as described in Klingseisen et al. (2008). An overview of the methodology is provided in Fig. 21.2 and outlined hereafter.

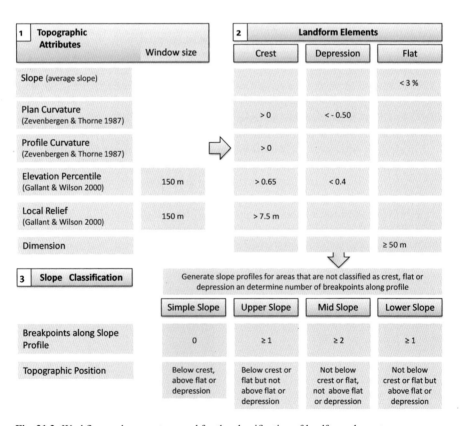

Fig. 21.2 Workflow and parameters used for the classification of landform elements

Firstly, general topographic attributes like slope, plan and profile curvature, as well as more regionalized attributes such as local relief and elevation percentile were derived from a DEM. The primary landform elements such as crests, depressions, and flats, were then generated using topographic attributes as defining parameters, following the definitions of Speight (1990). A combination of thresholds on plan and profile curvature, elevation percentile, and local relief defined the input for each landform element; the thresholds were based on work of Coops et al. (1998), and were modified to account for the higher resolution of the input data, as compared to the original study. These changes were mostly related to depressions and curvature measures, which are generally more affected by scale changes. Areas not initially classified as crests, depressions or flats were classified as slopes. Output primary landforms were input into the GIS as single layers, and combined through an overlay operation; singular cells or narrow strips that remained in the classification were removed using a low pass filter.

Subsequently, slope areas were subdivided into zones of upper, mid, lower or simple slope through three steps; first, slope profiles were derived from the DEM following the direction of the steepest slope from slope cells. Second, each slope profile was broken up into segments at significant changes in slope. As a last step, the cells along the profile were assigned a slope class, according to their relative position in a toposequence between crests and depressions. In this toposequence, upper slopes are the highest elements and occur underneath crests, followed by mid and lower slopes near the valley bottom. Areas lacking significant break in slope were classified as simple slopes.

As a form of validation, and mainly to compare the results against a landform map produced by 'traditional' methods of photo-interpretation, an expert classified the same area by photo-interpretation of colour aerial photographs at scale 1:25,000, mapping the same landform elements using the guidelines for photo-interpretation of geomorphic units described in Part I of this book. The outcomes of both techniques were compared for their similarity using a fuzzy set approach proposed by Hagen (2003) as a part of the Map Comparison Kit (MCK) software (Visser and de Nijis 2006; RIKS 2006). This software produces a category similarity matrix to highlight or disregard different types of similarity, taking into account that some landform categories are more alike (Hagen 2003; Hagen-Zanker et al. 2005). For example, mid and lower slopes are considered to be more similar between them than to crests and depressions. The result of comparing two maps is a third map, indicating for each location the level of agreement in a range from 0 (low similarity) to 1 (identical) between categories. Additionally, statistical values such as average similarity (e.g. the average similarity of all cells in the map), and a similarity index called fuzzy kappa are calculated (Hagen 2003).

21.2.4   Results

Using the aforementioned methodology, the study area was divided into landform elements as defined by Speight (1974, 1990). The resulting landform map covering the farm area of Muresk and beyond is presented in Fig. 21.3 (Map 1); side to side

368 B. Klingseisen et al.

with the expert classification of the same area (Map 2). These maps were input into the Map Comparison Kit (MCK) to determine their average similarity, and explore spatial differences amongst mapped landform categories.

To account for similarities between categories as perceived by the expert, a similarity matrix was defined (see Klingseisen et al. 2008). Using this matrix, the fuzzy comparison between the photo-interpreted and semi-automated classifications yielded an average similarity of 0.629 (Fig. 21.4). Areas of agreement between the semi-automated classification and the expert photo-interpretation show values close to one, whereas areas of total disagreement on category assignation take a value close or equal to zero.

The most significant differences are evidenced in the spatial structure of the landscape mapped. A high disagreement occurs in categories where human cognition is required (e.g. to generate a connected depression network), and in those categories that are derived from exact topographic attributes, like slope percentage, difficult for the expert to quantify. For instance, in Fig. 21.3 (Map 2) the expert recognises landforms as larger homogeneous areas, whereas the semi-automated approach generates smaller landform elements. One of the main reasons for this to happen is the difficulty of a photo-interpreter to gather an exact estimation of slope percentage, and thus there is a tendency to misclassify simple slopes. Table 21.3 substantiates this observation, showing simple slopes to be the category with the lowest similarity. The human interpreter appears unable to break simple slopes into lower, mid, and upper slopes, when subtle changes in slope percentage are present in the

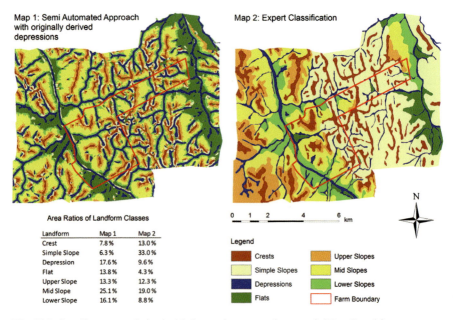

Fig. 21.3 Landform maps derived with the semi-automated approach (*Map 1*) and from an expert classification (*Map 2*), including a comparison of the area covered by each category

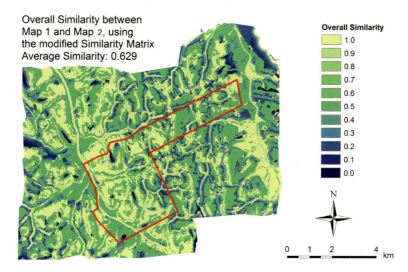

Fig. 21.4 Spatial assessment of similarity between maps *1* and *2*

Table 21.3 Average similarity per category

Map 1: automated approach			
Map 2: expert classification	Overall similarity	Commission	Omission
Crest	0.956	0.968	0.988
Simple slope	0.794	0.816	0.978
Depression	0.937	0.991	0.946
Flat	0.896	0.995	0.901
Upper slope	0.862	0.948	0.914
Mid slope	0.818	0.951	0.868
Lower slope	0.818	0.952	0.867

landscape. A similar situation appears to occur with the drawing of a clear boundary between lower slopes and depressions. Another effect of the false estimation of slope values is that areas with a slope smaller than 3 % are often not identified as flats by the expert. This explains the relatively low agreement of this category in Table 21.3.

The resulting landforms, together with slope and a Compound Topographic Index (CTI) were identified as important driving factors in soil formation, and the development of homogeneous land management units at Muresk Farm (Warren et al. 2005, 2006). Besides landform classification in the Western Australian landscape, the classification method has also been applied to an agricultural area in upper Austria, as described by Klingseisen et al. (2004). Further application areas beyond soil landscape mapping include geohazard modelling (Rauter 2006), seafloor mapping, and landscape studies in a planning context.

21.3 Case Study: Landform Information to Increase the Accuracy of Land Condition Monitoring in Western Australian Pastoral Rangelands

21.3.1 Introduction

Quality and quantity of information related to land condition is essential to monitoring pastoral leases in the rangelands of Western Australia. In this Australian State, pastoral leases cover large areas, with an average size of 1,850 km². A lease requires regular inspections by staff of the Department of Food and Agriculture of Western Australia, through field surveys, which are not always practical due to lack of accessibility in remote areas, cost and time considerations (Wilson and Corner 2011). If a lease has no identified land condition problems, then inspectors only perform a report once every 6 years. Current regional datasets lack resolution to provide data and information relevant for assessment of grazing impacts on ecosystem conditions. Aside of its relevance for the State Government Agencies charged with monitoring these leases, greater capability to provide accurate information on pastoral conditions of these rangelands could assist lessees improving management of their land, while increasing productivity.

Hereafter we present the main characteristics of the study area and describe the methodology adopted to derive landform units, including topographic position, based on the conceptual model of Speight (1990), described in Sect. 21.1.

21.3.2 Study Area and Data

The method was tested on an area of approximately 3000 km² corresponding to the Bow River Station (17° 01′S, 128°12′E) in the East Kimberly (Fig. 21.5). This study area was chosen since it had been the focus of previous attempts at higher resolution mapping and a land subsystem map had been prepared by fieldwork (Schoknecht 2003).

21.3.3 Methodology

Four main stages (presented in Fig. 21.6) were required to map landforms and use the output result to inform land condition reporting: (a) identifying landform features used to define landscape units at a land system (1:250,000 scale) and subsystem level (1:100,000 scale); (b) identifying best techniques for extracting landform elements and landform patterns from a DEM; (c) identifying relevant variables (e.g. geology, land systems, drainage, and land use data) used in other studies for landscape modelling; and (d) defining a landscape model that can be used to

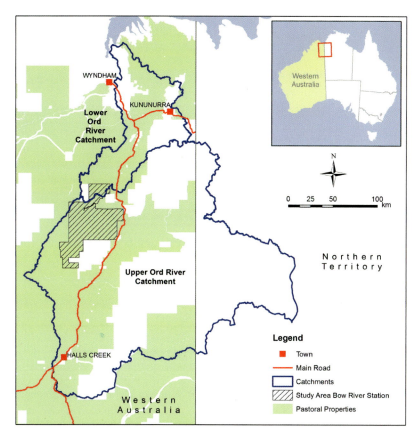

Fig. 21.5 Study area Bow River Station in the Kimberley, Western Australia

extrapolate landform features to other pastoral rangelands of Western Australia. The approach is fully described in Wilson and Corner (2011), and hereafter aspects relevant to mapping landform patterns and landform elements for monitoring land condition are described.

The classification of landforms was based on one arc-second SRTM level 2 digital elevation model (DEM) processed by Geoscience Australia (Gallant et al. 2011). The LandSerf software (Wood 2009) was used to extract landform features. This software offers feature extraction tools that allow user specification of window scale, slope tolerance and curvature tolerance, and it performs a semi-automated classification of a digital elevation model into six classes – peaks, channels, plains, passes, pits, and ridges, which can then be exported in a range of data formats (see Table 21.2).

Because LandSerf was unable to dynamically alter curvature and slope tolerances, or change the processing model, alternative land surface parameters including the Compound Topographic Index (Speight 1974; Quinn et al. 1991), and relative relief parameters such as those used in the Hammond-Dikau method (Dikau et al. 1991)

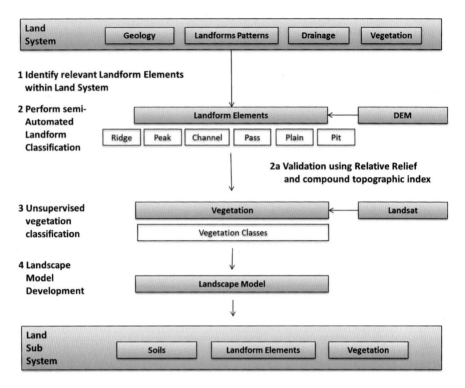

Fig. 21.6 Methodology for extraction of landforms to inform land condition reporting (After Wilson and Corner 2011)

were investigated using ArcGIS spatial analyst tools. The 'deviation from the mean' was found to best describe local relief, with three classes defined (Wilson and Corner 2011).

eCognition was also analysed for classification of landforms and other datasets. eCognition is an Object Based Image Analysis (OBIA) software that has the ability to use many different and disparate image layers in the analysis and classification processes (Dikau et al. 1991). Object based classification takes account not only of the attribute information in the layers (analogous to spectral content in a remotely sensed image) but also considers the spatial arrangement of those attributes.

21.3.4 Results

GIS and remote sensing techniques showed potential to enable the downscaling of "regional level" land system data, to derive localised land subsystem level data, with landforms being one of the key parameters. The LandSerf software initially produced six landform classes. However, as with the Muresk study area, some

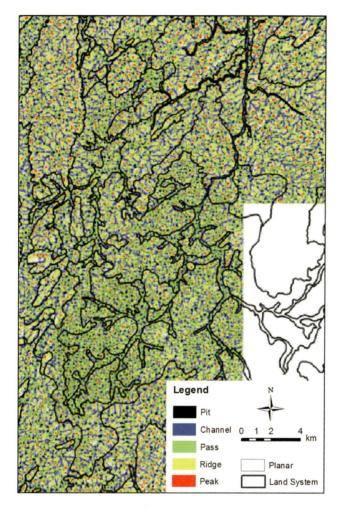

Fig. 21.7 Landform classes overlaid with land system boundary data

landforms were incorrectly labelled. For example, areas classified as channels were in fact plains or low hills. Errors and imperfection in the DEM also led to the misclassification of peaks as low lying isolated points. An iterative process that integrated LandSerf and eCognition was found to provide the best landform classifications over the study area, shown in Fig. 21.7 (Wilson and Corner 2011; Wilson et al. 2012).

A predictive model using Weighted Overlay and Weighted Sum tools in ArcGIS was used to derive information at land unit level. The six landform classes from LandSerf, as well as existing vegetation, geology, and soil data provided evidence layers for a predictive model of land subsystem units for pastoral rangelands. Additionally, relative relief and elevation were included to overcome some of the limitations of the landform classification mentioned above, and to improve the predictions. While the final accuracy assessment is still pending, model results suggest that land unit boundaries can be mapped accurately at a subsystem level, if all input datasets have a resolution of 30 m or better.

21.4 Conclusions

Two case studies were presented that investigated the development and application of tools for landform classification based on topographic attributes. The results are highly relevant for soil landscape studies to for instance, establish relationships between the landform elements and soil properties on agricultural areas; to produce homogeneous land map units that can be managed efficiently through targeted application of inputs (e.g. fertilizers, herbicides) and crop varieties best suited for a specific location, as reported by Warren et al (2006).

The second case study presented an approach to improve the accuracy of land subsystem mapping of the Western Australian rangelands, increasing the quality and quantity of data relevant to the assessment of land condition (e.g. degradation status and extent). Information for early warning of land degradation occurrence within leased pastoral areas could be used to trigger adaptive land management response, such as lowering stock numbers or complete de-stocking of degraded areas to avoid long-term, irreparable damage of pastoral rangelands. The understanding of landforms and their relation with soils and vegetation as presented in the case study of the Bow River Station ultimately leads to better monitoring of pasture degradation in rangeland, as reported by Wilson and Corner (2011).

While both studies were conducted in Australia, they covered very distinct landscapes shaped by different climatic, geological, and environmental conditions. This demonstrates the requirement for customisable semi-automated approaches that can be adapted to match data requirements of commonly used classification schemes, including surrounding conditions and the intended mapping scale.

The tools utilised in the case studies enable efficient, objective, and repeatable mapping of landforms based on their morphological type (crest/ridge, depressions, flat, and slope). These morphological types, although described by Speight (1990) in relation to their topographic position, can be derived from a DEM relatively independently of other elements in the neighbourhood, solely based on topographic attributes. Identification of more specific or compound landform elements such as mesas or terraces, where e.g. a flat may be above a slope, requires an approach that takes into account spatial relationships. For the latter, image segmentation techniques based on object-oriented analysis (Drăguţ and Blaschke 2006) may be used to further refine the classification methodology.

However, refining semi-automated classification based on an established landform and soil classification framework may only be one way to a solution. Typically, these classification schemes (e.g. Speight 1990) were defined prior to the widespread availability of software tools for automated landform classification, with the traditional soil surveyor in mind (e.g. 1960s–1990s). It has therefore been acknowledged that a new system for characterising landform is needed that takes full advantage of the GIS based tools, while retaining the link to geomorphic processes (The National Committee on Soil and Terrain 2009).

Acknowledgements The research underlying the two case studies was funded by a grant of the Australian Research Council, within the ARC-Linkage programme. The authors also acknowledge SpecTerra Services, the Muresk Institute of Agriculture, and the Department of Agriculture and Food of Western Australia for providing data and expert knowledge as well as assistance with field operations.

References

Blaszczynski JS (1997) Landform characterization with geographic information systems. Photogramm Eng Remote Sens 63(2):183–191

Caccetta PA (2000) Technical note. A simple approach for reducing discontinuities in digital elevation models (DEMs), Report 2000/231. CSIRO, Perth

Coops NC, Gallant JC, Loughhead AN, Mackey BJ, Ryan PJ, Mullen IC, Austin MP (1998) Developing and testing procedures to predict topographic position from digital elevation models (DEMs) for species mapping (Phase 1). Report to Environment Australia, Client report 271. CSIRO FFP, Canberra

Dikau R, Brabb EE, Mark RM (1991) Landform classification of New Mexico by computer, Open file report 91-634. US Geological Survey, Denver

Drăguţ L, Blaschke T (2006) Automated classification of landform elements using object-based image analysis. Geomorphology 81:330–344

Gallant JC, Wilson JP (2000) Primary topographic attributes. In: Wilson JP, Gallant JC (eds) Terrain analysis, principles and applications. Wiley, New York, pp 51–85

Gallant JC, Dowling TI, Read AM, Wilson N, Tickle P, Inskeep C (2011) 1 second SRTM derived digital elevation models user guide. Geoscience Australia, Canberra. Available at www.ga.gov.au/topographic-mapping/digital- elevation-data.html

Gessler PE, Moore ID, McKenzie NJ, Ryan PJ (1995) Soil-landscape modelling and spatial prediction of soil attributes. Int J Geogr Inf Syst 9(4):421–432

Hagen A (2003) Fuzzy set approach to assessing similarity of categorical maps. Int J Geogr Inf Syst 17(3):235–249

Hagen-Zanker A, Straatman B, Uljee I (2005) Further developments of a fuzzy set map comparison approach. Int J Geogr Inf Syst 19(7):769–785

Hengl T, Reuter HI (eds) (2009) Geomorphometry – concepts, software, applications, vol 33, Developments in Soil Science. Elsevier, Amsterdam

Klingseisen B, Metternicht G, Paulus G (2008) Geomorphometric landscape analysis using a semi-automated GIS-approach. Environ Model Software 23:109–121

Klingseisen B, Warren G, Metternicht G (2004) Landform: GIS based generation of topographic attributes for landform classification in Australia. In: Strobl J, Blaschke T, Griesebner G (eds) Angewandte Geoinformatik 2004, Beiträge zum 16 AGIT-Symposium, Salzburg. Wichmann, Heidelberg, pp 344–353

MacMillan RA, Shary PA (2009) Landforms and landform elements in geomorphometry. In: Hengl T, Reuter HI (eds) Geomorphometry – concepts, software, applications, vol 33, Developments in soil science. Elsevier, Amsterdam, pp 227–254

McKenzie NJ, Ryan PJ (1999) Spatial prediction of soil properties using environmental correlation. Geoderma 89:67–94

McKenzie NJ, Gessler PE, Ryan PJ, O'Connell DA (2000) The role of terrain analysis in soil mapping. In: Wilson JP, Gallant JC (eds) Terrain analysis, principles and applications. Wiley, New York, pp 245–265

Moore ID, Gessler PE, Nielsen GA, Peterson GA (1993) Terrain analysis for soil specific crop management. In: Robert PC, Rust RC, Larson WE (eds) Proceedings of first workshop, soil specific crop management. Soil Science Society of America, Madison, pp 27–55

Quinn P, Beven K, Chevallier P, Planchon O (1991) The prediction of hillslope flow paths for distributed hydrological modeling using digital terrain models. Hydrological Processes 5:59–80

Rauter M (2006) GIS-gestützte Analyse zur Berechnung potenzieller Lawinenabbruchgebiete. In: Strobl J, Blaschke T, Griesebner G (eds) Angewandte Geoinformatik 2006, Beiträge zum 18 AGIT-Symposium, Salzburg. Wichmann, Heidelberg, pp 569–578

RIKS (2006) Map comparison kit home page. Available from: http://www.riks.nl/mck/

Schoknecht N (2003) Land unit mapping for the Ord and Keep River catchments. Department of Agriculture Western Australia, Perth

Skidmore AK (1990) Terrain position as mapped from a gridded digital elevation model. Int J Geogr Inf Syst 4:33–49

Speight JG (1974) A parametric approach to landform regions. In: Brown EH, Waters RS (eds) Progress in geomorphology. Alden Press, London, pp 213–230

Speight JG (1990) Landform. In: McDonald et al (eds) Australian soil and land survey field handbook, 2nd edn. Inkata Press, Melbourne, pp 9–57

The National Committee on Soil and Terrain (2009) Australian soil and land survey field handbook, 3rd edn. CSIRO Publishing, Collingwood

Ventura SJ, Irvin BJ (2000) Automated landform classification methods for soil-landscape studies. In: Wilson JP, Gallant JC (eds) Terrain analysis, principles and applications. Wiley, New York, pp 245–294

Visser H, de Nijs T (2006) The map comparison kit. Environ Model Software 21(3):346–358

Warren G, Metternicht G, Speijers J (2005) Creating land management units on high resolution remote sensing data and DEM derived terrain attributes using spatially weighted multivariate classification. In: Proceedings of the Pecora 16 "Global Priorities in Land Remote Sensing", 23–27 October 2005, Sioux Falls, South Dakota. CD-rom

Warren G, Metternicht G, Speyers J (2006) Use of a spatially weighted multivariate classification of soil properties, terrain and remote sensing data to form land management units. ASPRS 2006 annual conference, Reno

Wilson D, Corner R (2011) Classification and use of landform information to increase the accuracy of land condition monitoring in Western Australian pastoral rangelands. ISDE 7, 23–25 August 2011, Perth

Wilson D, Corner R, Schut T (2012) Object based data fusion of landform and ancillary data for upscaling soil-landscape mapping in the Western Australian rangelands. IGARSS 2012, 23–25 August 2011, Munich

Wood J (2009) The LandSerf manual. User guide for LandSerf. Available from: http://www.soi.city.ac.uk/~jwo/landserf/

Zevenbergen LW, Thorne CR (1987) Quantitative analysis of land surface topography. Earth Surf Process Landf 12:47–56

Chapter 22
Digital Elevation Models to Improve Soil Mapping in Mountainous Areas: Case Study in Colombia

L.J. Martinez Martinez and N.A. Correa Muñoz

Abstract The demand for more detailed soil and relief information is steadily increasing. However, many countries have only general soil maps at 1:100,000 scale that do not satisfy the requirements needed for applications. This paper shows how geomorphometric analysis from digital elevation models (DEM) can contribute to improve information detail and accuracy and, thus, strengthen soil survey. The study was carried out in a mountainous area of Colombia where various geomorphometric parameters were calculated and a classification of landforms was created. The results can be useful to supplement existing soil studies and meet the information requirements of environmental spatial models, agriculture development, hydrology, land use and conservation.

Keywords DEM • Soil mapping • Geomorphometry • Geopedology • Relief parameters

22.1 Introduction

In Colombia, and elsewhere in developing countries, most of the territory is covered by general soil surveys at 1:100,000 scale. However, there is increasing demand for up to date, more detailed information that is costly and time-consuming to obtain. Topography is one of the soil forming factors that influences the variation of soil properties (Jenny 1994). It plays also an important role in soil degradation due to its effect on processes such as erosion, landslides, and flooding, among others.

L.J. Martinez Martinez (✉)
Facultad de Ciencias Agrarias, Universidad Nacional de Colombia,
Carrera 30 # 45-03, Bogotá, Colombia
e-mail: ljmartinezm@unal.edu.co

N.A.C. Muñoz
Facultad de Ingeniería Civil, Universidad del Cauca, Calle 5 # 4-70, Popayán, Colombia
e-mail: nico@unicauca.edu.co

© Springer International Publishing Switzerland 2016   377
J.A. Zinck et al. (eds.), *Geopedology*, DOI 10.1007/978-3-319-19159-1_22

Furthermore, relief characteristics are important for land use, management, and conservation (FAO 2007; Valbuena et al. 2008; Munar and Martínez 2014). Traditionally, relief was characterized qualitatively as part of soil surveys using visual interpretation of aerial photographs or other images for delineating soil-landscape units. Quantitative analysis of land surfaces is now central to different types of terrain modeling and applications. Before the 1990s, this was difficult to achieve because of the lack of data and appropriate methods for data analysis. Geomorphometry is the science of quantitative land-surface analysis (Pike 2000). It evolved from mathematics and earth and computer sciences with applications in several areas, such as soil, vegetation, agriculture, environment, and earth sciences, among others (Pike et al. 2008). Digital elevation models (DEM) are used to generate geomorphometric information and relate topographic parameters to the spatial variation of soil properties. The spatial distribution of soil moisture content is strongly related to slope gradient, aspect, curvature, and topographic position (Florinsky 2012). Horizontal and vertical curvatures are key topographic factors that determine overland and intra-soil water dynamics (Kirkby and Chorley 1976). It was found that DEM prediction increases with DEM resolution (Chaplot et al. 2000).

In a soil survey, relief properties are criteria used to define and characterize soil map units. Research has focused on the use of DEM data to delineate soil patterns at different levels of detail. The use of elevation, slope, aspect, and curvature as differentiating criteria can generate satisfactory results for soil characterization, particularly in reconnaissance soil surveys (Dobos and Montanarella 2007). However, in areas with complex physiography, DEM data should be complemented with other sources (Dobos et al. 2000). DEM and remote sensing image sensors were used to study soil drainage, finding a high correlation between the established drainage classes based on high-resolution images (Jiangui et al. 2008). Meanwhile, Dobos et al. (2000) found that soil mapping using feature selection algorithms applied to a DEM had an important role in soil characterization.

Smith et al. (2006) stated that terrain features such as slope, aspect, and curvature calculated from a DEM were key parameters for digital soil mapping; however, the accuracy of the results depended on the spatial resolution of the DEM and the size of the area under consideration (Wu et al. 2008). It was found that slope angles decrease and contributing area values increase constantly as DEMs are aggregated progressively to coarser resolutions. An investigation in Kansas that sought to differentiate soil classes found that the combination of SPOT imagery and DEM, using a canonical transformation of the data, was useful in second order soil mapping (Su et al. 1990).

In this paper, morphometric parameters obtained from an SRTM DEM are used to map and characterize landforms in mountainous areas of Colombia. The results can help improve soil survey, providing additional information with a higher level of detail and empowering the soil survey product.

22.2   Methods and Materials

22.2.1   Site Characteristics

The study area of 1200 km² is located in the Central Andes about 400 km southwest of Bogotá in the Cauca Department, between 76°40′25″W, 02°14′09″N and 76°24′13″W, 02°36′24″N (Fig. 22.1). The area was selected because it contains

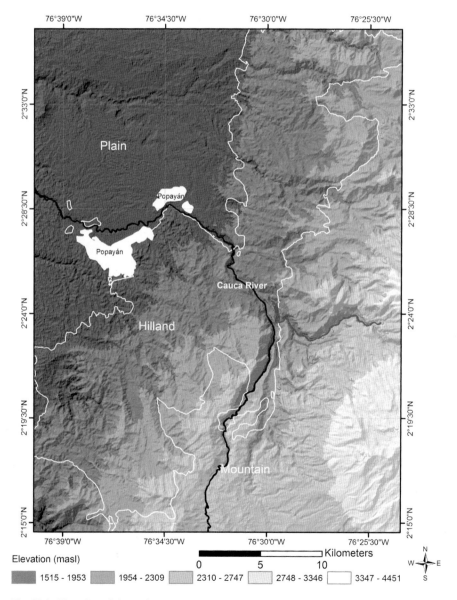

Fig. 22.1 Elevation of the study area

three contrasting landscape types: (1) mountains with slopes steeper than 30 % and relief amplitude higher than 300 m, (2) hills with slope gradients of 7–12 % and relief amplitude lower than 300 m, and (3) plains with slope gradients lower than 7 %. The climate is Am type according to Köppen-Geiger (Peel et al. 2007) with two short dry seasons in January and February and in July. The average annual rainfall varies between 1900 and 2800 mm throughout the area. The temperature ranges from 18 °C in the lowest part of the area and 10 °C in the highest one. The soil parent materials are mainly volcanic layers that overlay igneous and metamorphic rocks. Land use is dominantly pastures for cattle raising. Some areas are covered by montane and premontane rain forest (IGAC 2009).

22.2.2 Data Collection and Analysis

Three pilot areas with a total extent of 100 km^2 were selected to facilitate the development of the research and the validation of the results. In the field, 52 control points were selected and georeferenced using a GPS with metric precision. Location, height, and slope were recorded. A SRTM DEM and an ASTER-derived DEM with a spatial resolution of 30 m were compared in terms of accuracy to a self-produced model using an interpolated 1:25,000 topographic digital map. The accuracy assessment of the three DEMs was performed by means of the software DEMANAL (Jacobsen 2007) using the 52 control points. The SRTM DEM was chosen because it presented the best match with the heights measured in the field with RMSE of 10.6 m.

The following parameters were obtained from the DEM with the software SAGA (Böhner and Conrad 2012): elevation as the primary data given by the DEM; slope defined as the tangent of a plane relative to the surface topography; curvature calculated based on second derivatives for a topographic attribute that describes the convexity or concavity of a terrain surface (Romstad and Etzelmüller 2012); and topographic wetness index (TWI), calculated as a second-order derivative of the DEM and used as an indicator of water accumulation in an area of the landscape where water is likely to concentrate through runoff (Quinn et al. 1991).

Solar radiation was computed based on algorithms that consider atmospheric conditions, elevation, orientation of the surface, and topography (Ruiz Arias et al. 2009). The convergence index (CI) proposed by Köthe et al. (1996) uses the aspect values of neighboring cells to parameterize flow convergence and the respective divergence (Olaya and Conrad 2009). Valley depth is related to a plain multiresolution index of valley bottoms.

Correlation analysis among the geomorphometric parameters was performed and those with less collinearity ($r < 0.5$) were selected to identify the landforms. A geomorphometric delineation was done using k-means analysis with the R software (Venables and Ripley 2002). From an existing soil survey at scale of 1:100,000 for the Cauca Department (IGAC 2009) slope degrees and relief types were extracted and compared to the slope gradients and landform types obtained from the DEM by means of multinomial logistic regression.

22.3 Results

Only the parameters that had less collinearity, with correlation coefficient <0.5, were used for the classification of the landforms. These parameters are analyzed hereafter.

22.3.1 Elevation

Elevation is an important variable due to its relationship with temperature and, therefore, is the basis for climate models (Daly et al. 2008). In the study area, the elevation ranged between 1515 and 4451 m asl (Fig. 22.1); 36 % of the area had elevations between 1515 and 2000 m asl, corresponding to a warm climate with an annual mean temperature between 17 and 24 °C; in 51 % of the area elevations ranged between 2000 and 3000 m asl, with a cold climate and temperatures between 12 and 17 °C; and in 13 % of the area the elevation was higher than 3000 m asl, corresponding to a very cold climate with a mean annual temperature between 6 and 12 °C.

The measurement and mapping of the relief elevations are fundamental for computing other geomorphometric parameters and for the spatial modeling of several phenomena related to soil, land use, hydrology, ecology and environment in general. A W-E topographic profile (Fig. 22.2) of the study area depicts the change in elevation along the transect, allows the major landforms to be identified, and offers a synoptic view of the climatic zones.

Temperature and precipitation are used for soil map unit definition and are included in soil legends as a hierarchical level, although in a very general way. The

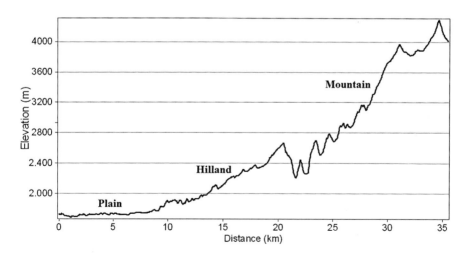

Fig. 22.2 Topographic west-east profile

elevation and its relationship to temperature have been used for life zone definition (Holdridge 1982) and, in Colombia, for the classification of climate and in land evaluation studies (Martínez 2006; Martínez and Munar 2010). A DEM provides an accurate representation of elevation and its spatial distribution, and may be a proxy for climatic data.

22.3.2 Slope

Table 22.1 shows the area occupied by each slope class based on the soil map (SCM) and on the DEM map (SCD). Large differences in the estimate of this parameter mostly in the ranges of 0–3 %, 12–25 %, and 50–75 % were found. In soil maps, the depicted slope usually corresponds to a general appreciation of the dominant slope in each soil unit. In contrast, the DEM allows for greater accuracy in estimating the slope gradients and their spatial distribution. The results were corroborated in the field and indicated that the DEM represented the slope in a more realistic way because each pixel had relatively good accuracy representation of the height.

Figure 22.3 shows the heterogeneity of the slope classes in the soil map. All slope classes shown in the soil map included important areas where the slope belonged to a different class. The percentage of the area where the slope classes from SCM and SCD coincided varied between 13 % (class b) and 38 % (class d). The remaining area was misclassified, with higher values in some cases and lower ones in others.

This indicates that the DEM allowed for a more accurate estimate of the slope, which is an important input to define the suitability, management, and conservation of the land. The detail provided by the DEM can be used to enhance the soil survey information. The use of DEMs to calculate the slope depends on the resolution of the DEM, the accuracy of the DEM data, and the algorithm used (Zhou and Liu 2004).

Table 22.1 Area covered by each slope class based on the soil map (SCM) and the DEM map (SCD)

Slope class and range (%)	Area (%) occupied in the SCM	Area (%) occupied in the SCD
a 0–3	1.8	12.4
b 3–7	10.7	7.8
c 7–12	13.4	12.1
d 12–25	12.9	33.7
e 25–50	28.7	34.4
f 50–75	27.8	7.4
g >75	4.8	2.2

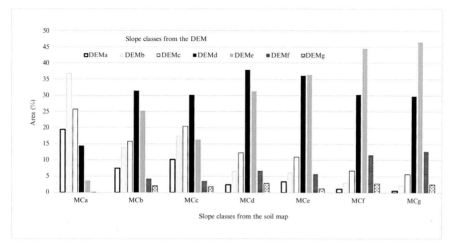

Fig. 22.3 Area (%) of slope classes from the soil map (MC) and from the DEM

22.3.3 Topographic Wetness Index

The topographic wetness index (TWI) (Beven and Kirkby 1979), also known as a compound topographic index (Quinn et al. 1991), relates an upslope area as a measure of water flowing towards a certain point, to the local slope, which is a measure of subsurface lateral transmissivity. In the study area, the TWI ranged from 4.6 to 21.3. Higher TWI values represent depressions on the landscape where water is likely to concentrate through runoff, while lower values represent crests and ridges.

The TWI has become a popular and widely used tool to infer information about the spatial distribution of wetness conditions (i.e. the position of shallow groundwater tables and soil moisture). TWI has been shown in some study areas to predict the solum depth (e.g. Gessler et al. 1995). In Fig. 22.4, important differences that distinguish the landscapes were observed. The NW part presents a pattern with gentle topography corresponding to the plain; in the south dominates the mountain landscape with a more parallel drainage pattern; and the middle part has mainly a dendritic drainage pattern.

22.3.4 Convergence Index

The convergence index (CI) (Köthe et al. 1996) uses the values of the aspect of the neighboring cells to calculate the convergence and divergence of the flow, which is similar to the curvature but not dependent on the absolute differences of heights. In the study area, the CI varied from −81 to 75. Positive values represent divergent areas, negative values represent convergence areas, and null values signify areas without curvature. Within each landscape some units can be differentiated by means

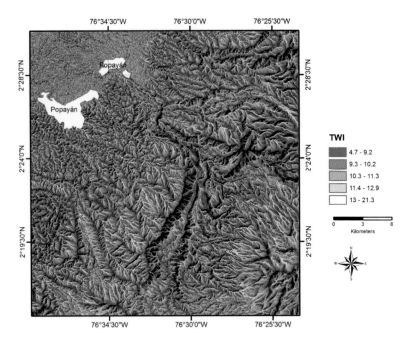

Fig. 22.4 Spatial distribution of TWI in the southern part of the study area

of CI values (Fig. 22.5) such as summits, small valleys, or bottom and backslope areas, which are important to identify and separate relief types.

22.3.5 Valley Depth

Valley depth (VD) is calculated as the vertical distance to a channel network representing a base level (Böhner and Conrad 2012). In the study area, the valley depth varied from 0 to 481 m (Fig. 22.6). The highest values represented the valley bottoms of the Cauca and San Francisco rivers; the lowest values corresponded to the mountain ridges. Valley depth is important in erosion studies to identify the location of gullies in the landscape and address the contribution of sidewall erosion, such as mass movement processes in gullies.

22.3.6 Landform Classification

For the classification of the landforms, various analyses with different parameters were tested. The correlation analysis showed that some parameters were highly correlated with each other: for instance, the horizontal and vertical curvatures with

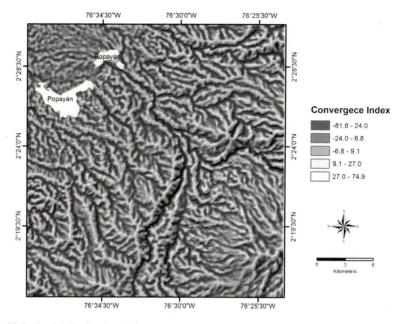

Fig. 22.5 Spatial distribution of the convergence index in the southern part of the study area

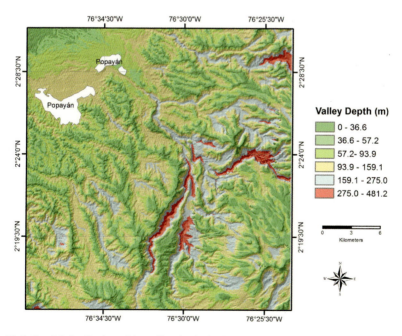

Fig. 22.6 Spatial distribution of the valley depths in the southern part of the study area

the general curvature (r: 0.87 and 0.86, respectively), the index of convergence with
the curvature (r: 0.71), and slope with insolation (r = −0.60). This indicates collinearity between some data, which will affect the variance and the final classification.
Therefore, the parameters that showed no significant correlations between them,
such as the index of convergence, the depth of the valleys, the topographic wetness
index, and slope were selected.

The k-means analysis with the R software (Venables and Ripley 2002) allowed
identifying and separating the landscapes of the study area: mountains in very
cold climate (landforms 2 and 6), mountains in cold climate (landforms 4 and 7),
hill-land in warm climate (landform 3), and plain in warm climate (landforms 1
and 5) (Fig. 22.7). With further level of detail, within these landscapes more
units can be separated which correspond to relief types like summits, ridges,
backslopes, and valley bottoms. This kind of analysis can improve the level of
detail of soil maps at 1:100,000 scale, for instance. According to the USGS
(U.S. Geological Survey 1993), the accuracy of the 30 m DEM data adequately
supports computer applications that analyze hypsographic features to a level of
detail similar to manual interpretations of information as printed at map scales
not larger than 1:63,360 scale.

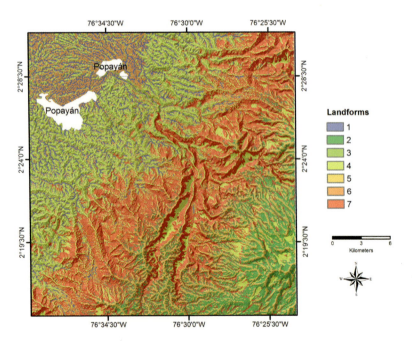

Fig. 22.7 Landform classification; landforms *2* and *6*: mountains in very cold climate; landforms
4 and *7*: mountains in cold climate; landform *3*: hilland in warm climate; landforms *1* and *5*: plain
in warm climate

22.4   Conclusions

In countries where relief information is scarce, especially in mountainous areas, DEMs with spatial resolution of 1 arc-sec constitute a suitable source to generate information with more detail through the calculation of quantitative geomorphometric parameters. The use of these parameters allows delineating and characterizing landforms as support to the traditional visual interpretation of aerial photographs. Slope, which is an important parameter in many models and applications, can be accurately calculated from DEM data. This relief information complements and strengthens existing soil surveys and can be integrated in new soil surveys to improve its quality and therefore to satisfy the needs of more users. It could be useful for applications in land suitability assessments, environmental studies, territorial planning, and watershed management, among others.

Acknowledgements The authors thank to the Research Division (DIB) of the Universidad Nacional de Colombia for financial support to carry out this research.

References

Beven KJ, Kirkby MJ (1979) A physically based, variable contributing area model of basin hydrology. Hydrol Sci 24(1):43–69

Böhner J, Conrad O (2012) System for automated geoscientific analyses. Available at: http://www.saga-gis.org/en/index.html

Chaplot V, Walter C, Curmi P (2000) Improving soil hydromorphy prediction according to DEM resolution and available pedological data. Geoderma 97(3–4):405–422

Daly C, Halbleib M, Smith J, Wayne P, Doggett M, Taylor G, Curtis J, Pasteris P (2008) Physiographically sensitive mapping of climatological temperature and precipitation across the conterminous United States. Int J Climatol 28:2031–2064

Dobos E, Montanarella L (2007) The development of a quantitative procedure for soilscape delineation using digital elevation data for Europe. In: Lagacherie P, McBratney AB, Voltz M (eds) Developments in soil science, vol 31. Elsevier, Amsterdam, pp 107–118

Dobos E, Micheli E, Baumgardner M, Biehl L, Helt T (2000) Use of combined digital elevation model and satellite radiometric data for regional soil mapping. Geoderma 97(3–4):367–391

FAO (2007) Land evaluation. Towards a revised framework. Land and water discussion paper 6. FAO, Rome

Florinsky IV (2012) Digital terrain analysis in soil science and geology. Elsevier, Amsterdam

Gessler PE, Moore ID, McKenzie NJ, Ryan PJ (1995) Soil-landscape modeling and spatial prediction of soil attributes. Int J GIS 9(4):421–432

Holdridge LR (1982) Life zone ecology. Tropical Science Center, San José

IGAC (2009) Estudio general de suelos y zonificación de tierras del departamento del Cauca. Instituto Geográfico Agustin Codazzi, Bogotá

Jacobsen K (2007) Manual of program system BLUH. Institute of Photogrammetry and Geoinformation, Leibniz University, Hannover

Jenny H (1994) Factors of soil formation: a system of quantitative pedology. Dover Publications, Toronto

Jiangui L, Patteya E, Nolin M, Miller JR, Kab O (2008) Mapping within-field soil drainage using remote sensing, DEM and apparent soil electrical conductivity. Geoderma 143(3–4):261–272

Kirkby MJ, Chorley RJ (1976) Through flow, overland flow and erosion. Bull Int Assoc Sci Hydrol 12:5–21

Köthe R, Gehrt E, Böhner J (1996) Automatische Reliefanalyse für geowissenschaftliche Anwendungen. Derzeitiger Stand und Weiterentwicklungen des Programms SARA. Arbeitshefte Geol 1:31–37

Martínez LJ (2006) Modelo para evaluar la calidad de tierras: caso del cultivo de papa. Agron Colomb 24(1):96–110

Martínez LJ, Munar O (2010) Digital elevation models as data source for land suitability analysis in Colombia. In: Reuter R (ed) Remote sensing for science, education, and natural and cultural heritage. EARSeL- European Association of Remote Sensing Laboratories, Paris, pp 641–647

Munar OJ, Martínez LJ (2014) Relief parameters and fuzzy logic for land evaluation of mango crops (Mangifera indica L.) in Colombia. Agron Colomb 32(2):246–254

Olaya V, Conrad O (2009) Geomorphometry with SAGA. In: Hengl T, Reuter H (eds) Geomorphometry: concepts, software, applications. Elsevier, Amsterdam, pp 293–308

Peel MC, Finlayson BL, McMahon TA (2007) Updated world map of the Köppen-Geiger climate classification. Hydrol Earth Syst Sci 11:1633–1644

Pike RJ (2000) Geomorphometry – diversity in quantitative surface analysis. Prog Phys Geogr 24:1–20

Pike RJ, Evans IS, Hengl T (2008) Geomorphometry: a brief guide. In: Hengl T, Reuter H (eds) Geomorphometry: concepts, software, applications. Elsevier, Amsterdam, pp 1–28

Quinn PF, Beven K, Chevallier P, Planchon O (1991) The prediction of hillslope paths for distributed hydrological modeling using digital terrain model. Hydrol Process 5:59–79

Romstad B, Etzelmüller B (2012) Mean-curvature watersheds: a simple method for segmentation of a digital elevation model into terrain units. Geomorphology 139:293–302

Ruiz Arias JA, Tovar J, Pozo D, Alsamamra H (2009) A comparative analysis of DEM-based models to estimate the solar radiation in mountainous terrain. Int J Geogr Inf Sci 23(8):1049–1076

Smith M, Zhu A, Burt J, Stiles C (2006) The effects of DEM resolution and neighborhood size on digital soil survey. Geoderma 137(1–2):58–69

Su H, Kanemasu E, Ransom M, Yang S (1990) Separability of soils in a tallgrasss prairie using SPOT and DEM data. Remote Sens Environ 33(3):157–163

US Geological Survey (1993) Digital elevation models – data users guide, 5th edn. Reston, Virginia

Valbuena C, Martínez LJ, Henao R (2008) Variabilidad espacial del suelo y su relación con el rendimiento de mango (Mangifera indica L.). Rev Bras Frutic 30(4):1146–1151

Venables WN, Ripley B (2002) Modern applied statistics with S. Springer, New York

Wu S, Lib J, Huang GH (2008) A study on DEM-derived primary topographic attributes for hydrologic applications: sensitivity to elevation data resolution. Appl Geogr 28:210–223

Zhou Q, Liu X (2004) Analysis of errors of derived slope and aspect related to DEM data properties. Comput Geosci 30(4):369–378

Chapter 23
Neuro-fuzzy Classification of the Landscape for Soil Mapping in the Central Plains of Venezuela

J.A. Viloria and M.C. Pineda

Abstract The application of geomorphology to soil survey has encouraged the study of genetic relationships between soil and geoforms. However, the qualitative classification of the landscape can be slow and expensive, and the outcome often depends on the perception of the classifier. This work applied a quantitative method based on artificial neural network and fuzzy logic to classify the landscape into land-surface units from a digital elevation model (DEM) of 5×5 m cells. The method helped explore the data to determine the optimal combination of number and fuzziness of classes. The classification output included the values of the geomorphometric parameters at the centre of each class, the memberships of the model cells to each class, and a map showing the spatial distribution of the land-surface classes. This output was transformed into a map of geoforms that was used as a framework for soil sampling and mapping. The resulting map disclosed the landscape structure consisting of a plateau dissected into mesas, hilltops, slopes, and valleys, with predominance of well-drained Alfisols in steep lands and imperfectly drained Vertisols in valleys. The method proved to be effective for establishing soil-landscape relationships in the study area.

Keywords Digital mapping • Digital elevation model • Geomorphometric parameters • Artificial neural network • Fuzzy logic

23.1 Introduction

Application of geomorphology to soil survey has encouraged the production of comprehensive maps which convey genetic relationships between soil and geoforms (Zinck 2013). These maps use a landscape classification as framework to determine soil map units. Conventional classification of landforms is usually based on a

---

J.A. Viloria (✉) • M.C. Pineda
Facultad de Agronomía, Instituto de Edafología, Universidad Central de Venezuela, Maracay, Venezuela
e-mail: jesus.viloria@gmail.com; maria.c.pineda@ucv.ve

© Springer International Publishing Switzerland 2016   389
J.A. Zinck et al. (eds.), *Geopedology*, DOI 10.1007/978-3-319-19159-1_23

qualitative characterization of the configuration of the land surface (McBratney et al. 2003). The method is applied to divide the landscape into successively smaller units, until simple and homogeneous landforms are obtained (Minár and Evans 2008). This procedure is lengthy and expensive, and the outcome depends on the experience and perception of the classifier, which influences the quality of the information produced (Debella-Gilo and Etzelmüller 2009).

New geomatic techniques allow making automated identification and quantitative description of basic land-surface forms from digital elevation models (DEM) and remote sensing (McKenzie and Ryan 1999; Scull et al. 2003; Bolongaro-Crevenna et al. 2005; Dobos et al. 2006; Minár and Evans 2008; Ehsani and Quiel 2008; Zhao et al. 2009). These land-surface forms can be grouped as basic elements of a system for characterization and classification of physiographic units (Bolongaro-Crevenna et al. 2005). They can be interpreted in terms of geomorphic genesis, dynamics, and chronology (Minár and Evans 2008) or used as a basis for representing relations between landscape and soil (Bolongaro-Crevenna et al. 2005) and for improving the mapping and modeling of soil and environment (Ehsani and Quiel 2008).

The methods used to identify land-surface classes are diverse; but often unsupervised classification procedures for grouping cells of a DEM in homogeneous classes with respect to the selected morphometric parameters are applied (e.g. Adediran et al. 2004; Iwahashi and Pike 2007; Ehsani and Quiel 2008). Bezdek et al. (1992) proposed the algorithm FKCN (Fuzzy Clustering Kohonnen Network) which combines two complementary classification techniques (i.e. artificial neural networks and fuzzy sets) in an integrated system. This allows benefitting from the strengths and overcoming the weaknesses of each one of these methods separately. Fuzzy sets show no sharp boundaries between classes. Each class is defined by its central concept and an array of membership functions of the individual to that class. The membership function is continuous with values ranging from 0 (no membership) to 1 (complete membership) (Zhu et al. 2010).

Viloria-Botello (2007) implemented the FKCN algorithm in a computer program for experimental classification of the landscape in a mountainous area of north-central Venezuela. The present work assesses the applicability of this method for modeling soil-landscape relationships in an area of the central plains of Venezuela.

23.2 Materials and Methods

23.2.1 Study Area

The study area of 3540 ha is located between 9°26.7′N and 67°11.4′W, in the Guárico state, center north of Venezuela. The area corresponds to a denudation plateau of flat to slightly undulating topography, with an average altitude of 160 m above sea level (Fig. 23.1). The substratum is mainly fine-textured rocks of sedimentary origin and Quaternary sediments. Average annual temperature is 26.5 °C and the mean annual precipitation is 1200 mm distributed mostly between

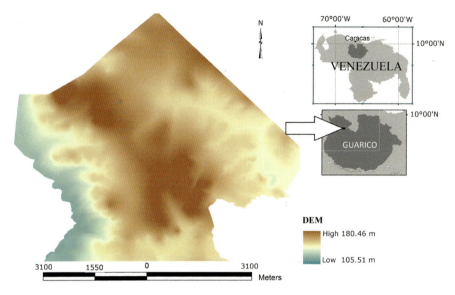

Fig. 23.1 Location of the study area

May and November. At the time of the study, much of the area was covered by a dense semi-deciduous forest.

23.2.2 Data

A DEM with a resolution of 5×5 m was generated with the algorithm ANUDEM (Hutchinson 1989) from contour lines separated 5 m obtained from a topographic survey. Table 23.1 shows the environmental covariates used for digital terrain classification. The geomorphometric parameters were computed from the DEM with SAGA (System for Automated Geoscientific Analyses, version 2.0.8) as descriptors of the land-surface morphology and/or the potential movement of water and materials over this surface. The environmental covariates were normalized to the same representation range [−1 to 1].

23.2.3 Optimal Values of Fuzzy Exponent (Φ) and Number of Classes

When a new and unknown area is classified into fuzzy classes of land surface, there is no prior information on what is the most favorable combination of number of classes and value of the fuzzy coefficient (Φ). To estimate the best combination of these parameters, the fuzziness performance index (FPI) was calculated for

Table 23.1 Environmental covariates used as input for the digital terrain classification

Covariate	Description	Reference
Relative height (RH)	Meters above the drainage network	
Slope gradient (Slope)	Magnitude in m/m of the steepest gradient in the X and Y directions	
Profile curvature (Profile C)	Curvature in m/m^2 in the direction of the slope	Zevenbergen y Thorne (1987)
Plan curvature (Plan C)	Curvature in m/m^2 perpendicular to the direction of slope	Zevenbergen y Thorne (1987)
Landscape curvature (Land C)	Ratio between plan curvature and profile curvature	
Catchment area (As)	Local upslope area in m^2 draining through a certain cell of the model	Quinn et al. (1991)
Topographic wetness index (TWI)	$\ln(As/\tan \beta)$, where As is the catchment area and β is the local slope in degrees	Willson y Gallant (2000)

different number of classes (6–11) and values of Φ (1.2–1.5 with successive increments of 0.1).

The FPI varies from 0 to 1. A value equal to 0 indicates that there is no fuzziness in the data and the classes are discrete. Conversely, when FPI equals 1, the fuzziness is maximal, and each individual has no clear membership to any class. A plot of FPI against number of classes, for different values of Φ, can be used as a guide for choosing the optimal combination of class number and fuzzy coefficient (Odeh et al. 1992). Following this procedure, a number of classes equal to 10 and fuzzy exponent equal to 1.3 were chosen for this study. Such a combination produces a value of FPI equal to 0.44, which assures that the land-surface classes are neither too crispy nor too fuzzy.

23.2.4 Analysis of the Land-Surface Classes Created by the Neuro-fuzzy Network

The FKCN output includes a set of fuzzy classes defined by: (a) the values of the input variables at the center of each class, and (b) a function of membership of every cell of the model to each class. To display the spatial distribution of the fuzzy classes, each cell of the model was allocated to the class with the greatest membership value. The resulting map of land-surface classes was used as a spatial framework to design a reconnaissance sampling which included 392 soil profiles. Each soil profile was georeferenced, described, and classified according to the USDA Soil Taxonomy (Soil Survey Staff 2010).

The class centers were analyzed together with the map of land-surface classes and the soil profiles to understand the geomorphic meaning of each class and its relationship with the soil variation in the study area.

23.3 Results and Discussion

Table 23.2 shows the centers of the land-surface classes produced by the neuro-fuzzy classification, while Fig. 23.2 displays the spatial distribution of these classes.

These results reveal details of the physiographic structure of the study area, which remained concealed under a thick forest cover. The area corresponds to a denudation plateau conformed by flat-topped mesas, hilltops, and hillocks separated by shallow valleys. The digital model of the land surface produced by the DEM classification divided the valleys into six different classes (1–6). All of them have relatively large wetness index (TWI $\geq$ 11), even slopes ($\leq$ 2 %), concave or lineal curvature ($\leq$ 0) (Table 23.2), and an elongated configuration (Fig. 23.2). Class 1 corresponds to flood channels. Class 2 with the lowest relative height and the largest values of catchment area and the wetness index, represents the valley floors along the drainage network. Class 3 corresponds to valley bottoms dominating the valley floors and dominated by the rest of the valley classes. Class 4 represents flat and relatively wide areas at both sides of the valley bottoms, while classes 5 and 6 correspond to relatively narrow incisions in mesas, hilltops, and slopes, occupying higher grounds than the rest of the valley classes. The remaining classes occupy the upper positions in the landscape, dominating the valleys by 3–20 m. They correspond to flat-topped mesas (class 9), hilltops (class 10), upper slopes (class 8), and lower slopes and hillocks (class 7).

Table 23.3 shows the soil variation among land-surface classes. Soil properties are strongly influenced by the parent material derived from shale and fine-textured sediments. The most important soil feature in the valley positions is the presence of cracks, slickensides, and wedge-shaped aggregates produced by the shrink and swell of expansive clay. In the valley bottom, soils are imperfectly drained but soil drainage improves as the relative height increases. In mesas, hilltops, hillocks and

Table 23.2 Centers of the land-surface classes produced by the neuro-fuzzy classification of geomorphometric parameters computed from the DEM

Land-surface class	RH (m)	Land_C ($\times 10^{-4}$)	TWI	As (km^2)	Slope (%)	Physiography
1	0.9	−1	14.2	12.5	0.6	Flood channels
2	0.1	−1	16.8	228.9	0.3	Valley floors
3	0.9	−2	12.6	6	1.1	Valley bottoms
4	1.2	−1	10.9	2	2	Lower valleys
5	3.6	0	14.8	19.5	0.5	Middle valleys
6	7.8	0	13.2	5.2	0.6	Upper valleys
7	3.1	0	10	1.5	4.1	Lower slopes and hillocks
8	7.6	2	9.4	1.1	7.4	Upper slopes
9	5.3	1	11.1	2.4	1.6	Mesas
10	10.8	6	9.8	1	3.3	Hilltops

RH (m) relative height, *Land_C* landscape curvature, *TWI* topographic wetness index, *As (km^2)* catchment area

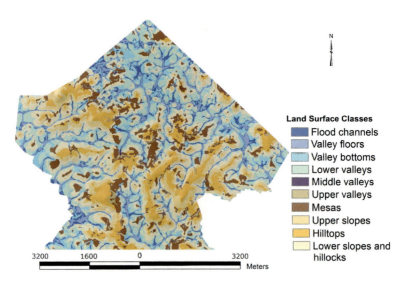

Fig. 23.2 Land-surface classes produced by the neuro-fuzzy classification of the landscape

Table 23.3 Variation of soil attributes among land-surface classes

Land-surface class	Physiography	Dominant soil class	Soil drainage class	Rock fragments at the surface %
1	Flood channels	(Not sampled)	Poorly drained	<3
2	Valley floors	Vertic Fluvaquents	Imperfectly drained	0
3	Valley bottoms	Ustic Epiaquerts	Imperfectly drained	<3
4	Lower valleys	Chromic Haplusterts	Moderately well drained	<3
5	Middle valley	Chromic Haplusterts	Moderately well drained	<3
6	Upper valley	Chromic Haplusterts	Well drained	>50
7	Lower slopes and hillocks	Vertic Haplustalfs	Well drained	15–50
8	Upper slopes	Vertic Haplustalfs	Well drained	15–50
9	Mesas	Typic Haplustalfs	Well drained	>50
10	Hilltops	Typic Haplustalfs	Well drained	>50

slopes, soils are well drained and have a subsurface horizon with illuvial clay. Rock fragments 2–10 cm large (pebbles) are frequent at the soil surface.

The digital model of the land-surface classes is continuous because (a) it consists in a grid in which each cell has been allocated to a given class, and (b) the membership of a cell to each class is defined by a continuous function in the interval [0, 1]. However, the spatial model obtained shows the study area divided into different physiographic units that appear clearly separated by natural boundaries (Fig. 23.2).

Thus, the procedure applied has produced an objective identification of natural discontinuities in the landscape of the study area. Such discontinuities result from the spatial differentiation caused by past and recent geomorphic processes (Minár and Evans 2008).

The morphology of the study area seems to be the result of three successive processes. First, a sedimentation process moved rock fragments from the mountain ranges in the north and deposited them along the flooding paths. Second, an erosion process affected the areas not protected by rock fragment cover and transformed them into valleys, leaving the areas protected by coarse fragments as mesas, hilltops, hillocks, and slopes. Third, a more recent process transported by gravity part of the coarse fragments from the steep lands into the valleys.

The digital classification by means of FKCN divided the landscape into land-surface classes which can be used as a basis to determine geomorphic units with different morphology, genesis, dynamic, and age. These units define particular environments of soil formation and control the spatial soil distribution, as shown in Table 23.3. As a result, the map of land-surface classes provides a geographical framework for soil sampling, soil-landscape mapping, and land-use planning.

23.4 Conclusions

The fuzzy neural network approach divided the study area into ten land-surface classes, which revealed details of the landscape structure that were hidden under a thick forest cover. The landscape corresponds to a denudation plateau conformed by flat-topped mesas, hilltops, and hillocks separated by narrow valleys.

Although the digital model of land-surface classes is continuous, it shows natural discontinuities resulting from geomorphic processes that modeled the landscape. Thus, the digital classification of the land surface was used as a basis to identify and map physiographic units that defined particular environments of soil formation.

The achieved model of physiographic units was used as a framework for soil identification and sampling to establish soil-landscape relationships. Soil drainage varies from imperfect to well-drained along a toposequence from the valley floors to the uplands. Dominant soil features result from shrink-swell process of expansive clay in the valleys and from clay illuviation into the subsoil in the uplands. Topsoils in mesas and hilltops are frequently covered by pebbles as a result of the processes that modeled the denudation plateau. Gravity transportation has extended the occurrence of rock fragments at the soil surface to some valley sectors.

Acknowledgments This research was supported by funds proceeding from the Venezuelan Organic Law for Science and Technology (LOCTI) and from the Consejo de Desarrollo Científico y Humanístico (Council of Scientific and Humanistic Development) of the Universidad Central de Venezuela (CDCH-UCV). We are also grateful to the International Centre for Theoretical Physics (Trieste, Italy) for financial support and fellowships.

References

Adediran AO, Parcharidis I, Poscolieri M, Pavlopoulos K (2004) Computer-assisted discrimination of morphological units in north-central Crete (Greece) by applying multivariate statistics to local relief gradients. Geomorphology 58:357–370

Bezdek JC, Tsao ECK, Pal NR (1992) Fuzzy Kohonen clustering networks. In: Proceednings of the IEEE international conference on Fuzzy Systems, San Diego, pp 1035–1043

Bolongaro-Crevenna A, Torres-Rodríguez V, Sorani V, Framed D, Ortiz MA (2005) Geomorphometric analysis for characterizing landforms in Morelos State, Mexico. Geomorphology 67:407–422

Debella-Gilo M, Etzelmüller B (2009) Spatial prediction of soil classes using digital terrain analysis and multinomial logistic regression modeling integrated in GIS: examples from Vestfold County, Norway. Catena 77:8–18

Dobos E, Carré F, Hengl T, Reuter HI, Tóth G (2006) Digital soil mapping as a support to production of functional maps. Office for Official Publications of the European Communities, Luxembourg. EUR 22123 EN, 68 pp

Ehsani HA, Quiel F (2008) Geomorphometric feature analysis using morphometric parameterization and artificial neural networks. Geomorphology 99:1–12

Hutchinson MF (1989) A new procedure for gridding elevation and stream line data with automatic removal of spurious pits. J Hydrol (Amst) 106:211–232

Iwahashi J, Pike RJ (2007) Automated classifications of topography from DEMs by an unsupervised nested-means algorithm and a three-part geometric signature. Geomorphology 86:409–440

McBratney AB, Mendonca Santos ML, Minasny B (2003) On digital soil mapping. Geoderma 117:3–52

McKensie MJ, Ryan PJ (1999) Spatial prediction for soil properties using environmental correlation. Geoderma 89:67–94

Minár J, Evans IS (2008) Elementary forms for land surface segmentation: the theoretical basis of terrain analysis and geomorphological mapping. Geomorphology 95:236–259

Odeh IOA, Chittleborough DJ, McBratney AB (1992) Soil pattern recognition with fuzzy-c-means: application to classification and soil-landform interrelationships. Soil Sci Soc Am J 56(2):505–516

Quinn PF, Beven KJ, Chevallier P, Planchon O (1991) The prediction of hillslope flow paths for distributed hydrological modelling using digital terrain models. Hydrol Process 5:59–79

Scull P, Franklin J, Chadwick OA, McArthur D (2003) Predictive soil mapping: a review. Prog Phys Geogr 27:171–197

Soil Survey Staff (2010) Keys to soil taxonomy, 11th edn. USDA–NRCS, Washington, DC

Viloria-Botello A (2007) Estimación de modelos de clasificación de paisaje y predicción de atributos de suelos a partir de imágenes satelitales y modelos digitales de elevación. Trabajo Especial de Grado. Universidad Central de Venezuela, Caracas

Wilson JP, Gallant JC (2000) Terrain analysis principles and applications. Wiley, Toronto

Zevenbergen LW, Thorne CR (1987) Quantitative analysis of land surface topography. Earth Surf Process Landf 12:47–56

Zhao Z, Chow TL, Rees HW, Yang Q, Xing Z, Meng F (2009) Predict soil texture distributions using an artificial neural network model. Comput Electron Agric 65:36–48

Zhu A, Moore A, Burt J (2010) Prediction of soil properties using fuzzy membership values. Geoderma 158:199–206

Zinck JA (2013) Geopedology. Elements of geomorphology for soil and geohazard studies. ITC Special Lecture Notes Series, Enschede

Part IV
Applications in Land Degradation and Geohazard Studies

Chapter 24
Gully Erosion Analysis. Why Geopedology Matters?

G. Bocco

Abstract Gully erosion dynamics is a complex phenomenon, usually man-induced, which cannot be described nor predicted using conventional soil erosion models such as USLE or similar. In the early 1980s it was stated that gully initiation and growth could be studied from a purely empirical perspective, because no deductive approach could serve the purpose. In this chapter, this premise is used to briefly summarize how gully erosion research has developed, what were the major achievements in the conceptual and methodological dimensions, and which may be potential courses of action for further research, with emphasis on the contribution of geopedology. Despite the advancements in the development of models and in RS and GIS techniques, gully erosion still remains a complex issue difficult to model and predict. In this sense, geopedology may play a role in its understanding and management. As other geomorphic processes, gullies occur in certain terrain, soil, and hydrology conditions which may be conveniently approached from a geopedologic perspective.

Keywords Gully erosion analysis • Erosion monitoring • Erosion modeling • GIS • Remote sensing

24.1 Introduction

Gully erosion initiation and development are complex phenomena which originate complex landforms (Bocco 1991). Gullies are usually man-induced and/or triggered by extreme rainfall events; they occur in different environments worldwide. Gully erosion cannot be described nor predicted using conventional soil erosion models such as the Universal Soil Loss Equation (USLE) (Wischmeier and Smith 1978) and similar tools. The initiation and development of gullies are the result of the activity

G. Bocco (✉)
Centro de Investigaciones en Geografía Ambiental (CIGA), Universidad Nacional Autónoma de México (UNAM), Morelia, Michoacán, Mexico
e-mail: gbocco@ciga.unam.mx

© Springer International Publishing Switzerland 2016   399
J.A. Zinck et al. (eds.), *Geopedology*, DOI 10.1007/978-3-319-19159-1_24

of a family of processes including those triggering inter-rill and rill erosion but also piping and shallow mass movement, involving soils and other surficial materials (Bocco 1993). Imeson and Kwaad (1980) stated that such phenomena as gully erosion could be studied from a purely empirical perspective, because no deductive approach was available. Has the situation changed after 35 years?

The purpose of this chapter is to analyze the potential contribution of geopedology to gully erosion research. After describing gully erosion in this introduction, research approaches to gully erosion studies are reviewed, with some emphasis on monitoring and modeling. Further the way how geopedology could serve gully erosion research is put forward. The idea is tested that the situation indicated by Imeson and Kwaad (1980) has probably not changed so far, because gullies are landforms originated by a variety of hydrologically-driven geomorphic and soil processes, and modeling cannot be based on deductive approaches. Gully erosion, in many instances, is a black box process.

Research on gully erosion is less developed than that on inter-rill and rill erosion. The hydrology of gully erosion is complex because it involves both surficial and subsurficial flows, which means that the hortonian-type of overland flow is only one of drivers of this type of soil erosion. In addition, there is no critical distance to water divide where incision and gullying would start, be it in large or small catchment. Subsurficial flows such as piping may be more important depending on the setting. As landforms, gullies are composed of head, slopes, bottom, channel, and sometimes fan. Incision usually takes place at the gully bottom, and slumps and slides affect mostly gully head and gully sides. Fan formation depends on the sediment delivery capacity of the gully. The upslope area contributing to gully head retreat, described as zero order, may be affected by inter-rill and rill erosion; but the sequence from laminar to gully erosion may not be present, and gullies may start independently of these processes. In addition, they may be triggered by micro-mass movements or because of human action (dirt roads, ill-defined culverts, boundaries between agricultural parcels, and others).

Gullies occur as individual features or as systems composed of several channels and thus of multiple heads and slopes. In the landscape, they are present as valley-side or valley-bottom types; they can be continuous or discontinuous. Usually, gullies develop on accumulative slopes, such as lower portions of footslopes or even plains. As a genetically erosional landform, this is a peculiar fact because gullying may upset the denudation chronology in a region.

Because of their initiation, development, and landscape position, gullies or gully systems were first analyzed from the standpoint of the so-called davisian cycle of erosion or "geographic cycle". In other words, the theory developed at the onset of the twentieth century by Davis (1905) for landscape development and denudation chronology was applied. Thus, gullies were described following the conventional stages of youth, maturity, and decline put forward by Davis to explain landscape evolution. Most of research in the US Department of Agriculture focused on developing typologies in this framework. The approach was criticized because the model assumed tectonic stability, temperate climate, hortonian-type of overland flow, a critical distance to the water divide to start incision, and a natural tendency of gully activity to become extinct. Empirical work showed, nonetheless, that gullies could start anywhere on a given slope, that the hydrologic flow could be complicated by the interference with subsurficial flows, and that gullies could be self-perpetuating systems. From the 1980s onwards, inductive empirical perspectives were established. The davisian approach was abandoned, and deductive models were no longer attempted.

Under these circumstances, empirical work in many contrasting regions showed that gullies could occur in a variety of climates and rainfall regimes, rock and soil types, slope facets and gradients, and land-cover and land-use types. Research progressed from establishing simple relationships between gully growth and time (i.e. gully erosion rates using sequential aerial photographs and photogrammetric means), to analyzing the contributing role of rainfall, soils, slopes, cover and practices, usually applying statistical approaches on the basis of field-verified remote-sensed data. In addition, comparisons between the severity of gully erosion and other types of soil erosion followed, with some emphasis on their respective sediment production and eventual delivery to streams or reservoirs. Gully erosion monitoring and modeling became increasingly popular in the scientific literature as well as in the grey literature produced by technical agencies (governmental and social) at various levels.

24.2 Research on Gully Erosion: Detection, Monitoring, and Modeling

Once the complexity involved in gully erosion was understood and the davisian model rejected, research moved with the beginning of this century towards several key interrelated topics to better understand the processes, their dynamics (monitoring) and simulation/prediction (modeling).

A thorough review of research on gully erosion is beyond the scope of this chapter. However, some lessons learned from the literature are convenient in particular to summarize limitations concerning the trends in gully erosion research and delineate some guidelines about the advantages of using geopedology as a tool in gully erosion.

24.2.1 Gully Erosion Processes and Modeling

Besides an early paper by the author of this chapter (Bocco 1991), one of the first published reviews was that of Bull and Kirkby (1997) who suggested that long-term rates of gully development were not well understood and that theoretical modeling could provide the way forward for more holistic investigations. Poesen (Poesen and Hooke 1997; Poesen et al. 2003) also called for the need of gully erosion modeling, in particular as related to environmental change. In both reviews, Poesen and collaborators indicate that most field measurements and modeling efforts had concentrated on water erosion processes operating at the runoff plot scale and not at smaller geographic scales (large areas). The implications of this limitation are not dealt with in the reviews. The authors conclude by suggesting research needs and priorities, including the quest for predictive models at various geographic scales and the study of the impact of gully development on hydrology, sediment yield, and landscape evolution. No specific practical reference is made as how to relate gully erosion development to landscape studies. Despite the calls for theoretical modeling, Desmet et al. (1999) highlighted the importance of slope gradient and contributing area for optimal prediction of the initiation and trajectory of ephemeral gullies at plot scale in the Belgian loess region. Moreover, Nachtergaele et al. (2001) when testing the physically-based Ephemeral Gully Erosion Model (EGEM) in the same region, where a robust collection of data depicting input parameters exists, determined a value for simple topographic and morphologic indices in the prediction of ephemeral gully erosion. They stressed the problematic nature of physically-based models, since they often require input parameters that are not available or can hardly be obtained. Capra et al. (2005) also used the EGEM in Sicily, but concluded that it seemed simpler using empirical relations between eroded volume and gully length in different environments, until more precise physically-based models were developed. This conclusion together with the data issue questions this type of modeling and its usefulness in potential practical applications in conservation. Likewise, Brazier (2004) stated that even for the UK, although provided with robust data, there is not enough information to validate soil erosion models (at large), especially when the goal is assessing the spatial heterogeneity of soil loss.

The trend from monitoring to modeling seems to be hampered by the lack of good quality data even in places where solid databases are available. As a consequence, the value of the models is challenged by the absence of validation through appropriate data. By the same token, this highlights the importance of the empirical data obtained from monitoring to face gully erosion analysis, even at large geographic scale in data-rich small areas. Brazier (2004) concluded that the paradox between data collection to improve models and erosion modelling to replace data collection must be addressed within the discipline if full use of datasets and improvement of models were to be made.

From a different perspective, Chaplot et al. (2005) quantified linear erosion (LE) at the catchment level, and found that some of the LE controlling factors could also be used for prediction over larger areas since topography and land-use data, closely correlated with LE, were easily accessible. Valentin et al. (2005) also referred to the relation between gully erosion and global topics and the need to develop research in areas larger than the cultivated plots, but without providing clear indication on the methods, techniques, and approaches to be used in this type of research in large areas, nor on how to extrapolate findings from the plot to the landscape level. However, the paper addresses soil and gully erosion triggered by agricultural land use in the Chinese loess plateau and suggests that land-use and land-use change are of a paramount importance, even more than climate change.

In a similar line of thinking, Boardman (2006) emphasized the limited value of existing soil erosion models, including those of gullies, in the real world as opposed to the academic sphere. He suggested that approaches should include socioeconomic variables, land-use concerns, and be less "data-rich and people-poor". DeVente et al. (2007) pointed at the absence of reliable spatially distributed process-based models for the prediction of sediment transport at the drainage basin scale and claimed that spatially distributed information on land use, climate, lithology, topography, and dominant erosion processes was required for modeling purposes, including that of gully erosion. Many papers have addressed those controlling variables at the landscape level and have detected relationships particularly between gullying, soil properties, and slope characteristics (Table 24.1). However, there is no reference to a given spatial frame such as terrain, landscape or other type of environmental units, which happen to be defined by the very same controlling variables mentioned above.

Nazari Samani et al. (2011) discuss the limitations of gully erosion models and emphasize, for land managers, the importance of gullies as sediment sources particularly as compared to inter-rill and rill erosion. By contrast, Porto et al. (2014) did not find significant differences in the contribution to soil loss between inter-rill, rill and gully erosion in a small cultivated plot in Sicily, but reported large inter-annual variability of this contribution. Capra (2013) also emphasized the need to study gully erosion in large areas. However neither of the two publications provided clear methods and techniques to tackle research at such scale.

Overall, modeling of gully erosion seems not to have moved beyond empirical approaches. No deductive model has been formalized. The attempts to move from plot to basin and from monitoring to modeling have not yielded the results expected

Table 24.1 Relationships between gully erosion and controlling variables in different environments

Author	Study area	Controlling factors
Gábris et al. (2003)	Hungary, temperate humid	Land-use (long term, decades)
Shrestha et al. (2004)	Nepal, temperate humid	Land-use, slope aspect
Chaplot et al. (2005)	Laos, tropical humid	Catchment surface area and perimeter, mean slope gradient
Zucca et al. (2006)	Sardinia, temperate Mediterranean	Rock type, slope gradient, land-use
Schmitt et al. (2006)	Poland, temperate humid	Land-use (long term, decades)
Lesschen et al. (2007)	SE Spain, sub-humid, semi-arid	Abandoned cropland
Moges and Holden (2008)	Ethiopia, sub-humid, semi-arid	Land cover change
Nazari Samani et al. (2009)	SE Iran, arid, semi-arid	Slope gradient, land-use
Van Zijl et al. (2013)	Lesotho, sub-humid, semi-arid	Duplex soils
Shrestha et al. (2014)	Thailand, tropical humid	Land-use, slope gradient

by scholars. An additional difficulty arises from the limited scientific results applied to hazard prevention and mitigation of gully erosion processes. Furthermore, the application of models and knowledge in practical management strategies is still a large issue. The call for "people-rich" approaches by Boardman (2006) is important. The question remains on to how to involve people, an issue that requires research on the social perception of erosion processes and land and landscape management. This would open a different avenue towards social and cultural research, particularly on rural land uses and local conservation knowledge.

24.2.2 Remote Sensing, GIS and Gully Mapping

Remote sensing and geographic information systems have been used to map gullies and controlling variables. The expectation is that research using these technical tools would contribute with a clear spatially-explicit framework to study gully erosion, but research objectives are in fact geared to a variety of topics. Sidorchuk et al. (2003) dealt with the identification of gully erosion forms and processes in the Mbuluzi River catchment (Swaziland) by using the Erosion Response Units (ERU) concept. Input data were obtained from remote sensing (API method) and GIS analyses.

Marzolff and Ries (2007) monitored headcut retreat in 12 gullies using detailed aerial photography taken from remote-controlled platforms to identify runoff patterns and infiltration behavior in the gully headcut surroundings. They emphasized the benefits of high-resolution aerial photography for monitoring and understanding gully erosion processes. Daba et al. (2003) and Ndomba et al. (2009) used sequential aerial photos to study the development and dynamics of gully erosion systems. Evans and Lindsay (2010) extracted gully maps from high-resolution digital elevation models (2 m LiDAR DEMs). Wang et al. (2014) proposed the use of object-oriented analysis (OOA) to quantitatively study small gullies. Shruthi et al. (2014) also used OOA and very high-resolution imagery to digitally detect gully systems. Peter et al. (2014) combined rainfall simulation, gully mapping, and volume quantification at local scale using unmanned aerial vehicle (UAV) remote sensing data. Gómez-Gutiérrez et al. (2014) used 3D photo-reconstruction methods (based on Structure from Motion (SfM) and MultiView-Stereo (MVS) techniques) to estimate gully headcut erosion. Results of this simulation, not surprisingly, pointed out to a clear decrease in the accuracy of the model when the photos were not acquired sequentially around the headcut.

Dube et al. (2014) went back to an empirical model based on seven environmental factors (land cover, soil type, distance from river, distance from road, sediment transport index (STI), stream power index (SPI), and wetness index (WI)) using a GIS-based weight of evidence modeling (WEM). The predictive capability of the weight of evidence model in this study suggests that land-cover, soil type, distance from river, STI, and SPI are useful but not sufficient to produce a valid map of gully erosion hazard. Conoscenti et al. (2014) also insisted in the use of GIS and multivariate statistical analysis in a small catchment. In contrast with the previous case study, they found "acceptable to excellent accuracies of the models" and good predictive skill.

Despite the advances in the use of new remote sensing techniques, sensors, and approaches (such as the OOA), and GIS models, the goals have not changed substantially with time. In addition, the question concerning how to conduct gully erosion research in areas larger than local plots, usually basins, is not addressed. How to stratify large areas for further sampling and extrapolation? Are topography as variable and DEM as tool enough to achieve this goal? Past and current research has focused more on technical issues than on essential topics such as testing spatial classification schemes of terrain, soils, and cover. Conceptual approaches based for instance on soil-landscape relationships can help support gully erosion research.

24.3 Using Geopedology in Gully Erosion Research

24.3.1 Why Is It Important?

Review of research on gullies suggests that (1) theoretical modeling is complex, requires usually unavailable data, and so far has not led to successful results; (2) empirical modeling, usually statistically-based and using remote-sensed data, seems to be the most common approach; (3) small catchments or runoff plots seem to be the most common type of study area; (4) no indications concerning methods and techniques to be applied in larger areas or catchments are provided. This includes the absence of proposals for the use of geographically-explicit models.

The first three points basically subscribe an idea which can be simply summarized as follows: landforms are difficult if not impossible to model; models may be conceptual but not operational. The last point suggests a lack of understanding of another simple fact: gullies occur in terrains, some of which are more susceptible than others to trigger this type of erosion. This calls for the need to stratify terrains to understand gully erosion processes and develop gully hazard models. Geopedology offers such a geographic approach. The basic assumption behind is that gullies are not randomly distributed but they develop in response to a combination of environmental factors (Vázquez and Zinck 1994). This assumption closely matches the need to stratify the landscape when working in relatively large areas. One would expect the occurrence of a combination of gully-prone factors per map unit, and predict that some units might be more susceptible than others to gully initiation and development.

These relatively simple relations have very seldom been referred to in the literature (see e.g. Bocco et al. 1990). One possible explanation is the development of purely quantitative approaches, which assume that a semi-quantitative approach is not scientifically sound. Something similar occurs with remote sensing, where digital interpretations are assumed to be more accurate than visual ones. Stand alone quantifications seem to have become more important than the understanding of natural processes, in many instances strongly influenced by human action. One example is the overuse of geomorphometric digital terrain modeling and the underuse of visual landform surveying and mapping as research tools in gully erosion analyses. Geopedology provides an alternative basis for gully erosion research.

24.3.2 Why Geopedology Matters?

The geopedologic approach allows differentiating spatial units at variable geographic scales, which are relatively homogeneous in terms of terrain and soil properties. These properties can be assessed as to their suitability for land use and crop production purposes, or as to their susceptibility to different types of land degradation processes, among others, gully erosion. Vázquez and Zinck (1994) used such an approach to model gully development in Central Mexico. They developed a stepwise procedure to explore the spatial relationships between gullies and six selected environmental factors assumed to control and explain gully formation and distribution (i.e. landscape type, relief type, slope gradient, slope shape, soil types and properties, and land use). They first analyzed the cartographic coincidence between factors and gullies, and derived threshold values signaling most favorable conditions for gully formation. Further they built rule-based models in a GIS where class boundaries were the selected threshold values. Then they tested and validated the models by evaluating their efficiency in reproducing existing gullied areas. Finally they applied the validated models to predict areas potentially favorable to gully formation, and derived recommendations for selecting priority areas for soil conservation.

The above research developed a semi-quantitative method which in practice offered a solution to many of the problems revealed in the literature review on gully erosion modelling previously discussed in this chapter. This approach can be used in fairly large areas at landscape level. It analyzes gully erosion factors considering soil and terrain attributes and provides a practical appraisal for soil conservation. The basis are data derived from a geopedologic survey, using both qualitative and quantiative methods. The survey does not involve complex data collection, instead requires elevation data, aerial photography or very high resolution satellite imagery, and a GIS platform. But it does require expert knowledge in terrain and soil survey, as well as in the hydrological processes that trigger gully erosion. This knowledge and field expertise cannot be replaced by algorithmic approaches or sophisticated data manipulations. It is a robust albeit simple approach to gully erosion or other conspicuous erosion phenomena, which can be extrapolated to any area where the above described knowledge and data are available or can be collected.

24.4 Conclusions

Gully erosion analysis, monitoring, and modeling are complex issues because gullies are complex landforms, usually polygenic, highly dynamic, and man-induced. Numerous variables have been tested using a variety of approaches. Terrain, soil, and cover harbor most of the properties from which variables are derived. These factors also control the effect of the hydrologic processes that trigger gully erosion. Relatively poor comprehension of the relationship between terrain, soil and cover, and strong emphasis on quantitative analyses have probably contributed to disregard conceptual models able to provide guidelines for stratifying land, soil, and cover, and allow sensitive sampling procedures to understand, monitor, and model gully erosion processes. Geopedology or analogous approaches have made substantial though less popular contributions to this end. A matching between quantitative analysis and a thorough spatial and conceptual framework to gully erosion seems to be a path to be further tested.

Acknowledgements   María Lira helped in the preparation of the manuscript. CONACYT project 247048 and PAPITT project 301914 offered support to this research.

References

Boardman J (2006) Soil erosion science: reflections on the limitations of current approaches. Catena 68(2–3):73–86. In: Helming K, Rubio JL, Boardman J (eds) Soil erosion research in Europe 68(2–3):71–202

Bocco G (1991) Gully erosion: processes and models. Prog Phys Geogr 15(4):392–406

Bocco G (1993) Gully initiation in quaternary volcanic environments under temperate sub-humid seasonal climates. Catena 20(5):495–513

Bocco G, Palacio JL, Valenzuela C (1990) Gully erosion modelling using GIS and geomorphic knowledge. ITC J 3:253–261

Brazier R (2004) Quantifying soil erosion by water in the UK: a review of monitoring and modelling approaches. Prog Phys Geogr 28(3):340–365

Bull LJ, Kirkby MJ (1997) Gully processes and modeling. Prog Phys Geogr 21(3):354–374

Capra A (2013) Ephemeral gully and gully erosion in cultivated land: a review. In: Lannon EC (ed) Drainage basins and catchment management: classification, modelling and environmental assessment. Nova Science Publishers, New York, pp 109–141

Capra A, Mazzara LM, Scicolone B (2005) Application of the EGEM model to predict ephemeral gully erosion in Sicily, Italy. Catena 59(2):133–146

Chaplot V, Coadou le Brozec E, Silvera N, Valentin C (2005) Spatial and temporal assessment of linear erosion in catchments under sloping lands of northern Laos. Catena 63(2):167–184

Conoscenti C, Angileri S, Cappadonia C, Rotigliano E, Agnesi V, Märker M (2014) Gully erosion susceptibility assessment by means of GIS-based logistic regression: a case of Sicily (Italy). Geomorphology 204:399–411

Daba S, Rieger W, Strauss P (2003) Assessment of gully erosion in eastern Ethiopia using photogrammetric techniques. Catena 50(2):273–291

Davis WM (1905) The geographical cycle in an arid climate. J Geol 13(5):381–407, http://www.jstor.org/stable/30067951

Desmet PJJ, Poesen J, Govers G, Vandaele K (1999) Importance of slope gradient and contributing area for optimal prediction of the initiation and trajectory of ephemeral gullies. Catena 37(3):377–392

deVente J, Poesen J, Arabkhedri M, Verstraeten G (2007) The sediment delivery problem revisited. Prog Phys Geogr 31(2):155–178

Dube F, Nhapi I, Murwira A, Gumindoga W, Goldin J, Mashauri DA (2014) Potential of weight of evidence modelling for gully erosion hazard assessment in Mbire District – Zimbabwe. Phys Chem Earth 67–69:145–152

Evans M, Lindsay J (2010) High resolution quantification of gully erosion in upland peatlands at the landscape scale. Earth Surf Process Landf 35(8):876–886

Gábris G, Kertész Á, Zámbó L (2003) Land use change and gully formation over the last 200 years in a hilly catchment. Catena 50(2):151–164

Gómez-Gutiérrez Á, Schnabel S, Berenguer-Sempere F, Lavado-Contador F, Rubio-Delgado J (2014) Using 3D photo-reconstruction methods to estimate gully headcut erosion. Catena 120:91–101

Imeson AC, Kwaad FJ (1980) Gully types and gully prediction. KNAG Geografisch Tijdschrift 14(5):430–441

Lesschen JP, Kok K, Verburg PH, Cammeraat LH (2007) Identification of vulnerable areas for gully erosion under different scenarios of land abandonment in Southeast Spain. Catena 71(1):110–121

Marzolff I, Ries JB (2007) Gully erosion monitoring in semi-arid landscapes. Z Geomorphol 51(4):405–425

Moges A, Holden NM (2008) Estimating the rate and consequences of gully development, a case study of Umbulo catchment in southern Ethiopia. Land Degrad Dev 19(5):574–586

Nachtergaele J, Poesen J, Steegen A, Takken I, Beuselinck L, Vandekerckhove L, Govers G (2001) The value of a physically based model versus an empirical approach in the prediction of ephemeral gully erosion for loess-derived soils. Geomorphology 40(3):237–252

Nazari Samani A, Ahmadi H, Jafari M, Boggs G, Ghoddousi J, Malekian A (2009) Geomorphic threshold conditions for gully erosion in Southwestern Iran (Boushehr-Samal watershed). J Asian Earth Sci 35(2):180–189

Nazari Samani A, Wasson RJ, Malekian A (2011) Application of multiple sediment fingerprinting techniques to determine the sediment source contribution of gully erosion: review and case study from Boushehr province, southwestern Iran. Prog Phys Geogr 35(3):375–391

Ndomba PM, Mtalo F, Killingtveit A (2009) Estimating gully erosion contribution to large catchment sediment yield rate in Tanzania. Phys Chem Earth A/B/C 34(13–16):741–748

Peter KD, D'Oleire-Oltmanns S, Ries JB, Marzolff I, AitHssaine A (2014) Soil erosion in gully catchments affected by land-levelling measures in the Souss Basin, Morocco, analysed by rainfall simulation and UAV remote sensing data. Catena 113:24–40

Poesen JWA, Hooke JM (1997) Erosion, flooding and channel management in Mediterranean environments of southern Europe. Prog Phys Geogr 21:57–199

Poesen J, Nachtergaele J, Verstraeten G, Valentin C (2003) Gully erosion and environmental change: importance and research needs. Catena 50(2):91–133. In: Poesen J, Valentin C (eds) Gully erosion and global change. Catena 50(2–4):87–564

Porto P, Walling DE, Capra A (2014) Using 137Cs and 210 Pbex measurements and conventional surveys to investigate the relative contributions of interrill/rill and gully erosion to soil loss from a small cultivated catchment in Sicily. Soil Tillage Res 135:18–27

Schmitt A, Rodzik J, Zgłobicki W, Russok C, Dotterweich M, Bork HR (2006) Time and scale of gully erosion in the Jedliczny Dol gully system, south-east Poland. Catena 68(2–3):124–132

Shrestha DP, Zinck JA, Van Ranst E (2004) Modelling land degradation in the Nepalese Himalaya. Catena 57(2):135–156

Shrestha DP, Suriyaprasit M, Prachansri S (2014) Assessing soil erosion in inaccessible mountainous areas in the tropics: the use of land cover and topographic parameters in a case study in Thailand. Catena 121:40–52

Shruthi RBV, Kerle N, Jetten V, Abdellah L, Machmach I (2014) Quantifying temporal changes in gully erosion areas with object oriented analysis. Catena. doi:10.1016/ j.catena. 2014.01.010

Sidorchuk A, Märker M, Moretti S, Rodolfi G (2003) Gully erosion modelling and landscape response in the Mbuluzi River catchment of Swaziland. Catena 50(2–4):507–525

Valentin C, Poesen J, Li Y (2005) Gully erosion: impacts, factors and control 63(2–3):132–153. In: Valentin C, Poesen J, Li Y (eds) Gully erosion: a global issue. Catena 63(2–3):129–330

Van Zijl GM, Ellis F, Rozanov DA (2013) Emphasising the soil factor in geomorphological studies of gully erosion: a case study in Maphutseng, Lesotho. S Afr Geogr J 95(2):205–216

Vázquez L, Zinck A (1994) Modelling gully distribution on volcanic terrains in the Huasca area, Central Mexico. ITC J 3:238–251

Wang T, He F, Zhang A, Gu L, Wen Y, Jiang W, Shao H (2014) A quantitative study of gully erosion based on object-oriented analysis techniques: a case study in Beiyanzikou catchment of Qixia, Shandong, China. Sci World J. doi:10.1155/2014/417325

Wischmeier WH, Smith D (1978) Predicting rainfall erosion losses – a guide to conservation planning. USDA Agriculture Handbook, Washington, DC, p 537

Zucca C, Canu A, Della Peruta R (2006) Effects of land use and landscape on spatial distribution and morphological features of gullies in an agropastoral area in Sardinia (Italy). Catena 68(2–3):87–95

Chapter 25
Assessing Soil Susceptibility to Mass Movements: Case Study of the Coello River Basin, Colombia

H.J. López Salgado

Abstract Mass movement susceptibility and hazard maps are essential tools for mitigation and prevention of landslides. The upper Coello river basin in the Colombian Andes is a catchment where the conjunction of high rainfall, steep slopes, and unstable volcanic ash cover, among other factors, causes frequent mass movements. A qualitative causal model was developed using detailed geopedologic information collected in sample areas and validated outside for mass movement hazard zoning. It highlights the relationships between mass movement-promoting soil properties (mainly mechanical and physical) and resulting morphodynamic processes and features (mainly landslides, various solifluction forms, and terracettes). Soil properties were assessed in terms of their susceptibility to mass movements from an integrated soil-geomorphic map.

Keywords Geopedology • Mass movement mapping • Mass movement hazard • Hazard zonation • Soil properties

25.1 Introduction

In general, the application of earth-sciences information and technology has potential for long term reduction of natural hazards produced by surface and subsurface phenomena. However, slope failures upon mass movement, although sometimes triggered by regional causes, may occur suddenly, at short term or periodically when the equilibrium of the soil mass is exceeded by internal changes caused by abnormally high precipitations or earthquakes. As such, earth-sciences technology can play a valuable role in determining hazard areas.

H.J. López Salgado (✉)
Private Activity, Environmental Studies, Bogotá, Colombia
e-mail: hector.lopezs@hotmail.com

Among natural hazards, mass movements are the most frequent and have the largest geographic distribution. Landslides alone, for example, cause greater loss of human lives and economic goods than any other single natural hazard.

In the Andes, mass movements are of common occurrence as natural and/or human-induced phenomena. Singular conjunctions of conditioning and activating factors contribute to it, such as steep slopes, unconsolidated sedimentary and volcanic rocks, deeply weathered soil covers, heavy rainfalls, active seismicity and volcanism, and environmentally degrading land uses. The upper Coello river basin in the Colombian Andes is a catchment where all these factors are jointly present.

The Coello river basin is located in the department of Tolima, in the central range of the Colombian Andes, approximately 200 km west-southwest of Bogotá. It is one of the most important catchments for water supply to Ibagué, the capital city of the department, as well as to smaller towns, and to the Coello irrigation district, an intensively used agricultural area, from where high yield products are distributed all over the country. One of the most important communication axes crosses the area east-west, allowing intensive traffic with other regions of the country.

The general research objective was to establish, run, and validate a GIS-assisted data management model aiming at spatial prediction of soil-induced surficial mass movements. Specific operational objectives were set for data collection and transformation to identify the various types of mass movements and develop a qualitative causal model which explains the occurrence of past mass movements and allows prediction of hazard areas (López and Zinck 1991).

25.2 Mass Movements: Hazard and Zonation

Natural hazard refers to the probability of occurrence of a potentially damaging phenomenon within a specified period of time and within a given area. The aim of mass movement studies is to develop an understanding of the processes involved by answering the questions of why, how, when, and where they occur, because this permits prediction of susceptibility by extension of site information to larger areas. Stresses on a slope are rarely constant over long periods of time. The variations may be very slow due to uplift, gradual erosion or deposition; seasonal, reflecting fluctuations of the groundwater level; or rapid as caused by transitory seismic vibrations, construction activity, cuttings, reservoir fluctuations, or changes in land use practices (Varnes 1984).

Apart from natural variations in rainfall, long term unfavorable changes in groundwater levels may result from human actions, through conversion of grassland to residential development with septic sewerage systems, by clearing of vegetation, irrigation, or timber harvesting. The loss of vegetation cover, either grass or forest, by overgrazing, fire or clear-cut logging, not only alters the hydrologic conditions of the slope but is widely believed to promote rapid runoff and erosion and to increase the possibility of mass movements. Zinck (1986) established a relationship between soil consistency limits and soil moisture with the aim to determining the susceptibility of the soil cover to mass movements on steep slopes under tropical cloud forest.

According to Hansen (1984), natural hazards can be assessed in three basic ways: (1) the historical method, in which all previously recorded data about magnitude, frequency, and dates of the hazardous events are collected; (2) geomorphological analysis, where field evidence is collected about frequency and magnitude of the events in question, and (3) experimentation and calculation which can provide additional quantitative information about the processes responsible for any hazardous event.

The term "zonation" applies in general sense to the division of a land surface into areas and the ranking of these areas according to degrees of actual or potential hazard from mass movements on slopes (Varnes 1984). Many landslide hazard zonation schemes employ the concept of superposing and integrating spatial information maps showing individually the factors thought important in assessing slope stability, such as maps reporting slope ranges; landslide deposits and their geomorphological characteristics; groups of geologic units having common lithology, structure, geotechnical properties or behavior; hydrologic conditions, rainfall and climate; and sometimes seismic activity and expected seismic response.

In determining the risk for each type of movement to occur, the following factors are considered: lithology, nature of the soil and subsoil, structure, angle of natural slopes, water drainage, morphology, local history of land sliding, and hydrology. In seismically active areas, some of the most disastrous landslides have been triggered by seismic shocks. Particularly susceptible materials are those with a loose or open structure such as loess, volcanic ash on steep slopes, saturated sands of low density, fine-grained "sensitive" deposits of clay or rock flour, and cliffs of fractured rock or ice. Where these conditions are given in seismic regions, landslide hazard zonation must be intimately linked with seismic zonation through an evaluation of how the materials will respond to acceleration, amplitude, and duration of seismic motion together with an estimation of recurrence intervals (Varnes 1984).

Margottini et al. (2013) reviewed various investigations and case studies worldwide, and summarized basic analytical concepts and approaches for the inventory and mapping of landslides. The integration of data derived from remote sensing and landform and soil surveys, in a geographic information system (GIS), was recognized as an important component in most of the approaches. At regional level, Mora et al. (2002), among others, put forward the need to compile a set of indicators, usually obtained by observation and measurement. Further, they point to the evaluation of the relative weight of the indicators and their spatio-temporal distribution, enabling the opportunity to establish a zoning of susceptibility of the land to move downslope. Van Westen et al. (1997) used terrain analysis as a spatial model to evaluate the potential of every unit to experience mass movements. In addition, they presented a three-level approach to landslide susceptibility mapping: heuristic at recognition level, causative-statistical (empirical) at semi-detailed level, and deterministic at detailed level.

In Colombia, during the last 15 years, geopedologic surveys have been carried out to collect the basic terrain and soil data needed to develop the national susceptibility zoning to mass movements. In this context, a set of variables was defined and a hierarchy was established to allocate a weight to each variable for further modeling efforts in a heuristic framework (INGEOMINAS 2001; Servicio Geológico Colombiano 2013).

25.3   Method and Materials

25.3.1   The Study Area

The study area of 125,000 ha is bounded by the coordinates 4°15′–4°39′ north and 75°09′–75°36′ west. The topography is rugged and elevation ranges from 800 to 5,200 m asl. The general characteristics of the Coello river basin are shown in Table 25.1.

Lithology is relatively varied, including mainly metamorphic rocks (schist, gneiss, and amphibolite) and igneous rocks (quartz diorite). In addition, pyroclastic layers originated from the Tolima-Quindío-Machin volcanic complex cover discontinuously the igneous-metamorphic substratum, especially above 1,500 m elevation. Several fault systems cross the area in N-S and N-SE directions.

According to CIAF (1983), the following landform units are present in the area: (a) units of denudational origin that have developed from schist and phyllite of the Cajamarca Group, and from gneiss, amphibolite, and igneous rocks of the Ibague Batholith; (b) units of volcanic origin that have developed from lava flows, lava fields, and pyroclastic layers; and (c) units of depositional origin that form mesas, old terraces, glacis, fluvio-volcanic flows, alluvial fans, young alluvial terraces, and colluvial surfaces.

The overall mountainous landscape is highly dissected and incised by deep, steeply sloping, narrow valleys. The relief energy between valley floors and crest lines can be more than 1000 m. Alignments of longitudinal ridges and transversal rafters with asymmetric topographic profiles are controlled by rock schistosity, layer dipping, and faults. The landscape and relief types, their elevation distribution, and their relationships with rock and soil units are shown in the soil map legend (Table 25.1). The variety of rock types and facies occurring in the Coello river basin can be grouped into four main lithologic sets: metamorphic, intrusive igneous, volcanic, and sedimentary rocks.

Historic seismic activity related to volcanic manifestations or not, has been recorded in the area. The Nevado del Tolima and Machín volcanoes are now in a latent stage (fumarolas), but they have undergone several explosive eruptions during the Holocene. Ballistic coarse-grained pumices and lapilli, as well as wind-transported ash, have been spread over the upper and middle catchment several times since 14,000 years BP (Cepeda 1989). As a consequence, pyroclastic mantles cover more than 70 % of the basin.

Table 25.1 General characteristics of the study area

Climatic elevation belts (m asl)	Life zones	Lithology	Soils (soil units in Fig. 25.2)	Slope gradient %	Morpho- dynamics	Land use
Páramo >3500	Subalpine moist scrub	Volcanic rocks pyroclasts	Cryandepts (1)	12 to >70	Creep landslide	Conservation
Subpáramo 3000–3500	Montane wet forest transitional to montane rainforest	Pyroclasts over metamorphic and volcanic rocks	Dystrandepts Placandepts Tropohemists (2)	12 to >70	Creep landslide solifluction terracettes	Conservation, grazing and some agriculture
Cold 2000–3000	Lower montane wet forest	Pyroclasts over metamorphic rocks colluvial material	Dystrandepts Troporthents Vitrandepts Haplustolls (3–4)	25 to >70	Creep solifluction landslide sheet erosion terracettes	Grazing forest along rivers and on slope summits
Temperate 1000–2000	Premontane moist forest transitional to premontane wet forest	Pyroclasts over igneous and metamorphic rocks	Dystrandepts eutropepts (5)	12 to >70	Creep solifluction landslide terracettes	Grazing agriculture (perennial and annual crops)
Temperate 1000–2000	Same vegetation as above	Igneous rocks	Eutropepts Dystropepts Troporthents (6)	12 to >70	Terracettes	Same land use as above
Warm <1000	Premontane moist forest transitional to tropical dry forest	Fluvio-volcanic material	Haplustalfs Natrustalfs Ustorthents (7–11)	<12	Sheet erosion gullies terracettes	Rainfed agriculture and cattle-raising

25.3.2   Methodologic Approach

25.3.2.1   Data Collection

Data were collected in two consecutive stages (Fig. 25.1). First, a general inventory
was carried out covering the entire basin (125,000 ha). Existing thematic maps
(topography, geology, geomorphology, and soils) were compiled, controlled by field
traverses and aerial photo-interpretation at scale of 1:50,000, and finally adjusted to
a common scale of 1:100,000. Maps were digitized and the attribute data were captured through a tabular database for further processing and analysis. Second, on the
basis of this general information, five sample areas were selected for detailed aerial
photo-interpretation (scale 1:20,000) and field research, taking into consideration
such factors as topography, climate, lithology, soils, morphodynamics, and land use
(Fig. 25.1) The sample areas were concentrated in two elevation zones (temperate
and cold from 1000 to 3000 m asl, in Table 25.1), showing either higher frequency
of present or past mass movement occurrences (cold zone) or exhibiting greater
hazard potential (temperate zone). The spatial distribution encompassed a large
range of relevant characteristics common to the major part of the catchment, such
as: black and green micaschists and granodiorite, with or without volcanic ash cover
layers; slopes between 12 and 75 %; variable climatic conditions from wet temperate to very wet cold; main morphodynamic processes related to terracettes and solifluction; soils belonging mainly to Andepts and Tropepts; cattle grazing as the
dominant land use.

   Fieldwork was carried out along transects for characterization of the landforms,
soil data collection, verification of the map units, soil classification to the family
level (Soil Survey Staff 1975), soil sampling, and collection of ancillary data (e.g.
climate, soil mechanics, among others). Soil samples were processed according to
the methods used by the IGAC laboratory in Bogotá (pH, CEC, organic carbon,
total bases, consistency limits, texture, bulk density, porosity, pF, infiltration).

25.3.2.2   Causal Model

A qualitative causal model was developed on the basis of the detailed information
collected in the sample areas. It highlights the relationships between mass
movement-promoting soil properties (chemical, physical, and mechanical) and
resulting morphodynamic processes and features (mainly landslides, various solifluction forms, and terracettes). Specific correlations were established, for example
between the liquefaction of quicksands (lapilli) and the formation of solifluction
lobes.

   Established cause-effect relationships were tested outside the sample areas for
broader validation of the relational model. Then the relationships were aggregated
from soil property level to taxonomic units and cartographic units to allow for generalization from sample areas to the basin area. The causal model was used as a
conceptual framework for channeling the extrapolation process through soil-
controlled homogeneous map units (Fig. 25.1).

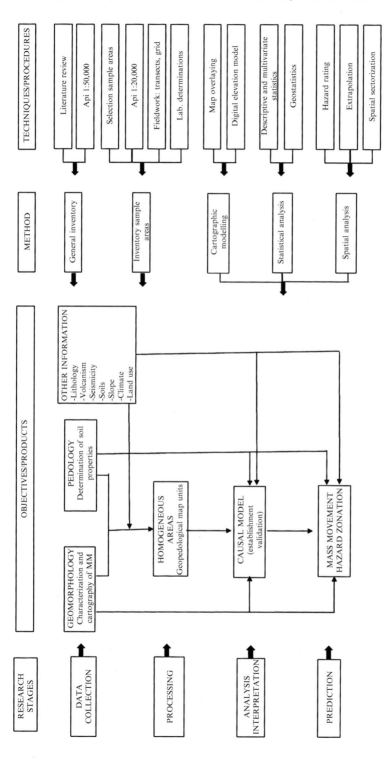

Fig. 25.1 Methodologic frame (*Api* aerial photo-interpretation, *MM* mass movements) (López and Zinck 1991)

The soil map was considered as a cartographic document integrating the four main mass movement conditioning factors: soil properties, relevant rock properties, slope gradients, and surficial morphodynamics. As a consequence, it was assumed that soil map units (i.e. geopedologic units) would be sufficiently homogeneous for the purposes of: (a) extrapolating detailed information from sample areas over the whole basin area; (b) incorporating additional information provided by other thematic maps (e.g. lithology, morphodynamics); and (c) allowing for mass movement hazard rating and zoning.

Cartographic modelling involved the integration of four thematic maps:

(a) Generalized soil map compiled from two original source maps;
(b) Morphodynamic map compiled from different sources and complemented with aerial photo-interpretation and field identification;
(c) Hybrid lithologic-geomorphic map obtained by extracting relevant information on rock types and cover formations from geologic and physiographic soil maps of the region;
(d) Slope gradient map derived from a digital elevation model (DEM). In a subsequent stage, the same DEM was implemented together with scarce ground climatic data, to create a mean annual rainfall distribution map and a mean annual temperature distribution map.

Maps were overlaid stepwise by crossing pairs of maps to allow for sequential validation of results. Through three successive overlay steps, the basic information of the soil map was refined or quantified (e.g. slope gradients). This incorporation of additional data contributed mainly to improving the phase information of the soil map units (e.g. slope, parent material, and erosion phases).

The causal model was tested first in the sample areas and then extrapolated over the whole catchment basin. To allow for extrapolation of the cause-effect relationships established in the sample areas, soil variability was assessed in terms of population variability and spatial variability (López and Zinck 1991). For the purpose of structural analysis, a dataset of 90 soil profiles, of which 36 were located in sample areas and the remainder spread over the rest of the basin, was submitted to statistical treatment to determine correlation levels and population distribution trends. Geostatistical analysis to determine soil spatial variability focused on andic materials (mainly Vitrandepts), since pyroclastic ash and lapilli are pervasive covers over most of the catchment.

25.4   Results and Discussion

25.4.1   Mass Movement Processes and Resulting Geoforms

Current or recent mass movements are frequent between 1,400 and 3,000 m asl. They affect without distinction all types of rocks, weathering products, and geoforms. Major occurrence is between 1,800 and 2,500 m asl, corresponding to the

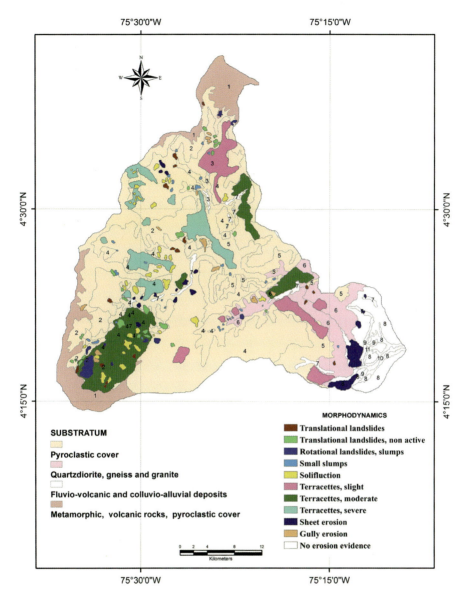

Fig. 25.2 Current and recent mass movements in relation to geologic substratum and soil cover; numbers correspond to soil units identified in Table 25.1

elevation zone with the highest rainfall (e.g. map unit 4 in Table 25.1) (Figs. 25.2 and 25.3).

Mass wasting processes result from the alteration of the mechanical properties of weathering materials by external factors. When the total and effective rainfall or the seismic activity deviates from its normal tendency, the behavior of soils and sapro-

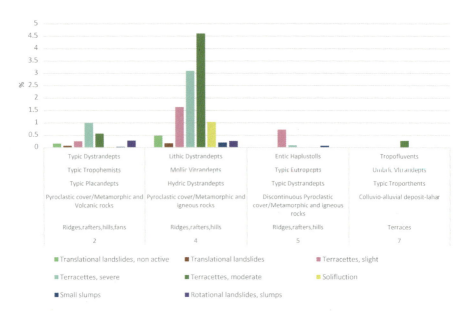

Fig. 25.3 Frequency of observed mass movements; numbers correspond to soil units identified in Table 25.1

lites results modified: they move slowly or quickly downslope in plastic or liquid solifluidal state, or by means of liquefaction. In general, soil properties together with topography and moisture conditions determine the susceptibility of unconsolidated cover formations to be affected by mass movements.

The main types of geoforms created by mass wasting processes present in the area are terracettes, solifluction tongues and lobes, and rotational and translational slides. Terracettes occur everywhere, without distinction of the type of rock substratum. In soil materials developed from granodiorite and gneiss, the landslide process takes place in plastic state. On volcanic ash, liquid solifluction is frequent. When the pyroclastic cover material is coarse-grained, liquefaction may occur.

Terracettes form parallel steps whose width and height vary between 30 and 80 cm and between 35 and 60 cm, respectively. Part of or the whole slope length may be affected. The grass cover is frequently broken along the microscarps separating consecutive treads, but the surface soil horizons rarely show visible discontinuity. Three degrees of severity of disturbance were established: (1) slight: weakly developed pattern and no rupture of the grass cover; (2) moderate: well developed pattern and less than 30 % of the grass cover disturbed; and (3) severe: well developed pattern and more than 30 % of the grass cover disrupted, often including local ruptures of the upper soil cover and even of deep-lying placic horizons.

Elongated, slightly undulating solifluction tongues are frequent in pyroclastic materials covering the Cajamarca Group, especially on concave slope facets. The forming process takes place in plastic state when the material is rich in clay, or in liquid state when the silt fraction dominates. Solifluction lobes, affected by radial crevices, are common in volcanic materials showing recurrent vertical variations in texture and consistence. Usually, the triggering effect is caused by the liquefaction of coarse-grained lapilli strata, flowing on impervious schist substratum (quicksand effect). Plastic solifluction occurs when the clay fraction is higher than 30 %. Liquid solifluction dominates when the clay fraction is below 20 % and the silt plus fine sand content is between 30 and 60 %. Finally, liquefaction affects materials with more than 60 % of medium and coarse sands.

25.4.2 Controlling Factors and Properties

Water dynamics in the soil cover strongly controls slope stability, because the water supply by effective rainfall exceeds the holding capacity of the soils during ten consecutive months per year. More than 80 % of the total pore space is occupied by water for long periods. Infiltration rates and hydraulic conductivity are low.

In general, soil horizon differentiation is weak, as reflected by the predominance of Inceptisols (Andepts and Tropepts), and does not have any significant function in inducing differential shear strength and material displacement, although this has been observed in other Andean regions (Zinck 1986). Only the placic horizon in Placandepts, when not deranged by previous mass movements acts as an impervious pan favoring water saturation and sliding of overlying strata.

Coarse pyroclastic cover materials, which are frequent all over the Coello river basin, originated from past explosive eruptions of the Machin volcano. Soils developed from those materials are younger than 9,000 years BP. Soil samples from buried A horizons taken at variable distance southwest of the Machín volcano provided the following dates: at 3 km, 3.6 m depth: $8,590 \pm 50$ BP; at 16 km, 2.2 m depth: $6,280 \pm 70$ BP; at 32 km, 0.6 m depth: $5,590 \pm 45$ BP (carbon-14 determinations at the Center for Isotope Research (CIO) of the University of Groningen, The Netherlands). It can be deduced from these data that volcanic eruptions are of relatively low recurrence.

In general, the chemical soil properties vary rather widely. However, cation exchange capacity and organic carbon content strongly correlate with liquid and plastic limits. The physical properties of 90 selected profiles, distributed over the whole catchment area, show normal distribution and therefore a high degree of homogeneity. The spatial variability of physical properties, belonging to soils derived from coarse-grained pyroclastic materials, as tested in a sample area of 30 observation points, resulted to be low, reinforcing the validity of the extrapolation procedure (López and Zinck 1991).

25.4.3 Mass Movement Hazard Zonation

The degrees of severity of mass movement hazards were estimated using a relative, qualitative rating scale of five levels. The factors taken into account were basically the intrinsic mechanical and physical properties of soil and saprolite materials, as modified by slope steepness and observed recent or past mass movement features. Transient factors such as abnormal rainfalls, earthquakes, volcanic activity, and human-induced risks, were not taken into account. The resulting model is thus more a mass movement susceptibility model than a dynamic hazard model (Fig. 25.4).

The following potential instability classes were established:

(a) None or uncertain: flat or slightly undulating areas where materials are intrinsically not susceptible and/or no mass movements were observed; areas with partially missing data were provisionally included in this class.
(b) Very low: materials slightly susceptible, slopes less than 7 %.
(c) Low: materials slightly susceptible, rare observed mass movement features, slopes 7–25 %.
(d) Moderate: materials moderately susceptible (mainly volcanic ash cover), common observed mass movement features, slopes 7–25 %.
(e) High: materials strongly susceptible (thick volcanic ash cover), frequent observed mass movement features, slopes 25–75 %.

Vitrandepts, Dystrandepts, and Eutrandepts, developed from volcanic ash and lapilli layers on slopes in the wet cold elevation zone, present the highest hazard for mass wasting. Andic Dystropepts and Eutropepts, of common occurrence where the volcanic ash and lapilli cover is discontinuous, in both cold and temperate life zones, are affected by moderate mass movement hazards. The same level of severity concerns the Dystropepts and Eutropepts formed on granodiorite and gneiss in the temperate zone. Dystropepts and Eutropepts without andic properties, Hapludolls, Troporthents, and Tropofluvents are much less susceptible to mass movements.

Geographically speaking, one of the areas most exposed to potential mudflows is the Coello river valley, directly threatened by recurring activity of the Nevado del Tolima volcano. In a similar setting in the same mountain range, a small eruption of the Nevado del Ruiz volcano in 1985 produced an enormous lahar that buried and destroyed the town of Armero in Tolima, causing an estimated 25,000 deaths.

25.5 Conclusions

Surficial mass movements, such as terracette formation, solifluction and landslides, are wide-spread in the Coello river basin, as well as elsewhere in the Colombian Andes. Most mass movements occur under grassland in the cold moist zones extending from 1,500–2,000 to 3,000–3,500 m elevation. Their spatial distribution is strongly controlled by the rock-soil-slope-land-use complex. Within this complex, soil materials play a fundamental role, especially through their physical and mechanical properties which determine degrees of mass movement susceptibility. Soil material is not only a conditioning factor; it is also the main natural resource damaged by mass movements.

By combining the properties and spatial distribution of individual soil (map) units, a mass movement hazard zoning map was generated. The resulting hazard

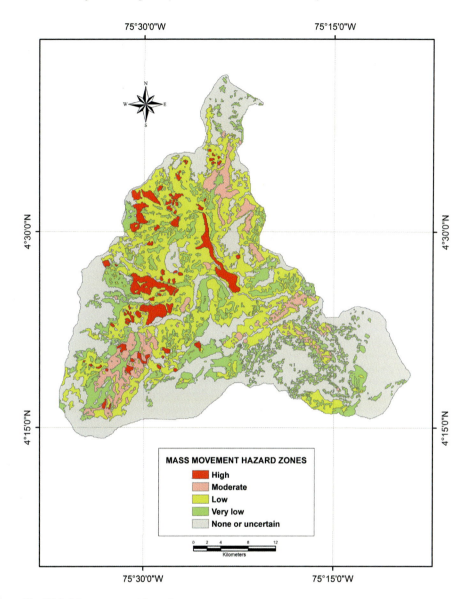

Fig. 25.4 Mass movement hazard zones

model focuses mainly on the spatial prediction of mass movements. Temporal prediction remains hypothetical because of the difficulty in determining the time recurrence of transient triggering factors such as abnormal rainfall, earthquakes, or volcanic eruptions.

Geopedologic maps can be reliably used for extrapolating mass movement hazards from point observations in sample areas, complemented by verifications outside, to whole catchment areas. Well defined soil map units may be homogeneous enough to allow cartographic extrapolation using the attributes of the taxonomic classes as information carriers.

References

Cepeda H, Murcia LA (1989) Mapa preliminar de amenaza volcánica potencial del Nevado del Tolima, Colombia Memoria V Congreso Colombiano de Geología tomo I. INGEOMINAS. Bogotá, pp 443–472

CIAF (1983) Estudio general integrado de geomorfología, suelos y socioeconómico de la cuenca superior del rio Coello. Centro Interamericano de Fotointerpretación, Bogotá

Hansen A (1984) Landslide hazard analysis. In: Brunsden D, Prior DB (eds) Slope instability. Wiley, New York, pp 523–602

INGEOMINAS (2001) Evaluación del riesgo por fenómenos de remoción en masa. Guía metodológica. Instituto Nacional de Investigaciones Geológico-Mineras, Escuela Colombiana de Ingeniería, Bogotá

López H, Zinck JA (1991) GIS-assisted modelling of soil-induced mass movement hazards: a case study of the upper Coello river basin. ITC J 1991(4):202–220

Margottini C, Canuti P, Sassa K (2013) Landslide science and practice, vol 1, Landslide inventory, susceptibility and hazard zoning. Springer, Berlin

Mora R, Chávez J, Vásquez M (2002) Zonificación de la península de Papagayo mediante la modificación del método Mora & Vahrson. Memoria del tercer curso internacional sobre microzonificación y su aplicación en la mitigación de desastres, Lima, pp 38–46

Servicio Geológico Colombiano (2013) Documento metodológico de la zonificación de susceptibilidad y amenaza por movimientos en masa escala 1:100.000, Bogotá

Soil Survey Staff (1975) Soil taxonomy. Agriculture handbook 436. US Government Printing Office, Washington, DC

Van Westen C, Rengers N, Terlien M (1997) Prediction of the occurrence of slope instability phenomena through GIS-based hazard zonation. Geol Rundsch 86:404–414

Varnes DJ (1984) Landslides hazards zonation, a review of principles and practice. UNESCO, Paris

Zinck JA (1986) Una toposecuencia de los suelos en el área de Rancho Grande. Dinámica actual e implicaciones paleogeograficas In: Huber O (ed) La selva nublada de Rancho Grande, Parque Nacional Henry Pittier. Fondo Ed Acta Cient Venez. Caracas, pp 67–90

Chapter 26
Geomorphic Landscape Approach to Mapping Soil Degradation and Hazard Prediction in Semi-arid Environments: Salinization in the Cochabamba Valleys, Bolivia

G. Metternicht and J.A. Zinck

Abstract Knowledge of the soilscape, i.e. the pedologic portion of the landscape, its characteristics, and composition helps understand the relationships between causes, processes, and indicators of land degradation. This chapter describes the application of the geopedologic approach to map land degradation caused by soil salinity, and to predict salinization hazard in the semi-arid environment of the Cochabamba valleys in Bolivia. In addition to providing a framework for generating soil information at sub-regional level, geopedology assisted in understanding top-soil spectral reflectance features of soil degradation, assessing soil salinity type and magnitude, and predicting salinity hazard.

Keywords Soil degradation • Salinization-alkalinization • Hazard prediction • Data synergy • Remote sensing

26.1 Introduction

Terrestrial life depends on land and yet, globally, land quality is declining, cropland is being lost through exhaustion, and soil fertility and productivity are decreasing (Bruinsma 2003; FAO 2011). According to the UNCCD (2011), 50 million people may be displaced within this decade as a result of land degradation in the world's

G. Metternicht (✉)
Institute of Environmental Studies, University of New South Wales, Sydney, NSW, Australia
e-mail: g.metternicht@unsw.edu.au

J.A. Zinck
Faculty of Geo-Information Science and Earth Observation (ITC), University of Twente, Enschede, The Netherlands

Institute of Environmental Studies, University of New South Wales, Sydney, NSW, Australia
e-mail: alfredzinck@gmail.com

© Springer International Publishing Switzerland 2016
J.A. Zinck et al. (eds.), *Geopedology*, DOI 10.1007/978-3-319-19159-1_26

drylands. Land degradation produces major externalities that are felt at regional and global scales, including dust storms, disruption of hydrological cycles, and greenhouse gas emissions. Furthermore, recent studies point to the interconnections between dryland degradation and climate change. Land degradation exacerbates and may itself be exacerbated by climate change. Land degradation depletes carbon stocks in vegetation and soils, increasing atmospheric carbon dioxide, while climate change is likely to increase temperatures and decrease rainfall in some regions such as Southern Africa (Boko et al. 2007; Cowie et al. 2011).

The state of knowledge about the extent, magnitude, and trends of land degradation globally is weak, and the lack of integrated approaches for monitoring and assessment has been identified as one of the main constraints to address land degradation (Vogt et al. 2011). Measurements on the extent of land degradation are uncertain, with estimates ranging from 15 to 63 % of Earth's land masses. Given the importance of the problem and its recognition as a global issue, it is surprising the lack of agreement to date on adequate methods for monitoring and assessment of this phenomenon, at global, regional and/or national scales.

Current practice in monitoring and assessment of land degradation at different scales were examined in Vogt et al. (2011). This chapter builds on findings of that review that called for more integrated, operational approaches to map and monitor land degradation. To this end, we discuss how the synergy of geospatial tools (i.e. remote sensing imagery and geographic information systems) and a geopedologic framework (Part I of this book) provide an effective approach to improve efficiency and accuracy in cartography and hazard prediction of soil degradation at regional and sub-regional scales. The Cochabamba, Sacaba, and Punata-Cliza valleys, in semi-arid central Bolivia, are used as case study. These valleys are affected by soil salinization, alkalinization, and accelerated erosion (Metternicht 1996). The chapter shows how geomorphic information can improve remote sensing-based mapping of land degradation, and guide strategic field observation and sampling for laboratory determination of soil properties and other landscape indicators relevant to map land degradation processes.

The methodology for deriving geopedologic information using the conceptual model described in Part I of this book is outlined. The chapter introduces key concepts related to soil salinity mapping and monitoring tools. It discusses the contribution of geopedology to: (a) improved understanding of topsoil spectral reflectance behavior, (b) enhanced understanding of soil salinity type and magnitude, and (c) salinity hazard prediction at regional and sub-regional levels.

26.1.1 Soil Degradation and Salinization

Soil degradation is the most critical component of land degradation, especially of irreversible land degradation leading to desertification. Soil degradation is a consequence of depletive human activities and their interaction with natural environments, resulting in soil quality decline. Three types of soil degradation are distinguished: physical

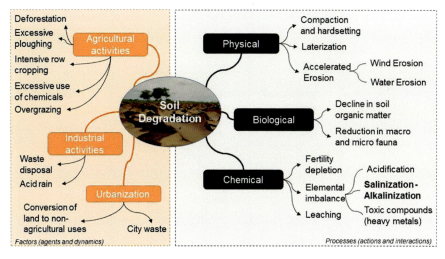

Fig. 26.1 Factors and processes of soil degradation (With information from Lal et al. 1989)

such as erosion and desertification by wind and water, biological like decline in soil organic matter, and chemical like soil acidification and salinization (Lal et al. 1989) (Fig. 26.1).

Salinization results from migration and dissemination of water-soluble salts from a source area to an area originally free of salt (Zinck and Metternich 2009). It causes accumulation of chlorides, sulphates, and carbonates of sodium, magnesium or calcium on the soil surface, in the subsoil, or in the groundwater. It can occur in soils, sediments, and porous rocks. A distinction is frequently made between primary salinization and secondary salinization. The former develops because of geologic materials (salt-bearing rocks), topography (low-lying areas, closed depressions, areas near the sea subjected to groundwater table fluctuations), and climate (low rainfall, high evaporation rates), whereas secondary salinization is mainly a human-induced process due to mismanagement practices. Human-induced salinization increases salt concentrations in soils already affected by natural salinity, or leads to contamination of salt-free soils because of inadequate water and land management (e.g. using brackish water for irrigation, or overgrazing).

Salt-affected soils form under the influence of ions in solid or liquid phase which alter the physical, chemical, and biological properties of soils. Water-soluble salts and compounds determine the processes leading to the formation of different types of salt-affected soils (Table 26.1).

26.1.2 Mapping Soil Salinity

Salinization alters physical, chemical, and biological properties of soils; therefore, changes affecting these properties can be used as indicators of land degradation status. For instance, deterioration of soil structure, soil surface crusting, salt content,

Table 26.1 Types of salt-affected soils and associated chemical and physical parameters

Salt-affected soils	Chemical indicators	Predominant anions	Predominant cations	Other properties	Main effects on the soil profile
Saline soils	EC >4dS/m SAR <13 pH <8.5	Chlorides, sulphates, and sometimes nitrates; small amounts of bicarbonates; carbonates absent	Na content not higher than half of the soluble cations; Ca and Mg in considerable amounts; K uncommon Sometimes gypsum and lime are also present	Generally flocculated; permeability equal or higher than that of similar non-saline soils white crusts on soil surface	Higher osmotic pressure
Alkaline (sodic) soils	EC <4dS/m SAR >13 8.5< pH <10	Chlorides, sulphates, and bicarbonates; carbonates in small amounts	Na dominant; K sometimes (exch and soluble); Ca and Mg in small amounts; at high pH and in the presence of carbonates, Ca and Mg precipitate	Organic matter dispersion and dissolution; clay deflocculation; columnar or prismatic structure	Change in structure and decrease in permeability and porosity; change in soil biological activity; pH increases beyond 9 or even 10
Saline-alkaline (sodic) soils	EC >4dS/m SAR >13 pH variable	If excess of salts: appearance and properties similar to saline soils (i.e. at pH <8.5, particles remain flocculated) If soluble salts are leached downwards: appearance and properties similar to sodic soils (i.e. at pH >8.5, soil particles are dispersed)			Soils become unsuitable for the entry and movement of water and for tillage

Modified from Richards (1954)

pH, sodium adsorption ratio, and soil electrical conductivity can be used to assess the state of land degradation caused by salinization. In the spatial context, information about the soilscape, i.e. the pedologic portion of the landscape, its characteristics and composition helps understand the relationship between causes, processes and best indicator sets to map and monitor land degradation, as discussed hereafter.

26.1.2.1 Indicators of Soil Salinization

Indicators are measurable physical, biological, or socio-economic characteristics that provide information about the state of a degraded landscape, drivers and pressures that determine current conditions, and impacts on soil quality. Common indicators of land degradation drivers and pressures are density of human settlements, deforestation, vegetation degradation, geomorphic position, and topography;

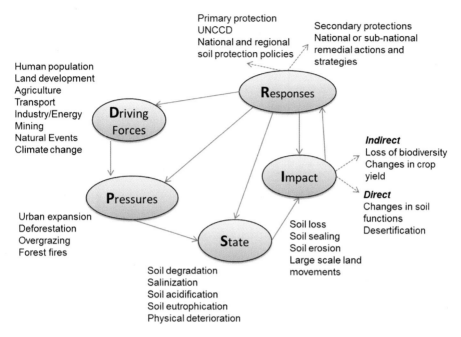

Human population
Land development
Agriculture
Transport
Industry/Energy
Mining
Natural Events
Climate change

Primary protection
UNCCD
National and regional
soil protection policies

Secondary protections
National or sub-national
remedial actions and
strategies

Responses

Driving
Forces

Indirect
Loss of biodiversity
Changes in crop
yield

Impact

Direct
Changes in soil
functions
Desertification

Pressures

Urban expansion
Deforestation
Overgrazing
Forest fires

State

Soil loss
Soil sealing
Soil erosion
Large scale land
movements

Soil degradation
Salinization
Soil acidification
Soil eutrophication
Physical deterioration

Fig. 26.2 Indicators associated to drivers, pressure, state, and impacts of soil degradation

whereas changes in soil organic carbon, surface crusting, saline efflorescence, soil structure, surface sealing, clay dispersion are indicators of status and impact of land degradation (Fig. 26.2).

Indicator definition is an essential step of any approach to map and monitor extent and rate of land degradation, independently of the mapping tool used to this end (i.e. remote sensing, field observations, laboratory determinations, or a combination thereof).

26.1.2.2 Tools for Mapping Soil Salinity Distribution

Conventional ways of determining and tracking changes in soil salinity in agricultural lands and rangelands are based on field observation and laboratory analysis, both of soils and vegetation. Remote sensing tools offer complementary data for assessing resource condition, and often constitute a less costly, more versatile and timely option to conventional monitoring and assessment, especially at regional and sub-regional scales (Metternicht and Zinck 2003).

Soils are mixtures of coarse primary minerals and organic fragments in the sand and silt fractions, and fine materials such as clay minerals and humus. The inclusion of rock fragments and folic material and different surface roughness conditions determine a complex reflectance response from the soil surface, with supplementary spatial and temporal variation patterns at multiple scales. Optical, infrared, and microwave remote sensing techniques exploit these patterns of energy interactions to derive relevant information about the physical and chemical characteristics of the soil surface (Metternicht 1996).

26.2   Methodology

This section introduces the main geologic, geomorphic, and environmental settings of the Cochabamba valleys, essential to frame the ongoing land degradation processes of the valleys, and to discuss the contribution of the geopedologic approach to mapping and hazard prediction.

26.2.1   The Study Area

The depressions of Cochabamba, Punata-Cliza, and Sacaba are located between 17°10′–17°40′S and 65°45′–66°30′W, at elevations ranging from 2500 to 3600 m asl (Fig. 26.3). The depressions are tectonic grabens filled with Quaternary sediments of lacustrine, glacio-lacustrine, and alluvio-lacustrine origin. The area belongs to the ecosystem of the mesothermic interandean valleys, with two distinctive morphologic environments: cordilleras and sedimentary basins formed during the Tertiary orogeny.

Landscape units are mountains, hillands, piedmonts, and valleys. Mountains are of structural or structural-denudational origin, and their topography reflects past

Fig. 26.3 The Cochabamba valleys: Central (*upper left*), Sacaba (*right*), and Alto (*lower right*). Landsat TM false color composite

glacial erosion. Moraine deposits, U-shaped valleys, and glacial lakes are relics of Quaternary glaciations. Slopes are very steep, promoting intense runoff, soil erosion, and deep river incision in the highlands.

Piedmonts are formed by glacis and fans of alluvial and colluvio-alluvial materials. Main relief types include: (a) dissected depositional glacis composed of coalescent alluvial fans; (b) old dissected fans of fluvio-glacial origin; (c) recent colluvio-alluvial fans; (d) active alluvial fans; (e) depositional glacis, a transition between the lagunary depressions in the center of the valleys and the recent alluvial fans of the piedmont; and (f) hills composed of marls, shales, siltstones, and sandstones, forming isolated relief types within the piedmont eroded by gullies.

Distal and terminal depositions, predominantly of lagunary facies, occur in the center of the depressions. Relief types and landforms include: (a) playas with saline soil surface during the dry season, that have formed in the lowest parts of the basins after drying up of Pleistocene lakes; (b) lagunary flats, slightly sloping to level landforms of alluvio-lagunary origin, located along the rims of the large depressions at the bottom of the valleys, forming transition between playas and alluvial fans or glacis; (c) badlands resulting from intensive dissection of the alluvio-lacustrine sediments composed of sandy, silty clay and clayey layers including lenses of coarser materials; and (d) alluvial terraces, located mainly along the Rocha river, in the Central and Sacaba valleys.

Climate is semiarid with mean annual temperature of 14–17 °C and mean annual rainfall of 400–600 mm. About 80 % of rainfall occurs from December to February, causing a moisture deficit during 8–9 months per year. The overall soil moisture regime is aridic, except in depressions and some flat areas that have ustic and sometimes aquic regimes evidenced by shallow groundwater table and mottling. In areas prone to salinization-alkalinization, salts tend to concentrate in the topsoil, and the leaching of soluble salts is restricted because of low rainfall.

The vegetation pattern varies according to the ecological conditions. The valleys belong to the subtropical lower montane thorn steppe ecosystem (Holdridge 1947). On mountain slopes with very shallow and stony soils, sparsely distributed native grasses and shrubs are found. On most alluvial fans, unfavorable to agriculture because of high amount of rock fragments on the soil surface and within the soil matrix, there are xerophytic species including cactaceae and shrubs such as locust and molle (*Schinus molle*) trees. Some alluvial fans of relatively recent formation, especially in the Central and Sacaba valleys, are being increasingly used for agriculture upon stone removal. Agricultural activities predominate on the lagunary flats and depositional glacis. Rainfed crops include corn, wheat, alfalfa and, to a lesser extent, beans, peas, onions, and quinua. Some of these crops, especially corn, are also produced under irrigation. Halophytic, salt-tolerant perennial vegetation grows in patches on the vast salt-affected areas of the Punata-Cliza valley, with distribution determined by the nature and degree of soil salinity (Metternich 1996).

26.2.2 Generating Geopedologic Information

Soil mapping combined three data sources: (1) visual interpretation of aerial photographs for cartography of geoforms and identification of surface degradation features, and digital processing of Landsat TM images to map land cover, land use, and indicators of soil degradation; (2) field observations and instrumental determinations of biophysical features, as well as observation of anthropic activities; and (3) laboratory determinations of mechanical, physical, and chemical soil properties. A soil geographic database was built up to store and process the data at different spatial scales. The methodology is described in Metternicht (1996), and summarized hereafter.

Because geomorphology is a relevant soil forming factor and soil mapping criterion, the geo-pedologic approach presented in Part I of this book guided field survey design using random-stratified sampling, to characterize different land degradation processes in piedmont and valley landscapes. Stratification was based on photo-interpreted geomorphic units, given that soil spatial distribution and variability are controlled by, among others, the geomorphic factor. Geomorphic units were the 'strata' within which soil samples were randomly located. Ancillary data such as existing soil observations, geological maps, vegetation reports, and meteorological records were assembled into the purpose-built soil geographic database aforementioned.

Field observations comprised the description of soil sites and surface dynamics, and reflectance measurements of the surface components. Point data were gathered from auger holes, mini-pits and some full pits. Modal profiles were described to check and improve existing soil maps. Horizon depth and designation, structure, color, texture, stoniness, porosity, and biological features were recorded according to the FAO guidelines for soil profile description (FAO 1990). Mini-pits were placed in the center of bare plots of approximately 90×90 m to account for location errors caused by positional inaccuracies of the GPS and/or the geometric correction of the Landsat TM image. Additional data were recorded concerning soil surface features such as crust color, texture and thickness, color and coverage percentage of surface rock fragments, and vegetation type and cover percentage.

Soil samples were collected at variable distances within selected geoform types exhibiting surface features related to salinization-alkalinization. A composite topsoil sampling scheme was adopted to account for soil spatial variability. Samples 5 cm deep, within a perimeter of 7 m from the central observation point, were collected for laboratory determinations (pH, electrical conductivity, soil particle size distribution, ion types and content) using a stratified-oriented sampling technique. These data were integrated with digital image classification of Landsat TM and JERS-1 SAR imagery acquired contemporaneously to fieldwork activities, to map saline-alkaline areas. Several image classification approaches were designed and tested using different regions of the spectrum (visible, infrared, thermal, microwaves), different classification schemes (statistical pattern recognition, fuzzy sets) and different soil parameters defining the information classes (pH, electrical conductivity, chloride to sulphate ratio, and carbonate to sulphate ratio), as described in Metternicht (1996).

26.3 Results: Value-Adding Geopedologic Information

The geopedologic maps of the three valleys are presented in Metternicht (1996). One example representing the Punata-Cliza valley is shown in Fig. 4.2 (Part I, Chap. 4) together with the corresponding geopedologic legend in Fig. 4.3 that summarizes the map unit types, pedotaxa, inclusions, and phases.

Because of the regionally dominant aridic regime, most soils classify as Aridisols or Entisols in the absence of diagnostic features (Soil Survey Staff 1996). Buried Alfisols and Mollisols, reflecting past more humid climatic conditions, occur in the Sacaba and Central valleys under loamy or silty loam recent alluvial covers. Details on soil types and characteristics can be found in Metternicht (1996).

Hereafter the value-added information provided by geopedology for mapping and monitoring soil degradation caused by salinization in the Cochabamba valleys is analyzed and discussed. In addition of providing a framework to generate sub-regional soil information, geopedology assisted in understanding topsoil spectral reflectance features of soil degradation, assessing soil salinity type and magnitude, and predicting salinity hazard.

26.3.1 Improved Understanding of Topsoil Spectral Reflectance Related to Soil Degradation

Understanding the relationship between the spectral reflectance of soil properties and surface components used as indicators of land degradation is essential to accurately identify and map areas prone to land degradation processes by means of remote sensing techniques.

Spectral similarity in the visible and infrared regions of the spectrum is a common criterion to group pixels in homogeneous areas during digital classification of air- or satellite-borne imagery. However, surface features related to degraded or non-degraded areas may exhibit similar spectral responses, causing inconsistencies in detection. For instance, organic matter, soil moisture, and soil surface roughness (e.g. stoniness) produce similar reflectance in the visible and near-infrared, that is, when these soil properties increase, reflectance decreases (Stoner and Baumgardner 1981).

The inclusion of geopedologic contextual information in the image classification process, for instance to define land cover categories, or mask out landforms that are not affected by land degradation, can resolve some issues of spectral similarity as shown in Fig. 26.4. Geomorphic units can be used to create a multi-layer classification of remotely sensed imagery, based on techniques like tree-classifiers, fuzzy-based classifications, and artificial neural networks. In a post-classification procedure, the

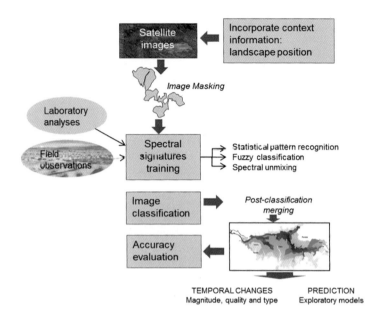

Fig. 26.4 Incorporation of landscape information, prior to image classification, as a means to reduce spectral confusion (Metternicht 1996)

different layers can be merged in a GIS to derive a final map. This approach was trialed in the Punata-Cliza valley and enabled reducing sources of spectral confusion between surface features related to different land degradation processes, improving the classification accuracy of salinity-affected areas.

26.3.2   Enhanced Understanding of Nature and Magnitude of Soil Salinity

The geopedologic approach, supported by field observations and laboratory determinations of anion percentages and composition, enabled to differentiate and map dominant salt types in the Punata-Cliza valley, and relate them to different landscape positions. Figure 26.5 shows that carbonates and bicarbonates dominate in higher lagunary flats. This might be attributed to the kind of alluvial sediments these geoforms received from surrounding calcareous uplands. Carbonates decrease towards the center of the valley, being replaced by sulphates in the lower lagunary flats, and by similar proportions of chlorides and sulphates in the lacustrine-lagunary playas. Playas have the highest salt concentration, while lagunary flats in relatively higher topographic position have much lower salt contents (i.e. 0.01–0.5 %).

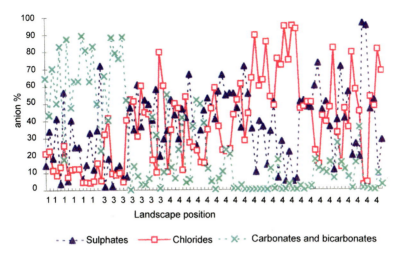

Fig. 26.5 Percentage anion composition of sample topsoils on different landscape positions: (*1*) higher lagunary flats, (*3*) lower lagunary flats, (*4*) playas (Metternicht 1996)

Multi-source cartographic information was overlaid, including a salt- and sodium-affected soil surface map derived from remotely sensed imagery (i.e. Landsat TM and JERS-1 SAR) and the geopedologic map of salt types based on field and laboratory data (e.g. SAR, pH, and EC). GIS-based data integration shows strong relationships between geomorphic positions and nature and magnitude of salinization (Fig. 26.6 and Table 26.2). Non to slightly saline-alkaline areas occur on the higher lagunary flats in the southern part of the valley, where coarser carbonate-rich alluvial deposits overlay lacustrine materials. The dominance of free carbonates and bicarbonates over chlorides and sulphates results in pH values of less than 8.5, as shown by laboratory analyses of soil samples.

Most degraded areas, classified as very strongly saline-alkaline, very strongly saline and slightly alkaline, very strongly alkaline and moderately saline, and strongly alkaline, occur in the clayey, low-lying playas. Moderately alkaline, non to slightly saline areas dominate on middle lagunary flats, the margins of high lagunary flats, and the distal parts of the piedmont glacis in the south of the valley.

Soil parent material, geomorphic position, and semi-arid climate that reduces salt-leaching and causes the migration of soluble salts towards the surface horizons, are the main factors controlling the spatial distribution of salt- and sodium-affected areas in the Punata-Cliza valley. Most degraded areas are located in low-lying playa landforms. Non to slightly saline-alkaline areas coincide with lagunary flats in relatively higher topographic position and with proximal piedmont glacis. Salinity levels are lowest in piedmont areas covered by alluvial deposits coming from eroded claystones, shales, and calcareous sandstones of the surrounding mountains.

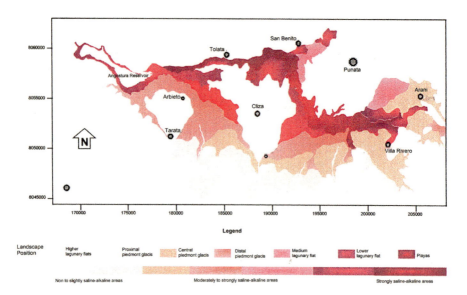

Fig. 26.6 Salinization types and magnitude, and their relation to landforms in the Punata-Cliza valley (Metternicht 1996)

Table 26.2 Distribution of soil salinity-alkalinity per geomorphic units

Salinity-alkalinity classes	Dominant geomorphic units	pH range	EC range
1. Non- to slightly saline-alkaline	Higher lagunary flats	7–8	0–8
2. Moderately and strongly saline-alkaline	Middle and lower lagunary flats	8.1–9.5	8.1–32
3. Very strongly saline-alkaline	Playas	>9.5	>32
4. Moderately and strongly saline, slightly alkaline	Middle and lower lagunary flats	7–8	8.1–32
5. Very strongly saline, slightly alkaline	Playas	7–8	>32
6. Very strongly saline, moderately and strongly alkaline	Playas	8.1–9.5	>32
7. Moderately alkaline, non- to slightly saline	Middle lagunary flats and piedmont glacis (distal)	8.1–8.5	0–8
8. Very strongly alkaline, moderately and strongly saline	Playas	>9.5	8.1–32
9. Strongly alkaline, non- to slightly saline	Playas	>9.5	0–8

26.3.3 Guiding Salinization Hazard Prediction

Geopedology and geospatial technologies were combined to design of a fuzzy knowledge-based exploratory model to predict salinization hazard (Fig. 26.7). The model assesses the nature, magnitude, rate, and reliability of soil salinity changes that took place in the period 1986–1994, and uses that information to predict hazard trends (Metternicht 2001). The model integrates the results of three maps depicting likelihood, nature, and magnitude of changes. It displays the areas exposed to increasing salinization as well as the areas already affected by the process. It incorporates 'prior knowledge' in the form of 9-year retrospective information on the dynamics of the degradation process derived from classified satellite imagery. Combining remote-sensed and geopedologic information, the model predicts likely salinization expansion to areas having similar parent materials and landform types.

GIS-based spatial analysis was conducted to ascertain spatial agreement between (1) salt concentration and composition in relation to landscape positions, (2) distribution of salt types and salinity degrees, and (3) cartography of anions, pH, and electrical conductivity derived through geostatistical interpolation of topsoil properties (described in Metternicht 1996). High multi-source data agreement simulated "expert elicitation" (i.e. a prediction factor) to infer likelihood of salinity expansion (Fig. 26.8).

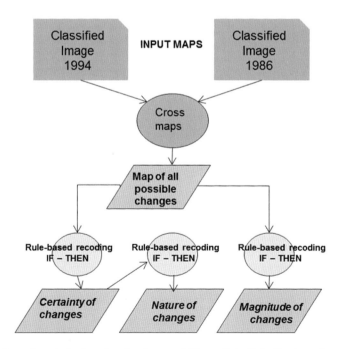

Fig. 26.7 Fuzzy knowledge-based exploratory model to predict salinization hazard

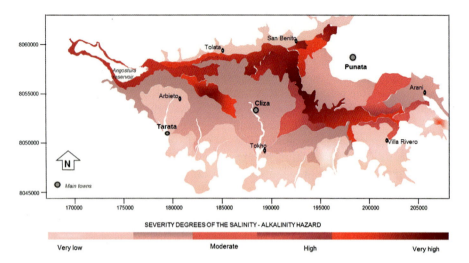

Fig. 26.8 Salinity hazard prediction in the Punata-Cliza valley (Metternicht 1996)

The exploratory model shows an overall trend towards increasing alkalinity, particularly in the higher lagunary flats (Fig. 26.8). This is attributed to the nature of the geologic deposits originating from the carbonate-rich southern mountains (calcareous sandstones). Soils of the higher lagunary flats are mainly composed of carbonate anions, although total anion concentrations are low. Therefore, middle as well as higher lagunary flats are threatened by potential expansion of salinization-alkalinization. The hazard refers not only to the geographic spreading of saline-alkaline areas, but also to its intensification in areas already degraded, particularly if dryness conditions accentuate in the basins.

26.4 Conclusions

This chapter shows how geopedology enables analyzing relationships between landscape position, lithology, and land degradation processes. Using the Punata-Cliza case study, it develops a framework that integrates salinity data derived from geopedologic survey and satellite imagery analysis in a GIS context to depict salinity distribution, distinguishing salt types and severity levels that revealed to be strongly related to specific geoforms and their soil contents. This information was used to identify spatial trends of soil salinization towards geomorphic positions potentially exposed to become salt-affected.

The information on soil degradation trends and hazards generated through the synergy of geospatial technologies and a geopedologic approach is important for land use planning and land management practices that can address and halt ongoing land degradation processes.

References

Boko M, Niang I, Nyong A, Vogel C, Githeko A, Medany M, Osman-Elasha B, Tabo R, Yanda P (2007) Africa. In: Parry ML, Canziani OF, Palutikof JP, van der Linden PJ, Hanson CE (eds) Climate change 2007: impacts, adaptation and vulnerability. Contribution of Working Group II to the fourth assessment report of the Intergovernmental Panel on Climate Change. Cambridge University Press, Cambridge, pp 433–467

Bruinsma J (2003) World agriculture: towards 2015/2030. An FAO perspective. Available at: http://www.fao.org/fileadmin/user_upload/esag/docs/y4252e.pdf. Accessed 18 May 2015

Cowie A, Penman T, Gorissen L et al (2011) Towards sustainable land management in the drylands: scientific connections in monitoring and assessing dryland degradation, climate change and biodiversity. Land Degrad Dev 22:248–260

FAO (1990) Guidelines for soil description, 3rd edn. FAO, Rome

FAO (2011) The state of the world's land and water resources for food and agriculture (SOLAW): managing systems at risk. Food and Agriculture Organization of the United Nations, Rome and Earthscan, London, 285 pp. Available at: http://www.fao.org/docrep/017/i1688e/i1688e.pdf

Holdridge LR (1947) Determination of world plant formations from simple climatic data. Science 105:367–368

Lal R, Hall G, Miller F (1989) Soil degradation: basic processes. Land Degrad Rehabil 1:51–69

Metternicht G (1996) Detecting and monitoring land degradation features and processes in the Cochabamba Valleys, Bolivia. A synergistic approach. PhD thesis, University of Ghent, Belgium. ITC publication 36, Enschede, The Netherlands. Available at: http://www.itc.nl/library/papers_1996/phd/metternicht.pdf

Metternicht G (2001) Assessing temporal and spatial changes of salinity using fuzzy logic, remote sensing and GIS. Foundations of an expert system. Ecol Model 144:163–179

Metternicht G, Zinck JA (2003) Remote sensing of soil salinity: potential and constraints. Remote Sens Environ 85:1–20

Richards L (ed) (1954) Agriculture handbook 60. US Department of Agriculture, Washington, DC

Soil Survey Staff (1996) Keys to soil taxonomy, 7th edn. US Department of Agriculture, Washington, DC

Stoner ER, Baumgardner MF (1981) Characteristic variations in reflectance from surface soils. Soil Sci Soc Am J 45:1161–1165

United Nations Convention to Combat Desertification (UNCCD) (2011) Desertification, land degradation and drought: some global facts and figures. Available at: http://www.unccd.int/Lists/SiteDocumentLibrary/WDCD/DLDD%20Facts.pdf. Accessed 3 June 2015

Vogt J, Safriel U, Von Maltitz G, Sokona Y, Zougmore R, Bastin G, Hill J (2011) Monitoring and assessment of land degradation and desertification: towards new conceptual and integrated approaches. Land Degrad Dev 22:150–165

Zinck JA, Metternicht G (2009) Soil salinity and salinization hazard. In: Metternicht G, Zinck JA (eds) Remote sensing of soil salinization: impact on land management. CRC Press, Boca Raton, pp 3–18

Chapter 27
Geopedology and Land Degradation in North-West Argentina

J.M. Sayago and M.M. Collantes

Abstract The subtropical region of north-western Argentina shows marked land degradation resulting from unrestricted use of ecosystems during the last centuries. From a geopedologic perspective, the land historical occupation and the distinctive features of the relief-soil relationship in a representative area, the province of Tucumán, are described in this work. For assessing the intensity and distribution of soil erosion in the region, soil potential loss is mapped at small scale, based on geomorphic sectorization and criteria of Universal Soil Loss Equation. Models of land erosivity are developed in two scenarios of future climate change from extreme rainfall values recorded over the last century. The assessment of erosion hazard at small scale using USLE, teledetection and geographic information system, helps develop programs oriented to the recovery of extensive degraded and desertified regions, through management systems adapted to current and future environmental conditions.

Keywords Relief-soil • Soil erosion • Erosion hazard • Climate change • Soil loss scenarios

27.1 Introduction

In a world exposed to socio-economic and environmental crisis and climate change, it seems appropriate to address the issue of land degradation in one of the least developed regions of a developing country. Paradoxically, the Argentine north-western region was the first territory colonized by the Spanish Crown, with flourishing economy during the first centuries of the Conquest.

The main ecosystems of the region, including the Yungas cloudy forest, the Chaco forest, the region of "El Monte" and the alto Andean steppe (Cabrera 1976),

J.M. Sayago (✉) • M.M. Collantes
Instituto de Geociencias y Medio Ambiente, Universidad Nacional de Tucumán,
S.M. de Tucumán, Argentina
e-mail: jmsayago@arnet.com.ar; mmcollantes@arnet.com.ar

© Springer International Publishing Switzerland 2016   441
J.A. Zinck et al. (eds.), *Geopedology*, DOI 10.1007/978-3-319-19159-1_27

are highly fragile as a consequence of a climate with strong seasonal variations, loess-developed soils, and an irregular relief with elevations higher than 5000 m.a.s.l. to the west and a flat plain to the east. Unrestricted and uncontrolled current land management is causing accelerated deterioration. This is in contrast with the careful management applied to natural ecosystems by native cultures that occupied the territory before the arrival of the Europeans (Caria et al. 2001). From a paleoclimatic perspective, the dry-wet subtropical landscape started developing in the early last millennium, contemporaneously with the extreme aridity of the Warm Medieval Period (Garralla 1999; Caria and Garralla 2003).

Since the arrival of the Spanish conqueror Francisco de Almagro in 1536, at the beginning of the colonial period (sixteenth to eighteenth centuries), the region was the annual wintering place of thousands of mules and horses (Niz 2003) used in mining activities in Chile, Bolivia (especially Potosí) and Perú. This was the beginning of intensive landscape degradation, first through deforestation, followed by the progressive disappearance of natural pastures (replaced by xerophytic species) in response to overgrazing. During the following centuries, deforestation expanded with the construction of a railway to the Andean mining area, particularly during World Wars I and II as the coal supply from Europe had to be replaced by the wood of carob trees (*Prosopis sp.*) and quebracho forests as fuel for steam locomotives (Sayago 1969). Native forests in the north-western region continued disappearing without interruption during the second half of the twentieth century because of rainfall increase that turned a large part of the region suitable for cereal cropping (Busnelli et al. 2009). Accelerated clearing of the western Chaco forest (about 2 million ha) for soybean and corn cultivation metamorphosed the primitive landscape.

Despite that in the late twentieth century three laws were enacted to protect natural forests, deforestation continued more intensively due to high international prices for soybean and corn. More than two million ha of native forest have been cleared in the last 10 years with variable impact (Secretaría de Medio Ambiente 2012). Assessment of environmental degradation is consequently imperative for the implementation of land conservation programs that will contribute to mitigate or neutralize such a situation.

This background accounts partly for the aim of this work, which is the application of geopedology to the inventory and assessment of land degradation in an extensive territory of the Argentine north-western region. The formation and evolution of landforms are described in their relationship with the soil cover in the Province of Tucumán. The erosion hazard under the current climate is evaluated and the impact of future climate changes on the surface geodynamics and land degradation is assessed through two scenarios based on extreme values of regional rainfall variability.

27.2 Materials and Methods

Geomorphology constitutes the structuring factor of the pedological landscape (Zinck 2012). In this sense, geomorphology covers a large part of the physical framework of soil formation through relief, morphodynamics, morphoclimatic context, non-consolidated or altered materials that serve as parental material to soils, and the time factor. Likewise, Jungerius (1985) states that the preparation of geomorphic and soil erosion maps substantially benefits from pedology contribution.

Zonneveld (1983) analyzes in depth the problem of the interaction of landscape elements and highlights the importance of applying geoecology concepts. He establishes a hierarchy of environmental factors based on the higher or lower capacity of some of them to influence unilaterally the others without being affected reciprocally by them. Although relief and climate have an independent position in the geoecological dynamics, the intensity and character of such an influence depend on the scale or level of perception at which it is considered. Climate is dominant at continental level (atmospheric circulation), but its influence is modified by the distribution of seas and mountain masses. At regional scale, although relief conditions climate (exposure, rainfall shadow, etc.), the latter influences directly or indirectly (arid, periglacial, subtropical morphogenesis, etc.). At local level, the influence of climate and relief depends on the landscape endogenous interrelationships. Thus, relief and climate show an ambivalent relationship with the remaining landscape elements and the prevalence of one or the other depends on the scale taken into consideration (Tricart 1982; Zonneveld 1983).

This work contributes methodologically to the inventory of water erosion processes, considering that mapping at a small scale of the relief-soil relationship constitutes the foundation for future actions. The inclusion of geomorphology as a conceptual and cartographic basis in erosion mapping involves the recognition of the essentially morphodynamic character of every degradation process. The relief classification applied in this work (Fig. 27.1, Tables 27.1 and 27.2) to obtain a partitioning of the Tucumán Province into geomorphic units, as a basis for the assessment of relief-soil relationships and potential erosion, is tentative, coinciding in its philosophy with Sayago (1982) and Zinck (2012).

The classification categories are as follows:

(a) Geomorphic province: it coincides with the generalized concept of geological province (Rolleri 1975), that is, an area characterized by a determined stratigraphic succession, a structural character of its own, and peculiar geomorphic features, the expression of a determined geological history (Puna, Cordillera Oriental, etc.) as a whole.

(b) Geomorphic region: territory characterized by a distinctive morphostructural style, defined by the recurrence of lithologic and morphogenic features developed during the Quaternary (Ancasti Range, Aconquija Range).

(c) Geomorphic association: defined as a part of a region, determined by the recurrence of typical morphogenic units conditioned by climate, which can be identified and mapped on aerial photographs and/or satellite images (Aconquija wet subtropical piedmont plain, constituted by fluvial valleys, erosion glacis, and alluvial paleo-fans).

(d) Geomorphic complex: it exhibits the same structure as the geomorphic association, but cannot be easily mapped due to the presence of dense vegetation cover or complex and spatially variable fluvial network (oriental slope of Calchaquí summits covered with vegetation, constituted by fluvial valleys, covered glacis, and covered slopes).

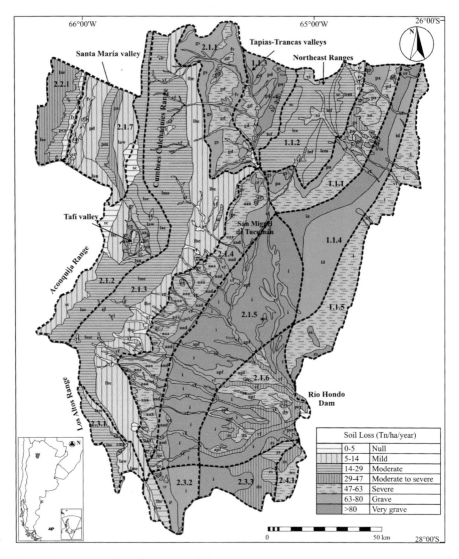

Fig. 27.1 Current soil erosion map (tn/ha/year), Tucumán province, Argentina (After Busnelli et al. 2009) Numbers (e.g. *1.1.1*) refer to the geomorphic map units as shown in Table 27.1

(e) Geomorphic unit: characteristic relief shapes, defined by a particular morphogenesis (alluvial fan, fluvial terrace, landslide, etc.) and mappable only at large scales (1:50,000 or more).

Erosion hazard was surveyed using GIS up to the level of geomorphic unit and delimited in every relief association. When considering the nature and causes of land degradation, different approaches influence the methodology used to represent what could be considered a complex environmental problem, although according to

Table 27.1 Geomorphic provinces, regions, and associations (legend of Fig. 27.1)

1. Geomorphic province subandean ranges
1.1. Geomorphic region northeast sierras (Medina, del Campo, and La Ramada)
1.1.1. Association humid subtropical eastern hillslope
1.1.2. Association humid subtropical mountainous sector
1.1.3. Association semiarid western hillslope
1.1.4. Association dry/wet subtropical floodplain
1.1.5. Association subtropical dry floodplain
2. Geomorphic province pampean ranges
2.1. Geomorphic region Calchaquíes and Aconquija ranges
2.1.1. Association subhumid to semiarid eastern hillslope of Calchaquíes range
2.1.2. Association dry-cold eastern hillslope of Aconquija and Calchaquíes ranges
2.1.3. Association humid subtropical eastern hillslope of Aconquija and Calchaquíes summits
2.1.4. Association humid subtropical piedmont plain of Aconquija range
2.1.5. Association subtropical dry-wet alluvial plain
2.1.6. Association subtropical dry alluvial overflow plain
2.1.7. Association western arid hillslope of Aconquija and Calchaquíes ranges
2.2. Geomorphic region Quilmes
2.2.1. Association arid eastern hillslope
2.3. Geomorphic region Ancasti/Los Altos range
2.3.1. Association cold subhumid summit
2.3.2. Association subtropical dry-wet subtropical hillslope
2.3.3. Association northern subtropical dry alluvial plain
2.4. Geomorphic region Guasayán range
2.4.1. Association western subtropical dry hillslope

Imeson (2012) what is more important are "people's actions, what they do and the relationships between them".

The use of quantitative criteria such as those of USLE (Universal Soil Loss Equation), developed by Wischmeir and Smith (1978) to assess erosion intensity at land parcel level, was updated by Sayago (1985) for the inventory of erosion hazard at small scale in the subtropics of Argentina, coinciding with Wischmeier (1984) in that USLE can be applied at small scale on the basis of proper geomorphic sectorization and the intensive use of visual and digital teledetection.

Quantification of soil erodibility by the potential loss in tn/ha/year is reflected in the behavior of all the USLE factors: rain erosivity, soil erodibility, slope length and gradient, vegetation cover, and conservation management (Wischmeier and Smith 1978; Renard et al. 1991). Rain erodibility is a key factor in the determination of soil loss by erosion because it relates to storm energy in a determined time span rather than to the total volume of rainfall. The difficulty to estimate the E130 index of the formula developed by Wischmeier and Smith (1978) in large areas, due to the lack of rainfall data, can be replaced by the index obtained by Arnoldus (1978) based on early work by Fournier (1960). This index has the advantage to use simple meteorological data (rainfall) and a good correlation with the measured values of

Table 27.2 Geomorphic
units (legend of Fig. 27.1)

Fluvial valley	vf
Fluvial terrace	tf
Valley bottom	fv
Interfluve area	i
Apical interfluve	ia
Distal interfluve	id
Interfluve with structural control	ice
Saline area	ed
Flooded depression	da
Perilake	per
Older floodplain	apf
Erosion glacis	ge
Covered glacis	gc
Upper	gs
Lower	gi
Dissected	gd
Glacis cone	cg
Undifferentiated piedmont	p
Apical	pa
Middle	pm
Distal	pd
Piedmont dominated by alluvial fans	pca
Alluvial fan	aa
Apical	aaa
Distal	aad
Residual hill	cr
Intermountain valley	vi
Stepped hillslope	rl
Summit surface	sc
Denudational hillslope	ld
Slopes: strong	lf
Moderate	lm
Gentle	ls
High sector	la
Middle sector	lm
Low sector	lb
North orientation	ln
South orientation	ls
West orientation	lo
East orientation	le
Dam	emb

the E130 index of USLE. Arnoldus (1978) established the general correlation equation $R = a \times FAO$ index + b, where R is the USLE rain erosivity factor and "a" and "b" are constants based on regional climatic conditions. This equation was tested in different parts of the world, showing high correlation with the USLE rain erosivity index in the USA, whose climatic characteristics range from arid/semiarid to humid subtropical.

Soil erodibility was assessed using the Wischmeier and Smith (1978) nomogram with soil data from laboratory and field to estimate the percentage of very fine sand and silt, organic matter content, permeability, and structure, the K factor values being obtained in every geomorphic unit. The LS factor, a combination of slope gradient and length, was estimated from a DEM (digital elevation model).

According to Imeson (2012), the vegetation cover (C) in the USLE equation is by far the most significant and critical quantitative term, but quite easy to estimate. The estimation of this factor was based on the separation of cultivated areas and natural vegetation. The cover in cultivated areas was measured in three stages, including plowing, emergence, and pre-harvest; the natural cover (mulch) of the wet and dry seasons was averaged.

The USLE management factor has major importance in the preservation and recovery of eroded areas, especially in regions of intensive agriculture in developed as well as in developing countries. The assessment of climate changes was carried out from extreme rainfall values recorded in the region (Torres Bruchmann 1977; Bianchi and Yañes 1992; Minetti 1999; Bianchi and Cravero 2010). Two scenarios were established: a wet scenario with rainfall 30 % higher than the average, and an arid scenario with 30 % less rainfall than average.

27.3 Results

27.3.1 Soil-Landscape Relationships

It has been said "every soil is a landscape" as an expression of the close relationship existing between relief and soil in their genesis and evolution (Birkeland 1999). Every geomorphic unit, delimited based on coherent taxonomic criteria, shows a spatial homogeneity given by the recurrence of shapes and endogenous processes, thus constituting itself a basic unit of soil/landscape (Sayago 1982).

In the undulating plain to the east of Tucumán province, Mollisols and Entisols predominate, whose moderate development on the loessic or detritic substratum responds to rainfall scarcity in winter, reflected in the Chaco forest vegetation, partly replaced by annual crops. To the southeast, in the depressed plain, the presence of Entisols on fluvial deposits and Mollisols with aquic and sodic characteristics reflects the persistence of past and current fluvial actions. Precipitation increase to the west due to a "rain shadow" process coincides with the appearance of more developed Mollisols, promoting significant agricultural activity in the alluvial plain without water deficit, in the Aconquija Range piedmont, and in the southern Sub-Andean Sierras. The wet subtropical climate of the eastern slope and part of the summit areas of the mountain ranges contributes to the development of Inceptisols under the Yungas perennial cloud forest. In the summit areas and western slopes of the Aconquija and Calchaquí ranges, cold climate and precipitation decrease intensify cryoclastism and mass movement, which accounts for the presence of Regosols and Entisols.

In the western piedmont of the mountain ranges, rainfall scarcity causes the presence of shrub communities with giant cacti, Regosols and Aridisols, and intense alluvial/torrential dynamics. At the bottom of the Santa María Valley, dryness accounts for the development of sodic and natric Aridisols, Entisols and Alfisols of fluvial origin, while on the eastern slope of the Quilmes Range, Entisols of lithic and detritic origin predominate. At last, in the Tapia-Trancas basin, the semiarid climate together with the subtropical lower mountain forest and the Chaco forest contributes to weakly developed Entisols and Mollisols that cover piedmont glacis and lower valleys. Similar soils occur also in the south-eastern tip of the province on the north-western piedmont of the Guasayán Range.

Modifications produced by deforestation during the last century have created a soil-landscape metamorphosis in the areas incorporated to intensive agriculture. In brief, it is a priority to assess land use and land occupation for determining the types of management best suited to the current and future environmental and socio-economic conditions.

27.3.2 Erosion Hazard Assessment

The meaning of the term "soil erodibility" differs from "soil erosion". The volume of soil loss through erosion may be controlled more by slope, cover or management than by the intrinsic soil properties. However, some soils erode more than others, even though all the other factors are similar (Bergsma 1986).

In the east of the Tucuman territory, rainfall erosivity is relatively low, but increases gradually towards west together with the "rainfall shadow" effect in the pre-Andean ranges. In the uplands, the adiabatic influence causes rainfall to progressively decrease together with higher elevation to the west. Orographic rain has as a consequence rainfall reduction and aridity in the western valleys. In the oriental plains, sheet erosion predominates due to unrestricted cultivation or overgrazing in relatively poor soils, previously covered by the Chaco forest (Fig. 27.1).

In areas with irregular relief, gully erosion predominates and soil loss is mostly related to concentrated overland flow and sediment transportation to riverbeds, canals or dams. Extensive low-gradient slopes account for sheet erosion caused by hortonian overland flow or top saturation overland flow (Bergsma 1986). Both surface erosion processes are closely related to soil characteristics. In the east of the Tucuman plain, soils show uniform permeability: the longer the slope, the higher the overland flow and, consequently, the higher the erosion. More developed soils have usually dense subsurface horizons (Bt, claypan, etc.) that influence the over-land flow/infiltration relationship and the surface horizon saturation time, causing stronger erosion hazard (Bergsma 1986).

The influence of the vegetation (C factor), either natural vegetation or crops, is directly correlated with soil use and cover values. The eastern side of the mountain region with subtropical cloud forest has dense cover, thus low C values (<0.1). In contrast, the intermountain valleys have relatively high values (0.21–0.25), reflecting intense and long-standing agricultural activity. To the west, deciduous forests on relief summits have slightly higher C values (0.11–0.15) than those of the Yungas cloud forest (0.01–0.05). To the east, in the undulating alluvial plain where the Chaco forest has not been totally cleared, C values are moderate (0.16–0.20). The simplicity and effectiveness of the vegetation cover measurement according USLE make it a useful tool to evaluate the "tipping point" process (Scheffer 2010) or eco-system landscape collapse due to extreme soil degradation. In the Santa María valley, located in the west of the study area, erosion hazard was determined using USLE in every relief unit. Vegetation in some relief units did not show changes of mulch cover between winter and summer due to heavy soil deterioration that prevented ecosystem recovery and resilience (Sayago et al. 2012).

North-western Argentina is an important agricultural region, without systematic and generalized use of soil conservation practices. The criteria used to assess the M factor (management) in dry areas may help evaluate the proximity to the landscape collapse point or "tipping point" or to test the effectiveness of changes in management systems to attenuate desertification. The erosion hazard values of the Argentine subtropical region are similar to the erosion classes established by El Swaify (1977) for Hawaii, where the erosion hazard values are higher than those normally measured in non-tropical regions.

27.3.3 Soil Loss Prediction Under Future Climate Changes

Climate changes resulting from "greenhouse effect" constitute one of the most distressing events in the history of humankind. The increase in greenhouse gases (carbon dioxide, methane, nitrous oxide, and chlorofluorocarbons, among others), as a consequence of industrial activity, deforestation, forest fires, etc., is responsible for global warming that might reach 1.5–6 °C in the next decades, with doubling of the CO_2 content in the atmosphere (Allen and Ingram 2002; IGPCC 2007).

Research programs dealing with the causes of climate change and mitigation and adaptation actions do not refer concretely to their influence on surface geodynamics (droughts, floods, erosion, sea level rise, etc.). The morphogenic and morphodynamic processes that model the earth surface result from the interaction between the geologic substratum and the morphoclimatic systems that influence soil development, surface and underground water distribution and, especially, the genesis and evolution of the main biomes and types of land occupation (Sayago and Collantes

2009). Future climate changes will influence the type, intensity, rhythm and duration of the processes that integrate surface geodynamics, whose effects on the landscape and living beings could only be mitigated from a thorough understanding of geomorphodynamics.

The severity of the erosion hazard in the subtropical region of north-western Argentina was assessed against two climate change scenarios determined on the basis of extreme values of rainfall variability during the last century. Erosion hazard was estimated for every map unit from the values of soil loss in the current conditions, obtained using the USLE criteria as defined in this work (Fig. 27.1, Tables 27.1 and 27.2). The relative difference in percent of soil loss between dry scenario (Fig. 27.2) and wet scenario (Fig. 27.3) was established, and both scenarios were compared with the current soil erosion loss (Fig. 27.1). The erosion values obtained for both scenarios are limited by the uncertainty of the future greenhouse effect on climate dynamics (rainfall intensity, extreme droughts, etc.). However, the strength of the USLE information (i.e. rainfall erosivity, soil erodibility, topography, vegetation cover, and management (Sayago 1985) allows an adaptation of the dominant conditions at least in the short term.

In the analysis, a modification of the K factor (soil erodibility) was taken into account in response to the two scenarios. Under wet conditions, higher organic matter content in the surface horizon can decrease the K factor, whereas the reverse would occur in arid conditions. Considering the relative stability of the relief as compared to climate variability, the LS factor (slope length and gradient) is assumed to vary little in both climate change settings. Variation rates of the K factor (soil erodibility) and C factor (cover) were estimated using the Langbein and Schumm (1958) curve for the both scenarios. In the wet setting, the C factor decreases in response to vegetation cover increase in cultivated areas and the Yungas forest. By contrast, in the western arid region, water erosion would increase despite cover increase because of larger bare soil areas susceptible to erode due to increasing R factor (rainfall erosivity).

Considering 30 % rainfall increase, the erosion hazard is assumed to increase in an equivalent percentage, although soil erodibility would decrease according to Langbein and Schumm (1958) because cover and organic matter content would also increase. Severely eroded units in dry environment would not experience any cover change, even with seasonal rainfall increase, because they have exceeded the threshold of the landscape resilience or "tipping point" (Sayago et al. 2012; Collantes and González 2012).

Summarizing, the maps of Figs. 27.2 and 27.3 show that, in both climate change scenarios, erosion would increase. Heavy rainfalls of the wet period would be reflected in the erosion intensity in the piedmont and eastern plains. In the desertified western regions, erosion increase would reflect intense landscape degradation, in many cases close to the ecosystem collapse point. During arid interruptions, the long slopes in the piedmont and eastern plain would be exposed to high erosion increase. In contrast, rainfall decrease in the arid western areas would account for the lowest erosion in the pre-Andean valleys and ranges.

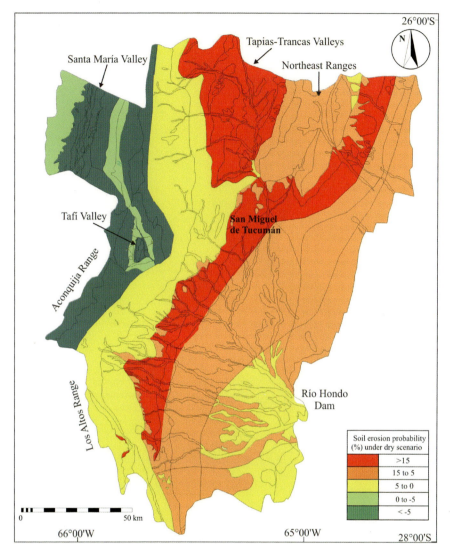

Fig. 27.2 Percentual differences in soil loss between the dry scenario and current soil erosion, Tucumán province, Argentina (Modified from Busnelli et al. 2009)

27.4 Discussion and Conclusions

The region shows, in general, high erosion hazard due to increasing anthropic pressure affecting especially areas still covered with natural vegetation. The maximal erosion hazard occurs in cultivated mountain areas where conservation practices are needed. On the contrary, mountain areas covered by cloud forest show low erosion hazard; potential deforestation would affect the soil integrity and the regional hydrologic balance.

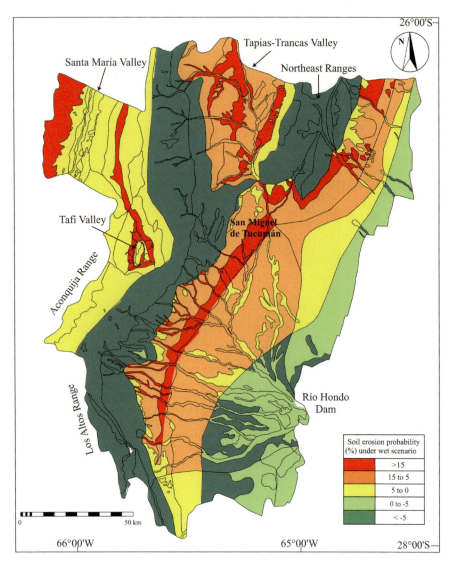

Fig. 27.3 Percentual differences in soil loss between the wet scenario and current soil erosion, Tucumán province, Argentina (Modified from Busnelli et al. 2009)

In plain areas, erosion hazard values are moderately high because of soil erodibility and rainfall aggressiveness. Soils derived from loess are naturally vulnerable to the impact of farming due to their unbalanced particle size distribution, with high silt and low clay contents causing weak structural stability, making loess soils prone to wind and water erosion and susceptible to sealing and crusting (Zinck 2006). The absence of conservation practices, despite the generalized use of "direct sowing" to neutralize soil erosion, the risk of soil compaction by heavy machinery, the nutrient loss due to soybean monoculture, and the drop of international corn prices create a worrying perspective for farmers. It would be advisable to reduce the intensive use of agrochemicals, not only due to their negative effect on health, but also because the return to simple conservation management (plowing, minimal tillage, rotations) would contribute to develop agro-forestry and secure pasture sustainability, with higher demand of local labor.

In the western arid regions, the risk of desertification is high. It is therefore relevant to assess in every landscape environment the proximity to the collapse point or "tipping point" to adapt the agricultural systems to land suitability and restrictions.

Due to their relative simplicity, the USLE methodological criteria (Wischmeier and Smith 1978), together with the use of teledetection and geographic information systems, are useful for erosion evaluation at small scale, on the condition of field validation of the evaluation results.

A map showing geomorphic regions or associations, with evaluation of erosion hazard values, can contribute to regional planning and development such as in the Argentine north-western region. Within this perspective, predictive models of erosion hazard, based on the historical periodicity of rainfall in a region, may guide the adaptation to the possible consequences of future climate changes.

Finally, although not less important, successful development of a land conservation program, especially in a region as fragile as the Argentine subtropics, demands the collaboration of producers, extensionists, and scientists as a necessary condition to achieve consistent progress.

Acknowledgements This chapter is dedicated to the memory of Dr. José Busnelli.

References

Allen MR, Ingram WJ (2002) Constraints on future changes in climate and the hydrologic cycle. Nature 419:224–232

Arnoldus HMJ (1978) An approximation of the rainfall factor in the universal soil loss equation. In: Boodt M, Gabriels D (eds) Assessment of erosion. Wiley, New York, pp 127–132

Bergsma E (1986) Aspects of mapping units in the rain erosion hazard catchments survey. In: Siderius W (ed) Land evaluation for land use planning and conservation in sloping areas. International Institute for Land Reclamation and Improvement, Wageningen, pp 84–105

Bianchi A, Cravero SAC (2010) Atlas climático digital de la República Argentina. Ediciones INTA, Salta

Bianchi AR, Yañez CC (1992) Las precipitaciones en el noroeste argentino. Instituto Nacional de Tecnología Agropecuaria. EER, Salta

Birkeland PW (1999) Soils and geomorphology. Oxford University Press, London

Busnelli J, Sayago JM, Collantes MM (2009) Riesgo erosivo ante diferentes escenarios de cambio climático en la provincia de Tucumán (Argentina). In: Sayago JM, Collantes MM (eds) Geomorfología y cambio climático. Instituto de Geociencias y Medio Ambiente, MAGNA Ediciones, Tucumán, pp 97–118

Cabrera AL (1976) Regiones fitogeográficas argentinas. Enciclopedia Argentina de Agricultura y Ganadería 2(1). Edn ACME. Buenos Aires, pp 1–85

Caria MA, Garralla S (2003) Caracterización arqueopalinológica del sitio Ticucho 1 (Cuenca Tapia-Trancas, Tucumán, Argentina). In: Collantes MM, Sayago JM, Neder L (eds) Cuaternario y geomorfología. Instituto de Geociencias y Medio Ambiente, MAGNA Ediciones, pp 421–428

Caria MA, Sampietro MM, Sayago JM (2001) Las sociedades aldeanas y los cambios climáticos. XIV Congreso Nacional de Arqueología, Rosario

Collantes MM, González LM (2012) Mecanismos del proceso de desertificación en el valle de Santa María, Provincia de Tucumán (Argentina). Acta Geol Lilloana 24(1–2):108–122

El-Swaify SA (1977) Susceptibility of certain tropical soils to erosion by water. In: Greenland DJ, Lai R (eds) Soil conservation and management in the humid tropics. Wiley, Chichester, pp 71–77

Fournier F (1960) Climat et érosion: la relation entre l'érosion du sol par l'eau et les précipitations atmosphériques. Presses Universitaires de France, Paris

Garralla S (1999) Análisis polínico de una secuencia sedimentaria en el Abra del Infiernillo, Tucumán, Argentina. Primer Congreso Argentino de Cuaternario y Geomorfología, Actas de Resúmenes, Comunicaciones y Trabajos, La Pampa, p 11

IGPCC (2007) Intergovernmental panel on climate change. Climate change 2007 – mitigation of climate change: working group III contribution to the fourth assessment report of the IPCC (Climate change 2007). Cambridge University Press

Imeson A (2012) Desertification, land degradation and sustainability. Wiley-Blackwell, London

Jungerius PD (1985) Soils and geomorphology. In: Jungerius PD (ed) Soils and geomorphology, vol 6, Catena supplement. Catena Verlag, Cremlingen, pp 1–18

Langbein WV, Schumm SA (1958) Yield of sediment in relation to mean annual precipitation. Am Geophys Union Trans 39:257–266

Minetti JL (1999) Atlas climático del Noroeste Argentino. Laboratorio Climatológico Sudamericano, Universidad Nacional de Tucumán, Tucumán

Niz AE (2003) Geomorfología del sector meridional del Dpto. Tinogasta, Provincia de Catamarca, Argentina. Tesis Doctoral Inédita, Facultad de Tecnología y Ciencias Aplicadas, UNCA, Catamarca

Renard KG, Foster GR, Weesies GA, Porte JR (1991) Revised universal soil loss equation. Soil Water Conserv 46(1):30–33

Rolleri E (1975) Las Provincias Geológicas Bonaerenses. Relatorio Geológico de la Provincia de Buenos Aires. VI Congreso Geológico Argentino

Sayago M (1969) Estudio fitogeográfico del Norte de Córdoba. Boletín de la Academia Nacional de Ciencias, Tomo XLVI, Córdoba

Sayago JM (1982) Las unidades geomorfológicas como base para la evaluación integrada del paisaje natural. Acta Geol Lilloana 16(1):169–180

Sayago JM (1985) Aspectos metodológicos del inventario de la erosión hídrica mediante técnicas de percepción remota en la región subtropical del noroeste argentino. Msc thesis, International Institute for Aerospace Survey and Earth Sciences (ITC), Enschede

Sayago JM, Collantes MM (2009) ¿Cambio climático o cambio geomorfodinámico? In: Sayago JM, Collantes MM (eds) Geomorfología y cambio climático. Instituto de Geociencias y Medio Ambiente, MAGNA Ediciones, Tucumán, pp 19–24

Sayago JM, Collantes MM, Niz AB (2012) El umbral de resiliencia del paisaje en el proceso de desertificación de los valles preandinos de Catamarca (Argentina). Acta Geol Lilloana 24(1–2):62–79

Scheffer M (2010) Foreseeing tipping points. Nature 467:363–494

Secretaría de Ambiente y Desarrollo Sustentable (2012) Monitoreo de la superficie del bosque nativo de la República Argentina. Periodo 2006–2012, Buenos Aires

Torres Bruchmann E (1977) Atlas agroclimático y bioclimático de Tucumán. Publicaciones Especiales 7 y 10. Universidad Nacional de Tucumán, Argentina

Tricart J (1982) Taxonomical aspects of the integrated study of the natural environment. ITC J 1982-3:344–348

Wischmeier WH (1984) The USLE: some reflections. J Soil Water Conserv 39(2):105–107
Wischmeier WH, Smith DD (1978) Predicting rainfall erosion losses – guide to conservation plan-
    ning, vol 537, Agriculture handbook. US Department of Agriculture, Washington, DC
Zinck JA (ed) (2006) Land use change and land degradation in the western Chaco. Tucumán prov-
    ince, Northwest Argentina, Burruyacú region, ITC publication 84. ITC, Enschede
Zinck JA (2012) Geopedologia. Elementos de geomorfología para estudios de suelos y de riesgos
    naturales, ITC special lecture notes series. ITC, Enschede
Zonneveld JIS (1983) Some basic notions in geographical synthesis. GeoJournal 72:121–129

Chapter 28
Adequacy of Soil Information Resulting from Geopedology-Based Predictive Soil Mapping for Assessing Land Degradation: Case Studies in Thailand

D.P. Shrestha, R. Moonjun, A. Farshad, and S. Udomsri

Abstract Soil is a natural body which delivers important ecosystem services apart from being a medium for plant growth. Soil mapping can be time consuming and expensive. During the 1960s and 1970s, introduction of air photo-interpretation in soil survey through element analysis, physiognomic and physiographic analysis, helped increase mapping efficiency. In the late 1980s, the geopedologic approach to soil mapping amplified the role of geomorphology. It helps understand soil variation in the landscape which increases mapping efficiency. In the present study, the adequacy of soil data resulting from geopedology-based predictive soil mapping for assessing land degradation in three locations in Thailand is assessed. The result shows that the geopedologic approach helps map soil in inaccessible mountain areas. However, for application in land degradation studies all the required soil properties may not be available in a soil map. The effect of land cover and land use management practices on soil properties, such as porosity and compaction having effect on hydraulic conductivity, a parameter used in modelling rainfall-runoff-soil erosion, is usually not reported in soil survey. These data have to be collected separately. For mapping areas susceptible to frequent flood, the geomorphic understanding of the river valley and soil characterization (Fluventic and Aquic) help identify susceptible areas. Similarly, the study shows how the geopedologic approach in combination with digital image processing helps in mapping soil salinity hazard.

Keywords Surface runoff • Erosion modelling • Flood-prone area • Soil salinity hazard • Geographic information systems

D.P. Shrestha (✉) • A. Farshad
Faculty of Geo-Information Science and Earth Observation (ITC), University of Twente, Enschede, The Netherlands
e-mail: d.b.p.shrestha@utwente.nl; abbasfarshad@gmail.com

R. Moonjun • S. Udomsri
Land Development Department, Ministry of Agriculture and Cooperatives, Bangkok, Thailand
e-mail: r.moonjun@utwente.nl; udomsri_sat@hotmail.com

© Springer International Publishing Switzerland 2016
J.A. Zinck et al. (eds.), *Geopedology*, DOI 10.1007/978-3-319-19159-1_28

28.1   Introduction

Soil is a natural body which delivers important ecosystem services apart from being a medium for plant growth. It can be considered the "skin of the earth" with interfaces between lithosphere, hydrosphere, atmosphere, and biosphere (Chesworth 2008). Soil mapping in general is time demanding and costly. The introduction of air photo-interpretation in soil survey in the 1950s helped increase mapping efficiency. During the 1960s and 1970s, several methods such as "elements analysis" (Buringh 1960), "pattern analysis" (Frost 1960), "physiognomic analysis" and "physiographic analysis" (Bennema and Gelens 1969; Goosen 1967) were developed. Gradually, it was noticed that understanding the relationship between landform and soil variation is crucial in drawing interpretation lines. Finally, in the late 1980s, the "geopedologic approach" to soil survey (Zinck 1989) presented in Part I of this book, amplified the role of geomorphology in understanding and mapping soil variations. Through a systematic and rather strict application of the geopedologic rules soil is mapped more efficiently (Farshad et al. 2013), although Esfandiarpoor et al. (2010) reported that a geopedologic map does not fully represent all the variability of soils. Recently, advances in digital soil mapping using an array of techniques including GIS, digital elevation models, multivariate statistics, geostatistics, neural network, fuzzy logic, among others, claim to increase mapping efficiency (McBratney et al. 2000; Behrens et al. 2005; Lagacherie et al. 2007; Grimm et al. 2008; Lagacherie 2008). These techniques are useful for mapping individual soil properties, but mapping the whole soil body (surface and depth) remains a challenge. A soil body incorporates not only solids, liquids, and air that cover the land, but it has also horizons and thus extends in depth. Mapping individual soil properties does not equate to mapping the whole soil body. In this respect, the geopedologic approach to soil survey can be considered very useful and efficient. However, the objective is not only to produce a soil map but to evaluate the value of such a map for various applications. In this chapter, three case studies of predictive soil mapping using the geopedologic approach are presented to assess the adequacy of soil data for applications in land degradation issues, namely soil erosion, flash flood, and soil salinity hazard studies in Thailand.

28.2   Method

Geopedologic photo-interpretation starts with delineating master lines across major landscape units such as mountain, hilland, valley, plain, among others, and drawing cross sections (Zinck 1989). Along the master lines follows the identification of sub-units (relief types) within major landscape units. Subsequently, main lithology types are identified for each unit. Lithological units can be derived from geological maps or inferred using expert knowledge, such as alluvial, colluvial or aeolian origin. Lastly, landform units are identified. This information is used to construct an interpretation legend where symbols are attributed to the landforms. After that, photo-delineation follows using a mirror stereoscope. Effective areas are determined within aerial photographs; these are marked perpendicular to flight lines, using the two transferred principle points, as explained in Paine and Kiser (2012).

Recently, the traditional way of interpreting aerial photos under mirror stereoscope has been replaced by on-screen digitizing of digital stereo pairs. The digital stereo pairs can be generated using either (a) scanning two overlapping air photos or ortho photos and creating a stereo pair, or (b) using a georeferenced image (i.e. ortho photo or satellite image) and the digital elevation data of the corresponding area. Many GIS software packages offer this facility. The OpenSource ILWIS software package helps generate stereo pairs which can be viewed using Red-Green or Red-Blue glasses in case of an anaglyph image or using a stereoscope in case of a stereo pair. For viewing a stereo image, a special stereoscope is needed which can be mounted on a computer screen. Interpretation of the stereo image can be done by directly digitizing on the screen. The advantage of using computer-assisted on-screen digitizing is that the interpretation lines are georeferenced with proper map projection parameters making the final map layout easier.

Once interpretation work is completed, sample areas are selected to facilitate fieldwork. A general rule for selecting a sample area is that it should include all the landform units. The area should also be easily accessible. Mini-pits are used for soil description and sampling. Soils are classified directly in the field following a classification system (e.g. FAO, USDA). Detailed soil study in the sample areas helps determine soil-landscape relationships and understand soil variability and patterns. This information is used to support extrapolation and mapping outside the sample areas. Laboratory determinations of soil samples are used to adjust soil classification. In this way the photo-interpretation map is converted into a soil map. With applying the above mentioned method, we demonstrate that soilscape knowledge enables analyzing cause-effect relationships between soil types, their distribution, and land degradation hazards as shown in the hereafter described case studies.

28.3 Soil Mapping for Assessing Soil Erosion in Inaccessible Mountain Areas

In mountainous areas, especially in the tropics, soil data are scarce mainly due to limited terrain accessibility. Sloping areas are often mapped as slope complexes. Soil survey and mapping have so far been carried out mainly in valleys, floodplains and other low-lying areas in the proximity of human settlements. Although mountain areas are usually considered low priority, they are important because of the ecosystem services they provide through rain water harvesting and storage, regulating weather conditions, supporting diversity of flora and fauna, offering scenic and panoramic views, among others. Inadequate management of watersheds can result in soil degradation in upland areas, which in turn affects the stream flow discharge causing offsite-effects in low-lying areas (e.g. stream avulsion, flooding). Conservation of watersheds and making effective management plans usually require data modelling and generating scenarios with detailed soil data.

In mountain areas, excess surface runoff as consequence of torrential rainfall can be the main causal factor of land degradation processes such as rill and gully erosion resulting in soil losses. Runoff generation is a function of rainfall volume and intensity during a rain event, interception by the vegetation cover, slope gradient, soil moisture storage capacity and infiltration into the soil, which depend on soil porosity, saturated hydraulic conductivity, and soil depth. Because of unavailability of detailed soil data, many soil properties, with exception of soil depth, are derived from soil particle size distribution using pedo-transfer functions. Interpolation techniques are commonly applied for mapping spatial variation of soil properties, disregarding in many instances the effect caused by topographic variation. The case study described hereafter attempts to map soils in an inaccessible mountain area of Thailand and assess the adequacy of the survey information for erosion estimation.

28.3.1   Study Area

The study area of 67 km² is located in the watershed of Nam Chun, in Petchabun province, about 400 km north of Bangkok, between 16°40′–16°50′N and 101°02′–101°15′E (Fig. 28.1a). The area has a tropical climate with distinct dry and wet seasons. Average annual precipitation is 1095 mm (Lomsak station) which fall mainly in the wet season (May–September). Average annual temperature is 28 °C, the hottest month being April (38 °C) and the coldest month being December (17 °C). Topography is rugged with mountain ridges of different heights separated by narrow valleys. Elevation varies from 185 to 1490 m asl. General accessibility is limited apart from a main road connecting Lomsak and Phitsanulok.

28.3.2   Soil Studies Along Roads

The rugged topography and lack of road access precluded stratified random sampling over the entire catchment area. Instead, observations were made along roads and nearby areas which could be reached on foot. A total of 219 soil samples was collected for laboratory analysis of particle size distribution, pH, organic matter content, bulk density, porosity, field capacity and saturated hydraulic conductivity. In addition to the conventional soil survey work, infiltration tests and shear-strength measurements of the topsoil in major land cover types were carried out to cope with the influence of human activities on soil compaction and cohesion. Such properties are necessary for assessing surface runoff and soil loss. Soils were classified at subgroup level according to USDA soil taxonomy (Soil Survey Staff 1999).

Fig. 28.1 Location of the three study areas in Nam Chun watershed (**a**), Pa Sak valley (**b**), and Nong Suang (**c**)

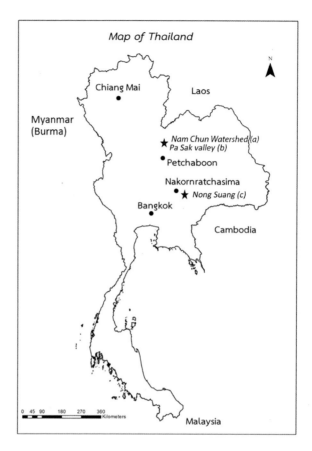

28.3.3 Results

Through visual interpretation of aerial photographs four main landscape units were identified as follows: (1) a high plateau (elevation 1,200 m) which borders the northwest of the watershed, (2) high mountain areas including very steep ridges (elevation 900 m) and lower dissected slope complexes, (3) low mountain areas (maximum elevation 600 m), and (4) a narrow valley cutting across the watershed. The cartography of these landscape units is presented in Fig. 28.2, and Table 28.1 contains information associated to the map legend.

The plateau landscape accounts for about 16 % of the watershed area. The main soils in this area are Typic Haplustalfs and the soil texture varies from loam to clay loam. The mountain landscape covers about 80 % of the watershed area. The landforms are narrow summits, mid-slopes including backslopes, and footslopes. Soils are mainly Lithic and Ultic Haplustalfs. In the low mountains Lithic Haplustolls occur on the narrow summits, while Lithic Haplustalfs and Typic

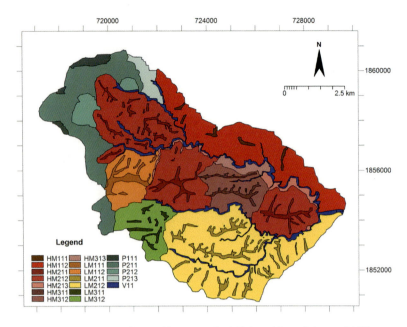

Fig. 28.2 Geopedologic map of Nam Chun watershed (Adapted from Solomon 2005)

Paleustalfs are common on mid-slopes. Typic Dystropepts and Typic Haplumbrepts are found on the footslopes. Soil texture varies from clay loam to silty clay loam. Soils in the area are characterized by rather high clay content and were mainly classified in the clay loam textural class. The area delineated as valley was very narrow and accounted for only 2 % of the total area. It consisted of the Hua Nam Chun river and a narrow floodplain that could not be differentiated as a separate unit. Soils are mainly Fluvents and Haplumbrepts and texture varies from sandy loam to clay loam.

Some of the soil properties especially in the topsoil, such as porosity, bulk density, and hydraulic conductivity, are very much related to land cover and land use practices. Highest soil porosity (53 %) and lowest bulk density (1.19 Mg m^{-3}) were found under forest cover (Table 28.2). Highest rate of saturated hydraulic conductivity was found in grassland followed by forest areas. Agricultural land had relatively compacted soil (average bulk density of 1.30 Mg m^{-3}) and reduced porosity. This contributes to increase surface runoff. Infiltration in grassland and forest land is higher than in other areas. Erosion assessment was carried out by applying the revised MMF model (Morgan 2001) which requires soil moisture content at field capacity, bulk density, cohesion, and erodibility. The result shows highest soil loss rates in agricultural land due to compaction and reduced soil porosity (Shrestha et al. 2014).

Table 28.1 Legend of the geopedologic map of Nam Chun watershed, Petchabun province

Landscape	Relief	Lithology	Landform	Map unit	Soil types USDA classification
Plateau	Cuesta	Sandstone	Undifferentiated	P111	
	Escarpment	Sandstone	Scarp	P211	
			Talus	P212	
			Undulating slope complex	P213	Typic Haplustalfs
High mountain	Ridge	Andesite	Summit	HM111	
			Slope complex	HM112	Ultic Haplustalfs
	Ridge	Andesitic tuff	Summit	HM211	Lithic Haplustolls
			Middle slope	HM212	Ultic Haplustalfs
			Footslope	HM213	Ultic Haplustalfs
	Erosional glacis	Andesitic and rhiolitic tuff	Summit	HM311	Lithic Haplustalfs
			Middle slope	HM312	Typic Paleustalfs
			Footslope	HM313	Lithic Haplustalfs
Low mountain	High ridges	Andesitic tuff	Summit	LM211	Ultic Haplustalfs
			Middle slope	LM212	Ultic Haplustalfs
	Moderately high ridges	Andesitic and rhiolitic tuff	Summit	LM111	Typic Haplustalfs
			Middle slope	LM112	Ultic Haplustalfs
	Low ridges	Andesitic and rhiolitic tuff	Summit	LM311	Typic Dystrustepts
			Middle slope	LM312	Ultic Haplustalfs
Valley		Alluvial Colluvial	Side slope/bottom complex	V111	Fluvents and Haplumbrepts

Table 28.2 Soil properties in different land cover types (Shrestha et al. 2014)

Land cover types	Hydraulic conductivity mm/h			Bulk density Mg m^{-3}		
	Mean	n	Std.Deviation	Mean	n	Std.Deviation
Forest	13.88	9	14.67	1.19	11	0.13
Degraded forest	7.06	8	9.19	1.28	10	0.12
Cornfield	4.36	9	2.79	1.30	11	0.10
Orchard	3.76	10	3.99	1.31	12	0.09
Grassland	15.43	11	15.70	1.26	11	0.11
Land cover types	Porosity %			Organic matter %		
	Mean	n	Std.Deviation	Mean	n	Std.Deviation
Forest	52.57	11	5.36	4.08	12	1.10
Degraded forest	49.04	10	4.95	3.15	8	1.38
Cornfield	48.12	11	4.08	2.24	19	0.84
Orchard	47.90	12	3.67	3.55	14	1.04
Grassland	49.61	11	4.50	2.99	13	1.12

28.4   Soil Mapping in Flood-Prone Areas

Lowlands in Thailand are usually easy to access because of good road network in flat areas. Abundant land for rice cultivation results in the increase of settlements and interconnecting roads. The area is subjected to frequent flooding during the rainy season. The study hereafter examines the adequacy of soil data for assessing flood hazard.

28.4.1   Study Area

The study area is located in the Pa Sak river valley, Petchabun province, about 400 km north of Bangkok. It served in the past for training ITC students in soil survey (ITC for International Institute of Geo-Information Science and Earth Observation, Enschede, The Netherlands) (Hansakdi 1998). The area is bounded in the east and west by mountain ranges that delineate the graben of the valley (Fig. 28.1b). Elevation varies from 120 m asl in the south to 170 m asl in the north from where the Pa Sak river flows. Floodplains are used for rice cultivation. In the surrounding foothills tamarind plantations are common. Other fruit crops are lychees and bananas. Main settlements are Lomsak and Lomkao.

28.4.2   Soil Studies in Sample Areas and Transects

Following visual interpretation of aerial photographs at the scale of 1:50,000, sample areas were selected so to include all the landform units and located on the basis of the local road network. The sample areas were surveyed in more detail using aerial photographs at scale of 1:15,000. Soils were described in (mini-)pits and auger holes following the FAO soil description manual.

28.4.3   Results

The geopedologic map (Fig. 28.3, Table 28.3) shows that the soils in the lowlands belong to Entisols, Inceptisols, and Alfisols (Soil Survey Staff 1999). Inceptisols are the most common soils, occurring in various landforms. Ustropepts, Eutropepts, and Tropaqepts are dominant. Entisols are mainly found on the sides and bottoms of narrow trench valleys. Alfisols include Haplustalfs and Tropaqualfs in low positions, and Paleustalfs in higher positions. Occurrence of Aquic, Fluventic, and Fluvaquentic subgroups is typical in the central valley, while Aeric and Ultic subgroups are

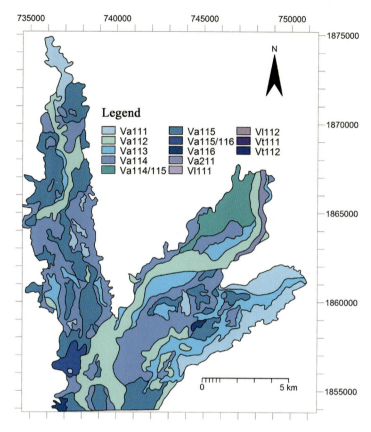

Fig. 28.3 Geopedologic map of the Pa Sak valley flood-prone area

dominant in the lateral valley (Table 28.3). In aquic moisture regime, the soil lacks dissolved oxygen for being saturated by ground water. Fluventic characteristics are typical of soils formed from alluvial sediments, stratified and showing frequent variations in texture and organic matter content. Fluvaquentic soils have both fluventic and aquic characteristics. Soil texture in the valley landscape varies from clay loam to silty clay loam. Although soil porosity is high, saturated hydraulic conductivity is very slow in the valley soils (less than 2 mm/h), meaning inability to percolate stagnated water fast to the groundwater. The soil characteristics and the landscape configuration make the area prone to frequent flooding. The soils in the lateral and trench valleys, higher in the landscape, are more developed, with argillic horizon, than the valley bottom soils, and they are not exposed to flooding. The result shows how geomorphic understanding of the river valley and characterization of soils (Fluventic and Aquic) help identify areas susceptible to frequent flooding.

Table 28.3 Geopedologic map legend of Pa Sak valley, Petchabun province

Landscape	Relief type	Lithology	Landform	Map unit	Soil map unit	Main soil types USDA classification
Valley	Terrace	Alluvium	Tread-riser complex	Va111	Association	Aeric Tropaquepts (40 %), Fluventic Ustropepts (30 %)
			Levee	Va112	Consociation	Fluventic Ustropepts (60 %), Typic Ustropepts (40 %)
			Levee/overflow mantle complex	Va113	Association	Aeric Tropaquepts (40 %) Fluventic Ustropepts (30 %) Typic Ustropepts (20 %)
			Overflow mantle	Va114	Association	Fluvaquentic Eutropepts (30 %) Typic Ustropepts (30 %) Fluventic Ustropepts (30 %)
			Overflow basin	Va115	Association	Fluvaquentic Eutropepts (30 %) Aquic Eutropepts (30 %) Typic Tropaquepts (20 %)
			Overflow mantle/basin	Va114/115	Association	Fluvaquentic Eutropepts Typic Ustropepts Fluventic Ustropepts
			Decantation basin	Va116	Association	Aquic Eutropepts (40 %) Aeric Tropaquepts (30 %) Fluventic Eutropepts (20 %)
			Overflow/decantation basin	Va115/116	Association	Fluvaquentic Eutropepts Aquic Eutropepts Typic Tropaquepts Aquic Eutropepts
	Flood plain	Alluvium	Levee/basin complex	Va211	Association	Fluventic Ustropepts (30 %) Aquic Ustropepts (30 %) Fluvaquentic Eutropepts (30 %)
Lateral Valley	Terrace complex	Colluvio-alluvium	Bottom/side complex	Vl111	Association	Aeric Tropaqualfs (40 %) Typic Haplustalfs (40 %)
	Depression		Bottom/side complex	Vl112	Consociation	Aeric Tropaquepts (60 %) Aeric Tropaqualfs (40 %)
Trench Valley	Terrace complex	Alluvium/residual	Bottom/side complex	Vt111	Association	Typic Ustifluvents (50 %) Typic Haplustalfs (30 %) Ultic Paleustalfs (20 %)
	High terrace		Bottom/side complex	Vt112	Consociation	Ultic Paleustalfs (50 %) Ultic Haplustalfs (50 %)

28.5   Soil Mapping in Areas Exposed to Soil Salinity Hazard

Soil salinity is a regional issue in the north-east of Thailand. The area is underlain by salt-bearing rocks affecting the groundwater. If the groundwater has high salt contents (conductivity of more than 15 dS m^{-1}), it is most likely that there is soil salinity hazard. The rate of water movement to the surface through capillary rise depends on soil particle size distribution, the finer the particle sizes, the higher the capillary action. The following section describes the application of the geopedologic approach to mapping salinity-prone areas.

28.5.1   Study Area

The study area is located in the Nong Suang region, Nakhon Ratchasima province, between 101°45′–102°E and 15°–15°15′N, with an elevation ranging from 160 to 175 m asl (Fig. 28.1c). The area is part of the Northeast Korat plateau landscape, and is underlain by salt-bearing rocks at about 80 m depth (Imaizumi et al. 2002). Average annual rainfall is 1035 mm (1971–2000), coming in mainly from May to October. Average annual potential evapotranspiration is 1817 mm, thus higher than the mean annual precipitation (1035 mm), indicating that climate is a potential driver of salinity in the area (Shrestha and Farshad 2008).

28.5.2   Soil Sampling and Data Analysis

Geopedologic interpretation was carried out to delineate the geomorphic units occurring in the lower parts of the landscape (i.e. floodplains, terraces, and vales), which have high potential for salinity development. Salt-affected areas give generally high reflectance values in the visible to near-infrared bands due to concentration of salts on the terrain surface and the formation of salt crusts. A Landsat TM image of the 2003 dry season was transformed using band rotation of near-infrared and red spectral bands to derive a soil line that enhances soil reflectance from saline areas (Shrestha et al. 2005). Level slicing of the resulting band (soil line) was carried out to generate salinity intensity classes within the geomorphic units of floodplain, terrace, and vale. A total of 126 samples was collected from three depths: 0–30 cm, 30–60 cm, and 60–90 cm to study soil salinity (Soliman 2004).

28.5.3   Results

At the landscape level, the area was separated into peneplain and valley that were further divided into corresponding relief type, lithology, and landform levels (Fig. 28.4, Table 28.4). Five soil order classes were distinguished (Soil Survey Staff

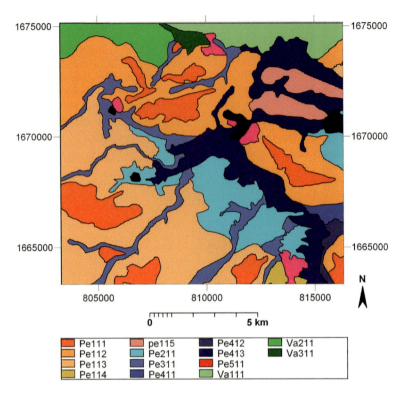

Fig. 28.4 Geopedologic interpretation of the Nong Suang area, Nakhon Ratchasima province

Table 28.4 Geopedologic interpretation legend of the Nong Suang area, Nakhon Ratchasima province

Landscape	Relief type	Lithology	Landform	Map unit
Peneplain	Ridge	Sedimentary rocks	Top complex	Pe111
		Korat Group	Side complex	Pe112
			Slope facet complex	Pe113
			Summit	Pe114
			Tread riser complex	Pe115
	Glacis	Sedimentary rocks	Tread riser complex	Pe211
		Korat Group		
	Vale	Sedimentary rocks	Slope complex	Pe311
		Korat Group		
	Lateral vale	Sedimentary rocks	Side complex	Pe411
		Korat Group	Bottom – side complex	Pe412
			Bottom complex	Pe413
	Depression	Sedimentary rocks	Basin	Pe511
		Korat Group		
Valley	Floodplain	Alluvial deposits	Levee – overflow complex	Va111
	Old terraces	Alluvial deposits	Overflow – basin complex	Va211
	New terraces	Alluvial deposits	Overflow – basin complex	Va311

1999). Ultisols occur on ridges, while Alfisols (Ustalfs and Aqualfs) are mainly in sloping areas adjacent to the ridges. Vertisols occur in the northern part of the area, along rivers and channels, where vertic features form due to the presence of swelling clays. Two suborders, namely Aquerts and Usterts, were distinguished based on the soil moisture regime. Inceptisols are common in the lower part of the lateral valleys that dissect the peneplain lobes. Inceptisols are mainly Aquepts due to poor drainage conditions that lead to the development of gleyic color, with no abrupt textural change. Wet Psamments, classified as Gleysols according to the FAO World Reference Base for Soil Resources (FAO 1998), occur in a few sloping spots on residual material derived from sandstone.

Geomorphic units such as floodplains, terraces, and vales have high potential for salinity development since they are located in the lower parts of the landscape and are thus most likely close to saline groundwater. They were masked out using digital elevation data in a GIS map overlay procedure in order to improve classification accuracy (Shrestha and Farshad 2008). The study also showed a good correlation between soil texture and salinity occurrence. Clayey soils are strongly saline because of higher capillary rise from the groundwater. Salinity is lower in coarse-textured soils. Since salt-affected areas usually present higher reflectance in all visible and near-infrared spectral bands when the soil is bare and dry, salinity variations can be mapped using remote sensing data and the enhancement technique aforementioned.

28.6 Conclusion

Visual interpretation of aerial photographs based on the geopedologic approach contributes efficiently in predicting and mapping soil types and their occurrence. Furthermore, soilscape knowledge enables analyzing cause-effect relationships between soil types, their distribution, and land degradation hazards as shown in the case studies with different land degradation problems. Fieldwork remains an essential component of soil mapping. Variability at the subgroup level, for instance between Typic and Ultic soil types, or at some of the intergrades (Alfisols-Ultisols) can only be discovered during fieldwork, by well-trained surveyors. On the other hand, the effect of land cover change and land use management on the changes in soil properties, such as porosity and compaction having effect on hydraulic conductivity, cannot be mapped solely with the geopedologic approach. Mapping soil properties influenced by human activities needs land cover and land use information which can be derived from remotely-sensed imagery. The use of DEM can provide sufficient information for mapping areas susceptible to frequent floods. For mapping salinity-affected areas, the combined use of geopedology, digital elevation data, and remote sensing techniques can be useful.

Through the case studies analyzed in this chapter we argue that the geopedologic approach helps in mapping soils very efficiently. For achieving the required adequacy of soil data for applications in different land degradation studies, incorporation of GIS-based spatial modelling and image processing techniques further improves soil mapping.

Acknowledgment Materials used in the case studies were derived from a joint research project of ITC, Enschede, the Netherlands, with the Land Development Department (LDD), Ministry of Agriculture and Cooperatives, Bangkok, Thailand. Contribution of ITC course participants, especially Anukul Suchinai, Ekanit Hansakdi, Harssema Solomon, and Aiman Soliman, is duly acknowledged.

References

Behrens T, Förster H, Scholten T, Steinrücken U, Spies ED, Goldschmitt M (2005) Digital soil mapping using artificial neural networks. J Plant Nutr Soil Sci 168(1):21–33

Bennema J, Gelens HF (1969) Aerial photo-interpretation for soil surveys, ITC lecture notes. ITC, Enschede

Buringh P (1960) The application of aerial photographs in soil surveys. In: Colwell RN (ed) Manual of photographic interpretation. American Society of Photogrammetry, Washington, DC, pp 633–666

Chesworth W (ed) (2008) Encyclopedia of soil science. Springer, Dordrecht

Esfandiarpoor BI, Mohammadi J, Salehi MH, Toomanian N, Poch RM (2010) Assessing geopedological soil mapping approach by statistical and geostatistical methods: a case study in the Borujen region, Central Iran. Catena 82(1):1–14

FAO (1998) World reference base for soil resources. World soil resources report 84, Rome, Italy

Farshad A, Shrestha DP, Moonjun R (2013) Do the emerging methods of digital soil mapping have anything to learn from the geopedologic approach to soil mapping and vice verse? In: Shahid SA, Taha FK, Abdelfattah MA (eds) Developments in soil classification, land use planning and policy implications. Springer, Dordrecht, pp 109–131

Frost RE (1960) Photo interpretation of soils. In: Colwell RN (ed) Manual of photographic interpretation. American Society of Photogrammetry, Washington, DC, pp 343–402

Goosen D (1967) Aerial photo interpretation in soil survey. Food and Agriculture Organization of the United Nations, Soils bulletin 6. FAO, Rome

Grimm R, Behrens T, Marker M, Elsenbeer H (2008) Soil organic carbon concentrations and stocks on Barro Colorado Island: digital soil mapping using random forests analysis. Geoderma 146:102–113

Hansakdi E (1998) Soil pattern analysis and the effect of soil variability on land use in the upper Pa Sak area, Petchabun, Thailand. Unpublished MSc thesis, ITC, Enschede

Imaizumi K, Sukchan S, Wichaidit P. Srisuk K, Kaneko F (2002) Hydrological and geochemical behavior of saline groundwater in Phra Yun, Khon Kaen, Thailand

Lagacherie P (2008) Digital soil mapping: a state of the art. In: Hartemink AE, McBratney A, Mendonca-Santos ML (eds) A state of the art: digital soil mapping with limited data. Springer, Dordrecht, pp 3–14

Lagacherie P, McBratney AB, Voltz M (eds) (2007) Digital soil mapping: an introductory perspective. Elsevier, Amsterdam

McBratney AB, Odeh IOA, Bishop TFA, Dunbar MS, Shatar TM (2000) An overview of pedometric techniques for use in soil survey. Geoderma 97(3–4):293–327

Morgan RPC (2001) A simple approach to soil loss prediction: a revised Morgan-Morgan-Finney model. Catena 44:305–322

Paine D, Kiser J (2012) Aerial photography and image interpretation, 3rd edn. Wiley, Hoboken, New Jersey

Shrestha DP, Farshad A (2008) Mapping salinity hazard: an integrated application of remote sensing and modeling-based techniques. In: Metternicht G, Zinck JA (eds) Remote sensing of soil salinization: impact on land management. CRC Press, Boca Raton, pp 257–272

Shrestha DP, Soliman AS, Farshad A, Yadav RD (2005) Salinity mapping using geopedologic and soil line approach. Asian Conference on Remote Sensing, Hanoi, Vietnam

Shrestha DP, Suriyaprasit M, Prachansri S (2014) Assessing soil erosion in inaccessible mountainous areas in the tropics: the use of land cover and topographic parameters in a case study in Thailand. Catena 121:40–52

Soil Survey Staff (1999) Soil taxonomy: a basic system of soil classification for mapping and interpreting soil surveys. US Department of Agriculture, Washington, DC

Soliman AS (2004) Detecting salinity in early stages using electromagnetic survey and multivariate geostatistical techniques: a case study of Nong Sung district, Nakhon Ratchasima province, Thailand. Unpublished MSc thesis, ITC, Enschede

Solomon H (2005) GIS-based surface runoff modeling and analysis of contributing factors: a case study of the Nam Chun Watershed, Thailand. Unpublished MSc thesis, ITC, Enschede

Zinck JA (1989) Physiography and soils, ITC lecture notes. ITC, Enschede

Part V
Applications in Land Use Planning and Land Zoning Studies

Chapter 29
Ecological Land Zonation Using Integrated Geopedologic and Vegetation Information: Case Study of the Cabo de Gata-Níjar Natural Park, Almería, Spain

P. Escribano, C. Oyonarte, J. Cabello, and J.A. Zinck

Abstract The aim of the present study is to determine zonation units geared towards balancing conservation and development in the Cabo de Gata-Níjar Natural Park, an arid environment located in south-eastern Spain. Ecosystems were identified selecting the attributes that exert the strongest influence on ecosystem dynamics at three different spatial scales. A multi-categorial geoform-soil classification system was used as base for the definition of the ecosystems hierarchy, including ecosection (1:100,000), ecoserie (1:50,000), and ecotope (1:25,000). Vegetation was used for the identification of ecosystems at ecotope level. The hierarchic structure of the geoform-soil database allowed maintaining the thematic and spatial coherence in which lower levels of the hierarchy inherit the attributes of higher levels. Geoform-soil and vegetation attributes provided the data needed to assess the conservation value and the vulnerability of the ecosystems to land use, crucial for the definition of zonation units.

Keywords Ecosystems hierarchy • Conservation value • Vulnerability • Management units • Ecological land zonation

P. Escribano (✉)
Estación Experimental de Zonas Aridas (CSIC), Campus Universitario, Almería, Spain
e-mail: paula.escribano@gmail.com

C. Oyonarte
Departamento de Agronomía, Universidad de Almería, Almería, Spain
e-mail: coyonart@ual.es

J. Cabello
Departamento de Biología y Geología, Universidad de Almería, Almería, Spain
e-mail: jcabello@ual.es

J.A. Zinck
Faculty of Geo-Information Science and Earth Observation (ITC), University of Twente, Enschede, The Netherlands

Institute of Environmental Studies, University of New South Wales, Sydney, NSW, Australia
e-mail: alfredzinck@gmail.com

© Springer International Publishing Switzerland 2016   475
J.A. Zinck et al. (eds.), *Geopedology*, DOI 10.1007/978-3-319-19159-1_29

29.1   Introduction

Protected areas cover around 15 % of the total earth surface. According to Dudley (2008), a protected area is a clearly defined geographical space, recognized, dedicated and managed, through legal or other effective means to achieve the long-term conservation of nature with associated ecosystem services or cultural values. Protected areas, when correctly managed, play a critical role in conservation and sustainability of natural resources. A particular type of protected areas concerns Natural Parks as defined in Spanish law 4/1989, which is comparable to the V category of the IUCN classification (Dudley 2008). The main goal is to make compatible the presence of humans and their activities in the park with the preservation of the environment through sustainable use of the natural resources. For this kind of protected areas, the Spanish law 4/1989 requires a management scheme for assuring the preservation, protection, and rational use of the natural resources. To facilitate decision-making, this scheme has to be based on the identification of zonation units that are geographical units with the same regulation needs in terms of protection and land use restrictions.

The classification of ecosystems constitutes the basic and initial step in the process of evaluation and analysis of the natural resources in an area. These environmental interpretations are usually needed at different scales for appropriate management of a territory. The ecosystems hierarchy is a way to break down complexity and render order to the natural complexity of ecosystems (Wu and David 2002). Nested hierarchic models emphasize both top-down and bottom-up approaches. The higher levels of the hierarchy control or exert constraints on the lower ones, while the lower levels provide the initial conditions (Wu and David 2002). Several studies have applied hierarchic models for the management of protected areas (Ortiz-Lozano et al. 2009), biodiversity studies (Noss 1990), and the study of ecological boundaries (Yarrow and Salthe 2008), among others.

A main issue affecting ecological analysis and environmental decision making is the complexity of ecosystems and landscape patterns. However, managers and policy-makers require information on the status, condition, and trends of the ecosystems. Providing a legal status of protection to a natural area is not necessarily a sufficient measure to protect the ecological integrity of the territory (Lajeunesse et al. 1995). Conservation requires the right measures to be applied to the right areas (Botrill and Pressey 2012). A method to understand and study the characteristics of ecological systems is the use of variables or indicators able to represent the most important features of the environmental state (Müller 1999). Several variables have been proposed to assess the ecological value of an ecosystem based on species and community traits related to conservation concern such as species richness, specificity, rarity, vulnerability, endemicity, or population connectivity (Bonn and Gaston 2005). The use of soil variables for conservation is less common but equally important. Ibáñez et al. (2012) stated that the pedosphere is part of our natural heritage. Soils should be considered both as biological and geological resources. As such, the use of soil singularity and diversity indices in the analysis of soil patterns would be similar to those used in biodiversity analysis of plant species or animals (Bockheim and Schliemann 2014). The use of geological and geomorphological values for assessing the conservation value of a protected area is also recommended (Dudley 2008). Incorporating vulnerability to land uses into conservation planning is a critical issue in protected areas. Yet few studies have tackled both conservation and vulnerability simultaneously (Wilson et al. 2005). Several variables have been proposed for the assessment of vulnerability, mainly related with soil properties such as texture, drainage, and organic matter content (Kosmas et al. 2013).

This study aims at determining and mapping zonation units in the Cabo de Gata-Níjar Natural Park to allow conservation and sustainable use of the natural resources to cohabit according to ecological suitability. An approach integrating geopedologic and vegetation information is used to this end, as described hereafter.

29.2 Materials and Methods

29.2.1 The Study Area

The Cabo de Gata-Níjar Natural Park (CGNP) is located in Almería province, southeast of Spain, with a continental surface of 38,000 ha. It is the most arid spot of Western Europe, being also one of the few protected subdesertic and steppe areas in Europe. Mean annual precipitation is 220 mm and mean annual temperature is around 18 °C, annual potential evapotranspiration is around 1390 mm, with an aridity index (Ia) below 0.2. Overall the park is mountainous with heterogeneous relief and lithology. Most of the area is of volcanic origin with an upper platform of detrital carbonate lithology (Fernández Soler 1996). The soil types follow geomorphic patterns. At landscape level, the major soil types are rendzic and eutric Leptosols, eutric and calcaric Regosols, calcaric Phaeozems, and Luvic Calcisols (Oyonarte 2004). Xerophytic scrubs of *Chamaerops humilis* and *Periploca laevigata*, and grass steppes of *Macrochloa tenacissima* are the dominant vegetation types in the less modified areas.

29.2.2 Determination of Ecosystems

An integrated geoform-soil database was built using a hierarchic geoform classification system. Geoforms and soils were fused in soil map units following the geopedologic approach described in Zinck (2013). Geomorphology provides the cartographic boundaries of the map units, while pedology provides the soil information content of the units. An existing vegetation database of the CGNP was used. Information about habitats was included to each vegetation community following the guidelines of the Interpretation Manual of European Union (European

Commission 2007). Vegetation provides the cartographic boundaries, while habitats are the attributes.

In the present case study, an ecosystem is considered as a geographic unit of interrelated biotic and abiotic components that can be identified and surrounded by boundaries (Bailey 1996). Ecosystems were identified on the basis of selected attributes, or key factors, that exert the strongest influence on the ecosystem at three spatial scales considered (Wu and David 2002). Information on landscape, relief/molding, and lithology was used to define the first two ecosystem hierarchic levels, namely ecosection and ecoserie (Fig. 29.1). Geomorphic patterns tend to structure the landscape at broad spatial levels, while vegetation plays a dominant role in the ecosystem dynamics at lower levels of the hierarchy. Therefore, the vegetation database was used to define ecosystems at ecotope level. Geomorphology and vegetation provided the cartographic units (spatial definition) and the soil profile and soil taxa/profile and habitat databases provided the attributes needed for the ecological assessment (Fig. 29.1).

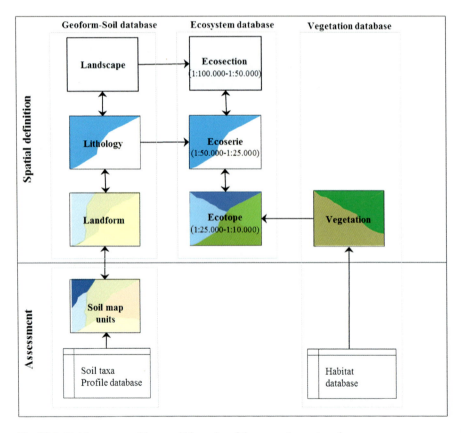

Fig. 29.1 Databases assemblage and hierarchy of the ecosystem categories

29.2.3   Ecological Assessment

29.2.3.1   Selection and Scoring of Ecological Variables

The conservation value of an ecosystem is understood here as a series of qualities that make that ecosystem of interest for conservation (Arponen et al. 2008). Seven attributes were used related to geomorphic, soil, and vegetation characteristics of the ecosystems that made them of conservation concern (Table 29.1). Geomorphic singularity refers here to geoforms that are unique in the regional context or display a particular scenic value. This variable was retrieved from the information on landscape, relief, molding, and lithology contained in the geoform database. Pedogenic singularity and pedodiversity (Ibáñez et al. 2008) were derived from the soil map units. Ecotopes were characterized considering their habitat, species richness, endemicity, specificity, and priority of conservation. These variables were obtained from the habitat attributes of the vegetation database, and are variables commonly used for assessing the conservation value of an area (Bonn and Gaston 2005).

Vulnerability is understood as the inherent fragility of a natural object or land unit when submitted to natural or human-induced disturbances. Four soil variables were selected to assess vulnerability, including organic matter content, texture, drainage, and soil depth. These are the most common variables used in the assessment of land use sustainability (Doran et al. 1994). The profile database of the geoform-soil attributes provided the data needed to assess the vulnerability of the ecosystems to land uses.

The ecological variables of conservation and vulnerability were evaluated in their original data sources (Fig. 29.2a), according to the score criteria of Table 29.1. Every variable was classified in three score classes: 1 (low), 2 (moderate), and 3 (high). For instance, unit MT31-16 in Fig. 29.3 is an ecosystem corresponding to a volcanic cone in a mountainous landscape, with a vegetation community of *Stipa tenacissima* and *Periploca angustifolia*. This ecosystem gets a high conservation value score when applying the score criteria of Table 29.1. Volcanic cone is a rare geomorphic type within the Andalusian region and is included in the Management Plan of Natural Resources (Decreto 37/2008) as a protected landscape (geomorphic singularity = 3). Dominant soils are associations of Argixerolls and Haploxerepts with inclusions of Xerothents (pedological diversity = 3). The presence of red soils is a factor of singularity (pedological singularity = 3). The habitat variables of species richness, specificity, endemicity, and priority of conservation were evaluated similarly on basis of the score criteria in Table 29.1. Regarding the vulnerability assessment, the soils of the ecosystem considered here have 1–3 % organic matter content, indicating moderate structural stability that makes them relatively resistant to degradation processes such as erosion (organic matter = 2). Soil texture varies between silt loam and clay loam, giving adequate hydraulic conductivity and structural stability to the soils (texture = 2). Soil depth of 25–75 cm is moderate (soil depth = 2).

Table 29.1 Ecological variables used to assess conservation value and vulnerability to land uses of the ecosystems in the CGNP

Ecological assessment	Data source	Ecological variables	Definition	Score criteria[a]
Conservation	Lithology	Geomorphic singularity	Geomorphic singularity degree within the Andalusian region	3 = Presence of protected landscapes included in PORN[b]
				1 = Absence of protected landscapes
	Soil map units	Pedological diversity	Number of soil taxa within the ecotope	3 = Two or more contrasting taxa
				2 = Two or more non contrasting soil taxa
				1 = One taxon
		Pedological singularity	Singularity of the pedogenic process	3 = Reds soils
				2 = Soils with mollic horizon
				1 = Other soil types
	Vegetation (habitat database)	Habitat species richness	Number of characteristic species of conservation concern	3 = Habitats with >3 characteristic species of conservation concern
				2 = Habitats with 1–3 characteristic species of conservation concern
				1 = Habitats without characteristic species of conservation concern
		Habitat specificity	Ecological restriction degree	3 = Habitats linked to rare ecological conditions
				2 = Habitats linked to narrow ecological conditions
				1 = Habitats of wide ecological conditions
		Habitat specificity	Geographic restriction degree	3 = Habitats with current distribution area restricted to the Natural Park
				2 = Habitats distributed over the arid southeast Iberian Peninsula
				1 = Habitats of wide distribution
		Habitat priority	Consideration in the Habitat Directive (HD)[c]	3 = Habitats considered priority for conservation in HD
				2 = Habitats considered in HD but not as priority for conservation
				1 = Habitats not considered in HD

Vulnerability	Soil map units (profile database)		
	Organic matter content	Surface stability/resistance to degradation	3 = <1 % Organic matter content
			2 = 1– 3 % Organic matter content
			1 = >3 % Organic matter content
	Texture	Hydraulic conductivity/surface stability	3 = Sand, loamy sand, clay
			2 = Sandy loam, sandy clay, silt, silt loam, silty clay loam
			1 = Clay loam, loam, sandy clay loam
	Drainage	Infiltration rate	3 = Excessively drained or poorly to very poorly drained
			2 = Somewhat or somewhat poorly drained
			1 = Well or moderately well drained
	Soil depth	Promote plant growth	3 = Soil depth <25 cm
			2 = Soil depth 25–75 cm
			1 = Soil depth >75 cm

[a]Conservation and vulnerability classes: 1 Low, 2 Moderate, 3 High

[b]Plan de Ordenación de los Recursos Naturales/Management Plan of Natural Resources (Decreto 37/2008 JA)

[c]Habitat Directive (European Commission 2007)

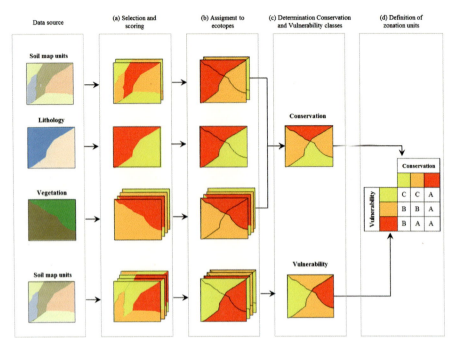

Fig. 29.2 Method used for ecological assessment and the determination of zonation units. (**a**) score criteria: 1 (low) *yellow*; 2 (moderate) *orange*; 3 (high) *red*; (**b**) *black lines* superimposed on vegetation, lithology, and soil map are ecotope boundaries; (**c**) data reduction and classification into high (*red*), moderate (*orange*), and low (*yellow*) classes; (**d**) example of a decision matrix for the determination of the zonation units

29.2.3.2   Assignment of the Score Values to Ecotopes

Ecological assessment was performed using data about lithology, soil, and vegetation. A common cartographic unit synthesizing all the information was established for management purposes by means of assigning the information to the ecotopes (Fig. 29.2b). The ecosystem and geoform databases have a hierarchic structure in which the lower levels of the hierarchy inherit the information of the upper levels (Fig. 29.1). Conversely, information contained at lower levels can be integrated into upper levels maintaining the spatial and thematic coherence. In the case of the variables derived from the lithology (e.g. geomorphic singularity), the score values were assigned to the ecoserie units, and then all ecotope units within a given ecoserie inherited these score values. For the variables derived from the soil profile database, the information was integrated into the soil map units, followed by a spatial aggregation to integrate the information in the corresponding ecoserie unit. The habitat variables, derived from the vegetation database, were directly assigned to the ecotopes, as vegetation was used to define the latter (Fig. 29.1).

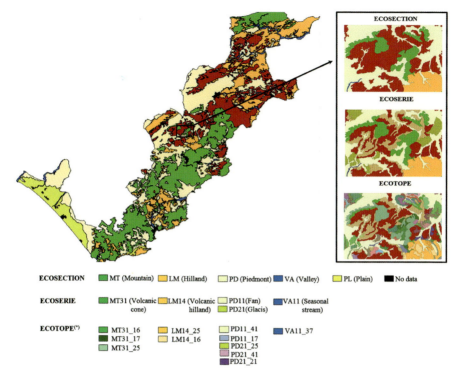

Fig. 29.3 Ecosection map of the Cabo de Gata-Níjar Natural Park; close-up of a selected area showing ecosystem units at ecosection, ecoserie, and ecotope levels; [*] numbers following the ecotope nomenclature indicate vegetation communities: 16 (Alpha grass steppe), 17 (Alpha grass steppe with sparse shrubs), 21 (Alpha grass steppe and low scrubland), 25 (Annual grassland), 37 (Thermo-Mediterranean broom vegetation), 41 (Rainfed cereals)

29.2.3.3   Determination of Conservation and Vulnerability Classes

A factor analysis was performed to integrate the information from the evaluated ecological variables, separately for conservation and vulnerability variables (SPSS 2000). The aim was to identify the underlying factors that explain the correlation patterns between the variables, avoiding redundancy. The application of this kind of analysis to semi-quantitative data is possible whenever the data are classified in meaningful ordered classes (Legendre 1993). A set of factors that explained at least 85 % of the total variability was selected. The output of this analysis provided the factor loadings which express the relationship of each variable to the underlying factor. The factor loadings for each conservation variable using the score of every variable per ecotope were summed up to obtain a global conservation value for each ecotope (Fig. 29.2c). The maximum possible value was calculated on the basis of all the variables scoring 3 and the minimum value on the basis of all the variables scoring 1. Subsequently, the data were normalized between 10 for maximum value and 0 for minimum value. Finally, the conservation values were distributed in three classes as follows: high (10–7), moderate (6–4), and low (<4). The same procedure was applied to the vulnerability variables (Fig. 29.2c).

29.2.4 Definition of Zonation Units

A decision matrix was prepared for the determination of the zonation units (Fig. 29.2d). It combines the conservation and vulnerability classes for decision-making and it is easy to adapt and interpret for management purposes. Zonation units are defined as geographic units with the same regulation needs in terms of protection and land use restrictions. In this case, three classes of zonation units were recognized following Ibáñez et al. (2007) as defined hereafter:

- *A areas*: protected areas with high conservation value and high to moderate vulnerability to land uses. Human activities are restricted.
- *B areas*: areas that have well preserved ecosystems with moderate vulnerability to land uses. B1 areas refer to traditional agricultural activities and B2 areas to forest activities.
- *C areas*: areas with lower ecological value and low vulnerability to land uses in which the regulations for agriculture or forest activities are less restrictive.

29.3 Results and Discussion

Figure 29.3 shows the ecosystems identified in the CGNP at ecosection level, together with an example of their hierarchic structure from ecosection to ecotope. Landscape diversity is high with the identification of five ecosection classes, including mountain, piedmont, hilland, plain, and valley. The polygons at this level are large and easily discernible at 1:100,000 scale, with exception of the valley class. Valleys in the CGNP are generally narrow, elongated units, as usual in arid environments, representing less than 2 % of the total area. Nevertheless, the high productivity and the ecological singularity of this type of landscape in the regional context justify its cartographic separation even at the small ecosection spatial scale.

At ecoserie level, 24 classes were recognized on the basis of differences in relief, molding, and lithology. At ecotope level, 203 classes were identified on the basis of vegetation diversity. From higher to lower hierarchic levels, the classes become narrower and the number of boundaries increases. Klijn and de Haes (1994) point out that when zooming in (i.e. downscaling in the hierarchy) the detail is steadily increasing (Fig. 29.3). While landscape, relief, and lithology are factors relatively constant over time, the vegetation cover changes due to natural processes or human modifications. Therefore the ecosystem hierarchy is at the same time a spatial and temporal hierarchy, being more stable at the upper level than at the lower (Zonneveld

1989). As a consequence, the management of a natural park should concentrate on the lower levels of the hierarchy to cope with local environmental changes that can be detected in time. The use of the ecosystem hierarchy for the determination of zonation units allows integrating information retrieved from different data sources into the delimited spatial units. Ecotopes are relatively homogeneous spatial entities with regard to land use regulation needs. Therefore, the application of the present method guarantees that the ecotope units will be managed as a whole.

The variables used for ecological assessment operate at different spatial scales. Soil map units, containing information on soil properties needed to assess vulnerability, were integrated at ecotope level following the spatial and thematic coherence of the hierarchic structure of the geoform database. Geomorphic features were integrated at ecoserie level because these features affect the ecosystem dynamics at meso-scale. The ecosystem hierarchy is a nested system where data incorporated at an upper level are transferred the lower levels. Thus, the ecosystem hierarchy demonstrates to be a good tool for the integration of variables at different spatial scales for a variety of management purposes, in accordance with experiences reported from other natural areas (Palik et al. 2000, Ortiz-Lozano et al. 2009). The zonation map is shown in Fig. 29.4. The comparison of the ecotopes (Fig. 29.3) and the zonation units (Fig. 29.4) reveals the underlying thematic and spatial generalization process which is crucial for management purposes. The zonation database maintains the original information of the ecosystem, geoform, and vegetation databases. Thus the detailed information contained in these databases can be recovered for management purposes.

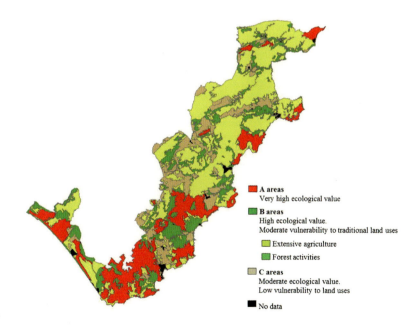

Fig. 29.4 Zonation units proposed for the Cabo de Gata-Níjar Natural Park

29.4 Conclusions

The hierarchic framework discussed in this chapter is able to assist the management of natural resources at different spatial levels in a consistent way. Information coming from the fine scale, at ecotope level, can be aggregated to broader scales for answering different resource management issues. In this way, the determination of ecological units at different spatial scales helps conduct the management of natural resources in a multi-scale frame.

The information contained in each ecotope class may focus on one conservation variable or a set of variables depending on management aims. Therefore, the flexibility of the method allows analyzing the variables in different ways depending on the management objective. Besides, the score classes simplified the process of translating the decision criteria, for the determination of the zonation units, from common language to technical implementation. Furthermore, the use of meaningful score classes helps managers explain people the reasons behind land use restrictions, gaining their adhesion and minimizing social conflicts.

The hierarchic geoform classification system used in the geopedologic approach provided the framework for classifying and assessing the ecosystems of the natural park, integrating biotic and abiotic data in a coherent spatial and thematic way. This study provides a protocol to assist managers in implementing recommended land use regulations. The protocol can be improved incorporating new ecological knowledge to the protection status of the ecosystems in the Cabo de Gata-Nijar Natural Park.

References

Arponen A, Moilanen A, Ferrier S (2008) A successful community-level strategy for conservation prioritization. J Appl Ecol 45(5):1436–1445
Bailey RG (1996) Ecosystem geography. Springer, New York
Boekheim J, Schliemann S (2014) Soil richness and endemism across an environmental transition zone in Wisconsin, USA. Catena 113:86–94
Bonn A, Gaston KJ (2005) Capturing biodiversity: selecting priority areas for conservation using different criteria. Biodivers Conserv 14(5):1083–1100
Botrill MC, Pressey RL (2012) The effectiveness and evaluation of conservation planning. Conserv Lett 5(6):407–420
Decreto 37/2008, de 5 de febrero, por el que se aprueba el Plan de Ordenación de los Recursos Naturales y el Plan Rector de Uso y Gestión del Parque Natural Cabo de Gata-Níjar y se precisan los límites del citado Parque Natural
Doran JW, Coleman DC, Bezdicek DF, Stewart BA (eds) (1994) Defining soil quality for sustainable environment, Special publication 35. Soil Science Society of America, Madison
Dudley N (ed) (2008) Guidelines for applying IUCN protected area categories. IUCN, Gland. http://www.iucn.org/dbtw-wpd/edocs/paps-016.pdf
European Commission, DG Environment (2007) Interpretation manual of European Union Habitats. EUR 27, Brussels, July 2007

Fernández Soler JM (1996) El volcanismo calco-alcalino en el Parque Natural Cabo de Gata-Níjar (Almería). Soc. Almeriense Historia Natural/Junta de Andalucía, Almería

Ibáñez E (2007) La planificación en los espacios naturales protegidos: aplicación de los PORN en las Cordilleras Béticas andaluzas. Investig Geográficas 44:103–127

Ibáñez JJ, Sánchez-Díaz J, Rodríguez-Rodríguez A, Effland WR (2008) Preservation of European soils: natural and cultural heritage. In: Dazzi C, Costantini E (eds) The soils of tomorrow. Advances in Geoecology 39. Catena Verlag-IUSS, pp 37–59. Reiskirchen, Germany

Ibáñez J, Krasilnikov PV, Saldaña A (2012) Review: archive and refugia of soil organisms: applying a pedodiversity framework for the conservation of biological and non-biological heritages. J Appl Ecol 49(6):1267–1277

Klijn F, de Haes UH (1994) A hierarchical approach to ecosystems and its implications for ecological land classification. Landsc Ecol 9(2):89–104

Kosmas C, Kairis O, Karavitis C, Ritsema C, Salvati L, Acikalin S, Qinke Y (2013) Evaluation and selection of indicators for land degradation and desertification monitoring: methodological approach. Environ Manage 54(5):951–970

Lajeunesse D, Domon G, Drapeau P, Cogliastro A, Bouchard A (1995) Development and application of an ecosystem management for protected natural areas. Environ Manage 19:481–495

Legendre P (1993) Spatial autocorrelation: trouble or new paradigm? Ecology 74(6):1659–1673

Ley 4/1989 de 27 de Marzo, de Conservación de las Espacios Naturales y de la Flora y Fauna Silvestres. Ley 42/07 del Patrimonio Natural y de la Biodiversidad. BOE 14 diciembre 2007, núm. 299

Müller C (1999) Modelling soil-biosphere interactions. CABI Publishing, Wallingford

Noss RF (1990) Indicators for monitoring biodiversity: a hierarchical approach. Conserv Biol 4(4):355–364

Ortiz-Lozano L, Grandaos-Barba A, Espejel I (2009) Ecosystemic zonification as a management tool from marine protected areas in the coastal zone: application for the Sistema Arrecifal Veracruzano National Park, Mexico. Ocean Coast Manage 52:317–323

Oyonarte C (2004) Cárcavas y regueros. In: Villalobos M, Salas R, Lastra J (eds) Cabo de Gata, un espacio de leyenda. ACUSUR/Junta de Andalucía, Madrid, pp 45–49

Palik B, Goebel P, Kirkman L, West L (2000) Using landscape hierarchies to guide restoration of disturbed ecosystems. Ecol Appl 10(1):189–202

SPSS for windows 10.0 (2000) SPSS Inc., 1989–1999

Wilson K, Pressey R, Newton A, Burgman M, Possingham H, Weston C (2005) Measuring and incorporating vulnerability into conservation planning. Environ Manage 35(5):527–543

Wu J, David JL (2002) A spatially explicit hierarchical approach to modelling complex ecological systems: theory and applications. Ecol Model 153:7–26

Yarrow MM, Salthe SN (2008) Ecological boundaries in the context of hierarchy theory. Biosystems 92:233–244

Zinck JA (2013) Geopedology. Elements of geomorphology for soil and geohazard studies. ITC special lecture notes series, Enschede

Zonneveld IS (1989) The land unit. A fundamental concept in landscape ecology and its applications. Landsc Ecol 3(2):67–86

Chapter 30
Design and Evaluation of an Afforestation Project Based on Geopedologic and Ecological Information in North-Western Patagonia, Argentina

M.C. Frugoni, A. Dezzotti, A. Medina, R. Sbrancia, and A. Mortoro

Abstract Forest plantations can positively influence ecosystem patterns and processes, but afforestation based on monocultures can also affect plant diversity and should therefore be soundly designed. In the Aguas Frías Forest Station, Argentina, planting areas were determined using a geopedologic and ecological approach, and project effects on vegetation and soil were assessed. In 2007, a fence was installed around the station for protection against herbivores and a fire control system was implemented. Vegetation cover in the study area includes natural steppes, meadows, forests, and a relict scrubland. Soils are Andisols derived from cinder and pumice of Holocene volcanic activity. Main soils are Humic (45 %) and Aquic Udivitrands (6 %), and Typic Endoaquands (14 %). Suitable land for forest tree planting comprised 160 ha, taking into account soil fertility and planting restriction on valuable ecosystems. Geopedology has proven to be useful for assessing land suitability for pine plantation. Ecological indicators related to plant diversity, forest regeneration, and soil protection show improvement after 7 years of project implementation. These variables should be carefully monitored so that the social, conservation, and economic objectives of the project can be sustainably achieved.

Keywords Soil-landscape relation • Volcanic ash soils • Biodiversity • Forest suitability analysis • *Pinus ponderosa*

M.C. Frugoni (✉) • A. Dezzotti • A. Medina • R. Sbrancia • A. Mortoro
Universidad Nacional del Comahue, San Martín de los Andes, Argentina
e-mail: crisfrugoni@gmail.com; dezzotti@infovia.com.ar;
andrepampa@yahoo.com.ar; renato@smandes.com.ar; terrafain@hotmail.com

© Springer International Publishing Switzerland 2016
J.A. Zinck et al. (eds.), *Geopedology*, DOI 10.1007/978-3-319-19159-1_30

30.1 Introduction

Anthropogenic activities have dramatically degraded natural forests (Hansen et al. 2010; Lindquist et al. 2012). During the twentieth century, Argentina lost 70 % of its forest cover (SAyDS 2007). Simultaneously, the demand for goods and services provided by forest ecosystems continues increasing (FAO 2014). The imbalance between demand and supply of forest resources partly explains ongoing creation of forest plantations. In Argentina, about 1 million ha have been planted, of which 64 % is with pine trees (DPF 2009). The province of Neuquén accounts for 54,000 ha representing 75 % of all forest plantations in Patagonia, with 89 % being *Pinus ponderosa* (Pinaceae) forest (CFI-MDT 2009). This species shows adequate growth and development under the semi-arid conditions of the region characterized by dry summer, with frequent and intense wind, and heavy winter snowfall.

Forest plantations contribute to reduce erosion (La Manna et al. 2013), conserve fragile and valuable habitats (Dezzotti et al. 2013), and capture CO_2 (Laclau 2003; Nosetto et al. 2006). However, dense monocultures of exotic trees frequently affect the diversity of plants (Paritsis and Aysen 2008), insects (Corley et al. 2006), and vertebrates (Lantschner et al. 2012; Nájera and Simonetti 2010; Simonetti et al. 2013). Therefore, this productive pine forest system needs to be adequately managed to ensure conservation through silviculture and to provide shelter and food for wildlife (CBD 2010).

Silvicultural practices require valuable soilscape information at semi-detailed or detailed scale to assess land suitability for forest trees. Geopedologic maps can provide this spatial information including physical and chemical properties (e.g. water and air holding capacity, root penetration resistance, effective soil depth, nutrient availability) and terrain form (slope gradient and aspect) (FAO 1984). The aim of the present study was to determine suitability areas for *P. ponderosa* plantation based on information derived from applying a geopedologic approach (Zinck 2013) in the Aguas Frías Forest Station (Neuquén), and assess afforestation effects on native vegetation and soil using current and historical data from 7 years after planting.

30.2 Materials and Methods

30.2.1 Study Area

Aguas Frías is a forest station owned by the national oil company Yacimientos Petrolíferos Fiscales S.A. (YPF), silviculturally managed by the Corporación Forestal Neuquina S.A. (CORFONE). It is located 38°46′S and 70°54′W at an altitude of 1510–1670 masl. Climate is humid and windy, with cold winters and warm summers. Average annual temperature is 7.6 °C and average annual rainfall is 1,266 mm (Dezzotti et al. 2013) (Fig. 30.1). The Köppen-Geiger climate classification qualifies the area as temperate with dry and warm summers (Csb) (Peel et al. 2007).

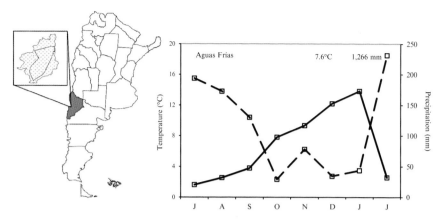

Fig. 30.1 Annual variation of mean temperature (*continuous line*) and precipitation (*dotted line*) and location of Aguas Frías Forest Station in the Neuquén province of Argentina (*grey*)

Rock substrata include basalts, andesites, breccias, volcanic agglomerates, and non-stratified glacial drift (Ferrer 1991). This lithology is covered by a thick mantle of Holocene tephra from the active Andean volcanoes that constitutes the soil parent material. Relief is mountainous with typical glacial morphology features such as glacial lakes, trough shoulders, cirques, hanging valleys, and erratic blocks (van Zuidam 1985; Gonzalez Díaz and Ferrer 1991). The area belongs to the Patagonic and Subantarctic ecoregions that exhibit a variety of vegetation types including forests, steppes, and meadows (Cabrera 1971). Historically, Aguas Frías was part of a route of nomadic and extensive ranching of goats and sheep, migrating from bottom valleys in winter to higher elevations in summer. In 2007, the station was fenced for protection against herbivores and trampling of domestic livestock and wildlife of medium and large size. A fire control system was also implemented.

30.2.2   Survey Approach and Land Suitability Assessment

Soil mapping method was based on a hierarchic geoform classification system (Zinck 2013), using visual interpretation of aerial photographs at scale 1:25,000 and field survey. Map units were determined taking into account changes in slope, aspect, shape, and elevation in relation to the surroundings. Profile descriptions and soil sampling for laboratory analyses were used to determine the composition of the soil map units (Schoeneberger et al. 1998). A geopedologic map was produced after ground-truthing the initial photo-interpretation map. Screen digitizing on ortho-rectified aerial photographs provided the baseline cartography. The mapped area of 1140 ha covered the whole catchment basin in which the Aguas Frias Forest Station is located. The mean soil map unit size was 22 ha and the smallest delineation was 0.2 ha. Soils were classified at subgroup level following the USDA Soil Taxonomy (Soil Survey Staff 1994).

Soil requirements for *P. ponderosa* are mainly related to physical properties, including effective soil depth, air and water holding capacity, and root penetration resistance (Girardín and Broquen 1995; Broquen et al. 1998; Suárez et al. 2012). These properties together with criteria related to conservation and restoration of natural forests and meadows helped determine suitability classes using GIS-based spatial analysis (e.g. map overlay and other map operations).

30.2.3   Vegetation and Ecological Indicators

Forest land units were identified on an ASTER satellite image (resolution 15 m, Gauss Krüger coordinates band 1, ellipsoid WGS 1984), using vegetation physiognomy, composition, and structure, and subsequently verified in the field. In 2014, the effect of fencing on species diversity and soil cover was evaluated within the units conformed by mixed forest and herbaceous-shrubby steppe, on the basis of ten random sampling sites inside (closed condition) and outside the fence (unclosed condition). From the centre of each sampling site, 10 m transects were laid in the four cardinal directions. On each transect, sampling points were located 1 m apart to determine the frequency of vascular species, using the point-intercept method (Kent 2011).

Plant species richness, diversity, and density were estimated according to Simpson (1949) and Rosenzweig (2003). Compositional similarity was based on presence/absence, using the Sørensen index (Diserud and Ødegaard 2007), and relative frequency data, using the Morisita-Horn index (Chao et al. 2008). Proportion of bare soil was also estimated using frequency data. The effect of enclosure on the natural forest was assessed in 2007 and 2014, using size and structure of the tree population. In each forested landscape unit, a permanent sampling plot of 2,000 m^2 was installed and all adult (diameter $d \geq 0.1$ m) and sapling trees ($d < 0.1$ m and total height $h > 0.1$ m) were measured for d (diameter tape and calliper) and h (hypsometer and tape). Composition and abundance of seedlings ($d < 0.1$ and $h \leq 0.1$ m) were evaluated in 20 subplots of 0.5 m^2 located inside the main plot. Past distribution of *Nothofagus antarctica* (Nothofagaceae) scrubland was identified and georeferenced on aerial photographs dated from 1970 (1:25000).

30.3   Results and Discussion

30.3.1   Vegetation Types

Main vegetation types present in the study area include herbaceous-shrubby steppe (HS, 69.4 % of the area), hygrophilous (HM) and xerophilous meadows (XM) (16.2 %), *Nothofagus pumilio* (Nothofagaceae) pure forest (PF), and *N. pumilio* and *Araucaria araucana* (Araucariaceae) mixed forest (MF) (8.6 %). The rest of the

Table 30.1 Landscape units of Aguas Frías based on present and past vegetation types

	Vegetation type	Code	Area (ha)	%
Present	Pure forest of *Nothofagus pumilio*	PF	35.9	7.6
	Mixed forest of *Nothofagus pumilio* and *Araucaria araucana*	MF	4.9	1.0
	Herbaceous-shrubby steppe dominated by *Chusquea culeou* and *Festuca pallescens*	HS	325.5	69.4
	Hygrophilous meadow with *Cortaderia egmontiana*, *Azorella trifoliolata*, and *Carex macloviana*	HM	71.1	15.1
	Xerophilous meadow on rocky outcrops with *Festuca pallescens*, *Colobanthus lycopodiodes*, and *Berberis microphylla*	XM	5.0	1.1
	Deflation area of pumice and cinder with very scarce vegetation cover	DA	27.3	5.8
Past	Pure scrubland dominated by *Nothofagus antarctica*	PS	76.5	16.3
Total			469.6	100

area corresponds to pumice and cinder deflation spots with scarce vegetation (DA, 5.8 %). The scrubland (PS) composed of isolated *N. antarctica* living and dead trees is a relict formation that covered 16.3 % of the area in the 1960s (Table 30.1, Fig. 30.2).

30.3.2 Soil Types

Well-developed allophanic Andisols have formed from pumice and cinder parent materials. They show low bulk density (<0.9 Mg/m^3), high water holding capacity, and high content of organic matter (50–100 g/kg). They are generally deep to very deep. Shallow and very stony soils are found mainly on eroded slopes. Soils show a positive reaction to the Fieldes and Perrots test, indicating presence of active aluminium (Mizota and van Reeuwijk 1989). Humic Udivitrands are the most widespread soils within the study area (Table 30.2). These are very deep (120 cm to more than 250 cm) and well drained soils. They occupy most of the backslopes and sideslopes, as for instance in the Mo121 and Mo214a map units. The presence of cobbles, stones, and boulders in Humic Udivitrands stony phase clearly reduces the effective soil depth, water holding capacity and, therefore, the physical fertility of the soils. They occupy 4 % of the study area and are dominant on strongly eroded slopes as for instance in unit Mo217 (Table 30.2, Fig. 30.3).

Aquic Udivitrands are very deep (>120 cm) and moderately well drained soils with redox concentrations from 90 cm down. The wetter soil bottom contributes an extra supply of water, which can balance the hydric stress during the dry season. They predominate in vale areas (Mo216) and on the lower backslopes (Mo122), and occupy 6 % of the study area (Table 30.2, Fig. 30.3).

Typic Endoaquands develop under aquic conditions with permanent water table, and occur in hanging valleys (Mo123) and alluvial plains (Mo210). They show

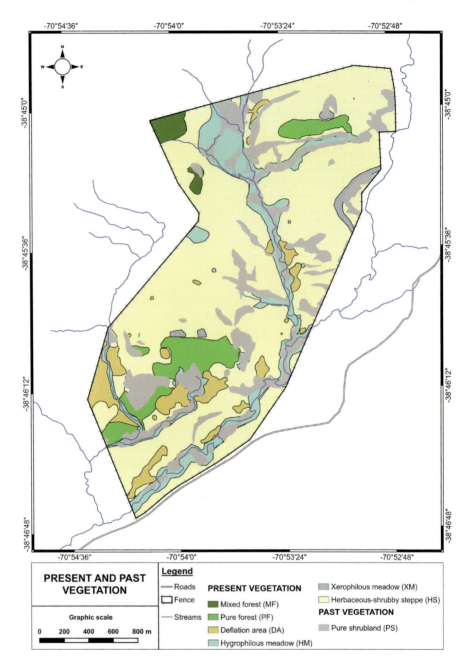

Fig. 30.2 Landscape units of Aguas Frías Forest Station based on present and past vegetation

Table 30.2 Legend of the geopedologic map of the Aguas Frías area

Landscape	Relief/molding	Lithology/facies	Terrain form	Map unit	Area (ha)	Main soils
Glacial modelled mountains Mo	Glacial erosion molding Mo1	Basalt, andesite Mo11	Summit and upper relief flank	Mo111	41	Detrital cover, rock outcrops
			Scarp	Mo112	30	Rock outcrops
			Moderately steep upper backslope	Mo113	8	Stony phase of HumicUdivitrands
			Cliff	Mo114	82	Rock outcrops
		Holocene lapilli and cinder Mo12	Steep slope	Mo121	370	Humic Udivitrands
			Gently sloping lower backslope	Mo122	57	Aquic Udivitrands
			Hanging valley	Mo123	52	Typic Endoaquands
			Small canyon	Mo124	42	Rock outcrops
	Hillsides and glacial valley bottom Mo2	Holocene lapilli and cinder over non stratified glacial drift Mo21	Interfluve	Mo211	20	Detrital cover, pockets of shallow soil
			Headslope	Mo212	102	Humic Udivitrands
			NE aspect side slope	Mo213a	32	Detrital cover, stony phase of HumicUdivitrands
			NW aspect side slope	Mo213b	85	Humic Udivitrands, detrital cover
			W aspect concave sideslope	Mo214a	29	Humic Udivitrands
			E aspect concave sideslope	Mo214b	27	Detrital cover, pockets of shallow soil
			Gently sloping upper backslope	Mo215	10	Humic Udivitrands
			Vale	Mo216	13	Aquic Udivitrands
			Moderately steep footslope	Mo217	7	Stony phase of HumicUdivitrands
			Incision	Mo218	92	Detrital cover, pockets of shallow soil
			Base slope	Mo219	24	HumicUdivitrands
			Alluvial plain	Mo210	5	TypicEndoaquands
	Deflation areas Mo3	Pumice and cinder Mo31	Pumice mantle	Mo311	12	Lapilli and cinder

Fig. 30.3 Geopedologic map of the Aguas Frías area

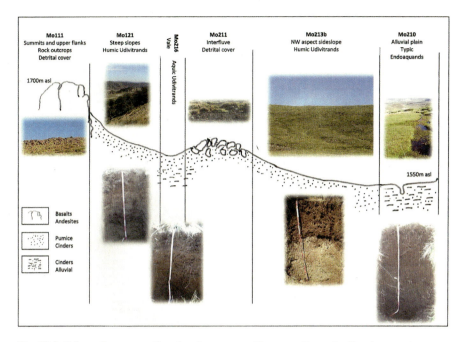

Fig. 30.4 Schematic cross-section showing geomorphic map units, and soil and vegetation types of Aguas Frías Forest Station

redoximorphic features from the topsoil downwards. Rock outcrops and detrital covers (29 %), with little soil development, are dominant in the higher and steepest portions of the landscape such as summits and upper relief flanks (Mo111), scarps (Mo112), cliffs (Mo114), and interfluves (Mo211) (Table 30.2, Fig. 30.3). Soil-landscape relationships along toposequences show a continuum from bare areas on summits, to moderately deep and very deep well drained soils on backslopes, to very deep moderately well drained soils on footslopes, and poorly drained soils on toeslopes (Fig. 30.4).

30.3.3   Forest Suitability

The prevailing volcanic ash soils determine areas of high physical fertility for *P. ponderosa* plantation. These are deep, well to moderately well drained and friable soils, with high water holding capacity, low bulk density, and high total porosity. They are particularly suitable for pine plantations on hillsides and in glacial valley bottoms (map units Mo121, Mo213b, and Mo216 in Fig. 30.4). Stoniness reduces effective soil depth and water holding capacity, therefore limiting land suitability. Rock outcrops, detrital covers, and poor drainage are excluding conditions for plantation because of insufficient effective soil depth, as in map units Mo111, Mo211, and Mo210 (Fig. 30.4). Slope gradient and aspect are not limiting conditions. Protection of valuable ecosystems prevailed over areas identified as highly suitable for silviculture, as in the case of unit Mo121 because of the presence of mixed natural forest. In total, 159 ha are deemed moderately and highly suitable for pine plantation, while 310 ha are unsuitable because of soil restrictions and conservation of highly valuable ecosystems (Table 30.3, Fig. 30.5).

30.3.4 Monitoring Afforestation Performance

In the herbaceous-shrubby steppe, total plant richness exhibits large differences between closed (n=47) and unclosed condition (n=24). In contrast, mixed forest shows similar values regardless of protection. Mean species richness is highest in the closed steppe $(\overline{x} = 10.7)$ and lowest in the closed mixed forest $(\overline{x} = 7.5)$. This variable differs significantly between closed and unclosed condition only in the herbaceous-shrubby steppe (t test, $p<0.05$, n=10). Plant species diversity ranges from 3.6 in the closed mixed forest to 4.1 in the herbaceous-shrubby steppe. Changes in protection condition within landscape units do not significantly affect mean diversity values (t test, $p\geq0.05$, n=10). Plant species evenness is highest in the closed mixed forest $(\overline{x} = 0.5)$ and lowest in fenced areas of the herbaceous-shrubby steppe $(\overline{x} = 0.4)$. However, differences between condition in a given vegetation type are not significant (t test, $p \geq0.05$, n=10) (Table 30.4).

Overall, the species density index is greater in closed than in unclosed areas. For instance, in the herbaceous-shrubby steppe, the value of this variable decreases from 4.6 in fenced to 1.9 in unfenced condition. In all cases, values of vegetation similarity between closed and unclosed areas are lower than 1. For example, in the herbaceous-shrubby steppe, similarity based on incidence data is 0.45 while based on relative frequency is 0.87 (Table 30.4). In both vegetation types, average bare soil is statistically much lower in closed than in unclosed areas (t test, $p<0.05$, n=10). For instance, in the mixed forest the value of this variable is 34.1 % for the closed and 51.4 % for the unclosed condition (Table 30.4).

In 2007, tree regeneration represented by seedlings and saplings was negligible or intensively browsed by domestic livestock. However, 7 years later after the fence was constructed around the forest station, young individuals within the pure forest comprised 150,000 ind/ha for *N. pumilio*, whereas within the mixed forest regeneration density was 1035 ind/ha for *A. araucana* and 240 ind/ha for *N. pumilio*. Between 2007 and 2014, the number of adult trees in the smallest diameter classes between 10 and 40 cm also increased markedly; for instance during this period the density of *A. araucana* within the mixed forest changed from 20 to 135 ind/ha (Fig. 30.6).

Table 30.3 Land suitability for forest plantation in map Aguas Frías based on soil properties and conservation criteria

Geomorphic Map unit	Dominant soil	Soil fertility	Vegetation cover	Forestry suitability (limitations)
Steep slope, headslope, W aspect concave sideslope, NW aspect side slope, gently sloping upper backslope, base slope	Humic Udivitrand	Deep to very deep, well drained, high water holding capacity, friable, mean carbon content 50 g/kg	Herbaceous-shrubby steppe	Highly suitable
Gently sloping lower backslope, vale	Aquic Udivitrand	Deep to very deep, moderately well drained, high water holding capacity, friable, mean carbon content 95 g/kg	Herbaceous-shrubby steppe	Highly suitable
Moderately steep upper backslope, moderately steep footslope	Stony phase of Humic Udivitrand	Moderately deep, well drained, less water holding capacity caused by stoniness	Herbaceous-shrubby steppe	Moderately suitable (effective soil depth, water holding capacity)
Steep slope	Humic Udivitrand	Deep to very deep, well drained, high water holding capacity, friable, mean carbon content 50 g/kg	Pure and mixed natural forest	Not suitable (conservation of natural forest)
Alluvial plain	Typic Endoaquand	Poorly drained, redoximorfic features from the topsoil downwards	Hygrophilous meadow	Not suitable (conservation of hygrophilous meadow, drainage)
Summit and upper relief flank	Rock outcrops and detrital cover	No soil	Xerophilous meadow	Not suitable (conservation of xerophilous meadow, soil development)
Scarp, cliff, small canyon, interfluve, NE aspect side slope, E aspect concave sideslope, incision	Rock outcrops and detrital cover	No soil	No vegetation	Not suitable (soil development)
Pumice mantle	No soil development	No soil	No vegetation	Not suitable (soil development)

Other excluded areas: strip of 15–50 m width around natural forest, meadow, alluvial plain, road, and fence

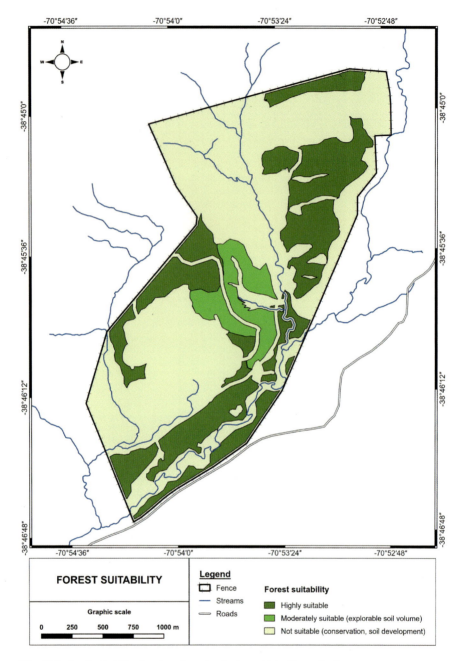

Fig. 30.5 Land suitability for pine plantation in Aguas Frías Forest Station

Table 30.4 Values of ecological variables in the mixed forest closed (MF$_c$) and unclosed (MF$_u$) and in the herbaceous-shrubby steppe closed (HS$_c$) and unclosed (HS$_u$) within the Aguas Frías Forest Station

Ecological variable		Vegetation type			
		MF$_c$	MF$_u$	HS$_c$	HS$_u$
Total plant richness (n)		24	23	47	24
Mean plant richness (n)	$\bar{x}$	7.5	8.6	10.7	8.5
	se	1.0	0.6	1.4	0.8
Plant evenness index	$\bar{x}$	0.5	0.4	0.4	0.5
	se	0.1	0.0	0.0	0.0
Plant diversity index	$\bar{x}$	3.6	3.7	4.1	4.1
	sc	0.4	0.3	0.4	0.2
Density index of plant species		2.4	1.2	4.6	1.9
Vegetation similarity index	Incidence	0.68		0.45	
	Frequency	0.87		0.87	
Bare soil (%)	$\bar{x}$	34.1	51.4	27.0	40.2
	se	9.2	3.5	4.4	5.8

Mean $\bar{x}$ and standard error se are indicated for n = 10

30.4   Conclusions

Detailed spatial information provided by the geopedologic survey, to characterise the landscape, soils and geomorphometry, has proven to be a valuable practical tool for determining suitability classes for pine plantation in the Aguas Frías afforestation project. In spite of parent material homogeneity, soil-landscape relationships created variability in particular as related to water dynamics. Toposequences from backslope to toeslope showed a sequence of well drained – moderately well drained – poorly drained soils. In concave slopes, accumulation of volcanic material was favoured allowing the development of very deep soils. Footslopes, which were expected to contain deep soils, showed moderately deep and very stony soils. Detection of these particular conditions helped establish soilscape units considering slope aspect and shape. At the working scale of 1:25,000, the minimum 0.2 ha delineation size was appropriate for the study purpose.

Trees are now successfully established. Productivity and its relation to land suitability should be monitored for at least 15 years from plantation onset, when trees are going to attain a commercial measurable size indicative of site quality. At present, ecological indicators related to plant diversity, regeneration of natural forest, and soil protection, particularly within the fenced steppe, show substantial improvement. Planting restriction in fragile and valuable areas, grazing exclusion, and fire control are playing a key role in landscape conservation. Direct positive impact on soil

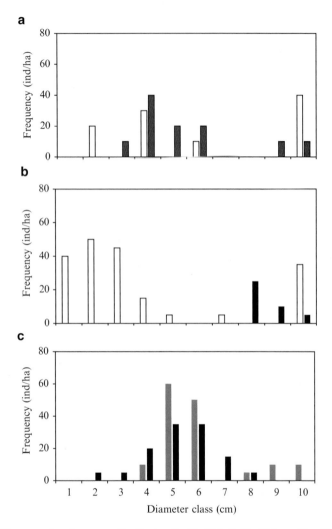

Fig. 30.6 Diameter-class frequency distribution (1: 10–19.9, 2: 20–29.9,…, 10: >100 cm) of *A. araucana* (*white bar*) and *N. pumilio* (*black bar*) in the mixed forest in 2007 (**a**) and 2014 (**b**), and of *N. pumilio* in the pure forest in 2007 (*grey bar*) and 2014 (*black bar*) (**c**)

protection is expected to increase as pine trees grow and develop. These indicators should continue to be monitored in the context of adaptive management to allow adjustments. This will help project objectives to be sustainably achieved as related to social (employment generation), economic (diversification of production), and conservation issues (increase CO_2 capture, protect hydrological basins, reduce soil erosion, and decrease anthropogenic pressure on fragile ecosystems).

Acknowledgements This study was funded by the Universidad Nacional del Comahue (UNCo) (project 04/S017), the Secretaría de Agricultura, Ganadería y Pesca de la Nación (grant 12004), and YPF S.A. Assistance of J Rizzo, C Calleja (YPF), L Sancholuz, L Chauchard, C Dufilho, R González Musso, H Attis, C Monte, M Catalán, H Mattes, S Tiranti, V Fontana, R Gönck, A Suárez, J Curruhuinca (UNCo), and the staff of CORFONE is gratefully acknowledged.

References

Broquen P, Girardín J, Falbo G et al (1998) Modelos estimadores de IS14 en *Pinus ponderosa* Dougl. Bosque 19(1):71–79

Cabrera A (1971) Fitogeografía de la República Argentina. Bol Soc Arg Bot 14:1–42

CBD (2010) Strategic plan for biodiversity 2011–2020 and the Aichi targets. Secretariat of the convention on biological diversity. Available via DIALOG. http://www.cbd.int/doc/strategic-plan/2011-2020/Aichi-Targets-EN.pdf. Accessed 30 Nov 2014

CFI-MDT (2009) Inventario del bosque implantado de la provincia de Neuquén. Consejo Federal de Inversiones, Ministerio de Desarrollo Territorial, Neuquén

Chao A, Jost L, Chiang S et al (2008) A two-stage probabilistic approach to multiple-community similarity indices. Biometrics 64:1178–1186

Corley J, Sackmann P, Rusch V et al (2006) Effects of pine silviculture on the ant assemblages (Hymenoptera: Formicidae) of the Patagonian steppe. For Ecol Manage 222:162–166

Dezzotti A, Sbrancia R, Acciaresi G et al (2013) Fortalezas ambientales y sociales de un programa forestal de desarrollo implementado en la Patagonia semiárida argentina. In: Pérez D, Rovere A, Rodríguez Araujo M (eds) Restauración de la diagonal árida de la Argentina. Larrea-INTA-CONICET, Buenos Aires, pp 468–477

Diserud OH, Ødegaard F (2007) A multiple-site similarity measure. Biol Lett 3:20–22

DPF (2009) Elaboración de un mapa de plantaciones forestales de la República Argentina de actualización permanente. Dirección de Producción Forestal, Ministerio de Agricultura, Ganadería y Pesca, Buenos Aires

FAO (1984) Land evaluation for forestry, FAO forestry paper 48. Food and Agriculture Organization, Rome

FAO (2014) El estado de los bosques del mundo: potenciar los beneficios socioeconómicos de los bosques. Food and Agriculture Organization, Rome

Ferrer JA (1991) Geología: estudio regional de suelos de la provincia del Neuquén. Consejo Federal de Inversiones – Consejo de Planificación y Acción para el Desarrollo, Neuquén

Girardín J, Broquen P (1995) El crecimiento de *Pinus ponderosa* y *Pseudotsuga menziesii* en diferentes condiciones de sitio. Bosque 16(2):57–69

González Díaz E, Ferrer JA (1991) Geomorfología: estudio regional de suelos de la provincia del Neuquén. Consejo Federal de Inversiones – Consejo de Planificación y Acción para el Desarrollo, Neuquén

Hansen M, Stehman S, Potapov P (2010) Quantification of global gross forest cover loss. Proc Natl Acad Sci U S A 107(19):8650–8655

Kent M (2011) Vegetation description and data analysis: a practical approach. Wiley-Blackwell, New York

La Manna L, Buduba C, Gigli A et al. (2013) Efecto de las plantaciones sobre la erosión hídrica potencial en suelos degradados de la Región Andino Patagónica. In: Abstracts of the II Jornadas Forestales de Patagonia Sur, Calafate, 16–17 May 2013

Laclau P (2003) Biomass and carbon sequestration of ponderosa pine plantations and native cypress forests in northwest Patagonia. For Ecol Manage 180:317–333

Lantschner MV, Rusch V, Hayes JP (2012) Habitat use by carnivores at different spatial scales in a plantation forest landscape in Patagonia, Argentina. For Ecol Manage 269:271–278

Lindquist EJ, D'Annunzio R, Gerrand A et al (2012) Global forest land-use change 1990–2005, Forestry paper 169. FAO, Rome

Mizota C, van Reeuwijk LP (1989) Clay mineralogy and chemistry of soils formed in volcanic material in diverse climatic regions, Soil monograph 2. Institute of Soil Reference and Information Centre, Wageningen

Najera A, Simonetti JA (2010) Enhancing avifauna in commercial plantations. Conserv Biol 24:319–324

Nosetto MD, Jobbágy EG, Paruelo JM (2006) Carbon sequestration in semiarid rangelands: comparison of *Pinus ponderosa* plantations and grazing exclusion in NW Patagonia. J Arid Environ 67:142–156

Paritsis J, Aizen M (2008) Effects of exotic conifer plantations on the biodiversity of understory plants, epigeal beetles and birds in *Nothofagus dombeyi* forests. For Ecol Manage 255:1575–1583

Peel MC, Finlayson BL, McMahon TA (2007) Updated world map of the Köppen-Geiger climate classification. Hydrol Earth Syst Sci 11:1633–1644

Rosenzweig ML (2003) Reconciliation ecology and the future of species diversity. Oryx 37:194–205

SAyDS (2007) Primer inventario nacional de bosques nativos. Secretaría de Ambiente y Desarrollo Sustentable, Buenos Aires

Schoeneberger PJ, Wysocki DA, Benham EC et al (1998) Field book for describing and sampling soils (ver 1.1). Natural Resources Conservation Service, USDA National Soil Survey Center, Lincoln

Simonetti JA, Grez A, Estades C (2013) Providing habitat for native mammals through understory enhancement in forestry plantations. Conserv Biol 27(5):1117–1121

Simpson EH (1949) Measurement of diversity. Nature 163:688

Soil Survey Staff (1994) Keys to soil taxonomy. US Department of Agriculture, Soil Conservation Service, Pocahontas Press, Blacksburg

Suárez A, Girardín J, Broquen P et al (2012) Potencialidad forestal de los suelos de una toposecuencia representativa de la región extrandina, SO de Neuquén, Argentina. In: Abstracts of the XXIII Congreso Argentino de la Ciencia del Suelo, Mar del Plata, 16–20 April 2012

van Zuidam RA (1985) Aerial photo-interpretation in terrain analysis and geomorphologic mapping. Smith Publishers, The Hague

Zinck JA (2013) Geopedology. Elements of geomorphology for soil and geohazard studies. ITC Special Lecture Notes Series, Enschede

Chapter 31
Territorial Zoning Based on Geopedologic Information: Case Study in the Caroni River Basin, Venezuela

P. García Montero

Abstract Geomorphology-based land surveys, founded on geoform-soil relationships, have been carried out in Venezuela at different scales and in a variety of environments using aerial photographs, radar and satellite images. The Caroni river basin is one of the most important watersheds in southern Venezuela, providing the largest part of the electric energy consumed in the country. To guarantee the sustainability of the hydroelectric production, a management plan of the natural resources of the watershed is needed. A territorial zoning of the catchment area based on geomorphic and soil information was undertaken as an initial step to propose land uses compatible with preserving the hydroelectric potential. Geomorphic units and their soil components, together with ancillary elements including the vegetation cover, were mapped at two scales using a multicategorial geoform classification system. From this information two zoning proposals were derived, one based on geomorphic landscape units at 1:250,000 scale and the other based on relief-type units at 1:100,000 scale. The zoning units were used for land evaluation and for establishing land use regulations required for the watershed management plan.

Keywords Geomorphology • Soils • Land use planning • Environmental planning • Caroni basin • Venezuela

31.1 Introduction

A watershed is a natural and functional biophysical unit for the management of natural resources. For each watershed there is a particular combination of biotic and abiotic interacting components. This allows differentiating watersheds according to their limitations, suitability, and capacity to provide environmental services, among others, water production. Not only is water one of the most important

The original version of this chapter was revised: The surname of the author has been corrected. The erratum to this chapter is available at: http://dx.doi.org/10.1007/978-3-319-19159-1_34

P. García Montero (✉)
Private Activity, Soil Survey and Environmental Planning, Caracas, Venezuela
e-mail: prgm2002@gmail.com

© Springer International Publishing Switzerland 2016
J.A. Zinck et al. (eds.), *Geopedology*, DOI 10.1007/978-3-319-19159-1_31

natural resources, it is also one of the most affected by human activities. As an economic system, a watershed encompasses resources that can be used to produce goods and services under diverse technological conditions. However, the lack of a holistic approach to characterize and evaluate the potentialities of the watersheds and the forms and operational ways of using the land units can cause the alteration of the hydrological cycle and that of other ecological functions commonly underestimated by planners, politicians, and users.

Watersheds are recognized as territorial planning frames for integrated management of natural resources. They represent a spatial arrangement of ecosystems with strong and complex relationships whose structure, functions, and environmental services must be taken into account when making management decisions.

Watersheds are under increasing anthropic pressure which generates serious problems of land degradation. There are many anthropogenic activities that cause environmental risks in watershed areas, including deforestation, inappropriate agricultural uses, overgrazing, degradation of wildlife habitats, loss of biodiversity, soil erosion, invading mining activities, water pollution, sediment load, river channel destruction, uncontrolled urban expansion, among others. Territorial zoning is a sound technical approach to prevent, control, and reduce watershed degradation. Zoning projects involve political decisions based on geospatial, environmental, social, and legal information that allows harmonizing land constraints and suitabilities with land uses and territorial occupation.

The Caroni river basin in southern Venezuela is a strategic watershed as it provides the largest part of the electrical energy consumed in the country. To guarantee the sustainability of the hydroelectric production, the formulation of a management plan of the natural resources of the watershed is needed. The objective of this study is to show how geomorphic and soil information can contribute to the partitioning of the watershed in zoning units as a basis for decision making on appropriate land uses and the preservation of the natural resources.

31.2   Land Inventories in the Neotropical Lowlands of South America

The neotropical lowlands of South America include a variety of large watersheds belonging mainly to the Amazon and Orinoco basins, with important natural resources that are increasingly threatened by uncontrolled exploitation. However systematic inventory of these resources is still lagging behind. One of the largest land inventory projects using remote sensing was the Radam Brazil Project carried out between 1970 and 1985 by the Ministry of Mining and Energy of Brazil (Projeto RADAM/RADAMBRASIL 1972–1978). This integrated natural resource inventory covered several regions of the vast and remote Brazilian territory, particularly the Amazon region. Radar images from SLAR were interpreted for the characterization and cartography of the physical and biotic environmental components. Geomorphic maps at 1:250,000 and 1:500,000 scales were generated and used as principal spatial information source for the cartography of soils, geology, vegetation, and potential land use. The integrated information was used for the ecological zoning of Amazonia. Similarly, in 1972, the Colombian Government initiated the "Proyecto Radargramétrico del Amazonas" (PRORADAM 1979) with the aim to carry out an exploratory study of the physical and human resources of the Colombian Amazonas as a basis for integrating the Amazon region into the development process of Colombia.

In the Venezuelan Guayana region, the Ministry of the Environment and Renewable Natural Resources (MARNR) in cooperation with ORSTOM, now IRD (Institut de Recherche pour le Développement, France) carried out during the period 1975–1981 the soil survey of the Venezuelan Amazonas territory. Geomorphic maps at 1:250,000 scale were generated including information related to geology, soils, and vegetation (MARNR 1981). Subsequently, one of the largest Venezuelan integrated land inventories was undertaken initially by the Venezuelan Guayana Corporation (CVG) and continued by CVG-TECMIN (1987). Under this project, an area of 468,000 km² in southern Venezuela (about 52 % of the national territory) was surveyed at 1:250,000 scale, making systematic use of remote sensing products (SLAR and Landsat images and aerial photographs at different scales). Field verification units were defined basically by geological and geomorphic attributes extracted by the multidisciplinary interpretation of the remote-sensed materials. Thematic geology, soil, and vegetation maps had as cartographic base the geomorphic delineations determined at the categorial level of geomorphic landscape according to the methodology developed by Zinck (1988).

31.3 The Study Area: National Importance of the Caroni River Basin

The Caroni river basin covers 92,170 km² in the Bolivar state, southern Venezuela, between 3°40′–8°40′ latitude north and 60°50′–64°10′ longitude west (Fig. 31.1). The Caroni river is the second largest of Venezuela with a length of 958 km and the principal tributary of the Orinoco river basin with an average discharge of about 4500 m³/s.

The Caroni river basin is one of the most important strategic territories of Venezuela. It represents a unique area in which converge remarkable natural and socioeconomic features. It is one of the oldest earth surfaces (Precambrian age), with a large reserve of hydrological, vegetal, and mineral resources. With an average rainfall of 2900 mm/year, including areas with >4000 mm/year, the basin contributes 13 % of the country's annual runoff and has one of the highest hydroelectric potential in Latin America (around 30,000 MW). It provides about 70 % of the

Fig. 31.1 Location of the Caroni river basin, Venezuela

country's electrical energy consumption, which is generated by four hydroelectric power plants representing an investment of around 15,000 million US$. Other assets include 65 % of forest lands with high biodiversity, 48 % of the country's fauna species, mineral reserves (e.g. gold, diamonds, iron, bauxite), and singular scenic resources associated to large and fast flow wing rivers and table-shaped highlands called tepuies, with high endemism and spectacular waterfalls (i.e. the Angel Fall, the highest worldwide).

It is still a largely uninhabited geographic space with an average population density of nine inhabitants per km^2. Indigenous people, around 3 % of the total basin population, live mainly upstream, while 90 % of the population concentrates in the lower basin. There are significant contrasts in cultural patterns, religions, ways of living, and modalities of using the natural resources. Spontaneous human settlements and aggressive exploitation of the natural resources are causing environmental threats and damages. This includes illegal mining leading to river channel destruction, pollution and sediment production; deforestation for expanding shifting cultivation, intensive agriculture, illegal timber exploitation, extensive cattle raising using fire for natural pasture management, and uncontrolled urban expansion in the rural and suburban areas. The lower basin area is the most impacted by agriculture, livestock, and urban expansion, while the sparsely populated middle and upper stretches are seriously affected by deforestation, vegetation fire, and illegal mining. Increasing land degradation and inappropriate use of the natural resources are threatening the integrity of the hydrological cycle and the sustainable generation of energy, while affecting at the same time other environmental services provided by the watershed ecosystems. This situation calls for the need to better control the settlement trends and rationalize the use of the natural resources by means of a territorial zoning plan.

31.4 Methodological Approach

31.4.1 Introduction

Two levels of zoning were proposed for the study area, based on available geo-environmental information and criteria such as land use intensity, degree of land degradation, land suitabilities, ecological features, and the presence of legally protected areas (i.e. national parks). For that purpose, geo-environmental information was collected and evaluated at two resolution levels, using as basic determinants of the map units the concepts of geomorphic landscape and relief type (Zinck 1974, 1988) at 1:250,000 and 1:100,000 scales respectively. Other environmental components and attributes resulted to be usually well correlated with the geomorphic background.

Geoforms are conspicuous and distinguishable natural terrain tracts that can be recognized by their external attributes at different levels of abstraction. They are comprehensive cartographic units which are usually correlated with other landscape components, some of them being easily observable (e.g. topography, vegetation, rockiness, morphodynamic activity) or less observable (e.g. soil, internal drainage).

The utility of remote sensing for geomorphic survey in such a large, remote, and poorly accessible area as the Caroni river basin was immense, allowing relatively rapid inventory of the natural resources of this vast territory and the monitoring of land use changes. The interpretation of radar (SLAR) and satellite (Landsat) images permitted the identification and mapping of relatively homogeneous geomorphic surfaces used as reference for the cartography of other landscape components, particularly geology, soil, and vegetation cover, as well as for land evaluation, environmental planning, and territorial zoning.

Territorial zoning can be considered as a multidisciplinary and integrated process of partitioning the landscape in areas that show clear spatial arrangement and internal coherence in their components. These spatial units are evaluated in terms of limitations, suitabilities, and use potentials to assess their level of tolerance to human interventions, environmental management, and conservation policies.

In different countries and morphogenic environments, regional planning and environmental studies have been carried out based on geodata (COPLANARH 1973, 1974; MARNR-ORSTOM 1979; Steegmayer and Bustos 1980; MARNR 1983; Zinck 1970; Santosa and Sutikno 2006; Santos et al. 2006; India National Institute of Hydrology 2010; Ferrando and de Lucas 2011; Prakasam and Biswas 2012; Islam et al. 2014). In these studies, terrain features from diverse geo-environments were used for land survey, land zonation, regional and environmental planning proposals.

In this chapter, geomorphic data and information are used as basic input for the zoning projects of the Caroni river basin at the scales of 1:250,000 and 1:100,000. The work reviews and analyses the information generated by CVG TECMIN (Venezuelan Guayana Corporation) and subsequently updated and used by EDELCA (Caroni Electrification Company) to formulate the environmental Master Plan of the

Caroni watershed (CVG EDELCA 2004a). The information provided by the geomorphic units is evaluated and used to prepare a zoning proposal for the environmental planning and management of the basin.

31.4.2 Collecting Geo-environmental Information at 1:250,000 Scale

Zoning units were derived from the geomorphic units generated by the integrated land survey carried out by CVG and CVG TECMIN in the Caroni river basin as part of the Natural Resources Inventory Project for the Guayana Region (PIRNG). Cartographic units were delineated by visual interpretation of radar (SLAR) and satellite images (Landsat), both at 1:250,000 scale, and aerial photographs at 1:100,000 and 1:50,000 scales. Map units at the categorial level of geomorphic landscape were identified using the following attributes: drainage pattern (type, density), local and regional geo-structures (faults, fractures), lineaments, relief, dissection intensity, image texture (greytones, roughness), all derived by interpretation of radar images and aerial photographs (Table 31.1). Black and white multispectral Landsat and false color (bands 4, 5, and 7) images were interpreted for delineating changes in the vegetation cover. Image interpretation was complemented by using aerial photographs at different scales to identify the relief types included in each geomorphic landscape and classify types of vegetation communities according to attributes such as life form, height, density, and intervention degree. These preliminary delineations were verified through multidisciplinary fieldwork, and the field information was correlated for each cartographic unit.

At the 1:250,000 resolution level, cartographic units are basically associations of landscapes and relief types identified and mapped following the criteria included in Table 31.1. Figure 31.2 shows a hilland landscape unit originally delineated on a SLAR image and later updated and validated using a digital elevation model (Instituto Geográfico de Venezuela 2003). Soil is considered one of the physical components of the landscape unit, together with other physical variables. Soils in each landscape unit represent the dominant taxa surveyed in each of the relief types that integrate that landscape unit. The composition of the cartographic units is a combination of soil taxa classified at great group level (Soil Survey Staff 1975) and their phases (soil thickness, rockiness, stoniness, slope, among others). Geomorphic map units and zoning units are combinations of several attributes.

31.4.3 Collecting Geo-environmental Information at 1:100,000 Scale

For the lower Caroni river basin, information was collected at 1:100,000 scale because of more intensive land use, severe land degradation, and the presence of the hydropower plants potentially impacted by inappropriate land use trends.

Table 31.1 Criteria for defining and mapping geomorphic landscape units at 1:250,000 scale

Attribute	Code	Mountain (Mo)	Plateau (Al)	Piedmont (Pm)	Hilland (Lo)	Peneplain (Pe)	Valley (Va)	Plain (Pl)
Altitude (masl)	1	Low	Low (<900)		Low	Low (<200)	Low	Low
	2	Medium	Medium (900–1,600)		Medium	Medium (200–500)	Medium	Medium
	3	High	High (1,600–2,600)		High	High (>500)	High	High
Topography (% slope)	1	Steep (30–60)	Flat (0–4)	Sloping (4–16)	Rolling (4–16)	Undulating (4–8)		
	2	Very steep (>60)	Undulating (4–16)	Mod sloping (16–60)	Hilly (16–30)	Rolling (8–16)		
	3		Steep (16–60)	Strong sloping (>60)	Steeply hilly (30–60)			
	4		Very steep (>60)					
Dissection	1		Slightly dissected	Slightly dissected				
	2		Moderately dissected	Moderately dissected				
	3		Strongly dissected	Strongly dissected				
Geo-substratum	1						Depositional	Depositional
	2						Residual	Residual
	3						Mixed	Mixed
	4						Rocky/stony	Rocky/stony
Drainage	1						Well drained	Well drained
	2						Poorly drained	Poorly drained
	3						Flooded	Flooded

CVG TECMIN (1987) modified by García

Dissection: remaining original terrain surface 1: >75 %, 2: 25–75 %, 3: <25 %

Code numbers indicate classes of attributes used to delineate cartographic units in each geomorphic landscape

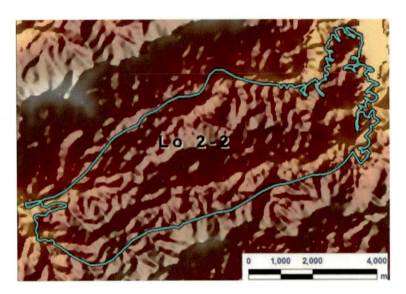

Fig. 31.2 Example of a hilland landscape (Lo 2-2) delineated for zoning at 1:250,000 scale (SLAR + DEM background)

Table 31.2 Examples of geomorphic units used for zoning in the lower Caroni river basin at 1:100,000 scale

Landscape	Relief type	Symbol
Hilland	High elongated hill	La (1–9)
	Medium elongated hill	Lm (1–11)
	Low elongated hill	Lb (1–17)
	Dome/inselberg	Di (1–7)
Peneplain	Medium rounded hill	Cm (1–7)
	Low rounded hill	Cb (1–9)

EDELCA (2004b)

The geomorphic units and types of vegetation cover were delineated by visual interpretation of aerial photographs at 1:50,000 and 1:100,000 scales, bands of radar images (SLAR) at 1:100,000 scale, and satellite images (Landsat) at 1:100,000 and 1:250,000 scales. At this resolution level, soil was the most important map unit component. The composition of the cartographic units is a combination of soil taxa classified at subgroup level (Soil Survey Staff 1993) and their phases (soil thickness, rockiness, stoniness, slope, among others). Table 31.2 shows the relief types mapped to construct the zoning units. The number added to the symbol indicates the different geomorphic units resulting from the combination of the attributes indicated in Table 31.1. For instance, Le 1-3 means "low, elongated hill developed on schist, 8–16 % slope".

Geomorphic units are combinations of relief types discriminated by criteria such as lithology, morphometric attributes (e.g. slope, relative height, configuration),

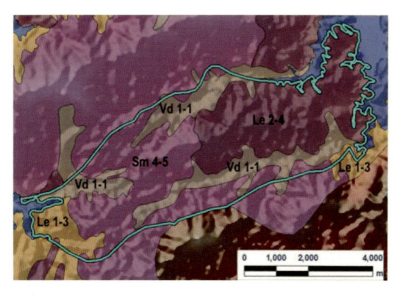

Fig. 31.3 Relief types identified in a hilland landscape (Lo 2-2 in Fig. 31.2) for zoning at 1:100,000 scale (Vd1-1: narrow Holocene floodplain, 0–4 % slope; Sm 4-5: ridges developed on metabasite, slope >30 %; Le 1-3: elongated hills developed on schist, 8–16 % slope; Le 2-4: elongated hills developed on schist, 16–30 % slope) (SLAR+DEM background)

surface features (rockiness, stoniness), and ancillary information particularly vegetation. Soil cover and actual erosion information was collected during fieldwork, when each geomorphic unit was surveyed following the hillslope model, specifically in undulating landscapes (CVG EDELCA 1990; CVG TECMIN 1995). Figure 31.3 shows four map units originally delineated on a SLAR image at 1:100,000 scale and later updated and validated using a digital elevation model (Instituto Geográfico de Venezuela 2003). This allowed disaggregating the landscape unit Lo 2-2 shown in Fig. 31.2. Each of the resulting cartographic units differs from the others in terms of soil cover, topography, limitations, suitabilities, and other land attributes. For the zoning proposal, the original soil classification was updated (Soil Survey Staff 1998).

At this resolution level, a zoning unit can be the summation of several geomorphic units having similar limitations and suitabilities, although they may have developed on different lithological substrata. For that reason, it is common to have different geoforms integrating a particular zoning unit.

31.5 Zoning Proposals

The use of geomorphic units at two levels of spatial resolution allowed the division of the Caroni river basin in zoning units suitable for different land uses and for supporting the creation of a system of protected areas subjected to specific environmental and administration regulations. The resulting territorial organization would be the basis for designing management programs allowing EDELCA and other national and regional public institutions to control the land uses, address the territorial occupation according to land limitations and potentialities, and promote environmental programs that guarantee the conservation of the natural resources and associated ecosystems, and the sustainability of the hydro-energy generation.

Table 31.3 shows the zoning units proposed at regional level (1:250,000). These units are associations of geomorphic landscapes with some similarity in bio-physical attributes, constraints, and potential uses. The geomorphic landscape units allowed the differentiation, description, and evaluation of 19 territorial zoning units at 1:250,000 scale for the whole Caroni river basin. They have been the basis for proposing a system of protected areas called Special Administration Regime Areas according to Venezuelan environmental laws, including two National Parks, one Natural Monument, two Protection Zones, one Public Work Protection Area, and five National Forest Reserves. Within this frame of protected areas there is a wide range of permitted land uses, including agriculture and forest exploitation.

The zoning proposal for the lower Caroni river watershed at 1:100,000 scale divided the territory of about 1268 km^2, representing 1.4 % of the total basin area, into 35 zoning units conformed by combinations of geoforms at the level of relief types. Two categories of areas with special administration regime were recognized: (1) a Public Protection Work Zone composed of 17 zoning units, and (2) the Guri Dam Protection Zone that assembles 18 zoning units including the Caroni River Zone and one Indigenous Community Zone. At this resolution level there is a relatively higher homogeneity in the zoning unit components and attributes. Zoning units have larger land use options because of the presence of soils suitable for multiple uses that have been better discriminated a 1:100,000 scale. At this scale, soil was one the most relevant factor for selecting land uses and determining the final vocation of the zoning units. Table 31.4 shows some zoning units proposed for the lower Caroni watershed. At this level, the geomorphic landscapes mapped at 1:250,000 scale are disaggregated in various relief types (Fig. 31.3) to conform the zoning units. In addition to the general objectives pursued by the territorial zoning for the entire basin, the zoning of the lower watershed focused especially on promoting environmental programs that guarantee the conservation and appropriate use of the land areas adjacent to the hydropower plants.

31.6 Discussion

Geomorphic studies are essential to understand the chorology of the land units as a basis for characterizing and evaluating the natural resources they encompass, with influence on the potential economic and social development of a region. Although geoforms were the most relevant criteria for delineating the zoning units in the study area, information on climate, geology, soils, vegetation, land suitability, land use, actual erosion, among others, was used to characterize and evaluate the land areas

Table 31.3 Zoning proposal at 1:250,000 scale for some areas of the Caroni river basin

Zoning unit	Area km²	Geomorphic landscape	Soils	Recommended uses
Integral Protection Zone (ZPI)	1094	Plateau, piedmont	Rock outcrops, Udorthents, Kanhapludults, Kanhaplohumults	Protection of unique and fragile ecosystems, biodiversity and genus reservoir, headwaters, riparian forests, wildlife habitats, peatlands and scenic sites. Carbon sinks. Environmental monitoring. Scientific research
Primitive Zone (ZP)	707	Mountain, plateau, piedmont, hilland, and valley	Udorthents, Kanhaplohumults, Haplohumults, Kanhapludults, Paleudults, Rock outcrops	Protection of fragile and high biodiversity forest ecosystems, wildlife habitats, and water sources. Carbon sinks. Ecotourism and research
Integral Protection Zone (ZPI)	2062	Mountain, plateau, hilland, piedmont, peneplain, and valley	Kanhaplohumults, Hapludults, Kanhapludults, Rock outcrops	Protection of unique and fragile ecosystems, biodiversity and genus reservoir, headwaters, riparian forests, wildlife habitats, peatlands and scenic sites. Carbon sinks. Environmental monitoring. Scientific research
Integral Protection Zone (ZPI)	107	Peneplain	Kanhaplohumults, Kandihumults, Kanhaplustults, Kandiustults, Rock outcrops	Protection of unique and fragile ecosystems, biodiversity and genus reservoir, headwaters, riparian forests, wildlife habitats, peatlands and scenic sites. Carbon sinks. Environmental monitoring. Scientific research
Preservation Zone for Intensive Agriculture (ZPAI)	32	Peneplain and hilland	Kanhaplohmults, Kandihumults, Kandiudults	Intensive agriculture, irrigated or rainfed, with soil conservation practices; agroforestry, forestry, and intensive cattle raising

EDELCA (2004a) modified by García

Table 31.4 Examples of proposed zoning units at 1:100,000 scale for the lower Caroni river basin

Landscape unit	Relief type % slope	Zoning unit	Area km²	Soils	Recommended uses
Peneplain	High elongated hills and low rounded hills; 4–8	Z2	38,698	Udic Kanhaplustults, Lithic Ustorthents; high rockiness and stoniness	Protection of stream headwaters, high biodiversity forest lands and wildlife habitats
Hilland and peneplain	High and low elongated hills and low rounded hills; 4–8	Z3	106,503	Udic Kanhaplustults, Typic Haplohumults, Kanhaplic and Udic Kandiustalfs; moderate rockiness	Cattle raising, agroforestry and locally forestry and intensive agriculture
Hilland	High elongated hills; >30	Z5	12,009	Lithic Kanhaplohumults, Kanhaplic Haplustalfs, Lithic Ustorthents, moderate to high rock outcrops	Protection of stream headwaters, high biodiversity forest lands and wildlife habitats
Hilland	Medium elongated hills; 16–30	Z9	22,805	Kanhaplic Haplustalfs, Udic Kanhaplustults, Ustic Kanhaplohumults; low to moderate rockiness	Cattle rising, agroforestry and forestry
Peneplain	Medium and low rounded hills; glacis; 4–8	A2	50,193	Typic and Lithic Kanhaplustults, Typic and Arenic Kandiustults, Lithic Ustorthents, high rockiness	Protection, Locally, agroforestry and forestry
Peneplain and hilland	Medium rounded hills and low elongated hills; 8–16	A6	161,215	Lithic Ustorthents, Typic and Lithic Kanhaplustults, Typic Kandiustults; low rockiness	Protection, Locally, agro forestry and forestry
Hilland	Medium and low elongated hills; 16–30	A8	20,714	Lithic Ustorthents, Typic and Lithic Kanhaplustults, high rockiness	Protection of stream headwaters, high biodiversity forest lands and wildlife habitats

EDELCA (2004b) modified by García
Z2, Z3, Z5, and Z9 represent zoning units mapped in the Protector Zone of the Guri Dam; A2, A6, and A8 are zoning units mapped in the Public Work Protection Zone of the hydropower plants downstream the Guri dam

included in each zoning unit. In all zoning areas, clear soil-geoform relationships were observed whatever the scale. Soil forming factors, pedogenic processes, and land suitabilities were identified in observable spatial limits corresponding to the boundaries of the zoning units. These land tracts possess some homogeneity in terms of characteristics, properties, conditions, and expected behavior in response to human activities. What is considered homogeneous depends on the use purpose, but generally each zone contains a mixture of environmental elements such as lithology, relief, slope, soilscape, vegetation, and other features, that allows for multi-purpose uses.

At 1:250,000 scale, attributes belonging to the geomorphic surfaces such as, for example, topography and level of dissection can be relevant to decide about land use options in large, relatively pristine, highly fragile, and poorly accessible areas like de Caroni river basin where the offer of land suitable for multiple uses is limited. At this resolution level, zoning for land use planning can be preliminarily supported by geodata provided by the geomorphic units together with some ancillary information (e.g. vegetation cover). In this situation, geomorphic information is very cost-effective, especially when the general objective is the protection of natural resources, the management of national parks, the control of areas with strong use restrictions or vulnerable to geo-hazards, among others. At 1:100,000 scale, the potential use of the zoning units is mostly based on soil suitability that sets the recommended uses of the different zones. In the Caroni case study, the territorial zoning generated on the basis of geopedologic information can be considered acceptable for land use planning purposes and for defining environmental policies and priority actions geared towards the protection of this strategic territory.

The mapping of geomorphic surfaces provides reliable spatial information for generating a variety of thematic maps. Cartographic units based on geomorphic criteria facilitate the landscape reading and understanding by non-specialists and planners. They provide means for correlating known and unknown areas, thus permitting terrain conditions to be reasonably predicted in the case of areas devoid of environmental information.

31.7 Conclusions

The morphometric elements of the geoforms are the result of morphogenic and morphodynamic forces acting on a particular geological substratum (rock or saprolite), under specific conditions and combinations of climate and vegetation cover, during a particular span of time. This is the main assumption to support using the geomorphic surface as a relevant and relatively stable source of data and information to map zoning units for environmental and land use planning purposes.

The geomorphic approach, in which the form and spatial distribution of terrain features are analyzed in an integrated manner, relates recurrent geomorphic surface patterns expressed by the interaction of environmental components, allowing the partitioning of the landscape into relatively homogeneous land units. Geoform classification and characterization may be used as part of the land-use planning decision making. In the examples of zoning presented here, classification schemes were based on knowledge of the bedrock geology, topography, surface formations, soils, and ancillary information such as vegetation cover and climate. Most of the information for geoform classification was derived from remote-sensed documents and complemented by low intensity fieldwork. From a practical point of view, territorial zoning based on geomorphic surfaces is a very useful approach in large, pristine, and remote areas like de Caroni river basin. It allows delineating land units for zoning objectives. At 1:250,000 scale where the key purpose may be formulating conservation proposals, as in the case of National Parks and Natural Monuments, land units founded on geomorphic surfaces can contribute to identify fragile areas susceptible to erosion, guide integrated and higher resolution land inventories, and formulate research projects, monitoring environmental programs, and preliminary regulations. At 1:100,000 scale where land use decisions are related to multiple uses, more intensive interventions, and higher risk of environmental impacts, zoning based only on geomorphic units introduces uncertainty because decisions on land use and planning require knowledge of other use determining factors such as soils. Geoforms, regardless of their abstraction level, are not only soilscape units; they are also useful multi-attribute territorial units for environmental planning and other land use decisions. Combining geoform and soil information, as it is intended in the geopedologic approach, offers a balanced way of tackling the territorial zoning issue.

References

COPLANARH (1973) Metodologías utilizadas en el Inventario Nacional de Tierras, Publicación 36. Comisión del Plan Nacional de Aprovechamiento de los Recursos Hidráulicos, Caracas

COPLANARH (1974) Inventario nacional de tierras. Estudio geomorfológico de los Llanos Orientales. Regiones 7 y 8, Sub-regiones 7C, 8A, 8B, Zonas 7C2, 8A2, 8A3, 8B1 y 8B2. Caracas, Venezuela

CVG EDELCA (1990) Estudio integral del área de influencia inmediata del embalse Guri. Dirección de Estudios e Ingeniería, División de Cuencas e Hidrología, Departamento de Estudios Básicos, Puerto Ordaz

CVG TECMIN (1987) Manual metodológico. Proyecto inventario de los recursos naturales de la Región Guayana. Gerencia de Proyectos Especiales, Puerto Ordaz

CVG TECMIN (1995) Caracterización de los recursos físico-naturales y aspectos socio- económicos del área de influencia de los futuros embalses de los desarrollos hidroeléctricos de Macagua, Caruachi, y Tocoma. Gerencia de Proyectos Especiales, Puerto Ordaz

EDELCA (2004a) Estudio plan maestro de la cuenca del río Caroní. Zonificación y ordenación espacial de la cuenca del río Caroní. Vol 3 Tomo 4. Gerencia de Gestión Ambiental. Caracas, Venezuela

EDELCA (2004b) Estudio plan maestro de la cuenca del río Caroní. Zonificación y reglamento de uso del área Bajo Caroní. Vol 4 Tomo 2. Gerencia de Gestión Ambiental. Caracas, Venezuela

Ferrando F, de Luca F (2011) Geomorfología y paisaje en el ordenamiento territorial: valorizando el corredor inferior del río Mapocho. Investig Geogr 43:65–86. Santiago, Chile

Instituto Geográfico de Venezuela (2003) Proyecto Cartosur Fase 2 SAR. Ministerio del Ambiente, Caracas

Islam S, Reza A, Islam T, Rahman F, Ahmed F, Haque N (2014) Geomorphology and land use mapping of northern part of Rangpur District, Bangladesh. Geosci Geomatics J 2(4):145–150

MARNR (1981) Sistemas Ambientales Venezolanos. Proyecto Ven/79/001. Dirección General Sectorial de Información e Investigación del Ambiente y Dirección General de Planificación y Ordenación del Ambiente. Caracas

MARNR (1983) Sistemas ambientales venezolanos. Proyecto Ven/79/001. Región Natural 27. Depresión de Unare. Documento 27. Caracas, Venezuela

MARNR-ORSTOM (1979) Atlas del inventario de tierras del Territorio Federal Amazonas. Dirección General Sectorial de Información e Investigación del Ambiente, Caracas

National Institute of Hydrology (2010) Geomorphological and land use planning for Danda watershed. A research report. http://www.indiawaterportal.org/sites/indiawaterportal.org/files/Geomorphological_and_Land use_Planning_for_Danda_Watershed.pdf

Prakasam C, Biswas B (2012) Evaluation of geomorphic resources using GIS technology: a case study of selected villages in Ausgram Block, Burdwam District, West Bengal, India. Int J Geol Earth Environ Sci 2:193–205

Projeto RADAM/RADAMBRASIL (1972–1978) Levantamento de recursos naturais. Vol 1–15. Ministério das Minas e Energia, Departamento Nacional da Produçao Mineral, Rio de Janeiro. http://pt.wikipedia.org/wiki/Projeto_Radambrasil. Accessed 29 Nov 2014

PRORADAM (Proyecto Radargramétrico del Amazonas, Colombia) (1979) La amazonia colombiana y sus recursos. 5 vols, figs., fotografías, mapas (1:500,000). Instituto Geográfico Agustín Codazzi (IGAC), Bogotá

Santos L, Duque M, Herrero AD (2006) Aspectos geomorfologicos en las Directríces de Ordenación Territorial de Segovia y Entorno (DOTSE). In: Geomorfología y Territorio. Actas de la IX Reunión Nacional de Geomorfología. Servicio de Publicación e Intercambio Científico, USC, Santiago de Compostela, pp 945–561

Santosa LW, Sutikno P (2006) Geomorphological approach for regional zoning in the Merapi volcanic area. Indones Geogr J 38(1):53–68. http://jurnal.ugm.ac.id/ijg/article/view/2235. Accessed 30 Nov 2015

Soil Survey Staff (1975) Soil taxonomy: a basic system of soil classification for making and interpreting soil surveys, Handbook 436. Govt Printing Office, Washington, DC

Soil Survey Staff (1993) Soil survey manual, Handbook 18. US Department of Agriculture, Washington, DC

Soil Survey Staff (1998) Keys to soil taxonomy, 8th edn. US Department of Agriculture, Natural Resources Conservation Service, Washington, DC

Steegmayer P, Bustos R (1980) Proposición metodológica para estudios de suelos en cuencas altas. MARNR, San Cristobal

Zinck JA (1970) Aplicación de la geomorfología al levantamiento de suelos en zonas aluviales. Ministerio de Obras Públicas (MOP), Barcelona

Zinck JA (1974) Definición del ambiente geomorfológico con fines de descripción de suelos. Ministerio de Obras Públicas (MOP), Cagua

Zinck JA (1988) Physiography and soils, Lecture notes. International Institute for Aerospace Survey and Earth Sciences (ITC), Enschede

Chapter 32
Contribution of Geopedology to Land Use Conflict Analysis and Land Use Planning in the Western Urban Fringe of Caracas, Venezuela

O.S. Rodríguez and J.A. Zinck

Abstract In areas where land use and land cover dynamics generates current and potential land use conflicts, geopedologic information derived from soil survey, landscape inventory, and land evaluation, together with data about economic and social conditions, contributes to guide the land use planning process. Geopedologic information is important in the first steps of the process to identify land use options for agricultural, engineering and sanitary purposes, among others, while social, economic, administrative, and political criteria are decisive in the final steps to select land use alternatives satisfying the stakeholders. A framework for land use planning is presented, where land use conflict analysis represents an essential tool for decision making on land use allocation, taking into account the stakeholders interests. An urban fringe case study carried out in the western periphery of Caracas (Venezuela) is discussed to illustrate the significant contribution of geopedology to strengthen land use planning.

Keywords Geopedologic unit • Land use policy • Land use scenario • Peri-urban area • Caracas

O.S. Rodríguez (✉)
Facultad de Agronomia, Universidad Central de Venezuela, Maracay, Venezuela
e-mail: osrp1958@gmail.com

J.A. Zinck
Faculty of Geo-Information Science and Earth Observation (ITC), University of Twente, Enschede, The Netherlands

Institute of Environmental Studies, University of New South Wales, Sydney, NSW, Australia
e-mail: alfredzinck@gmail.com

© Springer International Publishing Switzerland 2016   521
J.A. Zinck et al. (eds.), *Geopedology*, DOI 10.1007/978-3-319-19159-1_32

32.1 Introduction

Land use planning seeks to optimize land uses because land is a scarce and limited resource for particular uses. Multiple use of a piece of land is rarely attainable and may generate land use conflicts that can be managed and mitigated through land use planning and policy making. Land use conflicts occur when the same tract of land is appropriate for different uses and people with interest in the land disagree on what is the best use. Conflicts can also occur when one land user is perceived to infringe upon the rights, values, or amenity of a neighboring land user (NSW Government 2011). Two broad categories of land use conflicts can be recognized: the conflicts arising from incompatibilities caused by adjacency or proximity, and the conflicts associated with land conversion issues (Bryant et al. 1982). Conversion and incompatibility conflicts usually arise together in different proportions; it is therefore difficult to encase a specific land use conflict within one single class. Land use conflicts are not inherent to land resources but derived from the human behavior controlling land use decisions (Healy and Rosenberg 1979). Natural spatial variability of the landscape generates different values of the land units for specific uses according to distribution patterns, extent, qualities, and suitability for different purposes. To identify and assess this variability, land use inventory and land evaluation are needed. On the other hand, land use decision-making processes are driven by economic and social forces, being users participation important to reconcile divergent interests and ensure sustainability over time (de Groot 2006; Mann and Jeanneaux 2009; Nolon et al. 2013).

Land use conflicts can be functional, intensity-related, or generational. Functional land use conflicts refer to competitive land uses that lead to undesirable land use conversions and reduce the land available for specific land uses at regional or national levels, or cause incompatibilities with the surrounding land uses. They arise when land uses performing different functions are intended on a piece of land highly or moderately suitable for a variety of uses. Intensity land use conflicts occur in areas where the land qualities do not fulfill the agro-ecological, management, or environmental requirements of the actual or proposed land use. Generational land use conflicts include potential functional and intensity land use conflicts as they impact choices and needs of future generations. Any functional conversion today will be charged to future generations. Current land use allocations must anticipate and compromise with future land use demands (Rodríguez and Zinck 1998).

Peri-urban zones referring to the space between city and countryside are areas where dispersive urban growth creates fragmented landscapes. Such hybrid urban areas are particularly exposed to irreversible land use changes leading to the loss of agricultural land and threatening environmental amenities. These negative consequences result from uncontrolled and disorganized city expansion within the urban fringe as part of the global urbanization trend. Optimization of land use is needed to help solve land use conflicts via comprehensive land use planning. Land use conflict resolution has been commonly tackled from the political, economic, or legal points of view. Less attention has been given to the natural resource base where soil-landscape (i.e. geopedologic) information plays a significant role for assessing land suitabilities. Together soil information and land use conflict analysis constitute the basic input to policy formulation and scenario building as the main tools for land use allocation in a sustainable decision-making process. Mobilizing soil information in the search for suitable land use alternatives was set as central goal of this investigation. Scenario building was chosen as the main procedure for comparing different alternatives, identifying land use incompatibilities and generating discussion plans for analysis by expert groups and the community involved (Rodríguez 1995; Rodríguez and Zinck 1998).

The general objective of this research work was to establish, test, and implement a method of mobilizing soil-landscape information together with social, economic, and institutional data for detecting, assessing, and solving land use conflicts in urban fringes to assist planning and policy making. The western fringe of Caracas, the capital city of Venezuela, was used as case study because it presents multiple and severe land use conflicts generated by activities that compete for scarce land resources.

32.2 Study Area

The western periphery of Caracas is an area exposed to multiple peri-urban land use conflicts. The landscape is steeply mountainous, stretching from semiarid scrub at low elevation, just above sea level, to hyperhumid cloud forest at mountain summits around 2000 m elevation. Three main types of land use compete in this territory for the same, equally attractive locations, especially above about 500 m elevation: (1) intensive market-oriented agriculture specializing in vegetable, fruit, and flower production, adapted to highland conditions relatively scarce elsewhere in the region, (2) protection of the cloud forest environment for preservation of its high biodiversity, its role in supporting and regulating natural cycles (e.g. water cycle), and its vital water catchment sources, and (3) peri-urban expansion through residential subdivisions and intensive recreation activities, that tends to alleviate the pressure on the relief-constrained metropolitan area of Caracas.

Legally protected areas such as national parks, water production catchments, and peri-urban buffer zones are threatened by the uncontrolled, highly dynamic urban expansion. Water production catchments that collect water from the cloud forest areas (e.g. Petaquire river and many smaller rivers) benefit from a special administrative regime that theoretically regulates land utilization but is inefficient to prevent unpermitted land uses. Protected areas include the Macarao National Park, the Pico Codazzi natural monument, and the Caracas protection zone. Tracts of land falling within the Caracas metropolitan zone and other areas not specifically regulated were also included to show, in conjunction with protected areas, a broad spectrum of land uses and land use conflicts. On steep slopes, horticultural crops that require temperate climate and coffee plantations generate important farm production and complementary activities within the region, but their protection has not been sufficiently secured. The expansion of these agricultural systems to new areas is limited by the scarcity of suitable soil-landscape conditions and water resources and by the legal status of nature preservation imposed to large areas in this unique climatic environment. Conflicts between the main types of land use (e.g. agriculture versus nature conservation) and within specific land use areas (e.g. horticulture threatened by the invasion of intensive recreation resorts and second homes) derive from the competition for the same tracts of land. The location of the study area covering about 17,000 ha is shown in Fig. 32.1.

32.3   Research Method and Techniques

A conceptual model for land use planning was designed as a three-stage process involving land inventory, evaluation, and allocation (Fig. 32.2). Most of the basic information on land resources and land use context needed to implement the model was produced during the development of this work. The land inventory captured data on land use and land cover, geopedologic units, and institutional framework. Although highly variable in space, geopedologic units integrating soil and geoform information represent the most stable dataset through time so that other information about natural and cultural features can be georeferenced to them. They were also used as the main spatial units for land evaluation and allocation. Current land use was crossed with geopedologic units to support land use conflict analysis and carry out land evaluation and allocation, especially in areas where it was hypothesized that current land uses could not be reverted to former, less intensive kinds of land use.

The data provided by the inventory stage were integrated for land use conflict analysis and land evaluation. Land use policies were formulated and a policy weighting process was adopted for building alternative scenarios and land allocation discussion plans. After consultation with the concerned user groups operating in the area, a final land use allocation plan was proposed for implementation.

The collected data were georeferenced and introduced in a geographic information system for storing, processing, analysis, and interpretation. Three data bases were implemented: (1) a spatial database storing different maps within the ILWIS system (ITC 1993), (2) a soil profile database using a relational database management system (Arenas, FONAIAP, personal communication), and (3) a tabular database to manage farming systems data.

Available software was integrated including the ALES system (Rossiter and Van Wambeke 1993), designed to implement the FAO framework for land evaluation (FAO 1976), and the LUPIS system, an integrated package for land use allocation (Ive 1992). Terrain units were sorted and labelled according to a geoform classification system (Zinck 1988). The geopedologic map units were obtained by a sequential combination of thematic maps including life zones, relief, landform, slope, and current erosion. Geopedologic units and their individual delineations (polygons) were used as the basic map units throughout the land use planning process. Land

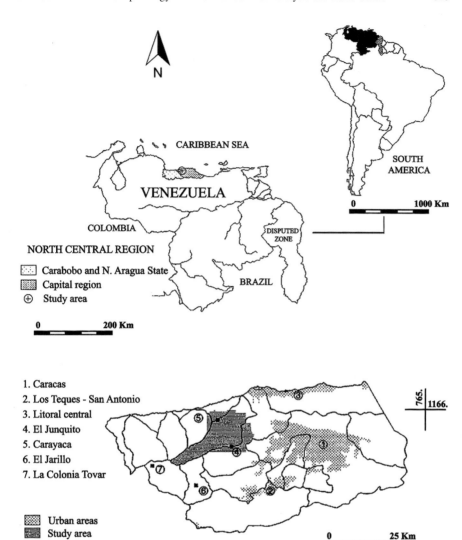

Fig. 32.1 Study area location

cover and land use were introduced as subdivisions of the geopedologic units when necessary for the detection and analysis of land use conflicts. In general, the biophysical and socioeconomic data were strongly correlated with the geopedologic context. Physical and economic suitabilities of the map units were determined for different land utilization types including agriculture, engineering, recreation, and nature conservation. Land use scenario building used LUPIS as the main tool to generate alternative plans. This was accomplished by varying the policy weights considered in land allocation for a given option.

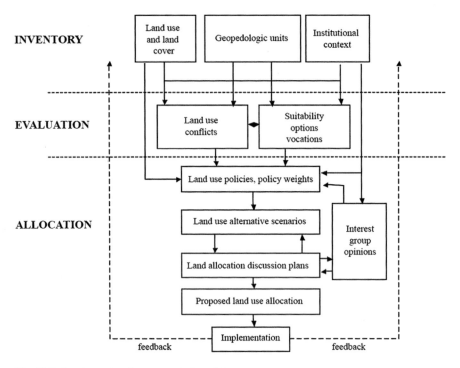

Fig. 32.2 Land use planning conceptual model

The core of LUPIS is a matrix-manipulation package, tailored to handle the tedious arithmetic required to repeatedly calculate suitability scores, summed across all policies for each land planning unit. A single LUPIS run, using a given set of policy-importance votes, determines the preferred land use option on a planning unit. This outcome is reached by ranking land uses in descending numerical order on the basis of the suitability scores obtained. Thomson's competitive solution (Thomson 1973) was used for the computation and aggregation of suitabilities, according to the following mathematical expression:

$$\text{Maximize } S_{ij} = E_{ij} \cdot \Sigma \, R_{ijk} \cdot V_k$$

where:

$i = 1, 2, \ldots n$: Planning unit (i.e. land unit)
$j = 1, 2, \ldots m$: Feasible land use options on a given planning unit
$k = 1, 2, \ldots p$: Preference policies whose satisfaction is to be maximized
E ($E = 0$ o 1): Exclusionary policies proscribing inadequate uses in given parcels
R ($0 \leq R \leq 1$): Policy satisfaction rating or degree to which a given use on a selected land unit satisfies a particular policy
V: Policy weight or "vote" for a policy by a given interest group
S: Aggregate policy satisfaction for a given use on a given parcel of land

Policies were defined as planning objectives concerned with the site requirements and environmental effects of particular land uses. Ratings were either performed from primary data items by a Qbasic code within the system, or entered directly from the keyboard, or calculated within a spreadsheet. The latter can also be used to interface the system with other software packages.

32.4   Results and Discussion

32.4.1   Land Inventory

The study area was divided into 540 terrain unit polygons, grouped hierarchically in landform and relief types. The association of primary and secondary summit relief facets dominates, and accounts for nearly 70 % of the study area. Within the incision type of landscape in the area, maximum separation into landform units was achieved at the scale of the available aerial photographs (1:20,000). The summit and shoulder associations, together with irregular slopes, account for 60 % of the study area. Slope ledges widely used for intensive horticulture cover only 1.6 % of the area, and slope summits, often used for urban settlements, correspond to only 1.9 % of the area. Slope ledges are small tracts of land, gently sloping in a general topographic context of steep slopes. They correspond to material displaced by local landslides and constitute among the best spots for cropping. The original 540 terrain unit polygons were clustered into 193 geopedologic units on the basis of associated landform and soil features (Fig. 32.3). Detailed taxonomic composition and general description of the geopedologic units can be found in Rodríguez (1995). Twenty eight soil subgroups were identified belonging to different soil orders, including Ultisols, Alfisols, Inceptisols, Mollisols, and Entisols according to Soil Taxonomy (USDA 1994). In general, there is good correlation between bioclimatic conditions and soil types, but no clear relationship could be identified between soils and relief or landform types at the selected map scale. An important feature is that most soils have developed from reworked materials displaced from weathered rock substratum or earlier soil materials by mass movements and water erosion. Very active morphodynamics has impeded the building of regular toposequences according to the classic slope facet model (e.g. Ruhe's model 1975). Eight bioclimatic zones were identified according to Holdridge (1967), including tropical very dry forest, tropical dry forest, premontane thorn woodland, premontane dry forest, premontane dry to moist forest, lower montane dry forest, lower montane dry to moist forest, and lower montane moist forest. Slope and current erosion information was relevant for the land evaluation analysis.

Land cover and land use evolution was analyzed between 1975 and 1991. By the time of 1975, the area was still characterized by a rural environment and natural cover despite the proximity to urban centers. However, the number of land cover and land use classes increased from 29 in 1975 to 56 in 1991, and the number of

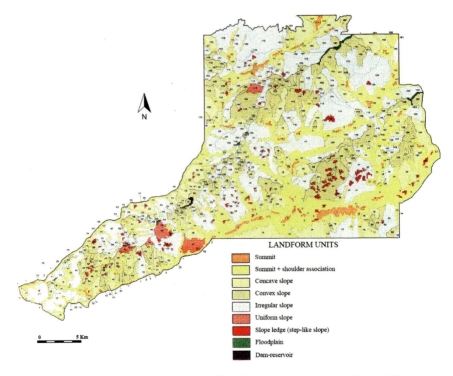

Fig. 32.3 Geopedologic units (integrated soil and landform units). Map includes 540 polygons clustered into 193 geopedologic units

delineations from 84 to 338. This reflects more diversification and fragmentation of the land use patterns resulting from increasing urban pressure, a typical adaptation response from land users in urban fringes.

Substantial land use increases between 1975 and 1991 include 700 ha horticulture, 500 ha urban areas, and 300 ha intensive recreation and rural residential facilities. These three activities cover only 15 % of the study area, but their expansion reflects the influence of urban forces on agricultural intensification and diversification, and the direct land consumption by urban uses. The most important area reduction concerns the coffee plantations from 1935 ha (8.17 % of the total area) in 1975 to 42 ha (0.24 % of the total area) in 1991. Many plantations reverted functionally to forest cover, representing a kind of social fallow waiting for land value increases, and to a lesser degree to more intensive land uses. The cloud forest and associated tree cover, highly valuable for their biodiversity and water catchment function, decreased by more than 500 ha during the same period. Natural cover decreased globally by about 7 %. Although these changes concern only 6 % of the total area, they affect mainly the cloud and dense forests, the most valuable natural resources within the area. Other information related to land use was developed as intermediate maps representing land market values, distance-accessibility conditions, legal and administrative status of the land, and farming systems.

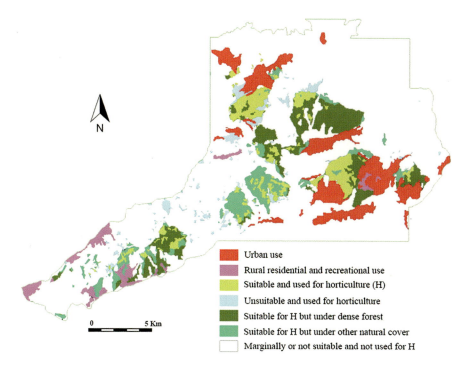

Fig. 32.4 Current intensity conflicts and potential functional conflicts for general horticulture

32.4.2   *Land Evaluation and Land Use Conflict Analysis*

Land evaluation maps were prepared using the automated land evaluation system of ALES (Rossiter and Van Wambeke 1993) to implement the FAO framework (FAO 1976) in the suitability assessment of the geopedologic units for selected land utilization types. The latter were grouped in three major land use types: (1) agriculture, (2) conservation, (3) intensive recreation and urban development. Ten specific agricultural land utilization types were evaluated and a generalized map for agriculture was also prepared. Adaptation of the FAO framework to evaluate land for purposes other than agriculture has been proposed by Baird and Ive (1989) in the case of recreation, and by Chapman et al. (1992) in the case of urban uses. For engineering and recreational applications, the values from the tables of the USDA-SCS (1983) were adjusted to local conditions. For nature conservation, an adaptation of criteria described by Sargent et al. (1991) was implemented.

Using the information from land evaluation and the current land use map, intensity conflicts and potential functional conflicts were detected among major land uses. An example of current intensity conflicts and potential functional conflicts affecting horticulture is presented in Fig. 32.4. It highlights areas currently used for horticulture but in fact unsuitable for that purpose. It also shows areas suitable for horticulture but now under other uses and thus exposed to future land conversion in the absence of protection policies.

32.4.3   Land Use Allocation

Three distinct scenarios, expressing the points of view and perceptions of the main land users (i.e. farmers, developers, and conservationists) were established. It was assumed that the land users respected the scenario conditions and the user opinions were taken into account. The primary inventory data and the transformed land evaluation data were used in a rating procedure to obtain the aggregate policy satisfaction for each specific use on a given parcel of land.

A basic set of planning conditions was established before running LUPIS to obtain particular land use scenarios. The conditions for this analysis include the following variables:

- Land use options: 4 (urban, rural residential/intensive recreation, agriculture, conservation)
- Planning parcels: 699
- Data items: 23 (e.g. biozone, slope, landform, land suitability, land market values)
- Land use policies:

  – Commitment policies: 5
  – Exclusion policies: 5
  – Preference/avoidance policies: 36

The land use scenarios were developed taking into account users' opinions through the weighting votes attached to each preference/avoidance policy by three selected interest groups: farmers, developers, and conservationists. Areas not suitable for or committed to particular land uses were respectively excluded or selected in advance.

Figure 32.5 shows the resulting maps, one for each particular scenario. In the farmer scenario, policies promoting agriculture were maximized. In that perspective, agriculture as well as urban and rural residential development uses would expand in the future at the expense of natural areas. The developer scenario encourages urban and rural residential development in any area with suitable engineering conditions at the expense of natural covers and agriculture. The conservationist scenario aims at restricting urban and rural development as much as possible: nature conservation is strongly emphasized, all forest areas will be preserved, and the role of the protected areas in nature preservation is maximized.

When the scenario maps of Fig. 32.5 are overlaid, areas of land use agreement and disagreement between the interest groups can be identified as shown in the discussion plan of Fig. 32.6. In the areas where the land use options selected by farmers, developers, and conservationists coincide, a planning consensus is reached. In contrast, opinion discrepancies arise in units showing multiple suitabilities. Such units must be submitted to a compatibilization process, leading to land use compromises. The process involves a trade between interest groups and a transfer of rights and power. The equity of the process assures a more sustainable plan.

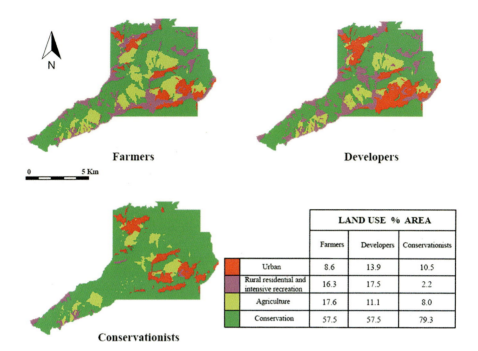

Fig. 32.5 Land use scenarios

A final land use plan was obtained after agreements were achieved among users. Consensus rules were developed to reach a compromise between different users' opinions and produced a proposed land allocation plan shown in Fig. 32.7. The proposed plan needs supporting planning strategies to be implemented. A legal framework, economic incentives, rights-in-land controls and compensations, farming systems adjustments, and institutional support are among some of the strategies that can be used to protect farmland and natural space.

32.5   Conclusions

- Land use conflicts are common consequences of the global process of urbanization. They are more acute in urban fringes because of insufficient open space, loss of suitable land for agriculture and nature conservation, and the high costs of services and waste disposal facilities. Peri-urban areas are volatile environments where land use changes take place rapidly and often unpredictably, causing different kinds of conflicts including intensity, functional, and generational conflicts.
- A combination of conventional and modern procedures for land resource inventory, land evaluation, land use assignment, and land allocation was used for data

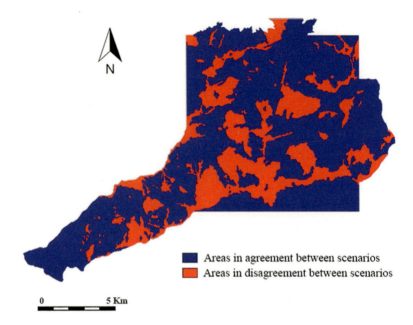

Fig. 32.6 Discussion plan

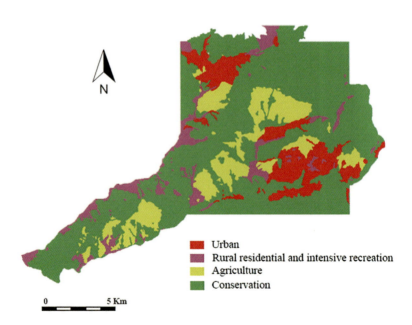

Fig. 32.7 Land use plan obtained by means of consensus rules

collection, processing, and analysis. Resource inventory, especially geopedo-logic mapping, and land use conflict analysis generated basic input data for policy formulation and scenario building. The geoform classification system played a fundamental role in the delineation of basic map units serving as cartographic containers for the mapping of the other natural resources. A set of data items with different levels of aggregation was selected from the resource inventory and land evaluation, and used to rate the planning units.

- A conceptual model integrating conflict analysis and land use planning was generated. Available software was combined and interfaced with a geographic information system for data handling. The incorporation of multiparty opinions as weighting votes for preference and avoidance land use policies is an innovative approach that encourages community participation in land use decisions. Although it is difficult to fully implement, participation generates support and adhesion to the proposed land use plan.

- Scenario building is a practical approach for identifying land use conflict areas and modeling potential land use plans. It enables the search for land use alternatives and the analysis of their possible negative and positive outcomes. It is also an efficient way of anticipating future land use conflicts and modeling the impact of land use policies in alleviating current and potential land use conflicts.

References

Baird IA, Ive JR (1989) Using the LUPLAN land use planning package to implement the recreation opportunity spectrum approach to park management planning. J Environ Manag 29:249–262

Bryant CR, Ruswurm LH, McLellan AG (1982) The city's countryside. Land and its management in the rural-urban fringe. Longman, New York

Chapman G, Hird C, Morse R (1992) A framework for assessment of urban land suitability for new South Wales. 7th ISCO Conference Proceedings, vol 1. Sidney, pp 42–51

de Groot R (2006) Function analysis and valuation as a tool to assess land use conflicts in planning for sustainable, multi-functional landscapes. Landsc Urban Plan 75:175–186

FAO (1976) A framework for land evaluation, Soils bulletin 32. FAO, Rome

Healy RG, Rosenberg JS (1979) Land use and states, 2nd edn. Resources for the Future Inc., John Hopkins University Press, Baltimore

Holdridge LR (1967) Life zone ecology, rev edn. Tropical Science Center, San José

ITC (1993) ILWIS 1.4 user's manual. International Institute for Aerospace Survey and Earth Sciences, Enschede

Ive JR (1992) LUPIS: computer assistance for land use allocation, Resource technology 92. Information Technology for Environmental Management, Taipei

Mann C, Jeanneaux P (2009) Two approaches for understanding land-use conflict to improve rural planning and management. J Rural Commun Dev 4(1):118–141

Nolon S, Ferguson O, Field P (2013) Land in conflict: managing and resolving land use disputes. Lincoln Institute of Land Policy, Puritan Press, New Hampshire

NSW Government (2011) Land use conflict risk assessment guide. Resource Planning and Development Unit, Department of Primary Industries. Factsheet, Primefact 1134, 1st edn. http://www.dpi.nsw.gov.au/factsheets

Rodriguez OS (1995) Land use conflicts and planning strategies in urban fringes. A case study of Western Caracas, Venezuela. Doctoral thesis, ITC publication 27. International Institute for Aerospace Survey and Earth Sciences, Enschede

Rodríguez OS, Zinck JA (1998) El ensamblaje de escenarios para la toma de decisiones ambientales sobre el uso de la tierra (Scenario building for environmental decision-making support on land use planning). In: Carrillo RJ (compilador) Memorias del IV Congreso Interamericano sobre el Medio Ambiente, Caracas, Venezuela, 1997. Colección Simposia, vol I, Editorial Equinoccio, Ediciones de la Universidad Simón Bolívar, Caracas, pp 337–342

Rossiter DG, van Wambeke AR (1993) ALES version 4 user's manual. Automated land evaluation system, SCAS teaching series T93-2. Cornell University, Department of Soil, Crop & Atmospheric Sciences, Ithaca

Ruhe RV (1975) Geomorphology. Geomorphic processes and surficial geology. Houghton Mifflin, Boston

Sargent FO, Lusk P, Rivera JA, Varela M (1991) Rural environmental planning for sustainable communities. Island Press, Covelo

Thomson W (1973) Middey Randstatd part 1: final report. Colin Buchanan, London

USDA-SCS (1983) National soil handbook. US Government Printing Office, Washington, DC

USDA-SCS (1994) Keys to soil taxonomy, 6th edn. USDA, Soil Conservation Service, Washington, DC

Zinck JA (1988) Physiography and soils, ITC Soil Survey Course Lecture Notes. International Institute for Aerospace Survey and Earth Sciences, Enschede

Part VI
Synthesis

Chapter 33
Synthesis and Conclusions

J.A. Zinck, G. Metternicht, H.F. Del Valle, and G. Bocco

Abstract The book contains a preface and 33 chapters that cover a large array of subjects including the basics of geopedology, implementation methods and techniques, and applications in land degradation and land use planning. Subjects addressed by the contributing authors are diverse but complementary. This shows that geopedology can be seen as a far-reaching discipline to support the inventory, scientific study, and practical management of natural resources. Geopedology aims at integrating soils and geoforms, two basic components of the earth's epidermis. Sets of examples that use different modalities or variants of geopedology are presented, from open soilscape approach for scientific research, to a more structured survey approach for mapping purposes.

Keywords Geopedology • Mapping • Geomorphic processes • Geomorphic landscape • Geomorphic environment

J.A. Zinck (✉)
Faculty of Geo-Information Science and Earth Observation (ITC), University of Twente, Enschede, The Netherlands

Institute of Environmental Studies, University of New South Wales, Sydney, NSW, Australia
e-mail: alfredzinck@gmail.com

G. Metternicht
Institute of Environmental Studies, University of New South Wales, Sydney, NSW, Australia
e-mail: g.metternicht@unsw.edu.au

H.F. Del Valle
Consejo Nacional de Investigaciones Científicas y Técnicas (CONICET),
Centro Nacional Patagónico (CENPAT), Instituto Patagónico para el Estudio de los
Ecosistemas Continentales (IPEEC), Puerto Madryn, Chubut, Argentina
e-mail: delvalle@cenpat-conicet.gob.ar

G. Bocco
Centro de Investigaciones en Geografía Ambiental (CIGA), Universidad Nacional Autónoma
de México (UNAM), Morelia, Michoacán, Mexico
e-mail: gbocco@ciga.unam.mx

© Springer International Publishing Switzerland 2016   537
J.A. Zinck et al. (eds.), *Geopedology*, DOI 10.1007/978-3-319-19159-1_33

33.1   Introduction

The book contains a preface and 33 chapters that cover a large array of subjects including the basics of geopedology, implementation methods and techniques, and applications in land degradation and land use planning. Subjects addressed by the contributing authors are diverse but complementary. This shows that geopedology can be seen as a far-reaching discipline to support the inventory, scientific study, and practical management of natural resources. Geopedology aims at integrating soils and geoforms, two basic components of the earth's epidermis. Commonalities between the subjects treated in the book allowed grouping them into five thematic parts; their relevant features are highlighted hereafter.

33.2   Part I: Foundations of Geopedology

The first part of the book written by Zinck deals with the basics of geopedology: its relationships with soil geomorphology, its focus and aims, its place in the pedologic landscape, and its supporting geomorphic framework.

After two initial chapters introducing the structure of the book, a brief review of the relationships between geomorphology and pedology is presented in Chap. 3. These relationships including the conceptual aspects and their practical implementation in studies and research have been referred to under different names, the most common expression being *soil geomorphology*. Definitions and approaches are reviewed distinguishing between academic and applied streams. There is consensus on the basic relationships between geomorphology and pedology: geomorphic processes and resulting landforms contribute to soil formation and distribution while, in return, soil development has an influence on the evolution of the geomorphic landscape. However, there is still no unified body of doctrine, in spite of a clear trend toward greater integration between the two disciplines. There are few references in international journals that provide some formal synthesis on how to carry out integrated pedogeomorphic mapping.

Chapter 4 outlines the essence of the geopedologic approach in conceptual, methodological, and operational terms. Geopedology is based on the conceptual relationships between geoform and soil which center on the earth's epidermal interface, is implemented using a variety of methodological modalities based on the three-dimensional concept of the geopedologic landscape, and becomes operational primarily within the framework of soil inventory, which can be represented by a hierarchic scheme of activities. The approach focuses on the reading of the landscape in the field and from remote-sensed imagery to identify and classify geoforms, as a prelude to their mapping along with the soils they enclose and the interpretation of the genetic relationships between soils and geoforms. There is explicit emphasis on the geomorphic context as an essential factor of soil formation and distribution. The geopedologic approach is essentially descriptive and qualitative. Geoforms and soils are considered as natural bodies, which can be described by direct observation in the field and by interpretation of aerial photographs, satellite images, topographic maps, and digital elevation models.

The pedologic component of geopedology is described in Chap. 5, with special consideration to the organization of the soil material in the pedologic landscape. Soil material is multiscalar with features and properties specific to each scale level. The successive structural levels are embedded in a hierarchic system of nested soil entities or holons known as the holarchy of the soil system. At each hierarchic level of perception and analysis of the soil material, distinct features are observed that are particular to the level considered. The whole of the features describes the soil body in its entirety. Each level is characterized by an element of the soil holarchy, a unit (or range of units) measuring the soil element perceived at that level, and a means of observation or measurement for identifying the features that are diagnostic at the level concerned. The holarchy of the soil system allows highlighting relevant relationships between soil properties and geomorphic response at different hierarchic levels. These relationships form the conceptual essence of geopedology.

The following three chapters refer to the geomorphic component of geopedology, considering successively the criteria for classifying geoforms, the classification of the geoforms, and the attributes of the geoforms. Chapter 6 describes how the combination of basic taxonomic system criteria with the hierarchic arrangement of the geomorphic environment determines a structure of six nested categorial levels. Geoforms have distinct physiognomic features that make them directly observable through visual and digital perception from remote to proximal sensing. Changing the scale of perception changes not only the degree of detail but most significantly the nature of the object observed. The geolandscape is a hierarchically structured and organized domain. Therefore, a multicategorial system, based on nested levels of perception to capture the information and taxonomic criteria to organize that information, is an appropriate frame to classify geoforms. Categorial levels are identified by their respective generic concepts, including from upper to lower level: geostructure, morphogenic environment, geomorphic landscape, relief/molding, lithology/facies, and the basic landform or terrain form.

Chapter 7 attempts to organize existing geomorphic knowledge and arrange the geoforms in the hierarchically structured system with six nested levels introduced in the foregoing Chap. 6. Geoforms are grouped thematically, distinguishing between geoforms mainly controlled by the geologic structure and geoforms mainly controlled by the morphogenic agents. It is thought that this multicategorial geoform classification scheme reflects the structure of the geomorphic landscape sensu lato. It helps segment and stratify the landscape continuum into geomorphic units belonging to different levels of abstraction. This geoform classification system has shown to be useful in geopedologic mapping, and it offers great potential for digital soil mapping.

Attributes are needed to describe, identify, and classify geoforms. These are descriptive and functional indicators that make the multicategorial system of the geoforms operational. Four kinds of attribute are used as outlined in Chap. 8: morphographic attributes to describe the geometry of geoforms; morphometric attributes to measure the dimensions of geoforms; morphogenic attributes to determine the origin and evolution of geoforms; and morphochronologic attributes to frame the time span in which geoforms originated and evolved. The morphometric and morphographic attributes apply mainly to the external component of the geoforms, are essentially descriptive, and can be extracted from remote-sensed imagery or derived from digital elevation models. The morphogenic and morphochronologic attributes apply mostly to the internal component of the geoforms, are characterized by field observations and measurements, and need to be substantiated by laboratory determinations.

33.3   Part II: Approaches to Soil-Landscape Patterns Analysis

Soil-landscape patterns can be analyzed in terms of spatial distribution, temporal evolution, or more advantageously a combination thereof. Part II presents a variety of approaches to establish and analyze relationships between soil and landscape in space and time. Information and knowledge can be obtained from field observation and landscape reading through systematic survey or transect description or a combination of both. A less common modality to identify patterns consists in translating the farmers' mental maps into soil-relief maps. Existing soil and soil-geomorphology maps are valuable sources of information; their interpretation reveals soil-landscape patterns not only in terms of geographic distribution but also in terms of temporal evolution. The concept of pattern suggests usually diversity: pedodiversity and geodiversity can be described using landscape ecology metrics.

In Chap. 9, Barrera-Bassols conveys findings from an integrated participatory soil-landscape survey in an indigenous community of central Mexico, combining ethnopedologic and geopedologic approaches. He describes the soil-landscape knowledge that local people use for selecting suitable agro-ecological settings, applying land management practices, and implementing soil conservation measures. Relief and soil maps generated by both procedures, the indigenous and the technical, are compared, and the level of spatial correlation of the map units is assessed. Commonalities, differences, and synergies of both soil knowledge systems are highlighted.

Diversity analysis of natural resources attempts to account for the variety of forms and spatial patterns exhibited by natural bodies, biotic and abiotic, at the earth's surface. Recently pedologists drew attention to soil diversity analysis and modelling using statistical tools similar to the ones used by ecologists, reporting insightful relations between spatial patterns of soil and vegetation. Geodiversity studies are mostly concerned with the preservation of the geological heritage, bypassing most of the aspects related to its spatial distribution. In Chap. 10, Ibáñez and Pérez Gómez explore a novel perspective of integrating soils, geoforms, climate, and biocenoses in a holistic approach to describe the structure and diversity of the earth surface systems.

Chapter 11 outlines a new sedimentological and geopedologic approach that explains more accurately soil development and spatial distribution in a sub-region of the Pampa plains in central Argentina. According to the traditional interpretation, vertic properties of the Pampa soils are due to a combination of finer parent materials resulting from granulometric selection during eolian transport from the southwestern Andean sources and intense smectite formation in a more humid eastern sector of the Pampa. In contrast to this view, Morrás and Moretti provide data that sustain a different soil-landscape evolution model. Smectitic sediments originating from northern sources in the Paraná basin were deposited in the Rolling Pampa and later covered by illitic loess sediments from south-western Andean sources. During a subsequent humid period in the Holocene, the illitic sediments were eroded and the smectitic sediments were exposed on the upper parts of the undulating relief. As a result, Typic Argiudolls developed on the illitic and volcanoclastic Andean sediments, while vertic soils evolved in higher positions of the landscape on the smectitic sediments of older age and different origin.

In Chap. 12, Pain et al. describe the landforms and soils in the arid region of the northern United Arab Emirates, and show how their form and evolution are closely related. Eight soil types were recognized at great group level, while 28 soil series were identified and grouped into 42 map units, each consisting of two or more soil series and a number of minor soil inclusions. At subgroup and family levels, these soils are related to specific landform morphologies and processes. Indeed, the example shows that although rainfall is scanty in this desert environment, the recognition of calcic and gypsic horizons clearly demonstrates that soil forming processes have been operating over a period of time.

Geopedology integrates an understanding of the geomorphic conditions under which soils evolve with field observations. In Chap. 13, Rossiter discusses examples from exhumed paleosol areas, low-relief depositional environments, and recent post-glacial landscapes, where simplistic digital soil mapping would fail but geopedology would succeed in mapping and explaining soil distribution. Mapping of soil bodies, not properties in isolation, is what gives insight into the soil landscape. Attempts at correlating environmental covariates from current terrain features, vegetation density, and surrogates for climate cannot succeed in the presence of unmapped variations in parent material, soil bodies, and landforms inherited from past environments.

In Chap. 14, Saldaña describes the effect of scale on the integration of landscape ecology with soil science principles, and emphasizes soilscape-pattern analysis complemented with the application of landscape ecology metrics. The approach is tested in the Jarama-Henares interfluve, central Spain, where all metrics showed to be scale-dependent, with higher values obtained at local scale. In addition, the number of indices required to describe appropriately the soilscape patterns was smaller at local than at regional scale.

Soils by virtue of their parent materials can provide key information about past sedimentologic or geologic processes and systems. Geologic and geomorphic processes substantially, but not solely, determine the materials from which soils are derived via the nature and redistribution of sediments. In Chap. 15, Schaetzl and

Miller focus on examples of studies or situations where careful examination of uniform parent material type and distribution can provide important information about the geomorphic attributes and history of the landscape. The relationship found between soils and their parent materials connects soil survey maps and geological maps. Different information collected for, and represented by, the respective maps – due to differences in purpose, focus, or resources – can assist the investigations of other disciplines. This multiple utility is especially true for studying soil-landform assemblages and soil-landscape evolution.

Chapter 16 of Yemefack and Siderius describes the application of a geopedologic approach for delineating and characterizing soil units and related soil fertility in tropical forest highlands of northern Thailand. A mathematical approach for analyzing relations between individual soil bodies was applied to study soil fertility variation as related to the categorial levels of a hierarchic geoform classification system. This relationship was displayed by means of numerical values of the Similarity Index (SI) and the Fertility Distance (FD), computed by integrating eight soil properties (pH, C, N, K, CEC-soil, CEC-clay, clay, and base saturation) assumed to influence soil fertility. The study revealed that the geopedologic approach for characterizing soils of this complex area was suitable and allowed the results obtained in sample areas to be extrapolated to similar areas. It has the advantage of being based on strong integration of geomorphology and pedology, and of considering the parent material at lower categorial levels of the system.

33.4 Part III: Methods and Techniques Applied to Pattern Recognition and Mapping

Part III comprises a set of chapters dealing with different spatial modelling techniques for soil pattern recognition and mapping, and the characterization of soil properties relevant for soil environmental risk management. A commonality between the case studies is the use of digital elevation models, remote-sensed imagery, digitally processed data using GIS, and spatial analysis and modelling techniques to transform data into usable information.

Soil classification deals with the systematic categorization of soils based on distinguishing characteristics as well as criteria that dictate choices in land use. In Chap. 17, Angueira et al. use DEM map derivatives, multi-spectral, multi-temporal and multi-spatial resolution satellite images, and visual interpretation techniques to enhance identification and classification of landscapes and soils. They describe major soil and land characteristics in a semiarid area of the Argentinean Chaco that has undergone intensive land use changes from forest to commercial agriculture over the last decade. These changes and the lack of reliable soil information at suitable scales are threatening sustainable development of the region and raising social conflicts. Map units were determined based on the integration of geoforms and soils, knowledge of landscape and soil forming factors, with the support of remote sensing data and modern survey techniques.

Quality of soil maps can be assessed from the producer's and the user's perspective. Modern methods can improve the quality of existent soil information systems in three ways: updating, upgrading, and corroborating. In Chap. 18, Bedendo et al. present an approach to improve a physiography-based soil map in Entre-Ríos province, Argentina, using digital soil mapping techniques. Continuous productivity-index (PI) classes were predicted from a number of environmental covariates, mostly digital elevation model (DEM) derivatives, using regression and geostatistical techniques. The PI land classification was used to adjust the soil-landscape/soil-series interpretation of the existing choropleth soil map by correlating discrete PI values obtained from a conventional mapping procedure with continuous PI values obtained by digital soil mapping procedures.

Limited research has been carried out on the potential of microwave remote sensing data for spatial estimation of different topsoil properties, except for soil moisture. In Chap. 19, del Valle et al. intend to narrow down this knowledge gap by assessing the potential of ALOS PALSAR image mosaics for identifying and mapping land covers, as a cartographic base for soil mapping or as a value-added layer for integration of multi-source thematic data. The chapter also analyses changes in L-band backscatter overtime, and their relation to land degradation processes. To this end, a test area covering the north-eastern Patagonia region, Argentina, was chosen for its diversity of geology, geomorphology, soil, and land use, as well as for existing soil expertise and an ongoing regional soil-mapping project.

In Chap. 20, Farshad et al. compare two approaches to prepare photo-interpretation maps that guide the location of field observations and serve as frames for soil cartography. The physiographic approach is mainly descriptive and aims at separating relief units based on their physiognomic appearance. The geopedologic approach highlights relationships between soils and geoforms and aims at predicting patterns of soil distribution prior to field survey. Both approaches have been applied in the Henares river valley, Spain. The two interpretation maps are compared in terms of soil patterns and density of delineations.

Chapter 21 of Klingseisen et al. examines geopedology in the context of soil-landscape studies in Australia. It discusses two cases where GIS-based geomorphometric tools were used for semi-automated classification of landform elements, based on topographic attributes like slope, curvature, and elevation percentile. The case studies illustrate how results of the geomorphic classification add value to management decisions related to rangelands, precision agriculture, spatial analysis, modelling of land degradation, and other spatial modelling applications where landscape morphometry is an influential factor in environmental processes.

Geomorphometric analysis from digital elevation models (DEM) can contribute to improve information detail and accuracy and, thus, strengthen soil survey. This topic is discussed in Chap. 22 by Martínez and Correa. The approach was tested in a mountainous area of Colombia. Several geomorphometric parameters were calculated and a classification of landforms was created. The outputs can supplement existing soil studies and meet the information requirements of environmental spatial models, agricultural development, hydrological studies, land use and conservation.

The application of geomorphology to soil survey has encouraged the study of genetic relationships between soils and geoforms. In Chap. 23, Viloria and Pineda applied a quantitative method based on artificial neural network and fuzzy logic to classify the landscape into land-surface units from a digital elevation model (DEM). The classification output included a map showing the spatial distribution of land-surface classes. The method proved to be effective for establishing soil-landscape relationships in the study area.

33.5 Part IV: Applications in Land Degradation and Geohazard Studies

Environmental deterioration, land degradation, and geohazard are of increased concern in many regions around the world. In this regard, understanding and quantifying the geopedologic processes that such regions are undergoing are fundamental towards promoting efficient solutions. Part IV is dedicated to applications in land degradation and geohazard studies that use geomorphic and pedologic analysis integrating spatial modelling and earth observation information.

Chapter 24 of Bocco summarizes how gully erosion research has developed, its major achievements in the conceptual and methodological dimensions, and potential courses of action for further research, with emphasis on the contribution of geopedology. It is claimed that despite the advancements in the development of models and in remote sensing and GIS techniques, gully erosion remains a complex issue difficult to model and predict. In this regard, the author argues that geopedology may play a role in its understanding and management. As other geomorphic processes, gullies occur in certain terrain, soil, and hydrology conditions, which may be conveniently approached from a geopedologic perspective.

In Chap. 25, López Salgado discusses a qualitative causal model to assess susceptibility to mass movements using detailed geopedologic information. The approach was applied in a Colombian Andean watershed. Data were collected in sample areas and validated outside for mass movement hazard zoning. The results highlight the relationships between mass movement-promoting soil properties (mainly mechanical and physical) and resulting morphodynamic processes and features (mainly landslides, various solifluction forms, and terracettes). Soil properties were assessed in terms of their susceptibility to mass movements from an integrated soil-geomorphic map.

Knowledge of the soilscape, i.e. the pedologic portion of the landscape, its characteristics, and composition helps understand the relationships between causes, processes, and indicators of land degradation. Chapter 26 by Metternich and Zinck describes the application of the geopedologic approach to map land degradation caused by soil salinity and predict salinization hazard in the semi-arid environment of the Cochabamba valleys in Bolivia. In addition of providing a framework to generate sub-regional soil information, geopedology assisted in understanding top-soil spectral reflectance features of soil degradation, assessing soil salinity type and magnitude, and predicting salinity hazard.

In Chap. 27, Sayago and Collantes report on significant land degradation in Tucumán province, Argentina, resulting from uncontrolled use of the Chaco ecosystem during the last centuries. Potential soil loss is mapped at small scale, based on geomorphic landscape sectorization and criteria of the Universal Soil Loss Equation (USLE). Models of land erosivity are developed in two scenarios of future climate change from extreme rainfall values recorded over the last century. The assessment of erosion hazard at small scale using USLE, remote sensing and GIS, helps develop programs oriented to the recovery of extensive degraded regions, through management systems adapted to current and future environmental conditions.

The geopedologic approach to soil mapping amplifies the role of geomorphology. It helps understand soil variation in the landscape, which increases mapping efficiency. In Chap. 28, Shrestha et al. show the adequacy of soil data resulting from geopedology-based predictive soil mapping for assessing land degradation in three locations of Thailand. The geopedologic approach helps map soil in inaccessible mountain areas, but for applications in land degradation studies all the required soil properties may not be available in a soil map. The effect of land cover and land use management practices on soil properties such as porosity and compaction having effect on hydraulic conductivity, a parameter used in modelling rainfall-runoff-soil erosion, is usually not reported in soil surveys. These data have to be collected separately. For mapping areas susceptible to frequent flood, the geomorphic understanding of the river valley dynamics and soil characterization help identify susceptible areas. Similarly, the study shows how the geopedologic approach in combination with digital image processing helps in mapping soil salinity hazard.

33.6   Part V: Applications in Land Use Planning and Land Zoning Studies

Part V is devoted to issues in land use planning and land zoning where geopedology plays a key role, both conceptually and in applied terms. These are important topics and are somehow neglected in the current scientific literature, more prone to purely digital mapping and pixel-based approaches. Semi-quantitative geopedologic studies aiming at the stratification of space for planning and zoning purposes are able to generate valuable scientific and practical information.

In Chap. 29, Escribano et al. determine zonation units geared towards balancing conservation and development in the Cabo de Gata-Níjar Natural Park, an arid region in south-eastern Spain. Ecosystems were identified selecting the attributes that exert the strongest influence on ecosystem dynamics at three nested spatial scales (ecosection, ecoserie, and ecotope). Geoform-soil and vegetation attributes provided the data needed to assess the conservation value and the vulnerability of the ecosystems to land use, crucial for the definition of a hierarchical zoning framework. The flexibility of the method allows analyzing the variables in different ways depending on the management objective. Furthermore, the use of meaningful score classes helps managers explain people the reasons behind land use restrictions, gaining their adhesion and minimizing social conflicts. This chapter fills a gap in conservation and development studies by characterizing ecosystems within a sound spatial framework. The approach is understandable by planners and other social actors lacking a thorough background in environmental studies; in addition it is solid and flexible, allowing for adaptive management purposes.

Silvicultural practices, including reforestation and afforestation are relevant to the provision of environmental services, to landscape rehabilitation, and in general to the sound conservation of natural resources, not only forests but also soils and water. One major challenge to these practices is the efficiency of the effort as measured in terms of successful plant viability and growth. In Chap. 30, Frugoni and Dezzoti show how geopedology has proven useful for assessing land suitability for pine plantation in north-western Patagonia, Argentina. The authors clearly indicate how silviculture requires valuable soilscape information at semi-detailed or detailed scale to assess suitability and monitor progress. Geopedologic maps provided spatial information on physical and chemical properties and terrain features. In spite of parent material homogeneity, soil-landscape relationships created variability in particular as related to water dynamics. In addition to supporting soil mapping, the approach served as a monitoring tool. Seven years after planting, ecological indicators related to plant diversity, forest regeneration, and soil protection showed improvement through project implementation. Lessons learned suggest that these variables should be carefully monitored so that the social, conservation, and economic objectives of the project can be sustainably achieved.

Land use planning and zoning frameworks are useful at all scales. However, area size (i.e. large, small) does not correlate with levels of difficulty associated to surveying and mapping. In fact, each scale offers challenges and solutions that can be addressed using geopedology. In Chap. 31, García describes a good example of territorial zoning at relatively low geographic resolutions (1:250,000 and 1:100,000 scales) in a fairly large fluvial basin in southern Venezuela, important in the provision of hydroelectricity at national level. The study is aimed at solving land management issues in the basin with the premise of long-term sustainability of power production. The study is strategic and involves far more than watershed management for suitable natural resource conservation. García proves that geopedology offers the backbone to these efforts. Territorial zoning of the catchment area, based on geomorphic and soil information, was undertaken as an initial step to propose land uses compatible with preserving the hydroelectric potential. Geomorphic units and their soil components, together with ancillary elements including the vegetation cover, were mapped at two scales using a multicategorial geoform classification system.

The zoning units were used for land evaluation and for establishing land use regulations required for the watershed management plan.

The closing chapter of Part V deals with land use planning in the western urban fringe of Caracas, the capital city of Venezuela. Urban fringes are special territories where understanding and managing social and natural processes in an integrated manner pose crucial challenges to settlers, politicians, planners, and administrators. Urban fringes in developing countries are dynamic areas that lack urban planning and usually harbor a majority of low income populations. In Chap. 32, Rodríguez and Zinck describe a framework for land use planning where land use conflict analysis represents an essential tool for decision making on land use allocation, taking into account the stakeholders interests. Resource inventory, especially geopedologic mapping, and land use conflict analysis generated basic input data for policy formulation and scenario building. The geoform classification system played a fundamental role in the delineation of basic map units serving as cartographic containers for the mapping of the other natural resources. A set of data items with different levels of aggregation was selected from the resource inventory and land evaluation, and used to rate the planning units.

33.7   Concluding Remarks

Geopedology is an approach to soil survey and other kinds of soil study. It combines pedologic and geomorphic criteria to establish soil map units in the practical-applied realm or analyze the relationships between soils and landscape evolution in the scientific realm. The geopedologic approach as described in Chap. 4 has been used primarily in soil mapping. In this context, geomorphology provides the contours of the map units ("the container"), while pedology provides the soil components of the map units ("the content"). Therefore, the units of the geopedologic map are more than soil units in the conventional sense of the term, since they also contain information about the geomorphic context in which soils have formed and are distributed. In this sense, the geopedologic unit is an approximate equivalent of the soilscape unit, but with the explicit indication that geomorphology is used to define the landscape. This is usually reflected in the map legend, which shows the geoforms as entries to the legend and their respective pedotaxa as descriptors.

Geopedology is mainly a conceptual framework, not a mapping technique in itself. It can be implemented with digital and convential survey techniques, apart or in combination, and using different survey norms and survey orders as shown in the various parts of the book. The geopedologic approach to soil survey and digital soil mapping are complementary, not mutually exclusive, and can be advantageously combined. The segmentation of the landscape sensu lato into geomorphic units provides spatial frames in which geostatistical and spectral analyses can be applied to assess detailed spatial variability of soils and geoforms, instead of blanket digital mapping over large territories. Geopedology provides information on the structure of the landscape in hierarchically organized geomorphic units, while digital techniques provide information extracted from remote-sensed imagery that help characterize the geomorphic units, mainly the morphographic and morphometric terrain surface features.

This book offers a set of examples that use different modalities or variants of geopedology from open soilscape approach for scientific research, to a more structured survey approach for mapping purposes. It shows the versatility and reach of geopedology thanks to the combination of pedology and geomorphology.

Chapter 31
Territorial Zoning Based on Geopedologic Information: Case Study in the Caroni River Basin, Venezuela

P. García Montero

© Springer International Publishing Switzerland 2016
J.A. Zinck et al. (eds.), *Geopedology*, DOI 10.1007/978-3-319-19159-1

DOI 10.1007/978-3-319-19159-1_34

The surname of the author has been incorrectly captured as P.G. Montero in the Table of Contents (p. no. xviii) and the chapter opening page (p. no. 505). The correct name should read as P. García Montero.

The online version of the original chapter can be found at
http://dx.doi.org/10.1007/978-3-319-19159-1_31

Index

Printed by Printforce, the Netherlands